evolution equations

PURE AND APPLIED MATHEMATICS

A Program of Monographs, Textbooks, and Lecture Notes

LECTURE NOTES IN PURE AND APPLIED MATHEMATICS

98. *P. Fischer and W. R. Smith,* Chaos, Fractals, and Dynamics
99. *W. B. Powell and C. Tsinakis,* Ordered Algebraic Structures
100. *G. M. Rassias and T. M. Rassias,* Differential Geometry, Calculus of Variations, and Their Applications
101. *R.-E. Hoffmann and K. H. Hofmann,* Continuous Lattices and Their Applications
102. *J. H. Lightbourne III and S. M. Rankin III,* Physical Mathematics and Nonlinear Partial Differential Equations
103. *C. A. Baker and L. M. Batten,* Finite Geometrics
104. *J. W. Brewer, J. W. Bunce, and F. S. Van Vleck,* Linear Systems Over Commutative Rings
105. *C. McCrory and T. Shifrin,* Geometry and Topology: Manifolds, Varieties, and Knots
106. *D. W. Kueker, E. G. K. Lopez-Escobar, and C. H. Smith,* Mathematical Logic and Theoretical Computer Science
107. *B.-L. Lin and S. Simons,* Nonlinear and Convex Analysis: Proceedings in Honor of Ky Fan
108. *S. J. Lee,* Operator Methods for Optimal Control Problems
109. *V. Lakshmikantham,* Nonlinear Analysis and Applications
110. *S. F. McCormick,* Multigrid Methods: Theory, Applications, and Supercomputing
111. *M. C. Tangora,* Computers in Algebra
112. *D. V. Chudnovsky and G. V. Chudnovsky,* Search Theory: Some Recent Developments
113. *D. V. Chudnovsky and R. D. Jenks,* Computer Algebra
114. *M. C. Tangora,* Computers in Geometry and Topology
115. *P. Nelson, V. Faber, T. A. Manteuffel, D. L. Seth, and A. B. White, Jr.,* Transport Theory, Invariant Imbedding, and Integral Equations: Proceedings in Honor of G. M. Wing's 65th Birthday
116. *P. Clément, S. Invernizzi, E. Mitidieri, and I. I. Vrabie,* Semigroup Theory and Applications
117. *J. Vinuesa,* Orthogonal Polynomials and Their Applications: Proceedings of the International Congress
118. *C. M. Dafermos, G. Ladas, and G. Papanicolaou,* Differential Equations: Proceedings of the EQUADIFF Conference
119. *E. O. Roxin,* Modern Optimal Control: A Conference in Honor of Solomon Lefschetz and Joseph P. Lasalle
120. *J. C. Díaz,* Mathematics for Large Scale Computing
121. *P. S. Milojević,* Nonlinear Functional Analysis
122. *C. Sadosky,* Analysis and Partial Differential Equations: A Collection of Papers Dedicated to Mischa Cotlar
123. *R. M. Shortt,* General Topology and Applications: Proceedings of the 1988 Northeast Conference
124. *R. Wong,* Asymptotic and Computational Analysis: Conference in Honor of Frank W. J. Olver's 65th Birthday
125. *D. V. Chudnovsky and R. D. Jenks,* Computers in Mathematics
126. *W. D. Wallis, H. Shen, W. Wei, and L. Zhu,* Combinatorial Designs and Applications
127. *S. Elaydi,* Differential Equations: Stability and Control
128. *G. Chen, E. B. Lee, W. Littman, and L. Markus,* Distributed Parameter Control Systems: New Trends and Applications
129. *W. N. Everitt,* Inequalities: Fifty Years On from Hardy, Littlewood and Pólya
130. *H. G. Kaper and M. Garbey,* Asymptotic Analysis and the Numerical Solution of Partial Differential Equations
131. *O. Arino, D. E. Axelrod, and M. Kimmel,* Mathematical Population Dynamics: Proceedings of the Second International Conference
132. *S. Coen,* Geometry and Complex Variables
133. *J. A. Goldstein, F. Kappel, and W. Schappacher,* Differential Equations with Applications in Biology, Physics, and Engineering
134. *S. J. Andima, R. Kopperman, P. R. Misra, J. Z. Reichman, and A. R. Todd,* General Topology and Applications
135. *P Clément, E. Mitidieri, B. de Pagter,* Semigroup Theory and Evolution Equations: The Second International Conference
136. *K. Jarosz,* Function Spaces
137. *J. M. Bayod, N. De Grande-De Kimpe, and J. Martínez-Maurica,* p-adic Functional Analysis
138. *G. A. Anastassiou,* Approximation Theory: Proceedings of the Sixth Southeastern Approximation Theorists Annual Conference
139. *R. S. Rees,* Graphs, Matrices, and Designs: Festschrift in Honor of Norman J. Pullman
140. *G. Abrams, J. Haefner, and K. M. Rangaswamy,* Methods in Module Theory

Additional Volumes in Preparation

evolution equations

edited by

Guillermo Ferreyra
Gisèle Ruiz Goldstein
Frank Neubrander
Louisiana State University
Baton Rouge, Louisiana

Marcel Dekker, Inc.   New York • Basel • Hong Kong

Library of Congress Cataloging-in-Publication Data

Evolution equations / edited by Guillermo Ferreyra, Gisèle Ruiz Goldstein, Frank Neubrander.
  p.   cm. — (Lecture notes in pure and applied mathematics; v. 168)
  Papers from the International Conference on Evolution Equations, held at Louisiana State University in Baton Rouge, Jan. 7-11, 1992.
  ISBN 0-8247-9287-4
  1. Evolution equations—Congresses. I. Ferreyra, Guillermo Segundo.
II. Goldstein, Gisele. III. Neubrander, Frank. IV. International Conference on Evolution Equations (1992 : Louisiana State University) V. Series.
QA377.E957 1994
515'.353—dc20   94-22879
                                                                            CIP

The publisher offers discounts on this book when ordered in bulk quantities. For more information, write to Special Sales/Professional Marketing at the address below.

This book is printed on acid-free paper.

MARCEL DEKKER, INC.
270 Madison Avenue, New York, New York 10016

Current printing (last digit):
10 9 8 7 6 5 4 3 2 1

PRINTED IN THE UNITED STATES OF AMERICA

Preface

The International Conference on Evolution Equations was held at Louisiana State University in Baton Rouge, Louisiana, January 7-11, 1992. It was an exciting and enjoyable conference with participants from Austria, Belgium, Canada, France, Germany, Hungary, Italy, Japan, Puerto Rico, Scotland, South Africa, Spain, Sweden, Switzerland, Taiwan, The Netherlands, and the United States. Entertainment was provided in true south Louisiana style with plenty of spicy seafood and original Cajun music.

This volume represents a cross section of the talks at the meeting. The scope of the conference reflected the areas of expertise of the Organizing Committee. The main local Organizing Committee consisted of J. R. Dorroh, Guillermo Ferreyra, Gisèle Ruiz Goldstein, Jerome A. Goldstein, Frank Neubrander and Peter Wolenski. Other members of the LSU faculty who assisted in the planning were Felipe Linares, Michael Tom and Lutz Weis. Nazar Abdelaziz, Wenyao Jia, and Thomas Schlumbrecht were also members of the Organizing Committee.

A variety of topics is covered in this volume. Both evolving and (to a lesser extent) stationary systems are studied. Partial differential equations (mostly nonlinear) are prominently featured including elliptic, hyperbolic, parabolic, dispersive, and functional differential equations. Both well-posedness and blow-up questions are considered. Volterra integral equations and reaction-diffusion systems are treated. In control theory, boundary control and identification problems are studied. Mathematical physics is present here, including Schrödinger operator theory and scattering theory. Bifurcation, asymptotics and other qualitative features are discussed. Linear operator theory is represented by semigroups of operators and related solutions of functional equations, spectral operators and spectral theory.

We gratefully acknowledge the financial support of the Department of Energy, the Porcelli Fund, the Vice-Chancellor for Research and Economic Development, Harvill Eaton, the College of Arts and Sciences and the Department of Mathematics of Louisiana State University. We sincerely thank Bob Dorroh and Jerry Goldstein for writing the proposals which enabled us to obtain funding for the conference. We thank Soula O'Bannon for her help with the logistics of the conference, and Eyvonne Coggins, Susan Oncal and Jacquie Paxton, for their help with the typing. The

Organizing Committee also would like to thank the graduate students (Boris Bäumer, Andrei and Aurora Breazna, Richard Edie, Genbao Shi, Bernd Straub, and Kunyang Wang) who helped us make this conference run smoothly.

Guillermo Ferreyra
Gisèle Ruiz Goldstein
Frank Neubrander

Contents

Contributors

Sergiu Aizicovici   Ohio University, Athens, Ohio

J. Albert   University of Oklahoma, Norman, Oklahoma

Wolfgang Arendt   Université de Franche-Comté, Besançon Cedex, France

Antonio Attalienti   Istituto di Matematica Finanziaria, University of Bari, Bari, Italy

F. Baptiste   University of Wisconsin-Milwaukee, Milwaukee, Wisconsin

Philippe Bénilan   UA CNRS 741, Université de Franche-Comté, Besançon Cedex, France

Jerry L. Bona   The Pennsylvania State University, University Park, Pennsylvania

C. Chicone   University of Missouri, Columbia, Missouri

Ioana Cioranescu   University of Puerto Rico, Rio Piedras, Puerto Rico

Wanda L. Conradie   Faculty of Sciences, University of Pretoria, Pretoria, South Africa

Richard Datko   Georgetown University, Washington, D.C.

Ralph deLaubenfels   Ohio University, Athens, Ohio

M. A. Demetriou   North Carolina State University, Raleigh, North Carolina

B. de Pagter   Faculty of Technical Mathematics and Informatics, Delft University of Technology, Delft, The Netherlands

Joachim Escher   Mathematisches Institut, University of Basel, Basel, Switzerland

W. E. Fitzgibbon   University of Houston, Houston, Texas

F. Gesztesy   University of Missouri, Columbia, Missouri

Evans M. Harrell II   School of Mathematics, Georgia Institute of Technology, Atlanta, Georgia

Mary Ann Horn   University of Minnesota, Minneapolis, Minnesota

L. Ke   University of Southwestern Louisiana, Lafayette, Louisiana

W. Lamb   University of Strathclyde, Glasgow, Scotland

Y. Latushkin   University of Missouri, Columbia, Missouri

Suzanne Lenhart   University of Tennessee, Knoxville, Tennessee

Yuan-Chuan Li   National Central University, Chung-Li, Taiwan, Republic of China

Laihan Luo   The Pennsylvania State University, University Park, Pennsylvania

A. C. McBride   University of Strathclyde, Glasgow, Scotland

D. F. McGhee   University of Strathclyde, Glasgow, Scotland

D. J. McLaughlin   University of Strathclyde, Glasgow, Scotland

Patricia L. Michel   School of Mathematics, Georgia Institute of Technology, Atlanta, Georgia

J. J. Morgan   Texas A&M University, College Station, Texas

Rainer Nagel   Mathematisches Institut, Universität Tübingen, Tübingen, Germany

B. Nagy   Faculty of Chemical Engineering, Technical University of Budapest, Budapest, Hungary

M. Parrott   University of South Florida, Tampa, Florida

G. F. Roach   University of Strathclyde, Glasgow, Scotland

Silvia Romanelli   Universita degli Studi di Bari, Bari, Italy

I. G. Rosen   University of Southern California, Los Angeles, California

W. M. Ruess   FB 6 Mathematik, Universität Essen, Essen, Germany

Niko Sauer   Faculty of Sciences, University of Pretoria, Pretoria, South Africa

Katarzyna Saxton   Loyola University, New Orleans, Louisiana

Ralph Saxton   University of New Orleans, New Orleans, Louisiana

Philip Schaefer   University of Tennessee, Knoxville, Tennessee

A. R. Schep   University of South Carolina, Columbia, South Carolina

Sen-Yen Shaw   National Central University, Chung-Li, Taiwan, Republic of China

Gieri Simonett   University of California at Los Angeles, Los Angeles, California

Hamidou Toure   UA CNRS 741, Université de Franche-Comté, Besançon Cedex, France, and Faculté des Sciences et Techniques, Université de Ouagadougou

J. M. A. M. van Neerven   Alfred P. Sloan Laboratory of Mathematics, California Institute of Technology, Pasadena, California

Xuefeng Wang   Tulane University, New Orleans, Louisiana

Dongming Wei   University of New Orleans, New Orleans, Louisiana

R. Weikard   University of Alabama-Birmingham, Birmingham, Alabama

Y. You   University of South Florida, Tampa, Florida

Conference Participants

N. Abdelaziz, Department of Mathematics, University of California, Berkeley, California 94720

S. Aizicovici*, Department of Mathematics, Ohio University, Athens, Ohio 45701

S. Alama, Departement of Mathematics, McMaster University, Hamilton, Ontario L8S 4K1, Canada

J. Albert*, Department of Mathematics, University of Oklahoma, Norman, Oklahoma 73019

F. Altomare, Dipartimento di Matematica, Universita degli Studi di Bari, Via E. Orbona 4, 70125 Bari, Italy

F. Andreu, Departamento Analisis Matematico, Universidad de Valencia, 46100 Burjassot (Valencia), Spain

W. Arendt*, Équipe dé Mathématiques, Université de Franche-Comté, 16 Route de Gray, 25030 Besançon Cedex, France

A. Y. Biyanov, Department of Mathematics, Caltech, Pasadena, California 91125

J. L. Bona*, Department of Mathematics, The Pennsylvania State University, University Park, Pennsylvania 16802

M. B. Bradley, Department of Mathematics, University of Louisville, Louisville, Kentucky 40292

---

*contributor

L. Bronsard,   Department of Mathematics, McMaster University, Hamilton, Ontario L8S 4K1, Canada

A. Castro,   Department of Mathematics, University of North Texas, Denton, Texas 76203

I. Cioranescu*,   Department of Mathematics, University of Puerto Rico, Rio Piedras, Puerto Rico 00931

C. Corduneanu,   Department of Mathematics, University of Texas at Arlington, Arlington, Texas 76019

R. Datko*,   Department of Mathematics, Georgetown University, Washington, D.C. 20057

R. deLaubenfels*,   Department of Mathematics, Ohio University, Athens, Ohio 45701

B. de Pagter*,   Delft University of Technology, Faculty of Technical Mathematics and Informatics, 2628 CD Delft, The Netherlands

J. R. Dorroh,   Department of Mathematics, Louisiana State University, Baton Rouge, Louisiana 70803

J. Escher*,   Mathematisches Institut, Rheinsprung 21, CH-45051 Basel, Switzerland

G. Ferreyra,   Department of Mathematics, Louisiana State University, Baton Rouge, Louisiana 70803

W. Fitzgibbon*,   Department of Mathematics, University of Houston, Houston, Texas 77204

F. Gesztesy*,   Department of Mathematics, University of Missouri, Columbia, Missouri 65211

A. Ghoreishi,   Department of Mathematics, Weber State University, Ogden, Utah 84408

T. L. Gill,   Department of Electrical Engineering, Howard University, Washington, D. C. 20059

G. R. Goldstein,   Department of Mathematics, Louisiana State University, Baton Rouge, Louisiana 70803

J. A. Goldstein, Department of Mathematics, Louisiana State University, Baton Rouge, Louisiana 70803

M. Grobbelaar, Faculty of Science, University of Pretoria, Pretoria 0002, South Africa

J.-S. Guo, Institute for Mathematics & Applications, University of Minnesota, Minneapolis, Minnesota 55455

D. Gurney, Department of Mathematics, Southeastern Louisiana University, Hammond, Louisiana 70402

M. Gyllenberg, Department of Applied Mathematics, Lulea University of Technology, S-95187 Lulea, Sweden

Y. Hamaya, Department of Applied Science, Okayama University of Science, Radai-cho 1-1, Okayama 700, Japan

E. Harrell*, School of Mathematics, Georgia Tech, Atlanta, Georgia 30332

J. V. Herod, School of Mathematics, Georgia Tech, Atlanta, Georgia 30332

M. Hieber, Mathematisches Institut, Universität Zürich, Winterthurerstrasse 190, CH-8057 Zürich, Switzerland

O. Hijab, Department of Mathematics, Temple University, Philadelphia, Philadelphia 19122

M. A. Horn*, Department of Mathematics, University of Minnesota, Minneapolis, Minnesota 55455

F. Howes, Appl. Math. Sci.-Sci. Comp. Staff, Office of Energy Research, Department of Energy, Washington, D.C. 20585

W. Jia, 5606 Bissonnet # 100, Houston, Texas 77081

R. Kauffmann, Department of Mathematics, University of Alabama-Birmingham, Birmingham, Alabama 35294

L. Ke*, Department of Mathematics, University of Southwestern Louisiana, Lafayette, Louisiana 70504

Y. Kurylev,   Department of Mathematics, Purdue University, West Lafayette, Indiana 47907

M. Lapidus,   Department of Mathematics, University of California-Riverside, Riverside, California 92521

Y. Latushkin*,   Department of Mathematics, University of Missouri, Columbia, Missouri 65211

L. Lefton,   Department of Mathematics, University of New Orleans, New Orleans, Louisiana 70148

F. Linares,   Instituto de Mathematica Pura e Aplicada, Estrada Dona Castorina, Rio de Janeiro, RJ 22460, Brazil

R. Logan,   Department of Mathematics, College of Charleston, Charleston, South Carolina 29424

G. Lumer,   Université de Mons-Hainaut, Institut de Mathématique et d'Informatique, 7000 Mons, Belgium

M. Mahdavi,   Department of Mathematics, Texas Christian University, Fort Worth, Texas 76129

J. M. Mazon,   Departamento Analisis Matematico, Universidad de Valencia, 46100 Burjassot (Valencia), Spain

A. McBride*,   Department of Mathematics, University of Strathclyde, Glasgow, G1 1XH, Scotland, UK

D. McGhee*,   Department of Mathematics, University of Strathclyde, Glasgow, G1 1XH, Scotland, UK

R. Nagel*,   Mathematisches Institut, Universität Tübingen, Auf der Morgenstelle, 72076 Tübingen, Germany

B. Nagy*,   Department of Mathematics, Fac. Chem. Eng., Technical University of Budapest, Budapest H-1521, Hungary

J. van Neerven*,   Mathematisches Institut, Universität Tübingen, Auf der Morgenstelle, 72076 Tübingen, Germany

J. W. Neuberger,   Department of Mathematics, North Texas State University, Denton, Texas 76203

F. Neubrander, Department of Mathematics, Louisiana State University, Baton Rouge, Louisiana 70803

M. Nkashama, Department of Mathematics, University of Alabama-Birmingham, Birmingham, Alabama 35294

S. Oppenheimer, Department of Mathematics & Statistics, Mississippi State University, Mississippi State, Mississippi 39762

M. Pang, Department of Mathematics, Cornell University, Ithaca, New York 14853

M. E. Parrott*, Department of Mathematics, University of South Florida, Tampa, Florida 33620

S. Piskarev, Department of Mathematics, Ohio University, Athens, Ohio 45701

Y.-W. Qi, Department of Mathematics, The Hongkong University of Science and Technology, Hong Kong, China

T. Randolph, Department of Mathematics, University of Missouri-Rolla, Rolla, Missouri 65401

M. Radzikowski, Department of Mathematics, Texas A & M University, College Station, Texas 77843

S. Rankin, American Mathematical Society, P.O. Box 6248, Providence, Rhode Island 02940

R. Rau, Mathematisches Institut, Universität Tübingen, Auf der Morgenstelle, 72076 Tübingen, Germany

G. Roach*, Department of Mathematics, University of Strathclyde, Glasgow, G1 1XH, Scotland, UK

S. Romanelli*, Dipartimento di Matematica, Universita degli Studi di Bari, Via E. Orbona 4, 70125 Bari, Italy

I. G. Rosen*, Department of Mathematics, University of Southern California, Los Angeles, California 90089

S. Rosencrans, Department of Mathematics, Tulane University, New Orleans, Louisiana 70118

W. Ruess*,   FB 6 Mathematik, Universität Essen, Universitätsstr. 1, 45141 Essen, Germany

N. Sauer*,   Faculty of Science, University of Pretoria, Pretoria 0002, South Africa

K. Saxton*,   Department of Mathematics, Loyola University, New Orleans, Louisiana 70118

R. Saxton*,   Department of Mathematics, University of New Orleans, New Orleans, Louisiana 70148

P. Schaefer*,   Deptartment of Mathematics, University of Tennessee, Knoxville, Tennesee 37996

W. Schappacher,   Institut für Mathematik, Universität Graz, Heinrich-strasse 36, A-8010 Graz, Austria

A. R. Schep*,   Department of Mathematics, University of South Carolina, Columbia, South Carolina 29208

G. Segalla-Pickett,   Department of Mathematics & Statistics, University of Southern Alabama, Mobile, Alabama 36688

S.-T. Shaw*,   Department of Mathematics, National Central University, Chung-Li, Taiwan 32054, Republic of China

R. Shivaji,   Department of Mathematics & Statistics, Mississippi State University, Mississippi State, Mississippi 39762

G. Simonett*,   Department of Mathematics, UCLA, Los Angeles, California 90024

Z. Sinkala,   Department of Mathematics & Statistics, Middle Tennessee State University, Murfreesboro, Tennessee 37132

P. E. Souganidis,   Applied Math Division, Brown University, Providence, Rhode Island 02912

J. Su,   Department of Mathematics, University of Texas at Arlington, Arlington, Texas 76019

W. H. Summers,   Department of Mathematical Sciences, University of Arkansas, Fayetteville, Arkansas 72701

D. Tataru, Department of Applied Mathematics, University of Virginia, Charlottesville, Virginia 22903

J. L. Vazquez, Departamento Matematicas, Universidad Autonoma de Madrid, C-XVI 28049 Madrid, Spain

X. Wang*, Department of Mathematics, Tulane University, New Orleans, Louisiana 70118

J. Ward, Department of Mathematics, University of Alabama-Birmingham, Birmingham, Alabama 35294

G. Webb, Department of Mathematics, Vanderbilt University, Nashville, Tennessee 37240

D. Wei*, Department of Mathematics, University of New Orleans, New Orleans, Louisiana 70148

R. Weikard*, Department of Mathematics, University of Alabama-Birmingham, Birmingham, Alabama 35294

L. Weis, Department of Mathematics, Louisiana State University, Baton Rouge, Louisiana 70803

P. Wolenski, Department of Mathematics, Louisiana State University, Baton Rouge, Louisiana 70803

X. Xu, Department of Mathematical Sciences, University of Arkansas, Fayetteville, Arkansas 72701

W. W. Zachary, Computational Science & Engineering Research Center, Howard University, Washington, D. C. 20059

S. Zhu, Department of Mathematics, University of Texas at Arlington, Arlington, Texas 76019

A Volterra Equation in Hilbert Space and a Related Control Problem

Sergiu Aizicovici   Ohio University, Athens, Ohio

1. INTRODUCTION

The topic of this paper is the abstract optimal control problem:
Minimize

$$\int_0^T [g(y(t)) + h(u(t))]dt$$

over the set of all $(y, u) \in C([0, T]; H) \times L^p(0, T; E)$ satisfying $\qquad$ (1.1)

$y'(t) + \int_0^t a(t - s)Ay(s)ds = Bu(t)$ on $(0, T)$,

$y(0) = y(T)$,

where H and E are real Hilbert spaces, $g : H \to \mathbf{R}$ and $h : E \to (-\infty, \infty]$ are convex functions, A is a linear densely defined operator in H, $B \in \mathcal{L}(E, H)$ (the space of all bounded linear operators from E to H), a : $(0, T] \to \mathbf{R}$ $(0 < T < \infty)$, and $1 \leq p < \infty$. Problems of this type can be considered in the context of heat control in materials with memory. (A simple example is obtained by combining the heat conduction model of Gurtin and Pipkin (1968) with the theory in Chapter II of Duvaut and Lions (1976)). Our main goal is to show that under suitable restrictions on data, (1.1) has a solution, called "optimal pair", and to establish necessary and sufficient conditions for optimality.

Related finite dimensional problems fall within the scope of Angell (1976a, 1976b, 1983) and Carlson (1987, 1990). See also Chapter 6 in Corduneanu (1991). In infinite dimensions, various authors studied the control of systems governed by Volterra integro-differential equations with initial, rather than end-point conditions. See, e.g., Avgerinos and Papageorgiou (1990) and Papageorgiou (1991). A periodic type optimal control problem involving a Volterra integro-partial differential equation has been discussed by Barbu and Pavel (1993) by a completely different method. Our approach combines the techniques used in Chapter 4 of Barbu and Precupanu (1986) for the control systems modeled by abstract differential equations, with methods specific to the theory of linear Volterra equations in Hilbert spaces. See Carr and Hannsgen (1979, 1982), Hannsgen and Wheeler (1984) and Prüss (1993).

The plan of the paper is as follows. In Section 2, we review some basic facts on abstract linear Volterra integrodifferential equations and include a result on a related end-point type problem. Section 3 contains the main results on the control problem (1.1), while a sketch of the proofs is presented in Section 4. We emphasize that although our methods can be carried over to a more general class of control problems with two-point boundary conditions, we have deliberately confined ourselves to the case of (1.1), in order to better illustrate the basic ideas.

2. LINEAR VOLTERRA EQUATIONS

Throughout this paper H denotes a real separable Hilbert space of norm $|\cdot|$ and inner product $(\ ,\ )$. We consider the following Volterra integrodifferential equation (in H)

$$y'(t) + \int_0^t a(t-s)Ay(s)ds = f(t), \quad 0 < t < T,$$

$$y(0) = y_0,$$

(2.1)

where $a \in L^1(0,T)$, A is a linear densely defined operator in H, $f \in L^1(0,T;H)$ and $y_0 \in H$.

By a *mild solution* to (2.1), we mean a function $y \in C([0,T];H)$ such that $b * y(t) \in D(A)$, $0 \leq t \leq T$, and

$$y(t) + A(b * y(t)) = y_0 + \int_0^t f(s)ds, \quad 0 \leq t \leq T.$$

Here "$*$" stands for convolution and $b(t) = \int_0^t a(s)ds$. See Prüss (1993) for details. Following Carr and Hannsgen (1979, 1982), we assume:

(H_A) A is self-adjoint and (strictly) positive definite,

(H_a) $a \in C(0,T) \cap L^1(0,T)$; $0 < a(0^+) \leq +\infty$; a is nonnegative, nonincreasing and convex, with $(-a')$ convex.

We extend a to all of $(0, +\infty)$ such that the conditions in (H_a) (nonnegativity, etc.) are satisfied, and let $z(t,\lambda)$, $\lambda \in \mathbf{R}$, denote the unique solution of the scalar integrodifferential equation

$$z'(t,\lambda) + \lambda \int_0^t a(t-s)z(s,\lambda)ds = 0 \quad (0 \le t \le \infty),$$

$$z(0,\lambda) = 1.$$
(2.2)

In view of (H_A), A admits a spectral family (resolution of the identity) $\{P_\lambda\}$ (see, e.g., Riesz and Nagy (1990)). Define the "resolvent" operator $U(t)(t \ge 0)$ on H, by

$$U(t) = \int_{-\infty}^{\infty} z(t,\lambda)d\,P_\lambda\,,$$
(2.3)

where $z(t,\lambda)$ is the solution of (2.2). It is well known (cf. Carr and Hannsgen (1979, 1982)) that $U(t)(t \ge 0)$ is a linear bounded, self-adjoint operator on H, such that $U(0) = I$ and $t \to U(t)x$ is continuous from $[0,\infty)$ to H for all x in H. Moreover, the theory of Carr and Hannsgen (1979, 1982) and Prüss (1993) ensures that Eq. (2.1) has a unique mild solution, given by the "variation of parameters" formula

$$y(t) = U(t)y_0 + \int_0^t U(t-s)f(s)ds, \quad 0 \le t \le T,$$
(2.4)

for any $f \in L^1(0,T;H)$.

We now consider the boundary value problem

$$y'(t) + \int_0^t a(t-s)Ay(s)ds = f(t) \quad \text{on} \quad (0,T),$$

$$y(0) = y(T),$$
(2.5)

where a and A satisfy (H_a) and (H_A), respectively, and $f \in L^1(0,T;H)$. The next definition makes sense because of (2.4).

DEFINITION 2.1 A *mild solution* to (2.5) is a function $y \in C([0,T];H)$ satisfying

$$y(t) = U(t)y(T) + \int_0^t U(t-s)f(s)ds, \quad 0 \le t \le T.$$
(2.6)

The following result will play a key role in the study of problem (1.1).

THEOREM 2.2 Let (H_A) and (H_a) be satisfied. In addition, assume that $a'(0^+) = -\infty$, and A^{-1} is compact. Then Eq. (2.5) has a unique mild solution.

Proof: A comparison of (2.4) and (2.6) shows that we must find $y_0 \in H$ such that

$$(I - U(T))y_0 = \int_0^T U(T-s)f(s)ds.$$
(2.7)

If $(I - U(T))$ is invertible, then (2.7) uniquely determines y_0 (specifically, $y_0 = (I - U(T))^{-1} \int_0^T U(T-s)f(s)ds$), and the corresponding $y(t)$, given by (2.4), is the unique mild solution of (2.5), in the sense of Definition 2.1.

By a result of Hannsgen and Wheeler (1984), the assumptions of Theorem 2.2 imply that $U(t)$ is compact for all $t > 0$. Consequently, proving the invertibility of $I - U(T)$ reduces to showing that $I - U(t)$ is one-to-one for all $t > 0$. To this end, we use a method that was kindly pointed out to us by K.B. Hannsgen (1992).

The spectral decomposition (2.3), which becomes an infinite sum here, yields for any $x \in H$ and $t > 0$:

$$| (I - U(t))x |^2 = \sum_{k=1}^{\infty} (1 - z(t, \lambda_k))^2 \, x_k^2, \tag{2.8}$$

where $\{x_k\}$ are the Fourier coefficients of x with respect to a basis in H consisting of eigenvectors of A, and $\lambda_k > 0$ $(k = 1, 2, \cdots)$ denote the corresponding eigenvalues. In view of (2.8), it is clear that $I - U(t)$ will be one-to-one if $z(t, \lambda)$ (the solution of (2.2) for a fixed, but arbitrary, positive λ) is never equal to 1 on $(0, +\infty)$. Let $V_\lambda(t)(\lambda > 0)$ be defined by

$$V_\lambda(t) = \frac{1}{2} z^2(t, \lambda) + \frac{1}{2}\lambda a(t) \left(\int_0^t z(s, \lambda)ds \right)^2 - \frac{1}{2}\lambda \int_0^t \left[\int_{t-\tau}^t z(s, \lambda)ds \right]^2 a'(\tau)d\tau. \tag{2.9}$$

Invoking a calculation of Levin (1963) in (2.9), yields

$$\frac{1}{2} z^2(t, \lambda) \le V_\lambda(t) = \frac{1}{2} + \int_0^t V_\lambda'(s)ds, \tag{2.10}$$

with

$$V_\lambda'(t) = \frac{\lambda}{2}a'(t) \left[\int_0^t z(s, \lambda)ds \right]^2 - \frac{\lambda}{2} \int_0^t \left[\int_{t-\tau}^t z(s, \lambda)ds \right]^2 da'(\tau), \quad \text{a.e. on } (0, \infty). \tag{2.11}$$

By hypothesis, $a' \le 0$, a.e. (with $a'(t) < 0$ for sufficiently small t), and the measure da' is nonnegative. As a consequence, (2.10) and (2.11) imply that $z^2(t, \lambda) < 1(t > 0)$, and we are done.

REMARK 2.3   A model kernel a, satisfying (H_a) and $a'(0^+) = -\infty$, is $a(t) = t^{-\alpha}$, $0 < \alpha < 1$. As regards the operator A, a typical choice in $H = L^2(\Omega)$ (where Ω is a bounded domain in $\mathbf{R}^N$ with smooth boundary), is $A = -\Delta$ with $D(A) = H_0^1(\Omega) \cap H^2(\Omega)$. In this case, the restrictions on A in Theorem 2.2 are satisfied.

3. MAIN RESULTS

Along with H, we consider another real Hilbert space E, of norm $\| \cdot \|$ and inner product $<, >$. In addition to (H_A) and (H_a), we assume

(H_B) $B \in \mathcal{L}(E, H)$,
(H_g) $g : H \to \mathbf{R}$ is convex and continuous,
(H_h) $h : E \to (-\infty, +\infty]$ is proper, convex and lower semicontinuous and satisfies

$\qquad h(u) \ge \omega(\|u\|^p - 1), \forall u \in E$, for some $p > 1$ and $\omega > 0$.

We are looking for pairs $(y, u) \in C([0, T]; H) \times L^p(0, T; E)$ (where p is the constant appearing in (H_h)) which satisfy the integrodifferential equation in (1.1) in a "mild" sense. Specifically, recalling Definition 2.1, yields

DEFINITION 3.1   A pair $(y, u) \in C([0, T]; H) \times L^p(0, T; E)$ is said to be *admissible* for problem (1.1), if

(i) $h(u) \in L^1(0, T)$,

$$(3.1)$$

(ii) $y(t) = U(t)y(T) + \int_0^t U(t - s)Bu(s)ds$, $0 \le t \le T$,

where U is given by (2.3).

The following is a direct consequence of Theorem 2.2.

THEOREM 3.2   Let conditions (H_A), (H_a), (H_B), (H_g) and (H_h) be satisfied. If also $a'(0^+) = -\infty$ and A^{-1} is compact, then (1.1) has admissible pairs.

DEFINITION 3.3   An admissible pair which solves problem (1.1) is called an *optimal pair*.

The existence of optimal pairs is established next.

THEOREM 3.4   Let the assumptions of Theorem 3.2 hold. Then (1.1) has at least one optimal pair.

REMARK 3.5. (i) Theorems 3.2 and 3.4 remain true for $p = 1$ as well. In this case, the space $L^p(0, T; E)$ in Definition 3.1 reduces to $L^1(0, T; E)$, while the inequality in assumption (H_h) should be changed to:

$$\frac{h(u)}{\|u\|} \to \infty \quad \text{as} \quad \|u\| \to \infty.$$

(ii) The above theory can be extended to the case when the function h is time dependent. Condition (H_h) should then be modified to include the restriction that h be a convex normal integrand from $[0, T] \times E$ into $(-\infty, +\infty]$ (cf, e.g., Barbu and Precupanu (1986)), satisfying for $p > 1$:

$h(\cdot, u_0(\cdot)) \in L^1(0, T)$ for some $u_0 \in L^p(0, T; E)$,

$h(t, u) \ge \omega\|u\|^p + \gamma(t)$, $\forall(t, u) \in [0, T] \times E$, with $\omega > 0$ and $\gamma \in L^1(0, T)$.

In particular, one may take $h(t, u) = h_1(u) + p^{-1}\|u - v(t)\|^p$, where $v \in L^p(0, T; E)$, and $h_1 : E \to (-\infty, +\infty]$ is proper, convex and lower semicontinuous.

We now turn our attention to the characterization of optimal pairs for problem (1.1). Let $(y^*, u^*) \in C([0, T]; H) \times L^p(0, T; E)$ be an admissible pair for (1.1), and assume that there exist $q \in C([0, T]; H)$ and $r \in L^\infty(0, T; H)$ such that

(i) $q(t) = U(T - t)q(0) - \int_t^T U(s - t)r(s)ds$, $0 \le t \le T$,

(ii) $r(t) \in \partial g(y^*(t))$,   a.e. on $(0, T)$,

$$(3.2)$$

(iii) $B^*q(t) \in \partial h(u^*(t))$,   a.e. on $(0, T)$.

In (3.2), "∂" stands for subdifferential, in the sense of convex analysis (see Brézis (1973)), and $B^* \in \mathcal{L}(H, E)$ denotes the adjoint of B.

REMARK 3.6   A simple computation shows that q, given by (3.2(i)), may be viewed as a mild solution (cf. Definition 2.1) of

$$q'(t) - \int_t^T a(s-t)Aq(s)\,ds = r(t), \quad 0 < t < T,$$

$$q(0) = q(T).$$

The following result can be considered as a "maximum principle" for the problem (1.1).

THEOREM 3.7   Suppose that (H_A) (with A^{-1} compact), (H_a) (with $a'(0^+) = -\infty$), (H_B), (H_g) and (H_h) (with $p \geq 2$) hold, and that $(y^*, u^*) \in C([0,T]; H) \times L^p(0,T; E)$ is admissible for problem (1.1). Then (y^*, u^*) is an optimal pair if and only if there exist $q \in C([0,T]; H)$ and $r \in L^\infty(0,T; H)$ satisfying (3.2).

REMARK 3.8   The sufficiency of conditions (3.2) can be established for any $p \in [1, \infty)$, See Remark 3.5(i) if $p = 1$.

4. PROOFS

In this section we outline the main steps in the proofs of Theorems 3.4 and 3.7. Note that the conclusion of Theorem 3.2 follows easily from Theorem 2.2. (It is sufficient to take $f(t) = B\bar{u}$, $0 \leq t \leq T$, where $\bar{u} \in D(h) := \{x \in E : h(x) < +\infty\}$, in Eq. (2.5), and apply Theorem 2.2.)

Proof of Theorem 3.4: Let

$$I(y, u) = \int_0^T [g(y(t)) + h(u(t))]\,dt, \tag{4.1}$$

where $(y, u) \in C([0,T]; H) \times L^p(0,T; E)$ satisfies (3.1(ii)). By Theorem 3.2, $I \not\equiv +\infty$; hence $d := \inf I < +\infty$. Consider a minimizing sequence $(y_n, u_n) \in C([0,T]; H) \times L^p(0,T; E)$, $n = 1, 2, \cdots$, satisfying

(i) $y_n(t) = U(t)y_n(T) + \int_0^t U(t-s)Bu_n(s)\,ds, \quad 0 \leq t \leq T$,

$$\tag{4.2}$$

(ii) $I(y_n, u_n) \to d$, as $n \to \infty$.

Invoking the properties of U (in particular, the invertibility of $I - U(T)$; cf. the proof of Theorem 2.2) and (H_B) in (4.2(i)), yields

$$| y_n(t) | \leq C(1 + \int_0^T \|u_n(s)\|\,ds), \quad 0 \leq t \leq T. \tag{4.3}$$

Here and in the sequel, C denotes a positive constant that may vary from line to line.

By (H_g) and (H_h), g and h are bounded from below by affine functions (see, e.g., Brézis (1973)). This, in conjunction with (4.1), (4.2(ii)) and (4.3), implies

$$\int_0^T h(u_n(t))dt \leq C\left(1 + \int_0^T \|u_n(t)\|dt\right). \tag{4.4}$$

Using (H_h) and Hölder's inequality in (4.4), we infer that

$$\{u_n\} \text{ is bounded in } L^p(0,T;E). \tag{4.5}$$

Therefore, (recall that $1 < p < \infty$), we can extract a subsequence $\{u_{n_k}\}$ of $\{u_n\}$ such that

$$u_{n_k} \to u^*, \quad \text{weakly in } L^p(0,T;E), \text{ as } n_k \to \infty. \tag{4.6}$$

(Note that (4.6) also holds when $p = 1$ and (H_h) changes as in Remark 3.5(i). In this case, one shows that (4.5) is satisfied with $p = 1$, and moreover $\{u_n\}$ is uniformly integrable. As a consequence, $\{u_n\}$ is precompact in $L^1(0,T;E)$). Employing (4.2(i)), (4.6), the properties of U, and the Dominated Convergence Theorem, we conclude that

$$y_{n_k}(t) \to y^*(t), \quad \text{weakly in } H, \quad \forall t \in [0,T],$$

$$y_{n_k} \to y^*, \quad \text{weakly in } L^1(0,T;H), \tag{4.7}$$

where $y^* \in C([0,T];H)$ satisfies

$$y^*(t) = U(t)y^*(T) + \int_0^t U(t-s)Bu^*(s)ds, \quad 0 \leq t \leq T. \tag{4.8}$$

Finally, remark that by (H_g), (H_h) (cf., e.g., Brézis (1973)), the functionals $x \to \int_0^T g(x(t))dt$ and $u \to \int_0^T h(u(t))dt$ are proper, convex and lower semicontinuous (hence weakly lower semicontinuous, as well) on $L^1(0,T;H)$ and $L^1(0,T;E)$, respectively. Thus, (4.1), (4.2), (4.7) and (4.8) lead to the conclusion that $I(y^*,u^*) = d$, and that $(y^*,u^*) \in C([0,T];H) \times L^p(0,T;E)$ is an optimal pair for (1.1), in the sense of Definition 3.3.

Proof of Theorem 3.7: "If" part. Let (3.2) hold. By the definition of a subdifferential (see Brézis (1973)) and (3.2(ii), (iii)), we have:

(i) $g(y(t)) - g(y^*(t)) \geq (r(t), y(t) - y^*(t))$, a.e. on $(0,T)$,

$$\tag{4.9}$$

(ii) $h(u(t)) - h(u^*(t)) \geq \langle B^*q(t), u(t) - u^*(t) \rangle$, a.e. on $(0,T)$,

for all $y \in C([0,T];H)$ and $u \in L^p(0,T;E)$. Combining (4.1) and (4.9), yields

$$I(y^*,u^*) \leq I(y,u) + \int_0^T [(r(t),\, y^*(t) - y(t)) + (q(t),\, B(u^*(t) - u(t)))]dt, \tag{4.10}$$

for all admissible pairs $(y,u) \in C([0,T];H) \times L^p(0,T;E)$. Recalling that (y,u) and (y^*,u^*) satisfy (3.1(ii)), and using the self-adjointness of $U(t)$, (3.2(i)) and Fubini's theorem, we obtain

$$\int_0^T (r(t),\, y^*(t) - y(t))dt = (\int_0^T U(t)r(t)dt,\, y^*(0) - y(0))$$

$$(4.11)$$

$$+ \int_0^T (B(u^*(t) - u(t)),\ U(T - t)q(0) - q(t))dt .$$

Next remark that

$$(\int_0^T U(t)r(t)dt, y^*(0) - y(0)) = (U(T)q(0) - q(0),\, y^*(0) - y(0)), \qquad (4.12)$$

while

$$(U(T)q(0),\, y^*(0) - y(0)) + \int_0^T (B(u^*(t) - u(t)),\, U(T - t)q(0))dt$$

$$(4.13)$$

$$= (q(0),\, y^*(0) - y(0)) .$$

Employing (4.12) and (4.13) in (4.11), we arrive at

$$\int_0^T (r(t),\, y^*(t) - y(t))dt = - \int_0^T (B(u^*(t) - u(t)),\, q(t))dt. \qquad (4.14)$$

From (4.10) and (4.14) it follows that $I(y^*, u^*) \le I(y, u)$, for all admissible pairs (y, u). This shows that (y^*, u^*) is an optimal pair, as desired.

"Only if" part. Assume that $(y^*, u^*) \in C([0, T]; H) \times L^p(0, T; E)$ is an optimal pair for (1.1). Let $\lambda > 0$, and consider the approximating problem.
    Minimize

$$\int_0^T [g_\lambda(y(t)) + h_\lambda(u(t)) + p^{-1}\|u(t) - u^*(t)\|^p]dt \qquad (4.15)$$

over all $(y, u) \in C([0, T]; H) \times L^p(0, T; E)$ satisfying (3.1(ii)).

In (4.15), g_λ and h_λ denote the regularizations of g and h (cf. Brézis (1973, Chapter 2)). By Remark 3.5 (ii) and the strict convexity of $u \to p^{-1} \int_0^T \|u(t) - u^*(t)\|^p dt$, (4.15) has a unique solution (y_λ, u_λ). Because of the smoothness of g_λ and h_λ, we have

$$\int_0^T [(\partial g_\lambda(y_\lambda(t)),\, z(t)) + \langle \partial h_\lambda(u_\lambda(t)) + F(u_\lambda(t) - u^*(t)),\, v(t)\rangle]dt = 0, \qquad (4.16)$$

for all $(z, v) \in C([0, T]; H) \times L^p(0, T; E)$ satisfying (3.1(ii)) (with z and v in place of y and u, respectively). Here $F(u) = \|u\|^{p-2}u$, and all integrands are easily seen to be L^1 functions. Note that (4.16) can be rewritten as

$$\int_0^T \big(\partial g_\lambda\,(y_\lambda\,(t)),\ U(t)z(T) + \int_0^t U(t-s)Bv(s)ds\big)dt$$

$$(4.17)$$

$$+ \int_0^T \langle \partial h_\lambda\,(u_\lambda\,(t)) + F(u_\lambda\,(t) - u^*\,(t)), v(t)\rangle dt = 0.$$

Next define $q_\lambda \in C([0,T]; H)$ by

$$q_\lambda\,(t) = U(T-t)q_\lambda\,(0) - \int_t^T U(s-t)\partial g_\lambda\,(y_\lambda\,(s))ds. \qquad (4.18)$$

Combining (4.17) and (4.18), and exploiting the properties of $U(t)$, we obtain

$$\int_0^T \langle -B^* q_\lambda\,(t) + \partial h_\lambda\,(u_\lambda\,(t)) + F(u_\lambda\,(t) - u^*\,(t)),\ v(t)\rangle dt = 0,$$

for all $v \in L^p\,(0,T;E)$. This yields

$$B^* q_\lambda\,(t) \in \partial h_\lambda\,(u_\lambda\,(t)) + F(u_\lambda\,(t) - u^*\,(t)), \quad \text{a.e. on } (0,T). \qquad (4.19)$$

By adapting the procedure used by Barbu and Precupanu (1986, Chapter 4), we conclude that

$$y_\lambda \to y^*, \quad \text{in } C([0,T]; H),$$

$$(4.20)$$

$$u_\lambda \to u^*, \quad \text{in } L^p\,(0,T;E),$$

as $\lambda \to 0^+$. It also follows that

$$\mid \partial g_\lambda\,(y_\lambda\,(t))\mid + \mid q_\lambda\,(0)\mid\ \leq\ C\ (0 \leq t \leq T,\ \lambda > 0),$$

which leads to (cf. (4.18))

$$q_\lambda\,(t) \to q(t), \quad \text{weakly in H},\quad \forall t \in [0,T],$$

$$(4.21)$$

$$\partial g_\lambda\,(y_\lambda) \to r, \quad \text{weakly-star in } L^\infty\,(0,T;H),$$

as $\lambda \to 0^+$, for some $q \in C([0,T]; H)$ and $r \in L^\infty\,(0,T;H)$. (The convergence in (4.21) actually holds for a subsequence $\{\lambda_n\}$ of $\{\lambda\}$, with $\lambda_n \to 0^+$ as $n \to \infty$, but we don't bother changing notation.) Finally (4.18)-(4.21) imply (3.2) upon passage to the limit, and the proof is complete.

Acknowledgments: This work was supported in part by the National Science Foundation under Grant No. DMS-91-11794. The author gratefully acknowledges stimulating conversations with Professors V. Barbu and K.B. Hannsgen.

REFERENCES

Angell, T.S. (1976a). On the optimal control of systems governed by nonlinear Volterra equations, J. Optim. Theory Appl. 19, 29-45.

Angell, T.S. (1976b). Existence of optimal control without convexity and a bang-bang theorem for linear Volterra equations, J. Optim. Theory Appl. 19, 63-79.

Angell, T.S. (1983). The controllability problem for nonlinear Volterra systems, J. Optim. Theory Appl. 41, 9-35.

Avgerinos, E.P., and Papageorgiou, N.S. (1990). Optimal control and relaxation for a class of nonlinear distributed parameter systems, Osaka J. Math. 27, 745-768.

Barbu, V., and Pavel, N.H. (1993). Optimal control problems with two point boundary conditions, J. Optim. Theory Appl., to appear.

Barbu, V., and Precupanu, T. (1986). *Convexity and Optimization in Banach Spaces.* D. Reidel, Dordrecht.

Brézis, H. (1973). *Opérateurs Maximaux Monotones et Semigroupes de Contractions dans les Espaces de Hilbert.* North Holland, Amsterdam.

Carlson, D.A. (1987). An elementary proof of the maximum principle for optimal control problems governed by a Volterra integral equation, J. Optim. Theory Appl. 54, 43-61.

Carlson, D.A. (1990). Infinite-horizon optimal controls for problems governed by a Volterra integral equation with state-and-control-dependent discount factor, J. Optim. Theory Appl. 66, 311-336.

Carr, R.W., and Hannsgen, K.B. (1979). A nonhomogeneous integrodifferential equation in Hilbert space, SIAM J. Math. Anal. 10, 961-984.

Carr, R.W., and Hannsgen, K.B. (1982). Resolvent formulas for a Volterra equation in Hilbert space, SIAM J. Math. Anal. 13, 459-483.

Corduneanu, C. (1991). *Integral Equations and Applications.* Cambridge Univ. Press, Cambridge.

Duvaut, G., and Lions, J.L. (1976). *Inequalities in Mechanics and Physics.* Springer-Verlag, Berlin.

Gurtin, M.E., and Pipkin, A.C. (1968). A general theory of heat conduction with finite wave speeds, Arch. Rational Mech. Anal. 31, 113-126.

Hannsgen, K.B. (1992). Personal communication.

Hannsgen, K.B., and Wheeler, R.L. (1984). Behavior of the solution of a Volterra equation as the parameter tends to infinity, J. Integral Equations 7, 229-237.

Levin, J.J. (1963). The asymptotic behavior of the solution of a Volterra equation, Proc. Amer. Math. Soc. 14, 534-541.

Papageorgiou, N.S. (1991). Nonlinear Volterra integrodifferential inclusions and optimal control, Kodai Math. J. 14, 254-280.

Prüss, J. (1993). *Linear Evolutionary Integral Equations in Banach Spaces and Applications.* Birkhäuser, Basel, to appear.

Riesz, F., and Nagy, B.Sz. (1990). *Functional Analysis.* Dover, New York.

Positivity Properties and Uniqueness of Solitary Wave Solutions of the Intermediate Long-Wave Equation

J. Albert   University of Oklahoma, Norman, Oklahoma

1. INTRODUCTION

The intermediate long wave (ILW) equation was first proposed in [9] and [12] as a model equation for long internal gravity waves at the interface between two fluids of different densities, each of finite depth H. Using the rescaled variables introduced in [12], the ILW equation can be written in the form

$$u_t + 2uu_x - (N_H u)_x + \frac{1}{H} u_x = 0, \tag{1.1}$$

where the "dispersion operator" N_H is the Fourier multiplier operator defined by

$$(N_H u)\hat{\ }(k) = (k \coth kH)\hat{u}(k).$$

(Here and throughout the paper, circumflexes are used to denote Fourier transforms in the x variable: thus $\hat{u}(k, t)$ denotes $\int_{-\infty}^{\infty} e^{ikx} u(x, t)\, dx$.)

An explicit family of solitary wave solutions of (1.1) was given by Joseph in [9]. These have the form $u(x, t) = \phi_{C,H}(x - Ct)$, where

$$\phi_{C,H}(z) = \left[\frac{a \sin aH}{\cosh az + \cos aH} \right].$$

Here $C > 0$ is arbitrary, and $a \in (0, \pi/H)$ is determined uniquely by the equation $aH \cot aH = (1 - CH)$. It was subsequently found that these solutions have solitonic properties like those of Korteweg-de Vries solitary waves ([10], [11],[13]). This was perhaps to be expected, in view of the role that ILW plays as an "interpolator" between two well-known soliton equations: the Korteweg-de Vries equation and the Benjamin-Ono equation, which are obtainable from ILW in the limits $H \to 0$ and $H \to \infty$ respectively ([1],[5],[14]).

Recently, it has been proved ([6]) that, for a given choice of $C > 0$ and $H > 0$, Joseph's solitary wave $\phi_{C,H}$ is, up to translations, the only L^2 travelling-wave solution of (1.1). The proof given in [6], which is based on Amick and Toland's proof of uniqueness of L^2 and periodic travelling wave solutions of the Benjamin-Ono equation ([7],[8]), treats the ILW operator as a Dirichlet-to-Neumann map for a strip in the complex plane, and makes crucial use of maximum-principle arguments. It is the purpose of this note to provide an alternate proof of uniqueness of ILW (and Benjamin-Ono) solitary waves which makes no use of complex analysis. Instead, it will be seen here that uniqueness of solitary waves can be deduced directly from two simple properties of the dispersion operator N_H: namely, the positivity of the resolvent $(N_H + \gamma)^{-1}$, and the well-known product identity stated below in Lemma 3.

Although Amick and Toland's original proof of uniqueness of Benjamin-Ono solitary waves is remarkable for its seemingly inexorable progression from one geometric idea to another, it is not easy to understand exactly which properties of the Benjamin-Ono equation are responsible for the success of the method. It is hoped that the alternate approach presented here will shed some light on this question. This approach also provides yet another illustration of the importance of positivity properties of the dispersion operator for the study of solitary waves. (See [3], [4], and [15] for applications of positive-operator theory to the questions of existence and stability of solitary waves.)

2. STATEMENT OF THE UNIQUENESS THEOREM

Attention in this paper will be confined to solutions of (1.1) which vanish in some sense as $|x| \to \infty$; thus excluding, for example, the periodic solutions discussed in [2]. In view of the results of [1] on the well-posedness of the initial-value problem for (1.1), the L^2 based Sobolev spaces H^s form a natural setting for the present study. (In this paper, H^s will denote the Hilbert space of all tempered distributions f whose Fourier transforms $\hat{f}$ are functions satisfying $\left(\int_{-\infty}^{\infty} |\hat{f}(k)|^2 (1 + |k|^2)^s \, dk \right)^{1/2} = \|f\|_s < \infty$. Use will also be made of the spaces $L^p = \{ f : \left(\int_{-\infty}^{\infty} |f|^p \right)^{1/p} < \infty \}$ for $1 \le p < \infty$ and $L^\infty = \{ f : \text{ess sup } |f(x)| < \infty \}$; and the Schwartz class $\mathbf{S}$ of functions f such that

$\lim_{|x|\to\infty} |x|^m \left(\frac{d^n f}{dx^n}\right)(x) = 0$ for all nonnegative integers m and n.)

By a "travelling wave" solution of (1.1) is understood any solution of the form $u(x,t) = \phi(x - Ct)$, where $C \in \mathbf{R}$. Usually a nontrivial (i. e., not identically zero) travelling wave solution which vanishes at $\pm\infty$ is referred to as a solitary wave. If ϕ is sufficiently regular and vanishes at $\pm\infty$ (say, $\phi \in H^1$), then substituting $u = \phi(x - Ct)$ into (1.1) and integrating once yields the equation

$$(N_H + \gamma)\phi = \phi^2 \tag{2.1}$$

where $\gamma = C - (1/H)$. This will be considered the defining equation for solitary waves. Since both sides of (2.1) are well-defined as tempered distributions for any $\phi \in L^2$ (note $N_H : H^s \to H^{s-1}$ for all $s \in \mathbf{R}$), one can extend the definition of solitary wave to include any nontrivial L^2 solution of (2.1).

The following uniqueness result for (2.1) is proved in [6].

THEOREM. Let $H > 0$ and $C > 0$ be given, and let $\gamma = C - (1/H)$. If $\phi \in L^2$ is a nontrivial solution of (2.1), then there exists $b \in \mathbf{R}$ such that $\phi(x) \equiv \phi_{C,H}(x + b)$, where $\phi_{C,H}$ is as defined in Section 1.

The goal of the present paper is to provide an alternate proof, which shows that the theorem is in fact an immediate consequence of the two special properties of the operator N_H stated in Lemmas 1 and 3 of the next section.

3. TWO PROPERTIES OF THE DISPERSION OPERATOR

For ease of notation, the subscript H will henceforth be dropped from N_H. Also, the notation $m(k)$ will occasionally be used for the multiplier function $k \coth kH$.

LEMMA 1. Let $\gamma > -(1/H)$ be given. Then the operator $(N + \gamma) : H^{s+1} \to H^s$ is invertible with bounded inverse for every $s \geq 0$. Moreover, for all $f \in H^s$, one has $\left[(N + \gamma)^{-1}f\right](x) = \int_{-\infty}^{\infty} K(x - y)f(y)\, dy$, where the kernel $K(x)$ belongs to L^p for every $p \in [1, \infty)$, and satisfies $K(x) > 0$ for all $x \in \mathbf{R}$.

PROOF: For every $f \in H^s$, the equation $(N + \gamma)g = f$ has the unique solution $g \in H^{s+1}$ given by $\hat{g}(k) = (m(k) + \gamma)^{-1}\hat{f}(k)$; and clearly $\|g\|_{s+1} \leq C\|f\|_s$. Therefore the kernel K of $(N + \gamma)^{-1}$ is given by

$$K(x) = \frac{1}{2\pi} \int_{-\infty}^{\infty} \left[\frac{1}{k \coth kH + \gamma}\right] e^{-ikx}\, dk.$$

A residue calculation using Jordan's Lemma shows that

$$K(x) = \sum_{j=0}^{\infty} e^{-\theta_j |x|/H} \left[\frac{2 \sin^2 \theta_j}{2\theta_j - \sin(2\theta_j)} \right]$$

where $\{\theta_j\}_{j=0,1,2,\dots}$ are the positive solutions of $\theta_j \cot \theta_j + \gamma H = 0$. It follows immediately that $K(x) > 0$ for $x \in \mathbf{R}$. Also, since $j\pi < \theta_j < (j+1)\pi$ for all $j \geq 0$, it follows that for some constants A, A_1, A_2 (independent of x) one has

$$
\begin{aligned}
|K(x)| &\leq A \left[e^{-\theta_0 |x|/H} + \sum_{j=1}^{\infty} \left(\frac{1}{j} \right) e^{-j\pi |x|/H} \right] \\
&= A \left[e^{-\theta_0 |x|/H} + \left| \log \left(1 - e^{-\pi |x|/H} \right) \right| \right] \\
&\leq \begin{cases} A_1 |\log |x|| & \text{for } |x| \leq 1 \\ A_2 \, e^{-\theta_0 |x|/H} & \text{for } |x| \geq 1 \end{cases}
\end{aligned}
$$

Therefore $K \in L^p$ for every $1 \leq p < \infty$. This completes the proof of the lemma.

Next we prove the product identity (3.1) for N. For Schwarz-class functions f and g, this identity follows immediately from a Fourier transform calculation. (It may also be derived as a consequence of the fact that N represents a Dirichlet-to-Neumann map for a horizontal strip in the complex plane, see e.g. [13].) However, here it will be required to establish the identity for functions f and g which do not necessarily vanish rapidly at infinity. For this purpose, the following lemma will be useful.

LEMMA 2. Suppose $f \in L^2$, $g \in H^2$, and ψ is in the Schwarz class **S**. Then

$$\int_{-\infty}^{\infty} f(x) \left(\int_0^x Ng(y) \, dy \right) \hat{\psi}(x) \, dx =$$

$$= \frac{-1}{2\pi i} \int_{-\infty}^{\infty} \coth kH \, \hat{g}(k) \left\{ \int_{-\infty}^{\infty} \left[\hat{f}(y-k) - \hat{f}(y) \right] \psi(y) \, dy \right\} \, dk.$$

PROOF: The left-hand side of the equation in the lemma may be rewritten as

$$\int_{-\infty}^{\infty} f(x) \left\{ \int_0^x \frac{1}{2\pi} \int_{-\infty}^{\infty} m(k)\hat{g}(k)e^{-iky} \, dk \, dy \right\} \hat{\psi}(x) \, dx.$$

Since $g \in H^2$, then $k \, \hat{g}(k) \in L^1(dk)$, and so Fubini's theorem may be applied to the integral in braces, yielding

$$\frac{1}{2\pi} \int_{-\infty}^{\infty} \left\{ \int_{-\infty}^{\infty} m(k)\hat{g}(k) \left[\frac{e^{-ikx} - 1}{-ik} \right] \, dk \right\} f(x)\hat{\psi}(x) \, dx.$$

Now, since

$$\left| f(x)\hat{\psi}(x) \left[\frac{e^{-ikx}-1}{-ik} \right] \right| = \left| xf(x)\hat{\psi}(x) \left[\frac{e^{-ikx}-1}{-ikx} \right] \right| \le C|xf(x)\hat{\psi}(x)| \in L^1(dx),$$

another application of Fubini's theorem yields

$$\frac{1}{2\pi} \int_{-\infty}^{\infty} m(k)\hat{g}(k) \left\{ \int_{-\infty}^{\infty} f(x)\hat{\psi}(x) \left[\frac{e^{-ikx}-1}{-ik} \right] \, dx \right\} \, dk.$$

Finally, upon applying Parseval's identity to the inner integral in braces, one obtains the right-hand side of the equation in the statement of the lemma, thus completing the proof.

LEMMA 3. If $f, g \in H^2$ and $Nf, Ng \in L^1 \cap L^2$, then

$$f \cdot g' + f' \cdot g + N \left[f \left(\int_0^x Ng \right) + g \left(\int_0^x Nf \right) \right] - Nf \left(\int_0^x Ng \right) - Ng \left(\int_0^x Nf \right) = 0. \quad (3.1)$$

PROOF: Let $h(x) = f(x) \left(\int_0^x Ng(y) \, dy \right)$. From the assumptions on f and g it follows that $Nh \in L^2$. Therefore, for every $\psi \in \mathbf{S}$, one has

$$\int_{-\infty}^{\infty} Nh(x)\hat{\psi}(x) \, dx = \int_{-\infty}^{\infty} h(x)\hat{\theta}(x) \, dx,$$

where $\theta \in \mathbf{S}$ is defined by $\theta(x) = m(x)\psi(x)$. So from Lemma 2 it follows that

$$\int_{-\infty}^{\infty} N \left(f \cdot \left(\int_0^x Ng \right) \right)(x) \, \hat{\psi}(x) \, dx = \int_{-\infty}^{\infty} f(x) \left(\int_0^x Ng(y) \, dy \right) \hat{\theta}(x) \, dx =$$

$$= \frac{-1}{2\pi i} \int_{-\infty}^{\infty} m(k)\hat{g}(k) \left\{ \int_{-\infty}^{\infty} \left[\hat{f}(y-k) - \hat{f}(y) \right] m(y)\psi(y) \, dy \right\} \, dk.$$

Also, by Lemma 2,

$$\int_{-\infty}^{\infty} Nf(x) \left(\int_0^x Ng(y) \, dy \right) \hat{\psi}(x) \, dx =$$

$$= \frac{-1}{2\pi i} \int_{-\infty}^{\infty} \coth kH \, \hat{g}(k) \left\{ \int_{-\infty}^{\infty} \left[m(y-k)\hat{f}(y-k) - m(y)\hat{f}(y) \right] \psi(y) \, dy \right\} \, dk.$$

Hence

$$\int_{-\infty}^{\infty} \left[N \left(f \cdot \left(\int_0^x Ng \right) \right) - Nf \left(\int_0^x Ng \right) \right] \hat{\psi}(x) \, dx =$$

$$= \frac{-1}{2\pi i} \int_{-\infty}^{\infty} \psi(y) \int_{-\infty}^{\infty} \hat{g}(k)\hat{f}(y-k) \coth kH \, \{m(y) - m(y-k)\} \, dk \, dy,$$

where the change in the order of integration in the last expression is justified by the absolute integrability of the integrand in $\mathbf{R}^2$.

Since $\psi \in \mathbf{S}$ was arbitrary, the preceding computation shows that, for all $y \in \mathbf{R}$,

$$\left[N\left(f \cdot \left(\int_0^x g \right) \right) - Nf \left(\int_0^x Ng \right) \right]^\wedge (y) =$$

$$= \frac{-1}{2\pi i} \int_{-\infty}^{\infty} \hat{g}(k)\hat{f}(y-k) \coth kH \left\{ m(y) - m(y-k) \right\} \, dk.$$

Now switching the roles of f and g and making the change of variables from k to $\tilde{k} = y - k$ yields the result

$$\left[N\left(g \cdot \left(\int_0^x f \right) \right) - Ng \left(\int_0^x Nf \right) \right]^\wedge (y) =$$

$$= \frac{-1}{2\pi i} \int_{-\infty}^{\infty} \hat{g}(k)\hat{f}(y-k) \coth(y-k)H \left\{ m(y) - m(k) \right\} \, dk.$$

Adding the two previous equations to the equation

$$[fg' + f'g]^\wedge(y) = \frac{1}{2\pi i} \int_{-\infty}^{\infty} \hat{g}(k)\hat{f}(y-k)y \, dk,$$

one finally obtains that the Fourier transform of the left-hand side of (3.1) is given as a function of y by

$$\frac{1}{2\pi i} \int_{-\infty}^{\infty} \hat{g}(k)\hat{f}(y-k) \, Q(y,k) \, dk,$$

where

$$Q(y,k) = y - \coth kH \, (m(y) - m(y-k)) - \coth((y-k)H) \, (m(y) - m(k)).$$

But an elementary computation shows that $Q(y,k) = 0$ for all y and k. Thus the proof of the lemma is complete.

Remark: The existence of some identity such as (3.1) seems to be required in order for an equation of the form (1.1) to have a "completely integrable" structure; see [1], [2], [11], [13].

4. A PROOF OF THE UNIQUENESS THEOREM

In this section, the uniqueness theorem stated in section 2 is derived as a consequence of Lemmas 1 and 3. The plan of the proof is as follows: first, in Lemma 4, regularity and positivity properties are derived for solutions of (2.1). Next, the identity (3.1) and the Krein-Rutman theorem for positive operators is used to reduce (2.1) to the ordinary differential equation (4.1). Standard uniqueness theory for ordinary differential equations then implies the desired result.

LEMMA 4. Let $\phi \in L^2$ be a nontrivial solution of (2.1). Then ϕ and $N\phi$ and all their derivatives are in L^p for $1 \le p \le \infty$. Also, $\phi(x) > 0$ for all $x \in \mathbf{R}$.

PROOF: First it will be proved by induction on k that $(\frac{d}{dx})^k(\phi) \in L^p$ for $1 \le p < \infty$. From (2.1) and Lemma 1, one has $\phi = K * \phi^2$, where $K \in L^p$ for $1 \le p < \infty$. Since $\phi \in L^2$, then $\phi^2 \in L^1$, and so it follows from Young's convolution inequality that $\phi \in L^p$ for $1 \le p < \infty$. Now assume that all the derivatives of ϕ up to order k are in L^p for $1 \le p < \infty$. Then from the Leibniz differentiation formula and the Cauchy-Schwarz inequality, it follows that $(\frac{d}{dx})^k(\phi^2) \in L^2$. Hence, by (2.1) and Lemma 1, $(\frac{d}{dx})^{k+1}(\phi) = \frac{d}{dx}(N+\gamma)^{-1}(\frac{d}{dx})^k(\phi^2) \in L^2$. Another use of Leibniz' formula now yields that $(\frac{d}{dx})^{k+1}(\phi^2) \in L^1$. Since $(\frac{d}{dx})^{k+1}(\phi) = K * (\frac{d}{dx})^{k+1}(\phi^2)$, it follows from Young's inequality that $(\frac{d}{dx})^{k+1}(\phi) \in L^p$ for $1 \le p < \infty$. So the inductive argument is complete.

It follows from what has just been proved that for all k, $(\frac{d}{dx})^k(\phi)$ is in $H^1(\mathbf{R})$ and hence in $L^\infty(\mathbf{R})$ as well. Also, the assertions of the Lemma concerning $N\phi$ follow from the fact that $N\phi = \phi^2 - \gamma\phi$.

Finally, to prove that if ϕ is not identically zero then $\phi(x) > 0$ for all $x \in \mathbf{R}$, one merely notes that, by Lemma 1, $\phi(x) = \int_{-\infty}^{\infty} K(x - y)\phi^2(y)\, dy$, where the kernel K is a strictly positive function on $\mathbf{R}$. This completes the proof of the Lemma.

To continue now with the proof of the main result, let ϕ be an arbitrary nontrivial L^2 solution of (2.1). By Lemma 4, $\phi > 0$ on $\mathbf{R}$, and so a Hilbert space Y may be defined to consist of all real-valued measurable functions g on $\mathbf{R}$ such that $\int_{-\infty}^{\infty} |g(x)|^2\phi(x)\, dx < \infty$, with norm furnished by the inner product $\langle g, h \rangle_Y = \int_{-\infty}^{\infty} g(x)h(x)\phi(x)\, dx$. The norm of g in Y is denoted by $\|g\|_Y$.

Next, define a linear operator $T : Y \to Y$ by

$$Tg = (N + \gamma)^{-1}(\phi g).$$

The spectral properties of T on Y have already been studied in some detail in [3] and [4]. Here we require only the following fact, which is a consequence of Lemmas 1 and 4 and the classical Krein-Rutman theorem on positive operators: T is a self-adjoint compact operator on Y, and has a positive simple eigenvalue λ_0 with a corresponding strictly positive eigenfunction $f_0(x)$. (For a proof, see Proposition 2.5.b of [3] and Lemma 8.a. of [4].)

On the other hand, the strictly positive function ϕ is also an eigenfunction of T, for the eigenvalue 1, as can be seen by rewriting (2.1) in the form

$$\phi = (N + \gamma)^{-1}(\phi^2) = T\phi.$$

Therefore λ_0 must equal 1; for otherwise, it would follow from the orthogonality of eigenspaces of self-adjoint operators that $\langle \phi, f_0 \rangle_Y = 0$, which is impossible since ϕ and f_0 are both strictly positive on $\mathbf{R}$.

Now consider the function $\psi \in Y$ defined by $\psi(x) = \phi'(x) + \phi(x) \cdot \int_0^x N\phi(y)\, dy$. We claim that ψ is also an eigenfunction of T for the eigenvalue $\lambda_0 = 1$. To see this, write

$$(N + \gamma)\psi = (N + \gamma)\phi' + (N + \gamma)(\phi \int_0^x N\phi)$$

$$= \frac{d}{dx}[\phi^2] + N(\phi \int_0^x N\phi) + \gamma\phi \int_0^x N\phi$$

$$= \phi\phi' + (N\phi)\int_0^x N\phi + \gamma\phi \int_0^x N\phi$$

$$= \phi\phi' + \phi^2 \int_0^x N\phi = \phi\psi,$$

where use has been made of (2.1) and Lemma 3 (with $f = g = \phi$). Hence

$$\psi = (N + \gamma)^{-1}\phi\psi = T\psi,$$

as claimed.

From the simplicity of the eigenvalue $\lambda_0 = 1$ of T, it now follows that ψ is a scalar multiple of ϕ: that is, there exists $\beta \in \mathbf{R}$ such that $\psi = \beta\phi$. Hence $\phi' + \phi \int_0^x N\phi = \beta\phi$. Solving for $N\phi$, one obtains that $N\phi = (-\phi'/\phi)'$. Substituting in (1.2) then yields the ordinary differential equation

$$\gamma\phi - \phi^2 = \left(\frac{\phi'}{\phi}\right)'. \tag{4.1}$$

To analyze equation (4.1), multiply by ϕ'/ϕ and integrate to obtain

$$(\phi')^2 = \phi^2(D + 2\gamma\phi - \phi^2), \tag{4.2}$$

where D is the constant of integration. Now since $\phi \in H^1$ by Lemma 4, then $\hat{\phi} \in L^1$ and, by the Riemann-Lebesgue lemma, $\phi \to 0$ as $|x| \to \infty$. It follows then, from taking the limit of (4.2) as $|x| \to \infty$, that $D \geq 0$.

Consider first the case $D > 0$. Then by (4.2),

$$\phi' = \pm\phi\sqrt{(P_+ - \phi)(P_- - \phi)}, \tag{4.3}$$

where $P_\pm = \gamma \pm \sqrt{D + \gamma^2}$, $P_+ > 0$, and $P_- < 0$. Let $\phi(0) = r$. From (4.3) it is clear that $r \in (0, P_+]$. If $r < P_+$, then an explicit integration of (4.3), together with the basic uniqueness theorem for ordinary differential equations, implies that, at least for x near 0,

$$\phi(x) = \left[\frac{a\sin aH}{\cosh a(x \pm b) + \cos aH}\right], \tag{4.4$_\pm$}$$

where $a = \sqrt{D}$, $H \in (0, \pi/a)$ is chosen so that $a \cot aH = -\gamma$, and $b > 0$ is uniquely determined by $\phi(0) = r$, with the positive or negative sign being used in (4.4)$_\pm$ according as to whether $\phi'(0) < 0$ or $\phi'(0) > 0$ respectively.

If $\phi'(0) > 0$, then the solution $(4.4)_-$ is valid at least from $x = -\infty$ to the point $x = b$, where $\phi = P_+$ and the solution of (4.3) ceases to be unique. However, at this point (4.3) gives $\phi' = 0$, and so (4.1) gives $\phi'' = \gamma - P_+ < 0$; showing that ϕ has a strict local maximum at $x = b$. Therefore $\phi(x) < P_+$ for $x > b$, and again uniqueness of solutions of (4.3) shows that ϕ is given by $(4.4)_-$ for all $x \in (b, \infty)$ as well. The same argument shows that $(4.4)_+$ holds for all $x \in \mathbf{R}$ if $\phi'(0) < 0$. Finally, if $\phi(0) = r = P_+$ and $\phi'(0) = 0$, then again (4.1) implies $\phi''(0) < 0$, and (4.4) is seen to hold for all $x \in \mathbf{R}$ with $b = 0$.

It remains to consider the case $D = 0$. But here the same argument as in the preceding paragraph shows that any positive C^2 solution of (4.1) and (4.2) must have the form

$$\phi(x) = \frac{2\gamma}{1 + \gamma^2 (x + b)^2} \tag{4.5}$$

for some $b \in \mathbf{R}$.

It has now been shown, therefore, that every solution of (4.1) must be given by either (4.4) or (4.5), where the constants $a > 0$ and $H > 0$ in (4.4) are related by $a \cot aH = -\gamma$. It is easily verified that the function in (4.5) does not satisfy (2.1) (but see the remark below). Therefore, for given $H > 0$ and $\gamma = C - (1/H)$, the only solutions of (2.1) are given by (4.4) with b arbitrary and a determined by $a \cot aH = C - (1/H)$. Hence $\phi(x) = \phi_{C,H}(x \pm b)$, and the proof of the theorem is complete.

Remark: A similar proof can be used to recover Amick and Toland's result in [7] on the uniqueness of Benjamin-Ono solitary waves. The Benjamin-Ono equation is

$$u_t + 2uu_x - (N_\infty u)_x = 0$$

where $(N_\infty u)\hat{\ }(k) = |k| \hat{u}(k)$. From the identity

$$\text{sign } k \,[1 - \text{sign } y \,\text{sign } (y - k)] = [\text{sign } y - \text{sign } (y - k)]$$

(valid for all real y and k), one easily derives the well-known product formula

$$fg + H(f \cdot Hg + g \cdot Hf) - Hf \cdot Hg = 0 \tag{4.6}$$

(valid for all f and g in L^2), where H denotes the Hilbert transform: $(Hf)\hat{\ }(k) = (i \,\text{sign } k)\hat{f}(k)$. Since $\frac{d}{dx} H = N_\infty$, differentiation of (4.6) yields (3.1) with N replaced by N_∞. It may be further seen that all the assertions of Lemma 1 hold with $N + \gamma$ replaced by $N_\infty + \gamma$, where $\gamma > 0$ is arbitrary. Hence the same argument as above leads to the conclusion that every L^2 solution of the Benjamin-Ono solitary-wave equation $(N_\infty + \gamma)\phi = \phi^2$ is also a solution of (4.1) and (4.2). From the above analysis of (4.2) it then follows that ϕ must in fact be given by (4.5).

REFERENCES

[1] L. Abdelouhab, J.L. Bona, M. Felland and J.-C. Saut, Nonlocal models for nonlinear, dispersive waves, *Physica D, 40*: 360-392 (1989).

[2] M.J. Ablowitz, A.S. Fokas, J. Satsuma and H. Segur, On the periodic intermediate long wave equation, *J. Phys. A, 15*: 781-786 (1982).

[3] J.P. Albert, Positivity properties and stability of solitary wave solutions of model equations for long waves, *Comm. Part. Diff. Equations, 17*: 1-22 (1992).

[4] J.P. Albert and J.L. Bona, Total positivity and the stability of internal waves in stratified fluids of finite depth, *IMA J. Appl. Math., 46*: 1-19 (1991).

[5] J.P. Albert, J.L. Bona and D. Henry, Sufficient conditions for stability of solitary-wave solutions of model equations for long waves, *Physica D, 24*: 343-366 (1987).

[6] J.P. Albert and J.F. Toland, On the exact solutions of the intermediate long-wave equation, to appear in *Differential and Integral Equations.*

[7] C.J. Amick and J.F. Toland, Uniqueness of Benjamin's solitary-wave solution of the Benjamin-Ono equation, *IMA J. Appl. Math., 46*: 21-28 (1991).

[8] C.J. Amick and J.F. Toland, Uniqueness and related properties of the Benjamin-Ono equation: A nonlinear boundary-value problem in the plane, *Acta Math., 167*: 107-126 (1991).

[9] R.I. Joseph, Solitary waves in a finite depth fluid, *J. Phys. A, 10*: L225-L227.

[10] R.I. Joseph and R. Egri, Multi-soliton solutions in a finite depth fluid, *J. Phys. A, 11*: L97-L102 (1978).

[11] Y. Kodama, J. Satsuma and M.J. Ablowitz, Nonlinear intermediate long wave equation: Analysis and method of solution, *Phys. Rev. Lett., 46*: 687-690 (1981).

[12] T. Kubota, D. Ko and L. Dobbs, Weakly-nonlinear long internal gravity waves in stratified fluids of finite depth, *J. Hydronautics, 12*: 157-165 (1978).

[13] D.R. Lebedev and A.O. Radul, Generalized internal long waves equations: construction, Hamiltonian structure, and conservation laws, *Commun. Math. Phys., 91*: 543-555 (1983).

[14] L.G. Redekopp, Nonlinear waves in geophysics: Long internal waves, *Fluid Dynamics in Astrophysics and Geophysics* (N. Lebovitz, ed.), Amer. Math. Soc., Providence: 59-77 (1983).

[15] M. Weinstein, Existence and dynamic stability of solitary wave solutions of equations arising in long wave propagation, *Comm. Part. Diff. Eqns., 12*: 1133-1173 (1987).

Spectrum and Growth of Positive Semigroups

Wolfgang Arendt   Université de Franche-Comté, Besançon Cedex, France

0. INTRODUCTION

Let A be the generator of a C_0-semigroup, $T = (T(t))_{t\geq 0}$ on a Banach space E.
We denote by $\sigma(A)$ the spectrum of A, by

$$s(A) = sup\{Re\ \lambda : \lambda \in \sigma(A)\}$$

the *spectral bound* of A and by

$$\omega(A) = inf\{w \in \mathbb{R} : \sup_{t\geq 0}\ e^{-tw}\|T(t)\|\ < \infty\}$$

the *growth bound* or *type* of T.
One always has $-\infty \leq s(A) \leq \omega(A) < \infty$, but in general, $s(A) < \omega(A)$.
It is of great interest to find conditions under which the identity $s(A) = \omega(A)$ is valid.
In fact, in that case, the growth of the semigroup is determined by the spectrum of the generator.
Several kinds of conditions may be considered.

A. *The nature of the Banach space* alone seems not to be helpful. A first counterexample
on ℓ^2 is due to Zabscyk (see [11, p. 62]). We show in Section 3 that the differential

operator

$$(Af)(x) = xf'(x) \tag{0.1}$$

generates a C_0-semigroup on $H^1(1,\infty)$ such that $s(A) \le -\frac{1}{2}$, $\omega(A) \ge \frac{1}{2}$.

B. *Regularity of the semigroup.* If T is eventually norm continuous (i.e. $\exists t_0 > 0$ such that $\lim_{h\downarrow 0} \|T(t_0 + h) - T(t_0)\| = 0$) then $s(A) = \omega(A)$ (see [8, p. 46] or [11, p.87]). This seems to be the most general condition of this type. Special cases are holomorphic and eventually norm continuous semigroups. Properties like compactness of the resolvent or having an extension to a C_0-group have no influence on the question, see [8, p. 44].

C. *Positivity.* Let E be an ordered Banach space and let T be positive (i.e. $T(t)E_+ \subset E_+$).
Then it is known that $s(A) = \omega(A)$ holds in three cases :

$$E = L^1, \ L^2 \text{ or } C_0(\Omega) \ (\Omega \text{ locally compact}),$$

see [11, p. 334].
It seems still be an open problem whether $s(A) = \omega(A)$ if $E = L^p$ ($1 < p < \infty$, $p \ne 2$). A well-known counterexample is given by Greiner–Voigt–Wolff [9] (see also [17], [11, p. 61], [12, p. 117]). It is valid on a mixed space $E = L^1((0,\infty), e^x dx) \cap L^p((0,\infty), dx)$.

In Section 2 we show that the operator (0.1) yields a counterexample in $L^p(1,\infty) \cap L^q(1,\infty)$ for $1 \le p < q < \infty$ (with the same measure space). The easy argument uses the fact that the realizations of the operator (0.1) in $L^p(1,\infty)$ have different spectra for different p.
In Section 3 we consider an individual version of the problem. Let T be a positive semigroup on a Banach lattice E. Then the spectral bound coincides with the abscissa of convergence of Laplace transform of T. It is natural to ask under which conditions

$$abs(T(.)f) = \omega(T(.)f) \tag{0.2}$$

for $0 \le f \in E$.
Whereas (0.2) holds for all $f \in E_+$ if $E = L^1$, we show that (0.2) fails in general, if $E = L^p$ ($1 < p < \infty$) or $E = C_0(\Omega)$.

1. SPECTRUM AND INTERMEDIATE SPACES

Let A be an operator on a Banach space E.

Assume that F is a Banach space such that $F \hookrightarrow E$ (i.e. $F \subset E$ and the injection is continuous) . Moreover, assume that there exists $\mu \in \rho(A)$ (the resolvent set of A) such that $R(\mu, A)F \subset F$ (where $R(\mu, A) = (\mu - A)^{-1}$). Let A_F be the part of A in F, i.e. :

$$A_F x = Ax, \quad D(A_F) = \{x \in D(A) \cap F : Ax \in F\}.$$

Then it is easy to see that $\mu \in \rho(A_F)$ and $R(\mu, A_F) = R(\mu, A)_{|F}$.

PROPOSITION 1.1. *In addition to the assumptions made above assume that* $D(A) \subset F$. *Then*

$$\sigma(A_F) = \sigma(A).$$

PROOF: Let $\lambda \in \rho(A)$. Then $R(\lambda, A)E \subset D(A) \subset F$. Hence $\lambda \in \rho(A_F)$. Conversely, let $\lambda \in \rho(A_F)$. Define $Q \in \mathcal{L}(F)$ by $Qx = R(\mu, A)x +$ $(\mu - \lambda)R(\lambda, A_F)R(\mu, A)x$. Then $Qx \in D(A)$ and $(\lambda - A)Qx = (\lambda - \mu)R(\mu, A)x +$ $(\mu - A)R(\mu, A)x + (\mu - \lambda)R(\mu, A)x = x$ for all $x \in E$. Since for $x \in D(A)$, $AQx = QAx$, it follows that $Q = (\lambda - A)^{-1}$. $\diamond$

2. EXAMPLES

a) Let $1 \leq p < q < \infty$ and let $F = L^p(1, \infty) \cap L^q(1, \infty)$ with norm $\|f\|_F = \|f\|_p + \|f\|_q$. Let $T(t) \in \mathcal{L}(F)$ be given by

$$(T(t)f)(x) = f(xe^t) \qquad (t \geq 0).$$

Then $T := (T(t))_{t \geq 0}$ is a C_0-semigroup on F. Denote by A the generator of T. Then

$$s(A) \leq -\frac{1}{p} < -\frac{1}{q} = \omega(A).$$

PROOF: Define the C_0-semigroup T_p on $L^p(1, \infty)$ by $(T_p(t)f)(x) = f(e^t x)$. Denote by A_p the generator of T_p . Then $T(t) = T_p(t)_{|F}$ and so $A = (A_p)_F$. One has

$$\|T_p(t)f\|_p = \left(\int_{e^t}^{\infty} |f(y)|^p dy\right)^{\frac{1}{p}} e^{-\frac{t}{p}} \leq e^{-\frac{t}{p}}\|f\|_p \quad (f \in L^p(1, \infty)). \tag{2.1}$$

Thus $s(A_p) \leq \omega(A_p) \leq -\frac{1}{p}$.

Observe that $(R(0, A_p)f)(x) = \int_0^\infty f(e^t x)\,dt = \int_x^\infty f(y)\dfrac{dy}{y}.$

Hence $R(0, A_p)f \in L^\infty(1, \infty)$. It follows that $\tilde{D}(A_p)) \subset L^\infty(1, \infty) \cap L^p(1, \infty) \subset F$. Hence by Proposition 1.2, $s(A) = s(A_p) \le -\frac{1}{p}$.

Observe that for $f \in L^p(1, \infty)$, $g(x) := (R(0, A_p)f)(x) = \int_0^\infty f(e^t x)\,dt = \int_x^\infty f(y)\dfrac{dy}{y}.$

Hence $|g(x)| \le \|f\|_1 \cdot \frac{1}{x}$ if $p = 1$ and

$$|g(x)| \le \|f\|_p \cdot \left(\int_x^\infty y^{-p'}\,dy \right)^{\frac{1}{p'}} = \|f\|_p \cdot \left(\frac{1}{p' - 1} \right)^{\frac{1}{p'}} x^{\frac{1}{p'} - 1}.$$

Thus $R(0, A_p)f \in C_0[1, \infty)$. It follows that $D(A_p) \subset L^p(1, \infty) \cap C_0[1, \infty) \subset F$. Hence by Proposition 1.1, $s(A) = s(A_p) \le -\frac{1}{p}$.

On the other hand, let $t > 0$. Let $f = \frac{1}{2}\,1_{[e^t, e^t + 1]}$. Then $\|f\|_F = 1$ and it follows from (2.1) that $\|T(t)f\|_F \ge \|T_q(t)f\|_q = \frac{1}{2}e^{-\frac{t}{q}}$. Consequently, $\omega(A) \ge -\frac{1}{q}$. $\diamond$

The example is based on the fact that the spectra of A_p are dependent on $p \in [1, \infty)$. This phenomenon is investigated in more detail in [1]. There it is shown, in particular, that the spectrum is p-independent for uniformly elliptic differential operators of second order with various boundary conditions.

b) Let $1 \le p < \infty$, $F = L^p(1, \infty) \cap C_0[1, \infty)$. Define the semigroup T on F by $(T(t)f)(x) = f(e^t x)$ and denote by A its generator. Then

$$s(A) \le -\frac{1}{p} < 0 = \omega(A).$$

The proof is similar to a) .

c) The first Sobolev space $H^1(1, \infty) = \{f \in L^2(1, \infty); f' \in L^2(1, \infty)\}$ is a Hilbert space for the norm $\|f\|_{H^1} = (\|f\|_2^2 + \|f'\|_2^2)^{1/2}$ (see [5, VIII]).

Define the C_0-semigroup T on $H^1(1, \infty)$ by

$$T(t)f)(x) = f(e^t x). \tag{2.2}$$

Denote by A its generator. Then

$$s(A) \le -\frac{1}{2} < \frac{1}{2} \le \omega(A).$$

PROOF: Consider the semigroup T_2 defined in a). Then T is the restriction of T_2 and so A is the part of A_2 in $H^1(1, \infty)$. Recall that $s(A_2) \le -\frac{1}{2}$. Let $f \in L^2$, $g = R(0, A_2)f$.

Then $g(x) = \int_x^\infty \dfrac{f(y)}{y}\,dy$. Thus $g \in H^1(1,\infty)$, $g'(x) = -\frac{1}{x}\,f(x)$. We have shown that $D(A_2) \subset H^1(1,\infty)$. It follows from Proposition 1.2 that $s(A) \leq -\frac{1}{2}$.

We show that $\omega(A) \geq \dfrac{1}{2}$.

Let $t > 0$. Choose $f \in H^1(1,\infty)$ such that $supp\, f \subset (e^t, \infty)$, $\|f\|_2 \leq 1$, $\|f'\|_2 = 1$ (one may take f linear on $(e^t, e^t + 1)$ and $(e^t + 1, e^t + 2)$ such that $f(e^t + 1) = \dfrac{1}{\sqrt{2}}$ and $f = 0$ on $[1, e^t] \cup [e^t + 2, \infty))$.

Then $\|f\|_{H^1} \leq \sqrt{2}$ and

$$\|T(t)f\|_{H^1} \geq \|(T(t)f)'\|_2 = \|e^t f'(e^t \cdot)\|_2 = e^t \left(\int_1^\infty |f'(e^t x)|^2 dx \right)^{\frac{1}{2}}$$

$$= e^t e^{-\frac{t}{2}} \left(\int_{e^t}^\infty |f'(x)|^2 dx \right)^{\frac{1}{2}} = e^{\frac{t}{2}} \|f'\|_2 = e^{-\frac{t}{2}}.$$

This proves the claim. $\diamond$

Example c) has interesting order properties. In fact, $H^1(1,\infty)$ is a vector lattice : $f \in H^1(1,\infty)$ implies $|f| \in H^1(1,\infty)$ and $|f|' = (sign\, f)f'$ (where $(sign\, f)(x) = 1$ if $f(x) > 0$, $= 0$ if $f(x) = 0$ and $= -1$ if $f(x) < 0$), see [7, chap. IV §7, p. 934]. In particular, $\|f\|_{H^1} = \| |f| \|_{H^1}$.

It follows from [4] that the lattice operations are continuous. However, $H^1(1,\infty)$ is not a Banach lattice since the norm is not monotone. In fact, the cone is not normal, i.e. order intervals $[-u, u] = \{f \in H^1(1,\infty) : -u \leq f \leq u\}$ are not norm bounded.

The example shows that positivity does not imply "$s(A) = \omega(A)$ " on such an ordered Hilbert space.

d) In an analogous way one sees that (2.2) defines a C_0-semigroup T on $W^{1,p}(1,\infty)$, where $1 \leq p < \infty$. Let A be its generator. Then

$$s(A) \leq -\frac{1}{p} < 1 - \frac{1}{p} = \omega(A).$$

See [5, VIII] for definition and properties of $W^{1,p}$. It is interesting to observe that the space $W^{1,p}(1, \infty)$ is isomorphic (as a Banach space) to a space L^p if $1 < p < \infty$ (see Wojtaszczyk [16, p. 47]).

3. INDIVIDUAL EXPONENTIAL STABILITY

Let T be a C_0-semigroup on a Banach space E with generator A.

First we show that T is exponentially stable whenever every orbit is exponentially stable (cf. [2]) :

PROPOSITION 3.1. *Assume that for every $f \in E$ there exist $\delta > 0$, $c \geq 0$ such that*

$$\|T(t)f\| \leq ce^{-\delta t} \quad (t \geq 0),$$

then there exist $\varepsilon > 0$, $M \geq 0$ such that

$$\|T(t)\| \leq Me^{-\varepsilon t} \quad (t \geq 0).$$

PROOF: For each $n \in \mathbb{N}$ the set

$$E_n := \{f \in E : \|T(t)f\| \leq ne^{-\frac{1}{n}t} \quad \forall t \geq 0\}$$

is closed and $E = \bigcup_{n \in \mathbb{N}} E_n$. By Baire's theorem there exist $n_0 \in \mathbb{N}$, $g \in E_{n_0}$, $\varepsilon > 0$ such that $f \in E_{n_0}$ whenever $\|f - g\| < \varepsilon$. Let $h \in E \backslash \{0\}$. Then $\dfrac{\varepsilon}{\|h\|} h + g \in E_{n_0}$. Thus

$$\frac{\varepsilon}{\|h\|} \|T(t)h\| \leq \left\|T(t)\left(\frac{\varepsilon}{\|h\|} h + g\right)\right\| + \|T(t)g\| \leq 2n_0 e^{-\frac{t}{n_0}} \quad (t \geq 0).$$

Consequently, $\|T(t)\| \leq 2\frac{n_0}{\varepsilon}.e^{-\frac{t}{n_0}} \quad (t \geq 0)$. $\diamond$

Next we compare abscissa (of the Laplace transform) and exponential growth bound for orbits.

Let $u : [0, \infty) \to E$. We denote by

$$\omega(u) = \inf \{w \in \mathbb{R} : \exists c \geq 0 \quad \|u(t)\| \leq ce^{wt} \quad (t \geq 0)\}$$

the exponential growth bound of u. In particular, one has $\omega(T) = \omega(A)$. If $u \in L^1_{loc}([0, \infty), E)$ we denote by

$$abs(u) = \inf\{w \in \mathbb{R} : \lim_{t \to \infty} \int_0^t e^{-ws}u(s)ds \quad \text{exists}\}$$

the *abscissa* of u (see [3] for a formula of $abs(u)$). It is clear that $abs(u) \leq \omega(u)$; but of course, one easily sees that for positive real-valued functions it can happen that $abs(u) < \omega(u)$.
If u is the norm of an orbit this is different :

PROPOSITION 3.2. *One has $abs(\|T(.)f\|) = \omega(T(.)f)$ for all $f \in E$.*

PROOF: This follows from Lemma [11 p. 102]. $\diamond$

We let $abs(T) := sup\{abs(T(.)f) : f \in E\}$. Then $\lambda \in \rho(A)$ and $R(\lambda, A)f = \int_0^\infty e^{-\lambda t} T(t)f \, dt$ whenever $Re \, \lambda > abs(T)$. But it may happen that $s(A) < abs(T)$ (see [11, A-IV]). However if E is a Banach lattice and T is positive, then

$$s(A) = abs(T) \tag{3.1}$$

(see [11, C-IV]).

Now if E is a space L^1, then for $u \in L^1_{loc}([0, \infty), E_+)$ one has $\| \int_0^t e^{-ws} u(s)ds \| = \int_0^t e^{-ws} \|u(s)\| ds \quad (w \in \mathbb{R})$ and so $abs(u) = abs(\|u\|)$. We conclude from Proposition 3.2 :

PROPOSITION 3.3. *Let E be a space L^1 and let T be positive. Then*

$$abs(T(.)f) = \omega(T(.)f) \quad (f \in E_+). \tag{3.2}$$

It follows immediately from (3.1) and (3.2) that $s(A) = \omega(A)$ if $E = L^1$ and T is positive. Thus one may consider (3.2) as an individual version of the identity "$s(A) = \omega(A)$ ". However, even though the latter identity is known to hold as well on L^2 and $C_0(\Omega)$ the individual version fails on these spaces. This is shown by the following example :

EXAMPLE 3.4. Let $(T(t)f)(x) = f(e^t x)$ on $E = L^p(1, \infty)$, $1 < p < \infty$, or $E = C_0[1, \infty)$. Then there exists $f \in E_+$ such that

$$abs(T(.)f) < \omega(T(.)f). \tag{3.3}$$

PROOF: Let $S(t) = T(t)_{|F}$, $F = E \cap L^1(1, \infty)$, and let B be the generator of S. Then by §2, $s(B) \leq -1$. Thus $abs(T(.)f) \leq -1$ for all $f \in F$. On the other hand we have shown in §3 that $\omega(B) = -\frac{1}{p}$ if $E = L^p$ and $\omega(B) = 0$ if $E = C_0[1, \infty)$. Hence, in view of Proposition 3.1, for $-1 < \lambda < -\frac{1}{p}$ there exists $f \in F_+$ such that $sup_{t \geq 0} \|e^{-\lambda t} T(t)f\|_F = \infty$. Since $\omega(\|T(t)f\|_1) \leq -1$, it follows that $sup_{t \geq 0} \|e^{-\lambda t} T(t)f\|_E = \infty$. One concludes that

$$\omega(T(.)f) \geq \lambda > -1 \geq abs(T(.)f). \qquad \diamond$$

REMARK 3.5. We have shown that on L^p, $(1 < p < \infty)$, and on $C_0(\Omega)$ it can happen that $abs(T(.)f) < abs(\|T(.)f\|)$ (by (3.3) and (3.2)) where T is a positive semigroup and $f \geq 0$.

It follows from a result by Schlotterbeck [14] (see [13, IV-2.1]) that a Banach lattice E is isomorphic to a space L^1 if and only if every measurable function $u : [0, \infty) \to E_+$ such that $\lim_{t \to \infty} \int_0^t u(s)dx$ exists satisfies $\int_0^\infty \|u(t)\|dt < \infty$.

REFERENCES

[1] W. Arendt, *p-independance of the spectrum for uniformly elliptic operators on L^p.* Differential and Integral Equations, to appear.

[2] W. Arendt, O. El Mennaoui and V. Keyantuo, *Local integrated semigroups : evolution with jumps of regularity,* J. Math. Analysis and Appl. To appear.

[3] W. Arendt, F. Neubrander and U. Schlotterbeck, *Interpolation of semigroups and integrated semigroups,* Semigroup Forum 45 : 26-37 (1992).

[4] J.M. Borwein and D.T. Yost, *Absolute norms on vector lattices,* Proc. Edinburgh Math. Soc. 27 : 215-222 (1984).

[5] H. Brézis, *Analyse fonctionnelle,* Masson, Paris (1983).

[6] Ph. Clément and al., *One-parameter semigroups,* CWI Monographs 5, North-Holland, Amsterdam (1987).

[7] R. Dautray and J.L. Lions, *Analyse mathématique et calcul numérique,* Masson, Paris (1987).

[8] E.B. Davies, *One-parameter semigroups,* Academic Press, London, (1980).

[9] G. Greiner, J. Voigt and M. Wolff, *On the spectral bound of the generators of semigroups of positive operators,* J. Operator Th. 5 : 245-256 (1981).

[10] E. Hille, R.S. Phillips, *Functional Analysis and semigroups,* AMS, Providence (1957).

[11] R. Nagel (ed.), *One-parameter semigroups of positive operators,* Springer LN 1184, Berlin (1986).

[12] A. Pazy, *Semigroups of linear operators and applications to partial differential equations,* Springer, Berlin (1983).

[13] H.H. Schaefer, *Banach lattices and positive operators,* Springer, Berlin (1974).

[14] U. Schlotterbeck, *Über Klassen majorisierbarer Operatoren auf Banachverbänden,* Dissertation, Tübingen (1969).

[15] G. Weiss, *Weak L^p-stability of a linear semigroup on a Hilbert space implies exponential stability,* J. Diff. Equ. 76 : 269-285 (1988).

[16] A. Wojtaszczyk, *Banach spaces for analysts.* Cambridge University Press 1991.

[17] M. Wolff, A remark on the spectral bound of the generator of semigroups of positive operators with applications to stability theory, ISNM. Functional Analysis and Approximation, Proc. Oberwolfach 1980, Birkhäuser, Basel (1981).

On Some Classes of Analytic Semigroups
on $C([a,b])$ Related to R or Γ-Admissible Mappings

Antonio Attalienti   Istituto di Matematica Finanziaria, University of Bari, Bari, Italy

Silvia Romanelli   Universita degli Studi di Bari, Bari, Italy

Abstract

Given $m \in \mathcal{C}^1([a,b])$, we prove that the operator $Af = m^2 f''$ defined on a suitable subspace of $\mathcal{C}([a,b])$, equipped with the sup-norm, generates an analytic semigroup. We apply this result to the case $m(x) = x^\alpha (1-x)^\alpha$ $(\alpha \geq 1, x \in [0,1])$.

Some one-dimensional diffusion processes are described by second order differential operators of the type $Af = pf''$, where p is a positive function and the operator A is defined on a suitable subspace of the space $\mathcal{C}([a,b])$ of all real-valued continuous functions on the bounded interval $[a,b]$, equipped with the sup-norm (see [6], [7], [2],[1], [10], [3], [4], [11]). Here we assume that $p = m^2$, where $m \in \mathcal{C}^1([a,b])$, and investigate for which domain $D(A)$ the strongly continuous semigroup generated by A is analytic.

Supported by M.U.R.S.T. 60% and 40%.

The main idea consists in interpretating A as a suitable perturbation of A_1^2, where $A_1 f = mf'$ is defined on the domain $D(A_1)$ consisting of all $f \in \mathcal{C}([a,b]) \cap \mathcal{C}^1(]a,b[)$ such that $f(a) = 0 = f(b)$ and

$$\lim_{x \to a^+} m(x)f'(x) = 0 = \lim_{x \to b^-} m(x)f'(x)$$

and in applying the same techniques of some perturbation results used in [9], 3.2 Theorem 2.1 (see also [5], Theorem 2.4). Thus, we are able to deal with the case of p vanishing at the endpoints of the interval.

In particular, some results concerning $p(x) = x^\alpha(1-x)^\alpha$ ($\alpha \in \mathbf{R}$, $x \in [0,1]$) can be derived.

Let us introduce the following

DEFINITION 1. *A function $m : [a,b] \to \mathbf{R}$, where $-\infty \le a < b \le +\infty$, is called $\mathbf{R}$-admissible (Γ-admissible, resp.), if the following holds*

(1) *m is continuous;*
(2) *$\forall x \in]a,b[: m(x) > 0$;*
(3) *there exists $z \in]a,b[$ such that*

$$\int_a^z \frac{1}{m(x)}\, dx = +\infty = \int_z^b \frac{1}{m(x)}\, dx$$

$$\left(\int_a^z \frac{1}{m(x)}\, dx < +\infty \qquad and \qquad \int_z^b \frac{1}{m(x)}\, dx < +\infty, \qquad resp. \right)$$

REMARK 2.

If m is $\mathbf{R}$-admissible and $a > -\infty$ ($b < +\infty$, resp.), then $m(a) = 0$ ($m(b) = 0$, resp.).

EXAMPLES.

(1) Every Lipschitz continuous function which vanishes at all finite endpoints of $]a,b[$ is $\mathbf{R}$- admissible.
(2) Let us consider $m : [0,1] \to \mathbf{R}$ defined by

$$m(x) := x(1-x) \text{ for every } x \in [0,1].$$

Then, for all $\alpha \ge 1$, m^α is $\mathbf{R}$-admissible, while for all $0 \le \alpha < 1$, m^α is Γ-admissible.

Let us introduce the spaces

$$\mathcal{C}_o([a,b]) := \{f \in \mathcal{C}([a,b]) \mid f(a) = 0 = f(b)\}$$

and

$$\mathcal{C}_p([a,b]) := \{f \in \mathcal{C}([a,b]) \mid f(a) = f(b)\},$$

where $a, b \in \mathbf{R}$, $a < b$.

We investigate how admissibility implies that certain first order differential operators generate analytic semigroups on the above spaces.

PROPOSITION 3. *Let us consider $m \in \mathcal{C}([a,b])$, m **R**- admissible (Γ - admissible, resp.). Then the operator $Af := mf'$ with domain $D(A)$ consisting of all $f \in \mathcal{C}_o([a,b]) \cap \mathcal{C}^1(]a,b[)$ ($f \in \mathcal{C}_p([a,b]) \cap \mathcal{C}^1(]a,b[)$, resp.) such that*

$$\lim_{x \to a^+} m(x)f'(x) = 0 = \lim_{x \to b^-} m(x)f'(x) \tag{1.1}$$

$$(\lim_{x \to a^+} m(x)f'(x) = \lim_{x \to b^-} m(x)f'(x) \in \mathbf{R}, resp.)$$

generates a strongly continuous group on $\mathcal{C}_o([a,b])$ (on $\mathcal{C}_p([a,b])$, resp.).

PROOF: Let us assume that m is **R**-admissible and observe that m is admissible in the sense of Definition 3.16 of [8, B-II]. Consequently, (see [8, B-II, Proposition 3.18]), the operator $Af = mf'$, with domain given by the subspace of $\mathcal{C}_o([a,b]) \cap \mathcal{C}^1(]a,b[)$ of all functions f satisfying (1.1), generates a strongly continuous group on $\mathcal{C}_o([a,b])$, which is similar to the translation group on the space $\mathcal{C}_o(\mathbf{R})$ of all real-valued continuous functions on $\mathbf{R}$, vanishing at infinity.

Now, let us consider m $\Gamma-$ admissible.

For every $x \in [a,b]$, let us define

$$\phi(x) := \int_z^x \frac{1}{m(y)} \, dy,$$

where $z \in]a,b[$ satisfies (3) of Definition 1. The Γ-admissibility of m assures that $\alpha := \phi(a)$ and $\beta := \phi(b)$ are real numbers.

Since

$$\phi'(x) = \frac{1}{m(x)} > 0 \qquad \text{for every } x \in]a,b[,$$

the mapping ϕ is strictly increasing in $[a,b]$. Hence ϕ is invertible and its inverse ψ is continuous in $[\alpha,\beta]$, differentiable in $]\alpha,\beta[$ with derivative different from 0.

Consequently, ϕ defines a homeomorphism from $[a,b]$ onto $[\alpha,\beta]$ and a diffeomorphism from $]a,b[$ onto $]\alpha,\beta[$.

Moreover, the operator

$$V: \mathcal{C}_p([\alpha,\beta]) \to \mathcal{C}_p([a,b])$$

defined by

$$Vf = f \circ \phi,$$

is a lattice isomorphism from $\mathcal{C}_p([\alpha,\beta])$ onto $\mathcal{C}_p([a,b])$.

If we set $B := V^{-1}AV$ with domain

$$\begin{aligned}
D(B) &= \{f \in \mathcal{C}_p([\alpha,\beta]) \mid Vf \in D(A), \, AVf \in \mathcal{C}_p([a,b])\} \\
&= \{f \in \mathcal{C}_p([\alpha,\beta]) \mid f \circ \phi \in D(A), \, A(f \circ \phi) \in \mathcal{C}_p([a,b])\} \\
&= \{f \in \mathcal{C}_p([\alpha,\beta]) \mid f \circ \phi \in \mathcal{C}^1(]a,b[), \, m[(f' \circ \phi)\phi'] \in \mathcal{C}_p([a,b])\} \\
&= \{f \in \mathcal{C}_p([\alpha,\beta]) \mid f \in \mathcal{C}^1(]\alpha,\beta[), \, f' \in \mathcal{C}_p([\alpha,\beta])\} \\
&= \{f \in \mathcal{C}_p([\alpha,\beta]) \cap \mathcal{C}^1(]\alpha,\beta[) \mid \lim_{x \to \alpha^+} f'(x) = \lim_{x \to \beta^-} f'(x) \in \mathbf{R}\},
\end{aligned}$$

then $Bf = f'$ for every $f \in D(B)$.

By virtue of [8, A-I, Example 2.5], B generates the group of periodic translations on $C_p([\alpha, \beta])$, i.e. the group $(T(t))_{t \in \mathbf{R}}$ such that

$$T(t)f(x) = f(y), \quad y = x + t \bmod (\beta - \alpha), \quad t \in \mathbf{R}.$$

Therefore, the operator $A = VBV^{-1}$ generates the strongly continuous group $(S(t))_{t \in \mathbf{R}}$ on $C_p([a, b])$ given by

$$S(t) = VT(t)V^{-1} \quad \text{for every } t \in \mathbf{R},$$

i.e., for every $f \in C_p([a, b])$ we have

$$S(t)f(x) = f(\psi(y)), \quad y = \phi(x) + t \bmod(\beta - \alpha)), \quad t \in \mathbf{R}. \quad \blacksquare$$

PROPOSITION 4. *Let us assume that* $m \colon [a, b] \to \mathbf{R}$ *is* $\mathbf{R}$-*admissible* (Γ- *admissible, resp.*) *and* $m \in C^1([a, b])$. *Then the operator* $Gf := m^2 f''$, *defined on* $C_o([a, b])$ *(on* $C_p([a, b])$, *resp.), with domain*

$$D(G) := \{ f \in C_o([a, b]) \cap C^2(]a, b[)|\ \lim_{x \to a^+} m(x)f'(x) = \lim_{x \to b^-} m(x)f'(x) = 0 =$$
$$= \lim_{x \to a^+} m^2(x)f''(x) = \lim_{x \to b^-} m^2(x)f''(x)\}$$
$$(D(G) := \{ f \in C_p([a, b]) \cap C^2(]a, b[)|\ \lim_{x \to a^+} m(x)f'(x) = \lim_{x \to b^-} m(x)f'(x) \in \mathbf{R},$$
$$\lim_{x \to a^+} m^2(x)f''(x) = \lim_{x \to b^-} m^2(x)f''(x) \in \mathbf{R}\},$$
$$\text{resp.)}$$

generates an analytic semigroup.

PROOF: Let $m \in C^1([a, b])$ be $\mathbf{R}$-admissible. If $(A, D(A))$ is the operator defined in Proposition 3, then from [8, A-II, Theorem 1.15], it follows that A^2 generates an analytic semigroup, where

$$D(A^2) = \{ f \in D(A)|\ Af \in D(A)\}$$
$$= \{ f \in D(A)|\ mf' \in D(A)\}$$
$$= \{ f \in C_o([a, b]) \cap C^1(]a, b[)|\ mf' \in C_o([a, b]) \cap C^1(]a, b[), m(mf')' \in C_o([a, b])\}.$$

This implies that

$$D(A^2) = \{ f \in C_o([a, b]) \cap C^2(]a, b[)|\ \lim_{x \to a^+} m(x)f'(x) = \lim_{x \to b^-} m(x)f'(x) = 0$$
$$= \lim_{x \to a^+} m^2(x)f''(x) = \lim_{x \to b^-} m^2(x)f''(x)\}.$$

Let us observe that

$$A^2 f = mm'f' + m^2 f'' = m'Af + m^2 f''$$

for every $f \in D(A^2)$. Consequently, the operator $Gf := m^2 f''$, with domain $D(G) = D(A^2)$ can be written as

$$Gf = (A^2 - m'A)f.$$

Since the semigroup generated by $(A^2, D(A^2))$ is analytic, [8, A-II, Theorem 1.14] implies that there exist $M > 0$ and $r \geq 0$ such that for every $\lambda \in \mathbf{C}$, with $Re\,\lambda > 0$ and $|\lambda| \geq r$ we have that $\lambda \in \rho(A^2)$ and

$$\|R(\lambda, A^2)\| \leq \frac{M}{|\lambda|}. \tag{1.2}$$

Moreover, $\lim_{|\lambda| \to \infty} \|R(\lambda, A)\| = 0$ implies that there exists $\mu \in \rho(A)$ such that

$$\|R(\mu, A)\| < \frac{1}{4(1 + M)(\|m'\| + 1)}.$$

Let

$$N := \|\mu A R(\mu, A)\|$$

and fix $\lambda \in \mathbf{C}$, with $Re\,\lambda > 0$ and $|\lambda| \geq r$.

Then, for every $f \in \mathcal{C}_o([a, b])$ we obtain

$$\begin{aligned}
\|AR(\lambda, A^2)f\| &= \|(\mu - A)R(\mu, A)AR(\lambda, A^2)f\| \\
&< N\,\|R(\lambda, A^2)f\| + \frac{\|A^2 R(\lambda, A^2)f\|}{4(1 + M)(\|m'\| + 1)} \\
&\leq (\frac{NM}{|\lambda|} + \frac{\|A^2 R(\lambda, A^2)\|}{4(1 + M)(\|m'\| + 1)})\|f\|.
\end{aligned}$$

From the inequality

$$\|A^2 R(\lambda, A^2)\| = \|\lambda R(\lambda, A^2) - I\| \leq M + 1,$$

it follows that

$$\|m'AR(\lambda, A^2)\| < \frac{NM\|m'\|}{|\lambda|} + \frac{1}{4}.$$

Consequently, there exists $R := \max\{4NM(\|m'\| + 1), r\} > 0$ such that

$$\forall \lambda \in \mathbf{C}, Re\,\lambda > 0, |\lambda| \geq R : \|m'AR(\lambda, A^2)\| < \frac{1}{2} \tag{1.3}$$

and thus the operator $I + m'AR(\lambda, A^2)$ has a bounded inverse in a suitable right half-plane.

Then every $\lambda \in \mathbf{C}$ with $Re\,\lambda > 0$ and $|\lambda| \geq R$ fulfills

$$\lambda - G = [I + m'AR(\lambda, A^2)](\lambda - A^2)$$

and hence, belongs to $\rho(G)$. From the estimates (1.2) and (1.3) we obtain also

$$\|R(\lambda, G)\| = \|R(\lambda, A^2)[I + m'AR(\lambda, A^2)]^{-1}\|$$

$$\leq \|R(\lambda, A^2)\| \sum_{n=0}^{\infty} \|m'AR(\lambda, A^2)\|^n$$

$$\leq \frac{M}{|\lambda|}\left(\frac{1}{1 - \|m'AR(\lambda, A^2)\|}\right)$$

$$< \frac{2M}{|\lambda|},$$

for any $\lambda \in \mathbf{C}$ with $Re\,\lambda > 0$ and $|\lambda| \geq R$. So G generates an analytic semigroup. If m is Γ- admissible, we can prove the assertion exactly in the same way. ∎

REMARK 5.

Let us observe that Proposition 4 can be applied to

$$m(x) := x^{\alpha}(1 - x)^{\alpha} \quad x \in [0, 1]$$

with $\alpha \geq 1$. In the case $0 \leq \alpha < 1$ Proposition 3 holds, but the assumptions of Proposition 4 are not fulfilled. Thus, it should be interesting to extend the result of Proposition 4 to more general classes of mappings m.

ACKNOWLEDGEMENT.

The authors would like to express their heartfelt thanks to Rainer Nagel for stimulating discussions on the subject.

REFERENCES

1. P.H. Clement and C. A. Timmermans, *On C_o - semigroups generated by differential operators satisfying Ventcel's boundary conditions*, Indag. Math. **89** (1986), 379-387.

2. G. Da Prato and P. Grisvard, *Equations d'evolution abstraites non lineaires de type parabolique*, Ann. Mat. Pura Appl., **70** (1979), 329-396.

3. J. A. Goldstein and C. Y. Lin, *Highly degenerate parabolic boundary value problems*, Differential and Integral Equations **2** (1989), 216-227.

4. _________________, *Degenerate parabolic problems and the Wentzel boundary condition*, in "Semigroup Theory and Applications , Ph. Clement, S. Invernizzi et al. (eds)," Lect. Notes in Pure and Appl. Mathem. 116, M. Dekker, 1989, pp. 189-199.

5. T. Kato, "Perturbation Theory for Linear Operators," Springer - Verlag, Berlin - Heidelberg -New York, 1966.

6. R. Martini, *A relation between semigroups and sequences of approximation operators*, Indag. Math. **35** (1973), 456-465.

7. R. Martini and W. L. Boer, *On the construction of semigroups of operators*, Indag. Math. **36** (1974), 392-405.

8. R. Nagel (ed), "One-parameter Semigroups of Positive Operators," Springer-Verlag New York, 1986.

9. A. Pazy, "Semigroups of Linear Operators and Applications to Partial Differential Equations," Springer-Verlag Berlin - Heidelberg - New York, 1983.

10. C. A. Timmermans, *On C_o - semigroups in a space of bounded continuous functions in the case of entrance or natural boundary points*, in "Approximation and Optimization, Proceedings of the International Seminar, held at the University of Havana, January 12 - 16, 1987," Lect. Notes in Mathematics 1354, 209 - 216, Springer-Verlag, 1988.

11. M. Ulmet, *Boundary conditions for one - dimensional positive semigroups*, Semigroup Forum **45** (1992), 92 - 119.

Sur l'Equation Générale
$u_t = a(\cdot,\, u,\, \varphi\,(\cdot,\, u)_x)_x + v$ dans L^1
I. Etude du Problème Stationnaire

Philippe Bénilan   UA CNRS 741, Université de Franche-Comté, Besançon Cedex, France

Hamidou Toure   UA CNRS 741, Université de Franche-Comté, Besançon Cedex, France, and Faculté des Sciences et Techniques, Université de Ouagadougou

1.Introduction

Nous étudions le problème général

$$(1) \qquad \begin{cases} u_t = a(.,u,\varphi(.,u)_x)_x + v \text{ sur } Q =]0,T[\times I \\ u = \ell \text{ sur } \Sigma =]0,T[\times \partial I \end{cases}$$

où I est un intervalle ouvert de $\mathbb{R}$, $v \in L^1(Q)$, $\ell : \partial I \to \mathbb{R}$, $a : (x,k,\xi) \in \mathbb{R}^3 \to \mathbb{R}$ continue et croissante (au sens large) en ξ, $\varphi : (x,k) \in \mathbb{R}^2 \to \mathbb{R}$ continue et croissante (au sens large) en k.

Mis à part des hypothèses techniques (que nous préciserons ci-dessous) sur la dépendance en x, l'hypothèse principale sera une coercivité de la fonction a par rapport à ξ, uniformément pour (x,k) borné.

En particulier, les résultats que nous développons ici généralisent ceux de [4] (cf. aussi [9]) pour l'équation

$$(2) \qquad u_t + f(u)_x = \varphi(u)_{xx} + v \text{ sur } Q, \ u = 0 \text{ sur } \Sigma$$

où f, φ sont des fonctions continues de $\mathbb{R}$ dans $\mathbb{R}$ avec φ croissante (au sens large); (2) correspond à $\ell = 0$, $a(x, k, \xi) = \xi - f(k)$, $\varphi(x, k) = \varphi(k)$.

Prenant $\varphi \equiv 0$, (2) se réduit à la loi de conservation

$$(3) \qquad u_t + f(u)_x = v \text{ sur } Q, \ u \equiv 0 \text{ sur } \Sigma$$

de telle sorte qu'il est clair que nous incluons dans (1) des problèmes hyperboliques du premier ordre, pour lesquels (même sous des hypothèses de régularité $\mathcal{C}^\infty$ sur les données) il n'y a aucun espoir d'avoir existence de solutions fortes globales ni de résultats d'unicité pour des solutions faibles. En d'autres termes nous devons introduire une notion de solution entropique pour le problème d'évolution du second ordre (1).

Suivant la démarche de [9] pour le problème (2), nous développons une notion de solution entropique pour (1) en deux étapes, grâce à la théorie des équations d'évolution dans un espace L^1.

Dans une première étape, qui est l'objet de ce premier article, nous étudions le "problème stationnaire "

$$(4) \qquad -a(., u, \varphi(., u)_x)_x = v \text{ sur } I, \ u = \ell \text{ sur } \partial I$$

pour lequel nous définissons une notion de solution entropique. Ceci permet d'associer aux données a, φ, ℓ un opérateur A de $L^1(I)$ et d'écrire formellement (1) sous la forme d'une équation d'évolution

$$(5) \qquad \frac{du}{dt} + Au \ni v \text{ sur }]0, T[\ .$$

Nous montrerons dans cet article que la fermeture dans $L^1(I)$ de l'opérateur A est m-accrétive à domaine dense (cf. Proposition 2.8 et Théorème 3.1 et 3.4). Ceci permettra, par application de la théorie des équations d'évolution dans un espace de Banach (cf. [3]) d'obtenir pour tout $v \in L^1(Q)$ et $u_0 \in L^1(I)$, l'existence et l'unicité d'une "bonne solution" $u \in \mathcal{C}([0, T]; L^1(I))$ de (5) avec $u(0) = u_0$.

Dans un deuxième article, nous préciserons en quel sens une "bonne solution" de (5) est solution de (1) ; entre autres, si $u_0 \in L^1(I) \cap L^\infty(I)$ et $v \in L^1(Q) \cap L^1(0, T; L^\infty(I))$, il sera clair que $u \in L^\infty(Q)$ et est solution entropique de (1) (cf. [4], [9] pour le cas de (2)).

La première section est consacrée à la définition d'une solution entropique de (4). Dans la deuxième section nous étudions l'opérateur A associé à a, φ, ℓ lorsque l'intervalle

I est borné. La troisième section est consacrée au cas I non borné ; nous considérons aussi, pour le cas $I =]0, \infty[$ un problème aux limites

$$(6) \qquad \begin{cases} -a(., u, \varphi(., u)_x)_x = v \text{ sur }]0, \infty[\\ a(., u, \varphi(., u)_x)_x(0) \in \gamma(u(0)) \end{cases}$$

où γ est un graphe maximal monotone à domaine borné.

Enfin dans la dernière section, nous étudions la dépendance de l'opérateur A associé aux différents problèmes stationnaires, par rapport aux données a, φ, ℓ, γ. Ces résultats permettront par application de la théorie générale, d'obtenir la dépendance des "bonnes solutions" de (5) par rapport aux données.

Section 1. Solution entropique du problème stationnaire

On se donne

$$(1.1) \qquad a : \mathbb{R} \times \mathbb{R} \times \mathbb{R} \to \mathbb{R}, \, \varphi : \mathbb{R} \times \mathbb{R} \to \mathbb{R} \text{ continues}$$

et on suppose

$$(1.2) \qquad \begin{cases} a(x, k, \xi) \text{ est croissante en } \xi, \, \varphi(x, k) \text{ est croissante en } k \\ \varphi(x, k) \text{ est continuement dérivable par rapport à } x. \end{cases}$$

On pose

$$(1.3) \qquad H(x, k) = a(x, k, \varphi_x(x, k)) \text{ pour } (x, k) \in \mathbb{R} \times \mathbb{R}.$$

On se donne d'autre part I un intervalle ouvert de $\mathbb{R}$; on note

$$(1.4) \qquad \alpha_- = Inf\, I, \, \alpha_+ = Sup\, I, \, \partial I = \{\alpha_-, \alpha_+\} \cap \mathbb{R}$$

On se donne enfin $v \in \mathcal{D}'(I)$, $\ell : \partial I \to \mathbb{R}$ et on considère le problème aux limites

$$(PS) \qquad -a(., u, \varphi(., u)_x)_x = v \text{ sur } I, \, u = \ell \text{ sur } \partial I$$

Il sera pratique de considérer une solution de (PS) comme une fonction u définie (p.p) sur $\mathbb{R}$.

Définition 1.1. On appelle *solution faible* de (PS) toute fonction $u \in L^\infty_{loc}(\mathbb{R})$ vérifiant

i) $w := \varphi(., u) \in W^{1,\infty}_{loc}(\mathbb{R})$, $a(., u, w_x) \in L^1_{loc}(\mathbb{R})$ et

$$-a(., u, w_x)_x = v \text{ dans } \mathcal{D}'(I)$$

ii) Si $\alpha_- > -\infty$ (resp.$\alpha_+ < +\infty$) alors

$$u = \ell(\alpha_-)(\text{ resp.}\ell(\alpha_+))p.p \text{ sur }]-\infty, \alpha_-[\text{ (resp.}]\alpha_+, +\infty[)$$

Lorsque $\varphi \equiv 0$, le problème (PS) se ramène au problème du premier ordre

$$(1.5) \qquad\qquad -a(., u)_x = v \text{ sur } I, \ u = \ell \text{ sur } \partial I$$

pour lequel il est bien connu, si $a(x, k)$ n'est pas monotone en k, qu'il n'y aura pas en général existence d'une solution u continue sur $\overline{I}$ ni unicité d'une solution faible (au sens de la Définition 1). Il suffit par exemple de considérer $a(x, k) = k^2$, $v = 0$: si I est borné et $\ell(\alpha_-) \neq \ell(\alpha_+)$, il n'existe pas de solution continue sur $\overline{I}$; pour tout $c > 0$ et $E \subset I$ mesurable, la fonction $c\chi_E - c\chi_{I\setminus E}$ se prolonge sur $\mathbb{R}$ en une solution faible.

Afin d'obtenir un problème (PS) bien posé dans le cas général, nous introduisons la notion de solution entropique de (PS) en adaptant celle introduite par O. Oleinik [7] pour une loi de conservation. Nous utiliserons pour cela des fonctions $u \in L^\infty_{loc}(\mathbb{R})$ réglées, c'est-à-dire vérifiant

$$(1.6) \qquad\qquad u(x+) = \lim_{h \downarrow 0} \text{ ess } u(x + h), \ u(x-) = \lim_{h \downarrow 0} \text{ ess } u(x - h)$$

existent pour tout $x \in \mathbb{R}$; on posera

$$(1.7) \qquad \underline{u}(x) = u(x+) \wedge u(x-), \ \overline{u}(x) = u(x+) \vee u(x-), \ \mathcal{I}(u, x) =]\underline{u}(x), \overline{u}(x)[$$

Définition 1.2. On appelle *solution entropique* de (PS) une solution faible réglée u vérifiant :

iii) il existe $h \in \mathcal{C}(\mathbb{R})$ telle que $h = a(., u, w_x)$ p.p sur I

et pour tout $x \in \mathbb{R}$ avec $u(x+) \neq u(x-)$

$$(1.8) \qquad\qquad (u(x+) - u(x-))(h(x) - H(x, k)) \geq 0 \quad \forall k \in \mathcal{I}(u, x).$$

Remarque 1.3.

a) La condition entropique (1.8) n'intervient qu'aux points $x \in \overline{I}$: la fonction h qui est parfaitement définie par u sur I peut être prolongée arbitrairement en dehors de $\overline{I}$ où elle n'intervient pas.

b) L'existence d'une solution entropique implique que v est la restriction à I de la dérivée d'une fonction continue sur $\mathbb{R}$; supposons maintenant que v soit ainsi. Alors toute solution faible continue est solution entropique. En particulier si

(1.9) $\varphi(x, k)$ est strictement croissante en k pour tout $x \in \mathbb{R}$

alors toute solution faible est solution entropique.

En d'autres termes la condition entropique n'apparaît que lorsque, pour certaines valeurs de $x \in \overline{I}$, le graphe de la fonction croissante $\varphi(x, .)$ présente des "parties plates" .

c) Supposons toujours que v soit la restriction à I de la dérivée d'une fonction continue sur $\mathbb{R}$ et considérons u une solution faible réglée de (PS), $h \in \mathcal{C}(\mathbb{R})$ telle que $h = a(., u, w_x)$ p.p. sur I et $x \in \overline{I}$ tel que $u(x+) \neq u(x-)$.

La fonction $w = \varphi(., u)$ étant continue, on a

$$\varphi(x, k) = w(x) \text{ pour tout } k \in \mathcal{I}(u, x)$$

et donc

(1.10) $$\varphi_x(x, k) = \varphi_x(x, u(x-)) = \varphi_x(x, u(x+)) \quad \forall k \in \mathcal{I}(u, x)$$

Pour se fixer les idées, supposons $u(x+) > u(x-)$ et fixons $k \in \mathcal{I}(u, x)$; il existe $\delta > 0$ tel que

$$u < k \text{ p.p sur }]x - \delta, x[, \ u > k \text{ p.p sur }]x, x + \delta[$$

et donc

$$w \leq \varphi(., k) \text{ sur }]x - \delta, x[, \ w \geq \varphi(., k) \text{ sur }]x, x + \delta[;$$

il en résulte

$$\lim_{y \uparrow 0} \text{ sup ess } w_x(y) \geq \varphi_x(x, k), \qquad \lim_{y \downarrow 0} \text{ sup ess } w_x(y) \geq \varphi_x(x, k)$$

et donc si $x > \alpha_-$ (resp. $x < \alpha_+$), alors $h(x) \geq H(x, u(x-))$ (resp. $h(x) \geq H(x, u(x+))$). Dans le cas $u(x+) < u(x-)$, les inégalités seront inversées ; on a donc dans le cas général

(1.11) $$\begin{cases} si & x > \alpha_-, \ (u(x+) - u(x-))(h(x) - H(x, u(x-))) \geq 0 \\ si & x < \alpha_+, \ (u(x+) - u(x-))(h(x) - H(x, u(x+))) \geq 0 \end{cases}$$

Il résulte de (1.10) et (1.11) que si

(1.12) $a(x, k, \xi)$ est croissante (resp. décroissante) en k

et si $\alpha_+ = \infty$ (resp. $\alpha_- = -\infty$), alors toute solution faible réglée de (PS) est solution entropique.

Enonçons maintenant le résultat principal de cette section :

Théorème 1.4. Supposons I borné. Soient u_1, u_2 des solutions entropiques de (PS) correspondant respectivement à v_1, $v_2 \in \mathcal{D}'(I)$ et ℓ_1, $\ell_2 : \partial I \to \mathbb{R}$.

On suppose

$$(1.13) \qquad v_1 \le v_2 + f \text{ dans } \mathcal{D}'(I) \text{ avec } f \text{ mesure de Radon sur } \mathbb{R}.$$

Alors

$$(1.14) \qquad \int_I sign_0^+(\underline{u}_1 - \overline{u}_2)f \ge \sum_{\alpha \in \partial I} \varepsilon(\alpha)sign_0^+(\ell_1(\alpha) - \ell_2(\alpha))(h_1(\alpha) - h_2(\alpha))$$

où $sign_0^+ r = 0$ si $r \le 0$, $sign_0^+ r = 1$ si $r > 0$, $\varepsilon(\alpha_-) = 1$, $\varepsilon(\alpha_+) = -1$ et h_1, h_2 sont les fonctions correspondantes à u_1, u_2 dans la Définition 1.2.

Preuve

$sign_0^+(\underline{u}_1 > \overline{u}_2)$ est la fonction caractéristique de $\{\underline{u}_1 > \overline{u}_2\}$ qui est un ouvert de $\mathbb{R}$. Si $(]c_n, d_n[)$ sont les composantes connexes de l'ouvert $I \cap \{\underline{u}_1 > \overline{u}_2\}$, compte-tenu de (1.13), on a

$$\int_I sign_0^+(\underline{u}_1 - \overline{u}_2)f \ge \sum_n (h(c_n) - h(d_n))$$

avec $h = h_1 - h_2$. Notons d'autre part que $\ell_1(\alpha_-) > \ell_2(\alpha_-)$ (resp. $\ell_1(\alpha_+) > \ell_2(\alpha_+)$) si et seulement si il existe une composante connexe de l'ouvert $\{\underline{u}_1 > \overline{u}_2\}$ contenant α_- (resp. α_+) ; et alors pour l'un des intervalles $]c_n, d_n[$ on aura $c_n = \alpha_-$ (resp. $d_n = \alpha_+$). Il suffit donc de montrer qu'étant donné $]c, d[$ une composante connexe de $\{\underline{u}_1 > \overline{u}_2\}$, on a

$$[c \in \overline{I} \Rightarrow h(c) \ge 0] \text{ et } [d \in \overline{I} \Rightarrow h(d) \le 0].$$

Nous allons prouver la première implication : on suppose $c \in \overline{I}$ et on montre que $h_1(c) \ge h_2(c)$; la seconde peut être vérifiée de manière identique.

Puisque $]c, d[$ est une composante connexe de $\{\underline{u}_1 > \overline{u}_2\}$ on a

$$(1.15) \qquad \underline{u}_1(c) = u_1(c+) \wedge u_1(c-) \le \overline{u}_2(c) = u_2(c+) \vee u_2(c-).$$

D'autre part $u_1 > u_2$ p.p sur $]c, d[$ et donc

$$(1.16) \qquad u_1(c+) \ge u_2(c+)$$

Distinguons plusieurs cas :

    cas 1 : $u_1(c+) > u_1(c-)$ et $u_2(c-) > u_2(c+)$

Alors d'après (1.15) et (1.16), $k = u_1(c-) \cap u_2(c+) \in \mathcal{I}(u_1, c) \cap \mathcal{I}(u_2, c)$; appliquant la condition d'entropie, $h_1(c) \ge H(c, k) \ge h_2(c)$.

cas 2 : $u_2(c+) < u_1(c+) \leq u_1(c-)$

D'après (1.15), $u_2(c+) < u_1(c+) \leq u_2(c-)$ et donc

$$H(c, u_1(c+)) \geq h_2(c).$$

D'un autre côté, $\varphi(c,.)$ est constante sur $\mathcal{I}(u_2, c)$; raisonnant comme dans la remarque 1.3-c), on en déduit

$$h_1(c) \geq H(c, u_1(c+)).$$

cas $2'$: $u_2(c-) \leq u_2(c+) < u_1(c+)$

De manière identique au cas précédent, en inversant les rôles de u_1 et u_2, on obtiendra $h_1(c) \geq H(c, u_2(c+)) \geq h_2(c)$.

cas 3 : hypothèse complémentaire des cas 1, 2 et $2'$.

Compte-tenu de (1.15) et (1.16), on a $u_1(c+) = u_2(c+)$ et donc $(w_1 - w_2)(c) = 0$; d'autre part, $u_1 > u_2$ p.p sur $]c, d[$ et donc $w_1 - w_2 \geq 0$ sur $]c, d[$; il en résulte

$$\lim_{x \downarrow c} \; \sup \; \text{ess} \; (w_{1,x}(x) - w_{2,x}(x)) \geq 0$$

Puisque $w_{1,x} \in L^\infty_{loc}(\mathbb{R})$,

$$\lim_{x \downarrow c} \; \text{ess} \; (h_1(x) - a(x, u_2(x), w_{1,x}(x))) = 0$$

et donc

$$h_1(c) \geq h_2(c). \; \diamond$$

Comme corollaire on obtient la caractérisation des solutions entropiques à l'aide d'inégalités correspondant à celles de S.N. Kruskov dans le cas d'une loi de conservation ([5], cf. aussi [1] pour le cas des conditions au bord) :

Corollaire 1.5. L'intervalle I étant quelconque, soient $v \in \mathcal{D}'(I)$, $\ell : \partial I \to \mathbb{R}$ et $u \in L^\infty_{loc}(\mathbb{R})$ réglée avec $w = \varphi(., u) \in W^{1,\infty}_{loc}(\mathbb{R})$ et vérifiant la condition (ii) de la Définition 1.1. Supposons

(1.17)   pour tout $k \in \mathbb{R}$, $H_x(., k) + v$ est la restriction d'une mesure de Radon sur $\mathbb{R}$.

Alors u est solution entropique de (PS) si et seulement si il existe $h \in \mathcal{C}(\mathbb{R})$ avec $h = a(., u, w_x)$ p.p sur I telle que pour tout $k \in \mathbb{R}$ et tout $\zeta \in \mathcal{D}(\mathbb{R})^+$ on ait les inégalités suivantes

$$(1.18-1) \quad \begin{cases} \int_I sign_0^+(\underline{u}-k)\{\zeta_x(H(.,k)-h)+\zeta(H_x(.,k)+v)\} \geq \\ \displaystyle\sum_{\alpha\in\partial I} \varepsilon(\alpha)sign_0^+(\ell(\alpha)-k)\zeta(\alpha)(h(\alpha)-H(\alpha,k)) \end{cases}$$

$$(1.18-2) \quad \begin{cases} \int_I sign_0^+(k-\overline{u})\{\zeta_x(H(.,k)-h)+\zeta(H_x(.,k)+v)\} \leq \\ \displaystyle\sum_{\alpha\in\partial I} \varepsilon(\alpha)sign_0^+(k-\ell(\alpha))\zeta(\alpha)(h(\alpha)-H(\alpha,k)) \end{cases}$$

Preuve

Montrons d'abord la condition nécessaire. Notons h la fonction correspondant à u dans la Définition 1.2, fixons $\zeta \in \mathcal{D}(\mathbb{R})^+$, $k \in \mathbb{R}$ et montrons les inégalités (1.18).

Pour celà, posons $\widetilde{I} = I\cap]-R,+R[$, où $R > 0$ est tel que $\partial I \cup supp\zeta \subset]-R,+R[$; pour $\alpha \in \partial\widetilde{I}$ posons $\widetilde{\ell}(\alpha) = \ell(\alpha)$ si $\alpha \in \partial I$, $\widetilde{\ell}(\alpha) = 0$ si $\alpha \notin \partial I$; enfin posons $\widetilde{a}(x,\widetilde{k},\xi) = \zeta(x)a(x,\widetilde{k},\xi)$ et $\widetilde{\varphi}(x,\widetilde{k}) = \widetilde{\zeta}(x)\varphi(x,\widetilde{k})$ où $\widetilde{\zeta} \in \mathcal{D}(]-R,R[)^+$, $\widetilde{\zeta} = 1$ sur $supp\zeta$. On vérifie immédiatement que la fonction $\widetilde{u}$ définie par $\widetilde{u} = u$ sur $\widetilde{I}$, $\widetilde{u} = \widetilde{\ell}(\widetilde{\alpha}_-)$ sur $]-\infty,\widetilde{\alpha}_-[$, $\widetilde{u} = \widetilde{\ell}(\widetilde{\alpha}_+)$ sur $]\widetilde{\alpha}_+,\infty[$ est solution entropique de

$$-\widetilde{a}(.,\widetilde{u},\widetilde{\varphi}(.,\widetilde{u})_x)_x = \zeta v - \zeta_x h \text{ sur } \widetilde{I}, \ \widetilde{u} = \widetilde{\ell} \text{ sur } \partial\widetilde{I}$$

avec $\widetilde{h} = \zeta h$ comme fonction correspondante. D'un autre côté, la fonction $\widehat{u} = k$ est solution entropique de

$$-\widetilde{a}(.,\widehat{u},\widetilde{\varphi}(.,\widehat{u})_x)_x = -(\zeta H(.,k))_x \text{ sur } \widetilde{I}, \ \widehat{u} = k \text{ sur } \partial\widetilde{I}$$

Appliquant le théorème 1.4 avec $(u_1,u_2) = (\widetilde{u},\widehat{u})$ (resp. $(\widehat{u},\widetilde{u})$), on obtient (1.18-1) (resp. (1.18-2)).

Réciproquement, supposons que les inégalités (1.18) soient satisfaites. Etant donné $\zeta \in \mathcal{D}(I)^+$, appliquant (1.18-1) (resp. (1.18-2)) avec $k < \underset{supp\zeta}{\text{Inf}} \text{ ess } u$ (resp. $k > \underset{supp\zeta}{\text{Sup}} \text{ ess } u$) on obtient

$$\int \zeta v - \zeta_x h \geq 0 \geq \int \zeta v - \zeta_x h$$

et donc

$$-h_x = v \text{ dans } \mathcal{D}'(I).$$

Pour prouver les inégalités entropiques, fixons $x \in \mathbb{R}$ tel que $u(x+) \neq u(x-)$; pour se fixer les idées, supposons $u(x+) > u(x-)$, le cas $u(x+) < u(x-)$ se traitant de manière identique. Etant donné $k \in \mathcal{I}(u,x)$, il existe $\delta > 0$ tel que

$$u < k \text{ p.p sur }]x-\delta,x[, \ u > k \text{ p.p sur }]x,x+\delta[;$$

choisissons $\zeta \in \mathcal{D}(]x - \delta, x + \delta[)^+$ avec $\zeta(x) = 1$ et appliquons les inégalités (1.18) en remarquant que

$$\zeta_x(H(.,k) - h) + \zeta(H_x(.,k) + v) = (\zeta(H(.,k) - h))_x.$$

Si $x \in I$, on peut toujours supposer $supp\zeta \subset I$ et on obtient en intégrant

$$(1.19) \qquad\qquad -(H(x,k) - h(x)) \geq 0 \geq H(x,k) - h(x)$$

c'est-à-dire deux fois $h(x) \geq H(x,k)$. Si $x \in \partial I$, par exemple $x = \alpha_-$, on peut toujours supposer $supp\ \zeta \subset] - \infty, \alpha_+[$ et on obtient encore (1.19) puisque $\ell(\alpha_-) = u(x-) < k$. $\diamond$

Remarque 1.6. La caractérisation du corollaire 1.5 fait clairement apparaître des notions de sous-solutions entropiques et sur-solutions entropiques ; nous ne développerons pas plus ici ces notions et renvoyons à [3] pour une étude détaillée de ces notions dans le cas d'une loi de conservation scalaire.

Remarque 1.7. Dans le cas où $a(x, k, \xi)$ est monotone en k (cf. (1.12) dans la Remarque 1.3-c), on peut étendre le théorème 1.4 aux solutions faibles de (PS) (cf. [9] dans le cas $a(x, k, \zeta) = \xi - f(k)$ avec $f : \mathbb{R} \to \mathbb{R}$ continue monotone) ; on peut même d'ailleurs sous des hypothèses de monotonie en k, considérer des applications $a(x, k, \xi)$ multivoques ; mais nous ne développerons pas plus ce cas particulier.

Remarque 1.8. Si $H_x(., k) + v \in L^1_{loc}(\overline{I})$ pour tout $k \in \mathbb{R}$ et $k \to H_x(., k) + v$ est continue de $\mathbb{R}$ dans $L^1_{loc}(\overline{I})$, alors (1.18-1) pour tout $k \in \mathbb{R}$ est équivalent à

p.p k, il existe $s \in L^\infty(I)$ tel que $s \in sign^+(u - k)$ p.p sur I et

$$\int_I s\{\zeta_x(H(.,k) - h) + \zeta(H_x(.,k) + v)\} \geq \sum_{\partial I} \varepsilon sign_0^+(\ell - k)\zeta(h - H(.,k))$$

(cf. [2], Lemme 2-2 p.II–11) ; on a évidemment la même chose pour (1.18-2).

Section 2. L'opérateur associé au problème de Dirichlet sur un intervalle borné

On reprend les données a, φ de la section 1 avec les hypothèses (1.1), (1.2). On se donne I un intervalle ouvert borné de $\mathbb{R}$ et $\ell : \partial I \to \mathbb{R}$. On va associer au problème stationnaire

$$(PS) \qquad\qquad -a(., u, \varphi(., u)_x)_x = v \text{ sur } I,\ u = \ell \text{ sur } \partial I$$

un opérateur dans l'espace de Banach $L^1(I)$.

On note $A = A_{a,\varphi,\ell}$ l'opérateur de $L^1(I)$ défini par

$$(2.1) \qquad \begin{cases} v \in Au \Leftrightarrow u, v \in L^1(I) \text{ , il existe } \tilde{u} \text{ solution} \\ \text{entropique de (PS) telle que } u = \tilde{u} \text{ p.p sur } I \end{cases}$$

Remarque 2.1. Bien que nous ayions employé une notation multivoque, l'opérateur A est clairement un opérateur univoque ; en fait il est même un "opérateur local" :

$$(2.2) \qquad \text{si } v_1 \in Au_1, v_2 \in Au_2 \text{ alors } v_1 = v_2 \text{ p.p sur } \{u_1 = u_2\}$$

En effet

$$\varphi(.,u_1)_x = \varphi(.,u_2)_x \text{ p.p sur } \{\varphi(.,u_1) = \varphi(.,u_2)\} \supset \{u_1 = u_2\}$$

et donc $\{u_1 = u_2\} \subset \{h_1 = h_2\}$ où $h_i \in \mathcal{C}(\mathbb{R})$ est telle que $h_i = a(.,u_i,\varphi(.,u_i)_x)$ p.p sur I ; alors $v_1 = -h_{1,x} = -h_{2,x} = v_2$ p.p sur $\{u_1 = u_2\}$.$\diamond$

Comme corollaire du Théorème 1.4, on obtient la "s-T-accrétivité" dans $L^1(I)$ de l'opérateur A ; plus précisément on a :

Proposition 2.2. Soit $\lambda > 0$ et, pour $i = 1, 2$, $u_i, v_i, z_i \in L^1(I)$ tels que

$$(2.3) \qquad z_i + \lambda A u_i \ni v_i.$$

Supposons

$$(2.4) \qquad (z_1 - z_2)(u_1 - u_2) \geq 0 \text{ p.p sur } I;$$

alors

$$(2.5) \qquad \int (z_1 - z_2)^+ \leq \int (v_1 - v_2)^+$$

Preuve

Posons $\tilde{v}_i = \dfrac{v_i - z_i}{\lambda} \in Au_i$ et $\mathcal{U} = \{z_1 > z_2\} \cup \{u_1 > u_2\}$.
Compte-tenu de (2.4),

$$(2.6) \qquad \int (z_1 - z_2)^+ = \int_{\mathcal{U}} (z_1 - z_2) \leq \int (v_1 - v_2)^+ - \lambda \int_{\mathcal{U}} (\tilde{v}_1 - \tilde{v}_2).$$

Notant $\widetilde{u}_i$ la solution entropique correspondant à $\widetilde{v}_i \in Au_i$ (cf. (2.1)), on a d'après (2.2)

$$\int_{\mathcal{U}} (\widetilde{v}_1 - \widetilde{v}_2) = \int sign_0^+(\underline{\widetilde{u}}_1 - \overline{\widetilde{u}}_2)(\widetilde{v}_1 - \widetilde{v}_2)$$

et donc d'après le Théorème 1.4, $\int_{\mathcal{U}} (\widetilde{v}_1 - \widetilde{v}_2) \geq 0$; d'où (2.5) en reportant cette inégalité dans (2.6).

On déduit également du Théorème 1.4, le "principe du maximum" pour l'opérateur A ; plus précisément on a :

Proposition 2.3. Avec la notation (1.3), supposons

$$(2.7) \qquad sign_0 \; k \; \frac{\partial H}{\partial x}(.,k) \leq c_0 + \omega|k| \text{ dans } \mathcal{D}'(I) \text{ pour tout } k \in \mathbb{R}$$

pour certaines constantes $c_0 \geq 0, \omega \in \mathbb{R}$. Etant donné $\lambda > 0$ avec $\lambda\omega < 1$ et $u, v \in L^1(I)$ avec $u + \lambda Au \ni v$, on a

$$(2.8) \qquad u^\pm \leq max\left(\sup_{\alpha \in \partial I} \ell(\alpha)^\pm, \sup_I ess \frac{(\lambda c_0 + v)^\pm}{1 - \lambda\omega} \right) \text{ p.p sur } I$$

Preuve

On démontre l'estimation de u^+, celle de u^- s'obtenant de manière identique. Notons k la constante du second membre de (2.8) ; la fonction $\widetilde{u} \equiv k$ est solution entropique de

$$\widetilde{u} - \lambda a(.,\widetilde{u},\varphi(.,\widetilde{u})_x)_x = \widetilde{v} \text{ sur } I, \; \widetilde{u} = k \text{ sur } \partial I$$

avec $\widetilde{v} = k - \lambda\dfrac{\partial H}{\partial x}(.,k)$. D'après (2.7) on a $\widetilde{v} \geq v$ dans $\mathcal{D}'(I)$ et donc

$$\frac{v - u}{\lambda} \leq \frac{\widetilde{v} - \widetilde{u}}{\lambda} + \frac{\widetilde{u} - u}{\lambda} \text{ dans } \mathcal{D}'(I)$$

Appliquant le Théorème 1.4 avec $(u_1, u_2) = (u, \widetilde{u})$ on obtient

$$\int -\frac{(u - \widetilde{u})^+}{\lambda} \geq 0$$

d'où $u \leq \widetilde{u}$. $\diamond$

Le résultat principal de cette section est le Théorème d'existence suivant :

Théorème 2.4. On suppose (2.7) et les deux propriétés suivantes :

i) $\dfrac{\partial H}{\partial x} \in BV_{loc}$ avec pour tout $R > 0$

$$(2.9) \qquad |\dfrac{\partial^2 H}{\partial x^2}| \leq \mu(R) \text{ dans } \mathcal{D}'(I \times] - R, R[)$$

$$(2.10) \qquad \dfrac{\partial^2 H}{\partial x \partial k} \leq C(R) \text{ dans } \mathcal{D}'(I \times] - R, R[)$$

où $\mu(R)$ est une mesure positive finie sur I et $C(R) \in \mathbb{R}$

ii) une hypothèse de coercivité de $a(x, k, \xi)$ par rapport à ξ :

$$(2.11) \qquad \lim_{|\xi| \to \infty} \inf_{I \times] - R, R[} |a(., ., \xi)| = +\infty \text{ pour tout } R > 0.$$

Alors pour tout $M > 0$, il existe $\lambda_M > 0$ tel que

$$R(I + \lambda A) \supset \{v \in BV(I); \|v\|_{L^\infty} \leq M\} \text{ pour } 0 < \lambda < \lambda_M.$$

Remarque 2.5. Notons que lorsque $a(x, k, \xi) = a(k, \xi)$ et $\varphi(x, k) = \varphi(k)$ sont indépendantes de x, les conditions (2.7), (2.9) et (2.10) sont automatiquement satisfaites. D'autre part, lorsque $a(x, k, \xi) = a_0(\xi) - f(x, k)$, la condition de coercivité (2.11) se réduit à

$$\lim_{|\xi| \to \infty} |a_0(\xi)| = \infty .$$

Pour prouver le Théorème 2.4 nous devons donc résoudre, au sens des solutions entropiques, le problème

$$(2.12) \qquad u - \lambda a(., u, \varphi(., u)_x)_x = v \text{ sur } I, \ u = \ell \text{ sur } \partial I$$

On peut faire diverses normalisations sur les données et supposer

$$(2.13) \qquad \varphi(x, 0) = 0, \ a(x, 0, 0) = 0, \lambda = 1$$

en remplaçant a, φ, v par

$$\widetilde{a}(x, k, \xi) = \lambda(a(x, k, \xi + \varphi_x(x, k)) - H(x, 0)),$$

$$\widetilde{\varphi}(x, k) = \varphi(x, k) - \varphi(x, 0), \ \widetilde{v} = v + \lambda H_x(x, 0)$$

Lorsque $\varphi(x, .)$ est un homéomorphisme de $\mathbb{R}$ sur $\mathbb{R}$ pour tout x, en faisant le changement d'inconnue $w = \varphi(., u)$ le problème (2.12) est équivalent à

$$(2.14) \qquad \beta(., w) - \lambda b(., w, w_x)_x = v \text{ sur } I, \ w = \varphi(., \ell) \text{ sur } \partial I$$

où

$$(2.15) \qquad \beta(x,.) = \varphi(x,.)^{-1} \text{ et } b(x,r,\xi) = a(x, \beta(x,r), \xi).$$

Il est clair, par des techniques classiques, (cf. par exemple [6]) que sous les hypothèses :

$$(2.16) \qquad \begin{cases} |b(x,r,\xi)| \le c(1 + |\xi|), \ b(x,r,\xi).\xi + c \ge c\,\xi^2 \\ |\beta(x,r)| \le c(1 + |r|) \end{cases}$$

alors pour tout $v \in L^2(I)$, il existe $w \in H^1(I)$ solution de (2.14) ;si de plus

$$(2.17) \qquad b(x,r,\xi), \ \beta(x,r) \text{ et } v(x) \text{ sont } \mathcal{C}^\infty, \ \frac{\partial b}{\partial x}(x,r,\xi) > 0,$$

alors $w(x)$ et $u(x) = \beta(x, w(x))$ sont $\mathcal{C}^\infty$.

Démontrons d'abord l'estimation de u dans $BV(I)$:

Lemme 2.6. Supposons que $\varphi(x,k)$ est $\mathcal{C}^\infty$, $\varphi(x,.)$ est un homéomorphisme de $\mathbb{R}$ sur $\mathbb{R}$ pour tout x et les fonctions b, β définies par (2.15) vérifient (2.16) et (2.17).

Soit $u \in \mathcal{C}^\infty(\overline{I})$ solution de (2.12). Alors

$$(2.18) \qquad \begin{cases} (1 - \lambda C(R))\|u_x\|_{L^1} \le \|v_x\|_{L^1} + \displaystyle\sum_{\partial I} (|v| + |\ell| + \lambda|H_x(.,\ell)|) \\ \qquad\qquad\qquad + \lambda \displaystyle\int \sup_{|k| \le R} |H_{xx}(.,k)| \end{cases}$$

où

$$R = \|u\|_{L^\infty} \text{ et } C(R) = \sup_{I \times]-R,R[} (H_{xk})^+$$

Preuve

Posons $z = \varphi(.,u)_x - \varphi_x(.,u) = \varphi_k(.,u)u_x$; on a

$$(2.19) \qquad u = d(.,u,z)_x + v$$

avec $d(x,k,\eta) = \lambda a(x,k,\varphi_x(x,k) + \eta)$.

Donnons-nous $p \in \mathcal{C}^\infty(\mathbb{R})$ avec $p' \ge 0$, $0 = p(0) \le |p| \le 1$.

Dérivant (2.19), multipliant par $p(z)$ et intégrant sur I, on a

$$(2.20) \qquad \begin{aligned} \int p(z)u_x &= \int p(z)v_x + \sum_{\partial I} \varepsilon p(z)(v - u) - \int p(z)_x \, d(.,u,z)_x \\ &\le \int |v_x| + \sum_{\partial I} (|v| + |\ell|) + I_1 + I_2 + I_3 \end{aligned}$$

avec

$$I_1 = -\int p^{'}(z)z_x\{d_x(.,u,z) - d_x(.,u,0) + d_k(.,u,z)u_x\}$$

$$\leq C\int |p^{'}(z)z|$$

$$I_2 = -\int p(z)_x d_x(.,u,0)_x$$

$$= \lambda\Big[\sum_{\partial I} \varepsilon p(z)H_x(.,u) + \int p(z)H_x(.,u)_x\Big]$$

$$I_3 = -\int p^{'}(z)d_\eta(.,u,z)z_x^2 \leq 0.$$

Faisant tendre $p(r)$ vers $sign_0\ r$, on a $I_1 \to 0$ et donc à la limite dans (2.20)

$$\int |u_x| \leq \int |v_x| + \sum_{\partial I} (|v| + |\ell| + \lambda|H_x(.,\ell)|)$$

$$+ \lambda\int |H_{xx}(.,u)| + H_{xk}(.,u)|u_x|$$

d'où (2.18) . $\diamond$

Preuve du Théorème 2.4

Fixons $\lambda_0 > 0$ avec $\lambda_0\omega < 1$. Etant donné $M > 0$ posons

$$R = max\Big(|\ell(\alpha_-)|, |\ell(\alpha_+)|, \frac{\lambda_0 c_0 + M}{1 - \lambda_0\omega} \Big) \text{ et } \lambda_M = min(\lambda_0,\ C(R)^{-1})$$

où $C(R)$ est définie par (2.10). Nous allons montrer que pour $v \in BV(I)$ avec $\|v\|_{L^\infty} \leq M$ et $\lambda \in]0, \lambda_M[$, il existe une solution de (2.12).

Pour cela, nous approchons le problème (2.12) par des "problèmes réguliers" de la manière suivante :

a) nous considérons $H^n \in \mathcal{C}^\infty(\mathbb{R} \times \mathbb{R})$ tels que $H^n \to H$ dans $\mathcal{C}(\mathbb{R} \times \mathbb{R})$ et pour tout n

$$(2.21) \qquad sign_0\ k\ H_x^n(x,k) \leq c_0 + \omega|k| \text{ pour tout } (x,k) \in I \times \mathbb{R}$$

$$(2.22) \qquad \int_I \Big(\sup_{|k|\leq R} |H_{xx}^n(x,k)| \Big) dx \leq \int_I \mu(R)$$

où $\mu(R)$ est définie par (2.10),

$$(2.23) \qquad H_{xk}^n \leq C(R) \text{ pour tout } (x,k) \in I\times]-R, R[$$

Ceci est toujours possible d'après les hypothèses (2.7), (2.9), (2.10) ;

b) nous considérons $\varphi^n \in \mathcal{C}^\infty(\mathbb{R} \times \mathbb{R})$ tels que $\varphi^n \to \varphi$, $\varphi_x^n \to \varphi_x$ dans $\mathcal{C}(\mathbb{R} \times \mathbb{R})$ et pour tout n, $\inf\limits_{\mathbb{R} \times \mathbb{R}} \varphi_k^n > 0$ de telle sorte que $\varphi^n(x,.)$ est un homéomorphisme de $\mathbb{R}$ sur $\mathbb{R}$ pour tout $x \in \mathbb{R}$ et $\beta^n(x,r) = \varphi^n(x,.)^{-1}$ est $\mathcal{C}^\infty$;

c) nous considérons $\widetilde{a}^n \in \mathcal{C}^\infty(\mathbb{R} \times \mathbb{R} \times \mathbb{R})$ tels que

$$\widetilde{a}^n(x,k,\eta) \longrightarrow \widetilde{a}(x,k,\eta) = a(x,k,\varphi_x(x,k)+\eta) - H(x,k)$$

dans $\mathcal{C}(\mathbb{R} \times \mathbb{R} \times \mathbb{R})$ et pour tout n

(2.24)
$$\inf_{\mathbb{R} \times \mathbb{R} \times \mathbb{R}} \widetilde{a}_\eta^n > 0, \ \widetilde{a}^n(x,k,0) = 0 \text{ pour tout } (x,k)$$

(2.25)
$$\inf_{\substack{|\eta|>m \\ x \in I \\ |k| \leq R}} |\widetilde{a}^n(x,k,\eta)| \geq \inf_{\substack{|\eta|>m \\ x \in I \\ |k| \leq R}} |\widetilde{a}(x,k,\eta)| \text{ pour } m \geq 1$$

d) nous posons

$$a^n(x,k,\xi) = \widetilde{a}^n(x,k,\xi - \varphi_x^n(x,k)) + H^n(x,k)$$

On a $a^n \to a$ dans $\mathcal{C}(\mathbb{R} \times \mathbb{R} \times \mathbb{R})$, $a^n(x,k,\varphi_x^n(x,k)) = H^n(x,k)$ et d'après (2.25) et (2.11)

(2.26)
$$\lim_{|\xi| \to \infty} \ \inf_{\substack{n \in \mathbb{N} \\ x \in I \\ |k| \leq R}} |a^n(x,k,\xi)| = +\infty \text{ pour tout } R > 0.$$

e) Enfin nous considérons $v^n \in \mathcal{C}^\infty(\overline{I})$ tels que $v^n \to v$ dans $L^1(I)$ et

(2.27)
$$v^n(\alpha_-) = v(\alpha_-^+), \ v^n(\alpha_+) = v(\alpha_+^-), \ \|v_x^n\|_{L^1} \leq var\, v, \ \|v^n\|_{L^\infty} \leq M.$$

Compte-tenu des hypothèses β^n, b^n, v^n vérifient (2.16) et (2.17) ; il existe donc pour tout n, une (unique) solution $u^n \in \mathcal{C}^\infty(\overline{I})$ de

$$u^n - \lambda a^n(., u^n, \varphi^n(., u^n)_x)_x = v^n \text{ sur } I, \ u^n = \ell \text{ sur } \partial I$$

D'après (2.21) et (2.27), la Proposition 2.3 donne

(2.28)
$$\|u^n\|_{L^\infty} \leq R$$

D'après (2.22), (2.23) et (2.27), le Lemme 2.6 donne :

$$\|u_x^n\|_{L^1} \leq (1 - \lambda C(R))^{-1} \left(var\, v + \sum_{\partial I} (|v| + |\ell| + \lambda|H_x^n(., \ell)|) + \lambda \int_I \mu(R) \right)$$

Notons maintenant que $(H_x^n(\alpha, \ell(\alpha)))$ est borné ; en effet

$$ sign_0 k\ H_x^n(x, k) \geq -\int_I \mu(R) - \frac{|H^n(\alpha_+, k) - H^n(\alpha_-, k)|}{\alpha_+ - \alpha_-} $$

et donc avec (2.21), H_x^n est borné uniformément sur $I \times]-R, R[$, il en résulte que

$$ (2.29) \qquad\qquad \|u_x^n\|_{L^1} \leq C $$

D'autre part, posant $w^n = \varphi^n(., u^n)$, $h^n = a^n(., u^n, w_x^n)$, on a d'après (2.28),

$$ (2.30) \qquad\qquad \|w^n\|_{L^\infty} + \|h_x^n\|_{L^\infty} \leq C $$

Utilisant (2.26), on en déduit

$$ (2.31) \qquad\qquad \|w_x^n\|_{L^\infty} + \|h^n\|_{L^\infty} \leq C. $$

Après extraction d'une sous-suite, on peut donc supposer $u^n \to u$ dans $L^1(I)$, $w^n \to w$, $h^n \to h$ dans $\mathcal{C}(\overline{I})$, et l'on a $u \in BV(I)$, $w = \varphi(., u) \in W^{1,\infty}(I)$, $h = a(., u, w_x) \in W^{1,\infty}(I)$ (en effet, $w_x^n \to w_x$ *-faiblement dans $L^\infty(I)$ et $a(x, k, \xi)$ est monotone en ξ). Il reste à prouver que u est solution entropique de (2.12) ; pour celà il suffit d'utiliser le Corollaire 1.5 (avec la Remarque 1.8) et de passer à la limite dans les inégalités (1.18-1), (1.18-2) appliquées avec u^n, h^n, H^n, v^n (en effet $H_x^n(., k) \to H_x(., k)$ dans $L^1(I)$ d'après (2.22) et $k \to H_x(., k)$ est continue de $\mathbb{R}$ dans $L^1(I)$ d'après (2.9)). $\diamond$

Remarque 2.7. D'après la preuve ci-dessus, sous les hypothèses du Théorème 2.4, étant donné M_1 il existe $\lambda_1 > 0$ et C_1 tels que pour $v \in BV(I)$ avec $\|v\|_{L^\infty} + var\ v \leq M_1$ et $0 < \lambda \leq \lambda_1$, $u = (I + \lambda A)^{-1} v \in BV(I)$ et $\|u\|_{L^\infty} + var\ u \leq C_1$.

On peut maintenant prouver le résultat qui permettra d'appliquer la théorie des équations d'évolution à l'opérateur A :

Proposition 2.8. Sous les hypothèses du Théorème 2.4,

1) $R(I + \lambda A)$ est dense dans $L^1(I)$ pour tout $\lambda > 0$

2) $D(A)$ est dense dans $L^1(I)$.

Preuve

Prouvons d'abord le point 2). Il suffit de montrer que $\overline{D(A)} \supset BV(I)$; fixons donc $v \in BV(I)$ et utilisant le Théorème 2.4 pour $\lambda > 0$ petit, considérons $u_\lambda = (I + \lambda A)^{-1} v$.

D'après la Remarque 2.7, $u_\lambda \in BV(I)$ et $\|u_\lambda\|_{L^\infty} + var\ u_\lambda$ est borné lorsque $\lambda \to 0$; posant $w_\lambda = \varphi(.,u_\lambda)$, $h_\lambda = a(.,u_\lambda,w_{\lambda,x})$ on a $u_\lambda = \lambda h_{\lambda,x} + v$; pour prouver que $u_\lambda \to v$ dans $L^1(I)$ lorsque $\lambda \to 0$, il suffit donc de montrer que $\lambda h_{\lambda,x} \to 0$ dans $\mathcal{D}'(I)$; puisque $\lambda h_{\lambda,x}$ est borné dans $L^\infty(I)$, il suffit donc de montrer que

$$(2.32) \qquad \lambda h_\lambda \to 0 \text{ en mesure sur } I \text{ lorsque } \lambda \to 0$$

Pour prouver (2.32), considérons $R > \|v\|_{L^\infty}$ et pour $r > 0$

$$C(r) = Inf\left\{ \frac{a(x,k,\xi).\xi}{|a(x,k,\xi)|} ; x \in I,\ |k| \le R,\ |a(x,k,\xi)| \ge r \right\}.$$

On a $\lim\limits_{r \to \infty} C(r) = +\infty$ et pour λ suffisamment petit,

$$(2.33) \qquad h_\lambda\, w_{\lambda,x} \ge C(|h_\lambda|)|h_\lambda|$$

Fixons $\delta > 0$ et $r_0 > 0$ tel que $C(r_0) \ge 0$. Pour $\lambda > 0$ suffisamment petit et vérifiant $\lambda r_0 \le \delta$, on a

$$
\begin{aligned}
C(\frac{\delta}{\lambda})\delta|\{\lambda|h_\lambda| > \delta\}| &\le \int_{\{\lambda|h_\lambda|>\delta\}} \lambda C(|h_\lambda|)|h_\lambda| \\
&\le \int_I \lambda h_\lambda w_{\lambda,x} - \int_{\{|h_\lambda|<r_0\}} \lambda h_\lambda w_{\lambda,x} \\
&= \sum_{\alpha \in \partial I} \varepsilon(\alpha)\lambda h_\lambda w_\lambda + \int_I (v - u_\lambda)w_\lambda - \int_{\{|h_\lambda|\le r_0\}} \lambda h_\lambda w_{\lambda,x}
\end{aligned}
$$

d'où

$$(2.34) \quad C(\frac{\delta}{\lambda})\delta|\{\lambda|h_\lambda| > \delta\}| \le \sum_{\alpha \in \partial I} \varepsilon(\alpha)\lambda h_\lambda w_\lambda + \int_I (v - u_\lambda)w_\lambda - \int_{\{|h_\lambda|\le r_0\}} \lambda h_\lambda w_{\lambda,x}$$

Notons maintenant que $\lambda\|h_\lambda\|_\infty$ est borné lorsque $\lambda \to 0$. Sinon, puisque $\lambda\|h_{\lambda,x}\|_\infty$ est borné, il existerait $\lambda_n \to 0$ tel que $|h_{\lambda_n}(x)| \to \infty$ uniformément pour $x \in \overline{I}$; on aurait alors d'après (2.33), $w_{\lambda_n,x} \to +\infty$ ou $w_{\lambda_n,x} \to -\infty$ uniformément sur I ; or ceci contredit le fait que w_λ est borné dans $L^\infty(I)$. Notons enfin que d'après l'hypothèse (2.11), $w_{\lambda,x}\chi_{\{|h_\lambda|\le r_0\}}$ est borné dans $L^\infty(I)$.

Il résulte donc de (2.34) que $C(\frac{\delta}{\lambda})\delta|\{\lambda|h_\lambda| > \delta\}|$ est borné et donc

$$|\{\lambda|h_\lambda| > \delta\}| \to 0 \text{ lorsque } \lambda \to 0.$$

Pour prouver le point 1), posons $k = \omega^+ + 1$. Il suffit de montrer que $R(kI + \overline{A}) = \overline{R(kI + A)} \supset BV(I)$. Fixons $v \in BV(I)$ et posons $A_1 = kI + \overline{A} - v$. Utilisant le

Théorème 2.4, pour tout $M > 0$, il existe $\lambda_M \in]0,1[$ tel que $R(I + \lambda A_1) \supset \{u \in L^\infty(I); \|u\|_\infty \leq M\}$ pour tout $0 < \lambda < \lambda_M$; de plus utilisant la Proposition 2.3, on a

$$\|(I + \lambda A_1)^{-1} u\|_\infty \leq max\left(\|\ell\|_\infty, \frac{\lambda(c_0 + \|v\|_\infty) + \|u\|_\infty}{1 + \lambda} \right);$$

on conclut alors de façon classique (cf. [8]) que $0 \in R(A_1)$.$\diamond$

Section 3. Cas d'autres problèmes aux limites

On reprend les données a, φ de la section 1 avec les hypothèses (1.1), (1.2).

On étudie d'abord le problème stationnaire

$$(3.1) \qquad\qquad -a(., u, \varphi(., u)_x)_x = v \text{ sur } \mathbb{R}.$$

On note ici $A = A_{a,\varphi}$ l'opérateur de $L^1(I)$ défini par

$$(3.2) \qquad v \in Au \Leftrightarrow v \in L^1(\mathbb{R}), u \in L^1 \cap L^\infty(\mathbb{R}) \text{ et } u \text{ est solution entropique de } (1)$$

Comme dans la section 2 (cf. Remarque 2.1), l'opérateur A est univoque et local.

Théorème 3.1. On suppose

$$(3.3) \qquad\qquad \lim_{(|x|,k)\to(\infty,0)} \frac{\varphi(x,k)}{x} = 0,$$

$$(3.4) \qquad\qquad \lim_{(x,k,\xi)\to(\pm\infty,0,0)} a(x,k,\xi) = h_\pm \text{ existe dans } \mathbb{R}$$

$$(3.5) \qquad\qquad x \longmapsto H(x,0) \text{ est absolument continue sur } \mathbb{R}$$

et que les hypothèses (2.7), (2.9), (2.10) et (2.11) sont satisfaites avec $I = \mathbb{R}$. Alors

1) L'opérateur A est s-T-accrétif dans $L^1(\mathbb{R})$

2) $R(I + \lambda A)$ est dense dans $L^1(\mathbb{R})$ pour tout $\lambda > 0$

3) $D(A)$ est dense dans $L^1(\mathbb{R})$.

Preuve du point 1)

Notons d'abord qu'étant donné $u \in D(A)$, il existe $w \in W_{loc}^{1,\infty}(\mathbb{R})$ et $h \in \mathcal{C}(\mathbb{R})$ tels que $w = \varphi(., u)$, $h = a(., u, w_x)$ p.p sur $\mathbb{R}$; puisque $h_x = -v \in L^1(\mathbb{R})$ on a $h \in AC(\mathbb{R})$

et donc en particulier $h \in \mathcal{C}([-\infty, +\infty])$; utilisant l'hypothèse de coercivité (2.11) (avec $I = \mathbb{R}$) on en déduit $w_x \in L^\infty(\mathbb{R})$. Nous allons montrer ci-dessous que

$$(3.6) \qquad h(\pm\infty) = h_\pm$$

On remarque également qu'étant donné un intervalle borné $I =]\alpha_-, \alpha_+[$, u est solution entropique de

$$-a(., u, \varphi(., u)_x)_x = v \text{ sur } I, u = \ell \text{ sur } \partial I$$

avec $\ell(\alpha_\pm) \in [\underline{u}(\alpha_\pm), \overline{u}(\alpha_\pm)]$.

Etant donné $u_1, u_2 \in D(A)$, $v_1 = Au_1$, $v_2 = Av_2$, utilisant le Théorème 1.4, pour tout intervalle borné $I =]\alpha_-, \alpha_+[$ on a

$$\int_{\alpha_-}^{\alpha_+} sign_0^+(u_1 - u_2)(v_1 - v_2) \geq -sign_0^+(\underline{u}_1(\alpha_+) - \overline{u}_2(\alpha_+))[h_1(\alpha_+) - h_2(\alpha_+)]$$
$$+ sign_0^+(\underline{u}_1(\alpha_-) - \overline{u}_2(\alpha_-))[h_1(\alpha_-) - h_2(\alpha_-)].$$

Mais d'après (3.6), $\lim_{\alpha \to \pm\infty} h_1(\alpha) - h_2(\alpha) = 0$ et donc

$$\int_{-\infty}^{+\infty} sign_0^+(u_1 - u_2)(v_1 - v_2) \geq 0;$$

utilisant la localité, ceci prouvera la s-T-accrétivité de A.

Nous prouvons maintenant (3.6). Notons d'abord, puisque $u \in L^1(\mathbb{R})$, qu'il existe $t_n \to \infty$ tel que $u(t_n x) \to 0$ p.p $x \in (1, 2)$, disons pour tout $x \in E \subset (1, 2)$ avec $(1, 2) \backslash E$ négligeable. Posons $p(x) = \lim_{n \to \infty} \inf w_x(t_n x)$, défini p.p $x \in (1, 2)$, disons pour tout $x \in E$ quitte à restreindre E.

Fixons $x_1, x_2 \in E$ avec $x_1 < x_2$; on a

$$\int_{x_1}^{x_2} w_x(t_n x) \, dx = x_2 \frac{\varphi(t_n x_2, u(t_n x_2))}{t_n x_2} - x_1 \frac{\varphi(t_n x_1, u(t_n x_1))}{t_n x_1}$$

et donc d'après l'hypothèse (3.3) et le lemme de Fatou (rappelons que $w_x \in L^\infty(\mathbb{R})$)

$$\int_{x_1}^{x_2} p(x) \, dx \leq 0.$$

On en déduit $p(x) \leq 0$ p.p $x \in (1, 2)$, disons à nouveau pour tout $x \in E$. Etant donné $x \in E$, il existe une suite extraite (n_k) et $\delta_k \to 0$ telles que $w_x(t_{n_k} x) \leq \delta_k$; on a

$$h(t_{n_k} x) \leq a(t_{n_k} x, u(t_{n_k} x), \delta_k)$$

et donc à la limite $h(\infty) \leq h_+$. On montrera de la même manière $h(\infty) \geq h_+$ (d'où $h(\infty) = h_+$) et $h(-\infty) = h_-$. $\diamond$

Pour la preuve des points 2) et 3) du théorème 1, nous utiliserons le résultat suivant

Proposition 3.2. On suppose satisfaite l'hypothèse (3.5), que les hypothèses (2.7), (2.8), (2.9) et (2.10) sont satisfaites avec $I = \mathbb{R}$ et que l'hypothèse (2.11) est satisfaite pour tout intervalle I borné de $\mathbb{R}$. Alors pour tout $M > 0$, il existe $\lambda_M > 0$ tel que

$$R(I + \lambda A) \supset \{v \in BV(\mathbb{R}) \cap L^1(\mathbb{R}); \|v\|_\infty \leq M\} \text{ pour tout } \lambda \in]0, \lambda_M[.$$

Preuve

Fixons $\lambda_0 > 0$ avec $\lambda_0 \omega < 1$ (où ω est la constante intervenant dans (2.7) avec $I = \mathbb{R}$). Etant donné $M > 0$, posons

$$R = \frac{\lambda_0 c_0 + M}{1 - \lambda_0 \omega} \text{ et } \lambda_M = min(\lambda_0, C(R)^{-1})$$

où $c_0, C(R)$ sont les données dans (2.7) et (2.10) respectivement (avec $I = \mathbb{R}$). Montrons que pour tout $v \in BV(\mathbb{R}) \cap L^1(\mathbb{R})$ avec $\|v\|_\infty \leq M$ et $\lambda \in]0, \lambda_M[$, il existe une solution u de

$$(3.7) \hspace{4cm} u + \lambda A = v$$

vérifiant

$$(3.8) \quad \|u\|_{L^1} \leq \|v + \lambda H_x(.,0)\|_{L^1}, \ \|u\|_\infty \leq \frac{\|v\|_\infty + \lambda c_0}{1 - \lambda \omega}, \ var\ u \leq \frac{var\ v + \lambda \mu(R)(\mathbb{R})}{1 - \lambda C(R)}$$

où $\mu(R)$ est donné par (2.9) (avec $I = \mathbb{R}$).

Pour cela, considérons pour tout $\alpha > 0$, $I_\alpha =]-\alpha, \alpha[$ et A_α l'opérateur de $L^1(I_\alpha)$ correspondant à I_α et $\ell = 0$ sur ∂I_α. D'après le Théorème 2.4 (voir la preuve), il existe $u_\alpha \in BV(I_\alpha)$ tel que

$$u_\alpha + \lambda A_\alpha u_\alpha = v \chi_{I_\alpha}.$$

Utilisant la proposition 2.2 (en notant que $A_\alpha 0 = -H_x(x,0)$), la Proposition 2.3 et le Lemme 2.6, la fonction u_α (prolongée par 0 sur $\mathbb{R} \backslash I_\alpha$) vérifient les estimations (3.8).

En particulier, $\{u_\alpha\}$ est relativement compact dans $L^1_{loc}(\mathbb{R})$. Utilisant d'autre part l'hypothèse de coercivité (2.11) pour tout intervalle I borné, les fonctions $\varphi(., u_\alpha)_x$ sont bornées dans $L^\infty_{loc}(\mathbb{R})$. Considérant une suite $\alpha_n \to \infty$ telle que $u_{\alpha_n} \to u$ dans $L^1_{loc}(\mathbb{R})$, on voit facilement que u est solution de (3.7) vérifiant (3.8). $\diamond$

Preuve des points 2) et 3) du Théorème 3.1

On suit pas à pas la preuve de la Proposition 2.8 grâce à la Proposition 3.2 remplaçant le Théorème 2.4. Pour le point 2), l'extension est immédiate. Pour le point 3), on adapte la preuve de la façon suivante.

Considérons $v \in BV(\mathbb{R}) \cap L^1(\mathbb{R})$ et pour λ suffisamment petit $u_\lambda = (I + \lambda A)^{-1} v$, $w_\lambda = \varphi(., u_\lambda)$, $h_\lambda = a(., u_\lambda, w_{\lambda,x})$. Puisque u_λ est borné dans $BV(\mathbb{R})$ et $\|u_\lambda\|_1 \leq \|v\|_1$, pour prouver que $u_\lambda \to v$ dans $L^1(\mathbb{R})$ lorsque $\lambda \to 0$, il suffit de montrer que $u_\lambda - v = \lambda h_{\lambda,x} \to 0$ dans $\mathcal{D}'(\mathbb{R})$. Il suffit donc de prouver que pour tout intervalle I borné et tout $\delta > 0$,

$$|\{x \in I; \lambda|h_{\lambda,x}| > \delta\}| \to 0 \text{ lorsque } \lambda \to 0.$$

Le raisonnement s'achève de manière identique. $\diamond$

Remarque 3.3. Les hypothèses (3.3), (3.4) et la coercivité globale (2.11) ne sont utilisées que pour prouver l'accrétivité de l'opérateur A.

On traiterait exactement de la même manière le problème aux limites

$$(3.9) \qquad -a(., u, \varphi(., u)_x)_x = v \text{ sur }]0, \infty[, \; u(0) = \ell_0$$

Nous allons plus généralement considérer un problème aux limites

$$(3.10) \qquad \begin{cases} -a(., u, \varphi(., u)_x)_x = v \text{ sur }]0, \infty[\\ a(., u, \varphi(., u)_x)_x(0) \in \gamma(u(0)) \end{cases}$$

où γ est un graphe maximal monotone de $\mathbb{R}$ avec

$$(3.11) \qquad D(\gamma) \text{ borné}$$

On note maintenant $A = A_{a,\varphi,\gamma}$ l'opérateur (univoque et local) de $L^1(0, \infty)$ défini par

$$(3.12) \qquad v \in Au \iff \begin{cases} v \in L^1(0, \infty), \; u \in L^1(0, \infty) \cap L^\infty(0, \infty) \\ \text{et il existe } \ell \in D(\gamma) \text{ et } \tilde{u} \text{ solution entropique de} \\ -a(., \tilde{u}, \varphi(., \tilde{u})_x)_x = v \text{ sur }]0, \infty[, \tilde{u}(0) = \ell \\ \text{tels que } h(0) \in \gamma(\ell), \tilde{u} = u \text{ p.p sur }]0, \infty[\\ (\text{ avec les notations de la Définition 1.2 }) \end{cases}$$

Théorème 3.4. On suppose (3.11) ,

$$(3.13) \qquad \lim_{(x,k) \to (\infty, 0)} \frac{\varphi(x, k)}{x} = 0,$$

$$(3.14) \qquad \lim_{(x,k,\xi)\to(\infty,0,0)} a(x,k,\xi) = h_+ \quad \text{existe dans } \mathbb{R}$$

$$(3.15) \qquad x \longmapsto H(x,0) \text{ est absolument continue sur }]0,\infty[$$

et que les hypothèses (2.7), (2.9), (2.10) et (2.11) sont satisfaites avec $I =]0,\infty[$. Alors

1) L'opérateur A est s-T-accrétif dans $L^1(0,\infty)$

2) $R(I + \lambda A)$ est dense dans $L^1(0,\infty)$ pour tout $\lambda > 0$

3) $D(A)$ est dense dans $L^1(0,\infty)$.

Le point 1) va résulter immédiatement du lemme suivant

Lemme 3.5. Supposons que (3.13), (3.14) ainsi que (2.11) avec $I =]0,\infty[$ sont satisfaites. Considérons $\lambda > 0$ et pour $i = 1,2$, $v_i \in L^1(0,\infty)$, $\ell_i \in \mathbb{R}$ et $\widetilde{u}_i$ solution entropique de

$$(3.16) \qquad -a(.,\widetilde{u},\varphi(.,\widetilde{u})_x)_x = \frac{v - \widetilde{u}}{\lambda} \text{ sur }]0,\infty[, \ \widetilde{u}(0) = \ell$$

correspondant à v_i, ℓ_i et vérifiant $u_i = \widetilde{u}_{i_{|]0,\infty[}} \in L^1(0,\infty) \cap L^\infty(0,\infty)$. Alors

$$(3.17) \qquad \int_0^\infty (u_1 - u_2)^+ + \lambda \, sign_0^+(\ell_1 - \ell_2)(h_1(0) - h_2(0)) \leq \int_0^\infty (v_1 - v_2)^+$$

où h_i est la fonction correspondant à $\widetilde{u}_i$ dans la Définition 1.2.

Preuve du Lemme 3.5

On a $h_i \in AC(]0,\infty[)$ et suivant la preuve du point 1 du Théorème 3.1, $h_i(+\infty) = h_+$; aussi pour tout $\alpha > 0$,

$$\int_0^\alpha sign^+(u_1 - u_2)\left(\frac{v_1 - u_1}{\lambda} - \frac{v_2 - u_2}{\lambda}\right) \geq$$

$$sign_0^+(\ell_1 - \ell_2)(h_1(0) - h_2(0)) - sign_0^+(\underline{u}_1(\alpha) - \overline{u}_2(\alpha))(h_1(\alpha) - h_2(\alpha)).$$

A la limite lorsque $\alpha \to +\infty$, on obtient (3.17). $\diamond$

Preuve du Théorème 3.4-1)

Si $u_i + \lambda A u_i = v_i$, on est dans les conditions du Lemme 3.5 avec $h_i(0) \in \gamma(\ell_i)$. Puisque γ est monotone, le deuxième terme de (3.17) est positif ou nul et donc

$$\int (u_1 - u_2)^+ \leq \int (v_1 - v_2)^+. \ \diamond$$

Pour la preuve des points 2) et 3) du Théorème 3.4, nous utilisons le résultat suivant

Proposition 3.6. On suppose satisfaite l'hypothèse (3.15), que les hypothèses (2.7), (2.8), (2.9) et (2.10) sont satisfaites avec $I =]0, \infty[$ et que l'hypothèse (2-11) est satisfaite pour tout intervalle ouvert borné $I \subset]0, \infty[$. Alors pour tout $M > 0$, il existe $\lambda_M > 0$ tel que pour $0 < \lambda < \lambda_M$, $v \in BV(0, \infty) \cap L^1(0, \infty)$ avec $\|v\|_\infty \leq M$ et $\ell \in [-M, M]$, il existe une (unique) solution entropique de (3.16) vérifiant $u = \widetilde{u}_{|]0,\infty[} \in L^1(0, \infty) \cap BV(0, \infty)$. De plus

$$(3.18) \qquad \|\widetilde{u}\|_\infty \leq max\left(|\ell|, \frac{\|v\|_\infty + \lambda c_0}{1 - \lambda \omega}\right)$$

$$(3.19) \qquad var\ \widetilde{u} \leq (1 - \lambda C(R))^{-1}[var\ v + |v(0)| + |\ell| + \lambda |H_x(0, \ell)| + \lambda \mu(R)(]0, \infty[)]$$

où $R = \|\widetilde{u}\|_\infty$ et ω, c_0, $\mu(R)$, $C(R)$ sont les données dans (2.7), (2.9), (2.10) (avec $I =]0, \infty[$).

Preuve de la Proposition 3.6

On suit exactement la preuve de la Proposition 3.2, considérant pour $\alpha > 0$, $\widetilde{u}_\alpha$ la solution entropique de

$$(3.20) \qquad -a(., \widetilde{u}_\alpha, \varphi(., \widetilde{u}_\alpha)_x)_x = \frac{v - \widetilde{u}_\alpha}{\lambda} \text{ sur }]0, \alpha[, \ \widetilde{u}_\alpha(0) = \ell, \ \widetilde{u}_\alpha(\alpha) = 0$$

On obtient les estimations de $\widetilde{u}_\alpha$ dans L^∞ et BV correspondant à (3.18) et (3.19) : le seul problème concerne l'estimation de $u_\alpha = \widetilde{u}_{\alpha|]0,\infty[}$ dans $L^1(0, \infty)$. On a clairement (cf. Lemme 3.5),

$$\|u_\alpha\|_1 \leq \|v + \lambda H_x(., 0)\|_1 + \lambda |h^\alpha(0)|;$$

puisque u_α est borné dans $L^\infty(0, \infty)$, il en est de même de h_x^α et $w_\alpha = \varphi(., u_\alpha)$; utilisant la coercivité (2.11), on en déduit que h^α est borné sur tout borné de $(0, \infty)$; donc $h^\alpha(0)$ est borné et u_α est borné dans $L^1(0, \infty)$.$\diamond$

Preuve du Théorème 3.4-2) et 3)

Suivant la preuve de la Proposition 2.8 en l'adaptant comme dans la démonstration des points 2) et 3) du Théorème 3.1, on se ramène à montrer qu'étant donné $M > 0$,

$$(3.21) \qquad \begin{cases} \text{il existe } \lambda_M > 0 \text{ tel que pour tout } 0 < \lambda < \lambda_M \\ R(I + \lambda A) \supset \{v \in BV(0, \infty) \cap L^1(0, \infty); \|v\|_\infty \leq M\} \end{cases}$$

En effet d'après (3.18), $D(\gamma)$ étant borné (hypothèse (3.11)) si $u + \lambda Au \ni v$ avec $\lambda > 0$, $\lambda\omega < 1$, $v \in L^\infty(0,\infty)$, alors

$$\|u\|_\infty \leq max\left(L_0, \frac{\|v\|_\infty + \lambda c_0}{1 - \lambda\omega}\right)$$

où $L_0 = sup\{|\ell|; \ell \in D(\gamma)\}$.

Pour démontrer (3.21) on peut toujours supposer $M \geq L_0$.

Considérons $\lambda_M > 0$ construit à la Proposition 3.6 et fixons $\lambda \in]0, \lambda_M[$ et $v \in BV(0,\infty) \cap L^1(0,\infty)$ avec $\|v\|_\infty \leq M$. Utilisant la Proposition 3.6, pour $\ell \in \overline{D(\gamma)}$ considérons $\widetilde{u}_\ell$ l'unique solution entropique de (3.16) et posons $T(\ell) = -h_\ell(0)$, où h_ℓ est la fonction correspondante.

Par définition de l'opérateur A, on a $v \in R(I + \lambda A)$ si et seulement si il existe $\ell \in D(\gamma)$ tel que $T(\ell) + \gamma(\ell) \ni 0$. Il suffit donc de montrer que T est une application croissante continue sur $\overline{D(\gamma)}$; en effet alors $T + \gamma$ sera un graphe maximal monotone à domaine borné et donc surjectif.

La croissance de T se déduit immédiatement du Lemme 3.5 ; en effet (3.17) donne pour $\ell_1 > \ell_2$

$$\int_0^\infty (\widetilde{u}_{\ell_1} - \widetilde{u}_{\ell_2})^+ - \lambda(T(\ell_1) - T(\ell_2)) \leq 0.$$

Notons que cette inégalité montre également que l'application $\ell \to \widetilde{u}_\ell$ est décroissante. Pour montrer la continuité de T donnons-nous (ℓ_n) une suite monotone de $\overline{D(\gamma)}$ de limite ℓ ; on a $\widetilde{u}_{\ell_n} \to \widetilde{u}$ dans $L^1(0,\infty)$, d'où $h_{\ell_n} \to h$ dans $C([0,\infty[)$ avec $-h(0) = \lim_n T(\ell_n)$, $\widetilde{u} - \lambda h_x = v$ dans $\mathcal{D}'(]0,\infty[)$. Raisonnant comme dans la fin de la preuve du Théorème 2.4 (dans un cas beaucoup plus simple), on voit que $\widetilde{u} = \widetilde{u}_\ell$, $h = h_\ell$; donc

$$T(\ell) = \lim_n T(\ell_n).\diamond$$

Remarque 3.7. Le point 1) du Théorème 3.4 reste évidemment valable pour un graphe maximal monotone γ quelconque. Le point 2) est par contre faux en général pour γ quelconque. Il suffit de prendre $\gamma \equiv 0$, $\varphi \equiv 0$, $a(x,k,\xi) = \xi + \dfrac{k^2}{2}$; l'opérateur A est alors défini par

$$v \in Au \Leftrightarrow v \in L^1(0,\infty), u \in L^1(0,\infty) \cap L^\infty(0,\infty),$$

u est solution entropique de $-(\dfrac{u^2}{2})_x = v$ sur $]0,\infty[$ et $u^2(0) = 0$; en d'autres termes, étant donné $v \in L^1(0,\infty)$ on a $u + Au \ni v$ si et seulement si u est solution entropique de

$$(3.22) \qquad\qquad -\left(\frac{u^2}{2}\right)_x = v - u \text{ sur }]0,\infty[, u(0) = 0$$

et $u(0+) = 0$; on vérifie directement pour $v = c\chi_{]0,x_0[}$ avec $c > 0$, $x_0 > 0$, que la solution entropique u de (3.22) existe et vérifie $u(0+) > 0$; on en déduit par comparaison, que $R(I + A)$ est disjoint de $\{v \in L^1(0,\infty); v \geq 0, \lim_{x \to 0} \inf \text{ ess } v(x) > 0\}$ et donc $R(I + A)$ n'est pas dense dans $L^1(0,\infty)$.

On peut développer de nombreux cas particuliers où le point 2) du Théorème 3.4) reste valable avec des graphes γ à domaine non borné : C'est par exemple classique (cf. [2] dans un cadre un peu différent) si $a(x,k,\xi) = a(\xi)$, $\varphi(x,k) = \varphi(k)$, $0 \in \gamma(0)$.

Le problème reste cependant ouvert de trouver des conditions satisfaisantes générales reliant a, φ et γ qui permettent d'obtenir le point 2) du Théorème 3.4.

Section 4. Dépendance continue de l'opérateur par rapport aux données

Dans cette section nous étudions la dépendance des opérateurs définis précédemment par rapport aux données $(a, \varphi, \ell, \gamma)$, ce qui impliquera par la suite la dépendance continue des "bonnes solutions " du problème d'évolution par rapport à ces mêmes données.

Considérons pour commencer l'opérateur $A_{a,\varphi,\ell}$ associé au problème de Dirichlet sur un intervalle ouvert borné de la section 2, avec les notations de cette section. Rappelons que I désigne dans ce cas, un intervalle ouvert borné quelconque de $\mathbb{R}$. Nous énonçons le résultat obtenu :

Théorème 4.1. Supposons I borné, soient pour tout $n \in \overline{\mathbb{N}}$, les fonctions a^n, φ^n, ℓ^n satisfaisant aux hypothèses (1.1), (1.2), (2.21), (2.22) et (2.23) (avec c_0, ω, $\mu(R)$, $C(R)$ indépendant de n) et

$$(4.1) \qquad \lim_{|\xi| \to \infty} \inf_{\overline{\mathbb{N}} \times I \times]-R,R[} |a^n(.,.,\xi)| = +\infty \text{ pour tout } R > 0$$

$$(4.2) \qquad \begin{cases} a^n \to a^\infty \text{ dans } \mathcal{C}(\mathbb{R} \times \mathbb{R} \times \mathbb{R}) \\ \varphi^n \to \varphi^\infty \text{ dans } \mathcal{C}(\mathbb{R} \times \mathbb{R}) \\ \dfrac{\partial \varphi^n}{\partial x} \to \dfrac{\partial \varphi^\infty}{\partial x} \text{ dans } \mathcal{C}(\mathbb{R} \times \mathbb{R}) \\ \ell^n \to \ell^\infty \text{ dans } \mathbb{R}^2 \end{cases}$$

alors, posant $A^n = A_{a^n,\varphi^n,\ell^n}$ pour tout $n \in \overline{\mathbb{N}}$ on a

$$(4.3) \qquad (I + \lambda \overline{A}^n)^{-1} v \to (I + \lambda \overline{A}^\infty)^{-1} v \text{ dans } L^1(I), \forall v \in L^1(I), \lambda > 0$$

Plus précisément, étant donné $M > 0$, si λ_M est la constante du Théorème 2.4 (associée aux données c_0, ω, $\mu(R)$, $C(R)$), on a

$$(4.4) \qquad (I + \lambda A^n)^{-1} v \to (I + \lambda A^\infty)^{-1} v \text{ dans } L^1(I)$$

pour tout $v \in BV(I) \cap L^1(I)$, $\|v\|_\infty \leq M$, $0 < \lambda < \lambda_M$.

Preuve du Théorème 4.1 Puisque $\overline{A}^\infty$ est la fermeture dans $L^1(I)$ de

$$\{((I + \lambda A^\infty)^{-1}v, \frac{v - (I + \lambda A^\infty)^{-1}v}{\lambda}); v \in BV(I), 0 < \lambda < \lambda_{\|v\|_\infty}\}$$

(cf. Proposition 2.8), il suffit de prouver (4.4). Fixons $v \in BV(I)$, $0 < \lambda < \lambda_{\|v\|_\infty}$ et posons $u^n = (I + A^n)^{-1}v$.

On a (u^n) borné dans $BV(I)$ (cf. (2.8), (2.18)) ; on peut donc supposer $u^n \to u$ dans $L^1(I)$ et p.p ; on montre alors que $u = (I + \lambda A^\infty)^{-1}v$ comme dans la preuve du Théorème 1.4. $\diamond$

Considérons maintenant le cas du problème sur $I = \mathbb{R}$; on a

Théorème 4.2. On suppose maintenant $I = \mathbb{R}$. Soient, pour tout $n \in \overline{\mathbb{N}}$, les fonctions a^n, φ^n satisfaisant les hypothèses du Théorème 4.1, ainsi qu'aux hypothèses (3.3), (3.4) et (3.5) avec

$$(4.5) \qquad\qquad \lim_{R \to \infty} \sup_n \int_{|x|>R} |H_x^n(x,0)| \, dx = 0$$

Alors on a la conclusion du Théorème 4.1 pour les opérateurs $A^n = A_{a^n, \varphi^n}$ correspondant à $I = \mathbb{R}$.

Preuve

Fixons $v \in BV(\mathbb{R}) \cap L^1(\mathbb{R})$, $0 < \lambda < \lambda_{\|v\|_\infty}$ et posons $u^n = (I + \lambda A^\infty)^{-1}v$. On a (u^n) borné dans $BV(\mathbb{R}) \cap L^1(\mathbb{R})$ et $u^n \to (I + \lambda A^\infty)^{-1}v$ dans $L^1_{loc}(\mathbb{R})$. Nous avons seulement à montrer que $u^n \to u^\infty$ dans $L^1(\mathbb{R})$.

Pour celà, nous utilisons la technique (maintenant classique) suivante; remarquons d'abord que si

$$(4.6) \qquad\qquad v \geq (\text{ resp. } \leq) - H_x^n(.,0) \text{ pour tout } n,$$

alors on a $u^n \geq 0$ (resp. $u^n \leq 0$) puisque $(I + \lambda A^n)^{-1}(-\lambda H_x^n(.,0)) = 0$; mais

$$\int u^n = \int v + \lambda(H^n(\infty,0) - H^n(-\infty,0))$$

et donc, utilisant (4.5),

$$\int u^n \to \int u^\infty;$$

on en déduit $u^n \to u^\infty$ dans $L^1(\mathbb{R})$.

Dans le cas général on utilise

$$v_+^n = \sup(v, -\lambda H_x^n(.,0)), \quad v_-^n = \inf(v, -\lambda H_x^n(.,0));$$

on a d'après (4.5), $v_\pm^n \to v_\pm^\infty$ dans $L^1(I)$ et, avec la preuve ci-dessus,

$$u_\pm^n = (I + \lambda A^n)^{-1} v_\pm^\infty \to (I + \lambda A^\infty)^{-1} v_\pm^\infty$$

dans $L^1(\mathbb{R})$. Puisque $u_-^n \le u^n \le u_+^n$, on en déduit

$$\lim_{R \to 0} \sup_n \int_{|x| > R} |u^n| = 0$$

et donc (u^n) converge dans $L^1(\mathbb{R})$. $\diamond$

Considérons enfin le cas $I =]0, \infty[$ avec conditions au bord en $x = 0$ non linéaire.

Théorème 4.3. On suppose $I =]0, \infty[$. Soient, pour tout $n \in \overline{\mathbb{N}}$, les fonctions a^n, φ^n satisfaisant les hypothèses du Théorème 1 ainsi qu'aux hypothèses (3-13), (3-14) et (3-15) avec

$$(4.7) \qquad \lim_{R \to \infty} \sup_n \int_{x > R} |H_x^n(.,0)| \, dx = 0$$

Soient d'autre part γ^n des graphes maximaux monotones de $\mathbb{R}$ telles que

$$(4.8) \qquad (I + \gamma^n)^{-1}(r) \to (I + \gamma^\infty)^{-1}(r) \quad \forall r \in \mathbb{R}$$

$(4.9) \qquad D(\gamma^n)$ est borné uniformément en n.

Alors on a la conclusion du Théorème 4.1 pour les opérateurs $A^n = A_{a^n, \varphi^n, \gamma^n}$.

Preuve

Soient $v \in BV(I) \cap L^1$, $0 < \lambda < \lambda_{\|v\|_\infty}$, $u^n = (I + \lambda A^n)^{-1} v$, $w^n = \varphi(., u^n)$, $h^n = a^n(., u^n, w_{n,x})$ et $\ell^n \in \mathbb{R}$ tel que $h^n(0) \in \gamma^n(\ell^n)$, u^n est solution entropique de

$$u^n - \lambda a^n(., u^n, \varphi^n(., u^n)_x)_x = v \text{ sur }]0, \infty[, \ u^n(0) = \ell^n.$$

Compte-tenu des hypothèses (en particulier (4.8)), on a (u^n) borné dans $BV(I) \cap L^1(I)$, (w^n) et (h^n) borné dans $W_{loc}^{1,\infty}([0, \infty[)$; on peut supposer $\ell^n \to \ell$, $u^n \to u$ dans

$L^1_{loc}([0, \infty[)$ et on a u solution entropique de $u - \lambda a^\infty(., u, \varphi^\infty(., u)_x)_x = v$ sur $]0, \infty[$, $u(0) = \ell$.

Maintenant $w^n \to w = \varphi^\infty(., u)$ et $h^n \to h = a^\infty(., u, w_x)$ dans $\mathcal{C}([0, \infty[)$. Puisque $h^n(0) \in \gamma^n(\ell^n)$, d'après (8) on aura $h(0) \in \gamma^\infty(\ell)$ et donc $u = (I + \lambda A^\infty)^{-1}v$.

On démontre que $u^n \to u$ dans $L^1(I)$ comme dans la preuve du Théorème 4.2 puisque

$$\int_0^\infty u^n = \int_0^\infty v + \lambda(H^n(\infty, 0) - h^n(0)). \diamond$$

Bibliographie

[1] Bardos (C.L)., Le Roux (A.Y)., Nedelec (J.C.)., *First order Quasilinear Equations with boundary conditions ,Comm. in partial differentiel equations,4(9),1017-1034(1979).*

[2] Benilan (Ph.)., *Equation d'Evolution dans un espace de Banach quelconque et Application. Thèse de Doctorat d'Etat. Orsay 1972.*

[3] Benilan (Ph.)., Crandall (M.G)., Pazy (A.)., *Evolution Equation governed by accretive Operators (à paraître)*

[4] Benilan (Ph.)., Touré (H.)., *Sur l'équation générale* $u_t = \varphi(u)_{xx} - \psi(u)_x + v$. *Note C. R. Acad. Sc. Paris t. 299, série I, n° 18, 1984.*

[5] Kruskov (S.N)., *First order quasilinear Equations with several independent variables. Math. Sb. 81 (123), 228–255 = Math USSR Sbornik (10) (1970), 217–243.*

[6] Lions (J.L)., *Quelques méthodes de résolution des problèmes aux limites non linéaires. Dunod-Gauthier-Villars, Paris (1969).*

[7] Oleinik (O.A)., *Discontinuous solutions of nonlinear differential Equations. Amer. Math. Soc. Transl. (2) 26 (1963), p. 95–172.*

[8] Pierre (M.)., *Un Théorème général de génération de semi-groupes non linéaires. Israël Journal of Mathematics. Vol. 23, N° 3-4 (1976).*

[9] Touré (H.)., *Etude des équations générales* $u_t - \varphi(u)_{xx} + f(u)_x = v$ *par la théorie des semi-groupes non linéaires dans* L^1. *Thèse de 3ème Cycle, 1982, Université de Franche-Comté.*

Introduction to the Porcelli Lecture

Pasquale "Pat" Porcelli was born in Chicago on December 2, 1926. He received a B.S. from the Illinois Institute of Technology in 1947 and a Ph.D. in Mathematics from the University of Texas in 1952. He was an instructor at the University of Texas (1952-53), Assistant Professor at DePaul University (1953-54), Associate Professor at Illinois Institute of Technology (1954-58), in residence at the U.S. Army Mathematics Research Center, University of Wisconsin (1958-59) and became a professor at Louisiana State University in 1959. In 1965, Pat was named Boyd Professor of Mathematics, the University's most prestigious honor.

Pat was a very active research mathematician and he had extraordinary success working with undergraduate honors students, master's students and, most importantly, doctoral students.

After a long battle with cancer, Pat died December 7, 1972 at the age of forty-six. The void left by his passing will never be filled.

The talk by Jerry Bona entitled *"Initial-Boundary Value Problems for Model Equations for the Propagation of Long Waves"* was designated as the Porcelli lecture in honor of Pasquale Porcelli.

Initial-Boundary Value Problems for Model Equations for the Propagation of Long Waves

Jerry L. Bona and Laihan Luo   The Pennsylvania State University, University Park, Pennsylvania

1. INTRODUCTION

This paper is concerned with initial- and boundary-value problems for evolution equations of the form

$$u_t + u_x + P(u)_x - \nu u_{xx} - \alpha^2 u_{xxt} = 0, \tag{1.1}$$

where $u = u(x,t)$ is a real-valued function of the two real variables x and t, and subscripts adorning u connote partial differentiation. Here P is a smooth, real-valued function of one real variable which will be suitably restricted later, and ν and α are non-negative real numbers. Such equations are often called pseudo-parabolic and the very particular form appearing in (1.1) arises in the modeling of unidirectional long waves in nonlinear dispersive systems. In case P is quadratic, (1.1) was studied by Peregrine [28] and Benjamin *et al.* [4] as an alternative to the well-known Korteweg-de Vries equation [18]

$$u_t + u_x + uu_x + u_{xxx} = 0. \tag{1.2}$$

Supported in part by the National Science Foundation and the Keck Foundation.

Equations of type (1.1) have significant advantages over those of type (1.2) when it comes to the imposition of the non-homogeneous boundary conditions arising in practice, especially when numerical techniques for the approximation of solutions are implemented. This point is discussed in some detail in Bona *et al.* [13] and Bona and Winther [17, 18] in the context of modeling surface waves in a flume generated by a wavemaker, and it arises again in the analysis presented by Albert and Bona [1] of the relation between a general class of models of type (1.1) and their Korteweg-de Vries-type analogues.

Analysis of initial- and boundary-value problems for equations like (1.1) began with the paper of Bona and Bryant [5] on the regularized long wave equation

$$u_t + u_x + uu_x - u_{xxt} = 0. \tag{1.3}$$

When (1.3) is used as a model for waves in a channel, the variable x is proportional to distance in the direction of propagation, t is proportional to time and u represents the deviation of the surface of the fluid from its rest position as it sustains the two-dimensional propagation of small-amplitude long waves. In [5], equation (1.3) was posed in a quarter plane $\{(x,t) : x \geq 0, t \geq 0\}$ with a single boundary condition at $x = 0$, and a theory of existence, uniqueness and continuous dependence established. Posing (1.1) in this form is of somewhat more practical interest than the more commonly considered pure initial-value problem in which u is specified for all x at some fixed time, say $t = 0$.

The theory in [5] was extended in [6] to initial- and two-point boundary-value problem for the equation

$$u_t + u_x + uu_x - \nu u_{xx} - \alpha^2 u_{xxt} = 0 \tag{1.4}$$

posed on a bounded interval with the solution specified at the right- and left-hand endpoints of the interval. These results are of especial interest in regard to the construction and analysis of numerical schemes for (1.4) since numerically feasible approximation schemes are necessarily applied on bounded intervals. The results in [5] and [6] are further developed in various ways in [19, 26, 27]. For example, in [19] Dang and Tran study the more general initial- and two-point boundary-value problem

$$u_t + P(u)_x + G(u) - F(u, u_x, x, t)_x - (b(x,t)u_{xt})_x = H(x,t), \quad x, t \in [0,1] \times [0,T],$$

$$u(x,0) = f(x), \qquad\qquad\qquad\qquad x \in [0,1], \tag{1.5}$$

$$u(0,t) = g(t), \qquad u(1,t) = h(t), \qquad\qquad t \in [0,T].$$

In all of the works just cited, either the nonlinear term P is allowed to grow at most quadratically or the boundary conditions are taken to be homogeneous.

Considerable interest has been shown recently in dispersive, dissipative evolution equations with nonlinearities that grow a rates higher than quadratic (cf. [7, 8, 21, 22, 23]). Perhaps the foremost reason for this attention stems from efforts to understand the interaction between these three, competing effects. In addition, mathematical issues arise for nonlinearities of order higher than quadratic that are not easily understood. Of course one can also broaden the perspective and consider at the same time more general dispersive and dissipative processes.

In this paper, we study well-posedness of equation (1.1) for the initial- and two-point boundary-value problem

$$
\begin{cases}
\qquad u(x,0) = f(x), & x \in [a,b], \\
u(a,t) = g(t), \qquad u(b,t) = h(t), & t \in [0,T],
\end{cases}
\tag{1.6a}
$$

and for the quarter-plane problem

$$
\begin{cases}
u(x,0) = f(x), & x \in \mathbb{R}^+, \\
u(0,t) = g(t), & t \in [0,T].
\end{cases}
\tag{1.6b}
$$

It will be shown that both of these problems for the equation (1.1) possesses at least locally in time a unique classical solution which depends continuously on ν in $\mathbb{R}^+$ and on variations of the data f, g, and h within their respective function classes. The local existence theory for the initial-boundary-value problems is relatively straightforward, and does not depend on the detailed structive of P. A theory that is global in time is more difficult, and depends upon the derivation of *a priori* bounds on local solutions. The provision of such bounds appears to need a growth condition on P, namely that its grow at infinity at a rate which is not more than quartic.

In addition to establishing the well-posedness of (1.1)-(1.6), we consider the degradation of the wave in case the parameter ν is actually positive. This aspect has already received some attention in the case of the pure initial-value problem on the whole line $\mathbb{R}$ (cf. [2, 9, 20]) and the periodic initial-value problem (see [7]). Especially the decay problem on the entire line is decidedly non-trivial, but all these results rely upon the homogeneity of the boundary conditions. Here we study decay in the more practically interesting setting of the two-point boundary-value problem (1.6a). Our results in this arena are interesting in their own right, but in addition they justify certain modelling considerations that arose in [13] in connection with water waves in channels. There is also interest in decay theory for the quarter-plane problem (1.6b), but the results in hand for this context appear not to be sharp, and consequently they will not be reported here.

This paper is organized as follows. In Section 2 basic notation is reviewed and the main theorem of well posedness is stated. In Section 3, the quarter-plane problem (1.1)-(1.6b) is studied. Local existence is proved by converting the differential equation (1.1) into an equivalent integral equation. *A priori* bounds that apply uniformly on compact subsets of the temporal variable are then derived in the presence of a growth condition on P, and these are used to extend the local solutions indefinitely. The continuous dependence of the solution on variations in the initial- and boundary-data follows readily from the proof of local existence. The two-point boundary-value problem (1.1)-(1.6a) is considered in Section 4. The theory for this problem is not dissimilar to that for the quarter-plane problem, and hence the presentation is abbreviated. In case $\nu > 0$, the decay theory for the two-point boundary-value problem (1.1)-(1.6a) is then discussed in the final part of Section 4.

2. NOTATION AND STATEMENT OF THE MAIN RESULTS

Throughout the paper, all functions will be real-valued. For any Banach space X, the associated norm will be denoted $\|\cdot\|_X$ except for a few abbreviations noted below. Spaces that arise in our analysis include the standard spaces $C^k(\bar{\Omega})$ for Ω a bounded open set in $\mathbb{R}^+$, $k = 0, 1, 2, \cdots$, $L_p(\Omega)$ for $1 \leq p \leq \infty$, and the L_2-based Sobolev spaces $H^m(\Omega)$ for $m = 0, 1, 2, \cdots$ (cf. Lions [25], Treves [30]). If Ω is an unbounded open set in $\mathbb{R}^+$, $C_b^k(\bar{\Omega})$ is defined exactly as $C^k(\bar{\Omega})$ except that the function and its first k derivatives are required to be bounded.

In the analysis of the initial- and boundary-value problem (1.1)-(1.6), the spaces $H^m(\Omega)$ will occur often with m a positive integer and $\Omega = \mathbb{R}^+ = (0, +\infty)$, $\Omega = (0, 1)$ or $\Omega = (0, T)$. Because of their frequent occurrence, it is convenient to abbreviate their norms thusly:

$$\|\cdot\|_m = \|\cdot\|_{H^m(\mathbb{R}^+)}, \qquad \text{or} \qquad \|\cdot\|_m = \|\cdot\|_{H^m(0,1)} \qquad \text{and} \qquad |\cdot|_{m,T} = \|\cdot\|_{H^m(0,T)}. \quad (2.1)$$

If $m = 0$, the subscript m will be omitted altogether, so that

$$\|\cdot\| = \|\cdot\|_{L_2(\mathbb{R}^+)}, \qquad \text{or} \qquad \|\cdot\| = \|\cdot\|_{L_2(0,1)} \qquad \text{and} \qquad |\cdot|_T = \|\cdot\|_{0,T}. \quad (2.2)$$

Let X be a Banach space, T be a positive real number and $1 \leq p \leq +\infty$. Then $L_p(0, T; X)$ denotes the Banach space of all measurable functions $u : (0, T) \to X$, such that $t \to \|u(t)\|_X$ is in $L_p(0, T)$. Similarly, by $C(0, T; X)$, we denote the subspace of $L_\infty(0, T; X)$ of all continuous functions $u : [0, T] \to X$.

For $\Omega = \mathbb{R}^+$ or $\Omega = [0,1]$, the abbreviation $\mathcal{B}_T^{k,l}$ will be employed for the functions $u : \Omega \times [0,T] \to \mathbb{R}$ such that $\partial_x^i \partial_t^j u \in C(0,T;C_b)$ for $0 \le i \le k$, and $0 \le j \le l$. This Banach space will carry the norm

$$\|u\|_{\mathcal{B}_T^{k,l}} = \sum_{\substack{0 \le i \le k \\ 0 \le j \le l}} \|\partial_x^i \partial_t^j u\|_{C(0,T;C_b)}.$$

The space $\mathcal{B}_T^{0,0}$ will be abbreviated simply $\mathcal{B}_T$ and its norm is just that of $L_\infty(\Omega \times [0,T])$.

In Sections 3 and 4, the following results are proved. For simplicity, and because all the interesting examples are thereby covered, it is assumed that P is a C^∞-function, though finite regularity assumptions suffice for most of our theory. The result is stated informally here, with precise versions provided later in the technical sections of the paper.

MAIN RESULTS. *Let $T > 0$, $f \in C_b^2(\mathbb{R}^+) \cap H^2(\mathbb{R}^+)$ and $g \in C^1(0,T)$ be given and suppose $f(0) = g(0)$. If the growth of P is no more than quartic at infinity, then there exists a unique solution u of (1.1) for the quarter-plane problem (1.6a) in the space $\mathcal{B}_T^{2,1} \cap C(0,T;H^2(\mathbb{R}^+))$. The solution depends continuously on the initial and boundary data, and on $\nu \ge 0$. Similar results hold for the initial- and two-point boundary-value problem (1.1)-(1.6b). If $\nu > 0$, and with appropriate decay assumptions on the boundary data g and h, the solution of (1.1)-(1.6b) tends to zero as t tends to infinity.*

3. WELL-POSEDNESS IN THE QUARTER-PLANE

In this section, interest will be focused on the initial- and boundary-value problem

$$u_t + u_x + P(u)_x - \nu u_{xx} - \alpha^2 u_{xxt} = 0, \qquad \text{for } x, t \ge 0, \tag{3.1a}$$

$$u(x,0) = f(x), \qquad \text{for } x \ge 0, \tag{3.1b}$$

$$u(0,t) = g(t), \qquad \text{for } t \ge 0. \tag{3.1c}$$

For consistency, the restriction

$$u(0,0) = f(0) = g(0), \tag{3.2}$$

will be imposed throughout the discussion and we take it that $\nu \ge 0$ and $\alpha \ne 0$.

By converting the differential equation (3.1a) with initial condition (3.1b) and boundary condition (3.1c) into an integral equation and applying the contraction mapping theorem to the integral equation, the existence of a local solution may be established. This local solution is extended to a global solution by appeal to an *a priori* estimation of smooth solutions of (3.1). The results of continuous dependence follow from the local theory.

3.1. Local Solution

To obtain a local existence theorem, we first convert the problem (3.1) into an integral equation. The argument closely parallels that given in detail in [5], and consequently many of the calculations are abbreviated.

Equation (3.1a) may be regarded as an ordinary differential equation for u_t by considering $\nu u_{xx} - u_x - P(u)_x$ as an external force. Solving this equation for u_t, formally integrating the solution by parts and then integrating from 0 to t, there appears the relation

$$u(x,t) = \exp(-t\nu/\alpha^2)f(x) + \tilde{g}(t)e^{-x/\alpha} + \mathbb{B}(u)(x,t), \qquad (3.3)$$

where

$$\mathbb{B}(u)(x,t) = \int_0^t \int_0^{+\infty} \exp(-\nu(t-\tau)/\alpha^2)K(x,\xi)\big[P(u(\xi,\tau)) + u(\xi,\tau)\big]d\xi d\tau$$

$$- \frac{\nu}{\alpha} \int_0^t \int_0^{+\infty} \exp(-\nu(t-\tau)/\alpha^2)L(x,\xi)u(\xi,\tau)d\xi d\tau, \qquad (3.4)$$

$$K(x,\xi) = \frac{1}{2\alpha^2}\big[\exp(-(x+\xi)/\alpha) + sgn(x-\xi)\exp(-|x-\xi|/\alpha)\big], \qquad (3.5)$$

$$L(x,\xi) = \frac{1}{2\alpha^2}\big[\exp(-(x+\xi)/\alpha) - \exp(-|x-\xi|/\alpha)\big], \qquad (3.6)$$

and

$$\tilde{g}(t) = g(t) - \exp(-t\nu/\alpha^2)g(0). \qquad (3.7)$$

Define the operator $\mathbb{A}$ by

$$(\mathbb{A}u)(x,t) = \exp(-t\nu/\alpha^2)f(x) + \tilde{g}(t)e^{-x/\alpha} + \mathbb{B}(u)(x,t). \qquad (3.8)$$

Then assuming that $f \in C_b(\mathbb{R}^+)$ and $g \in C(0,T)$, the operator $\mathbb{A}$ maps a function $u \in \mathcal{B}_T$ into itself since K is integrable. If T is chosen small enough, $\mathbb{A}$ is a contraction mapping of a ball centered at the origin in $\mathcal{B}_T$ into itself. This observation leads immediately to the following proposition.

PROPOSITION 3.1. *Let* $T > 0$, $f \in C_b(\mathbb{R}^+)$ *and* $g \in C(0, T)$. *Then there exists a positive constant*

$$T' = T'(\|f\|_{C_b(\mathbb{R}^+)}, \|g\|_{C(0,T)})$$

such that for any T_0 *with* $T_0 \leq \min\{T', T\}$, *there is a unique solution of (3.3) in* $\mathcal{B}_{T_0}$. *If* $f \in H^1(\mathbb{R}^+)$, *then there is a positive constant* $T'(\|f\|_1, \|g\|_{C(0,T)})$ *such that for any* $T_0 \leq \min\{T', T\}$ *there is a unique solution of (3.3) in* $C(0, T_0; H^1(\mathbb{R}^+))$. *In either case, for* T *sufficiently small, the mapping that associates to initial and boundary data* (f, g) *the solution* u *of (3.3) is continuous from* $C_b(\mathbb{R}^+) \times C(0, T)$ *into* $\mathcal{B}_T$ *or from* $H^1(\mathbb{R}^+) \times C(0, T)$ *into* $C(0, T; H^1(\mathbb{R}^+))$.

As mentioned, this proposition may be established by choosing positive values T' and R so that $\mathbb{A}$ is a contraction mapping of the ball of radius R centered at the origin in $\mathcal{B}_{T'}$. The crucial estimate is that for v, $w \in \mathcal{B}_{T'}$ with $\|v\|_{\mathcal{B}_{T'}}$, $\|w\|_{\mathcal{B}_{T'}} \leq R$, then

$$\|\mathbb{A}v - \mathbb{A}w\|_{\mathcal{B}_{T'}} \leq C(R)T'\|v - w\|_{\mathcal{B}_{T'}}$$

where the constant $C(R)$ is an absolute constant connected with norms of K and L times $\max_{|z| \leq R} |P'(z)|$. Once this inequality is in hand, the proof follows exactly the lines worked out in [1] or [5].

REMARK 3.2: The time interval T' for which $\mathbb{A}$ is inferred to be contractive depends inversely on $\|f\|_1$ and $\|g\|_{C(0,T)}$. If the boundary data g is given in $C(0, T)$ and it is known somehow that $\|u(\cdot, t)\|_1$ is bounded on bounded time intervals, then the contraction-mapping argument used to obtain Proposition 3.1 may be iterated to produce a solution of (3.3) defined for all t in $[0, T]$. This remark applies even if $T = +\infty$.

If $u \in \mathcal{B}_T$ is a solution of (3.3) and the boundary and initial data has regularity beyond just being bounded, continuous and consistent as in (3.2), it follows readily from the representation $u = \mathbb{A}u$ that u possesses additional regularity. The arguments leading to this conclusion parallel those spelled out in [4] and [5], and consequently we content ourselves with a statement of these useful results.

LEMMA 3.3. *Suppose* $f \in C_b^k(\mathbb{R}^+)$ *and* $g \in C^l(0, T)$ *with* $f(0) = g(0)$ *where* $k \geq 1$, $l \geq 0$ *and* $k > l$. *Let* $u \in \mathcal{B}_T$ *be a solution of (3.3). Then* $u \in \mathcal{B}_T^{k,l}$, *and, moreover, if* $k \geq 2$, $l \geq 1$, *then* u *is a classical solution of (3.1) on* $\mathbb{R}^+ \times [0, T]$. *Similarly, if* $f \in H^k(\mathbb{R}^+)$ *for some* $k > 1$ *and* $u \in C(0, T; H^1(\mathbb{R}^+))$ *is the solution of (3.3) guaranteed by Proposition 3.1, then* $u \in C(0, T; H^k(\mathbb{R}^+))$. *If* $f \in C_b^k(\mathbb{R}^+)$, $g \in C^l(0, T)$ *where* $k \geq 1$, $l \geq 0$, $k > l$ *and* $f^{(j)}(x) \to 0$ *as* $x \to +\infty$ *for* $0 \leq j \leq k$, *then the solution* u *of (3.1) has the property that*

$$\partial_x^j \partial_t^i u(x, t) \to 0 \qquad as \qquad x \to +\infty,$$

uniformly for $0 \leq t \leq T$, for $0 \leq j \leq k$, $0 \leq i \leq l$. In all cases, the mapping that associates the solution u to the initial and boundary data (f, g) is a continuous mapping between the function classes from which (f, g) and u are drawn.

With these preliminary results in hand, we are ready to undertake the derivation of *a priori* bounds that allow the local solution of (3.1) obtained via the contraction-mapping argument in Proposition 3.1 to be extended to arbitrary time intervals $[0, T]$.

3.2. Global Solutions

Suppose there is to hand a classical solution of (3.1) at least on a time interval $[0, T]$ for some $T > 0$. The following lemmas are aimed at extending this local solution to an arbitrary time interval.

The first foray will be into the situation with relatively weak assumptions on the initial data and with no dissipative effects. It will be handy in this result and later to define $\Lambda(s)$ by the specification

$$\frac{d\Lambda}{ds} = P(s), \qquad \Lambda(0) = 0. \tag{3.9}$$

LEMMA 3.4. *Let $T > 0$, $f \in C_b^2(\mathbb{R}^+) \cap H^1(\mathbb{R}^+)$ and $g \in C^1(0, T)$ be given with $f(0) = g(0)$. Suppose that $\Lambda(s)$ satisfies the one-sided growth condition*

$$\limsup_{|s| \to \infty} |s|^{-4} \Lambda(s) \leq 0. \tag{$*$}$$

Let $u(x, t)$ be the classical solution of (3.1) with $\nu = 0$ on $\mathbb{R}^+ \times [0, T']$ whose existence and regularity is guaranteed by Proposition 3.1 and Lemma 3.3. Then there exists a constant a_1 only depending on $\|f\|_1$ and $|g|_{1,T}$ such that $\|u(\cdot, t)\|_1 \leq a_1$.

PROOF: Multiply (3.1a) with $\nu = 0$ by $2u(x, t)$ and integrate the result over $\mathbb{R}^+ \times [0, t)$. After integrations by parts and using Lemma 3.3 to dismiss the boundary contributions at infinity, it appears that

$$\|u(\cdot, t)\|_1^2 = \|f\|_1^2 + \int_0^t \left[2Q(g(\tau)) + g(\tau)^2 - 2\alpha^2 g(\tau) u_{xt}(0, \tau) \right] d\tau, \tag{3.10}$$

where $Q(u) = \int_0^u \lambda P'(\lambda) d\lambda$. The Cauchy-Schwarz inequality then implies that

$$\|u(\cdot, t)\|_1^2 \leq C(\|f\|_1, |g|_{1,T}) + 2\alpha^2 |g|_T \left(\int_0^t u_{xt}^2(0, \tau) d\tau \right)^{\frac{1}{2}} \tag{3.11}$$

for $0 \leq t \leq T'$, where here and subsequently C will denote various constants that depend only on norms of the auxiliary data. Continue by multiplying (3.1a) with $\nu = 0$ by $2\alpha^2 u_{xt}(x,t) - 2P(u)$ and integrating the result over $\mathbb{R}^+ \times [0,t)$. After integrations by parts and using again Lemma 3.3, there obtains the relation

$$\alpha^2 \|u_x(\cdot,t)\|^2 + \alpha^4 \int_0^t u_{xt}^2(0,\tau)d\tau - 2\int_0^{+\infty} \Lambda(u(x,t))dx$$
$$= \alpha^2 \|f'\|^2 - 2\int_0^{+\infty} \Lambda(f(x))dx$$
$$+ \int_0^t \left[\alpha^2(g'(\tau))^2 + 2\alpha^2 P(g(\tau))u_{xt}(0,\tau) - P^2(g(\tau)) - 2\Lambda(g(\tau)) \right]d\tau. \tag{3.12}$$

It is deduced that

$$\alpha^4 \int_0^t u_{xt}^2(0,\tau)d\tau \leq C(\|f\|_1, |g|_{1,T}) + 2\|\Lambda(u)\|_{L_1}^2 \tag{3.13}$$
$$\leq C(\|f\|_1, |g|_{1,T}) + 2\|u\|^2 \tilde{E}(\|u\|^{\frac{1}{2}} \|u_x\|^{\frac{1}{2}}),$$

where

$$\Lambda(\lambda) = \lambda^2 E(\lambda), \qquad \tilde{E}(r) = \sup_{|\lambda| \leq r} E(\lambda),$$

by use of the elementary inequality

$$\|u(\cdot,t)\|_{L_\infty} \leq \sqrt{2}\|u(\cdot,t)\|^{\frac{1}{2}} \|u_x(\cdot,t)\|^{\frac{1}{2}}. \tag{3.14}$$

By using the assumption $(*)$, (3.13) may be further simplified to

$$\int_0^t u_{xt}^2(0,\tau)d\tau \leq C(\|f\|_1, |g|_{1,T}, \delta) + \delta\|u\|^2\|u\|_1^2, \tag{3.15}$$

for any $\delta > 0$. Substituting (3.15) into (3.11), it is concluded that there is a constant a_1 such that

$$\|u(\cdot,t)\|_1^2 \leq C(\|f\|_1, |g|_{1,T}) = a_1, \tag{3.16}$$

where a_1 only depends on $\|f\|_1$ and $|g|_{1,T}$. The lemma is proved. $\qquad \square$

REMARK 3.5: It is easy to see that if $\Lambda \leq 0$, for example when $P(u) = -u^{2m-1}$ where m is a positive integer, one immediately obtains H^1-bounds without resort to growth conditions. Even in case Λ is unrestricted in sign, it still follows from (3.11) and (3.13) that $\|u(\cdot,t)\|_1$ is bounded, independently of t, provided the initial data f and the boundary data g are small enough in $H^1(\mathbb{R}^+)$ and $H^1(0,T)$, respectively. This follows since (3.10) and (3.13) together imply that

$$\|u(\cdot,t)\|_1^2 \left(1 - \delta\tilde{E}(\|u\|_1)\right) \leq C(\|f\|_1, |g|_{1,T}),$$

where $\delta = |g|_T$. If δ is small enough relative to $\|f\|_1$ and if $\|f\|_1$ and $|g|_{1,T}$ are likewise not too large, then this last inequality provides a t-independent bound on $\|u(\cdot,t)\|_1$. Also note that estimates (3.11) and (3.15) yield a bound on $\|u(\cdot,t)\|_1$ that only depends on the norms of the auxiliary data and not explicitly on T. When boundary data g is specified in $H^1(\mathbb{R}^+)$, the solution $u(\cdot,t)$ of (3.1) with $\nu = 0$ is therefore bounded in $H^1(\mathbb{R}^+)$ independently of t. Finally, it is worth note that estimate (3.16), when unraveled implies that $\|u(\cdot,t)\|_1$ grows more or less linearly with the energy supplied by the wavemaker (cf. [5]). This satisfying state of affairs points to a certain consistency of the equation that lends credibility to its status as a model of real physical phenomena. $\qquad\square$

If stronger conditions on the initial value are given, the order of growth required of P can be as high as quartic while still maintaining a satisfactory theory of well-posedness. The following lemma applies even if $\nu > 0$.

LEMMA 3.6. *Suppose $f \in C_b^2(\mathbb{R}^+) \cap H^2(\mathbb{R}^+)$, $g \in C^1(0,T)$ for some $T > 0$ and $f(0) = g(0)$. It is presumed that P satisfies the growth condition*

$$\limsup_{|s| \to \infty} |s|^{-2}|P''(s)| \leq \varepsilon \qquad (**)$$

for some non-negative constant ε. Let $u(x,t)$ be a classical solution of (3.1) up to the boundary on $\mathbb{R}^+ \times [0,T']$ for some $T' \leq T$, as guaranteed in Lemma 3.3. Then for all $t \in [0,T']$, $u(\cdot,t) \in H^2(\mathbb{R}^+)$ and there exists a constant a_2 depending only on $T, \|f\|_2$ and $|g|_{C^1(0,T)}$ such that for $0 \leq t \leq T'$, $\|u(\cdot,t)\|_2 \leq a_2$.

PROOF: Multiply (3.1a) by $2u(x,t)$ and integrate the result over $\mathbb{R}^+ \times [0,t]$. After integration by parts and using Lemma 3.3, there obtains

$$\|u(\cdot,t)\|^2 + \alpha^2\|u_x(\cdot,t)\|^2 + \nu\int_0^t \|u_x(\cdot,\tau)\|^2 d\tau$$
$$= \|f\|^2 + \alpha^2\|f'\|^2 - 2\alpha^2 g(t)u_x(0,t) + 2\alpha^2 g(0)f'(0)$$
$$+ \int_0^t \left[2Q(g(\tau)) + g(\tau)^2 - 2\nu g(\tau)u_x(0,\tau) + 2\alpha^2 g'(\tau)u_x(0,\tau)\right] d\tau. \qquad (3.17)$$

By using some elementary inequalities, including (3.14) applied to u_x, we infer that

$$\left|\int_0^t -2\nu g(\tau)u_x(0,\tau)d\tau\right|$$
$$\leq \int_0^t 2\sqrt{2}\nu|g(\tau)|(\|u_x(\cdot,\tau)\|\|u_{xx}(\cdot,\tau)\|)^{\frac{1}{2}} d\tau$$
$$\leq 2^{\frac{3}{2}}\nu\left(\int_0^t |g(\tau)|^{\frac{4}{3}}\|u_{xx}(\cdot,\tau)\|^{\frac{2}{3}} d\tau\right)^{\frac{3}{4}} \left(\int_0^t \|u_x(\cdot,\tau)\|^2 d\tau\right)^{\frac{1}{4}} \qquad (3.18)$$
$$\leq \frac{\nu}{2}\int_0^t \|u_x(\cdot,\tau)\|^2 d\tau + C\nu|g|_T^{\frac{4}{3}}\left(\int_0^t \|u_{xx}(\cdot,\tau)\|^2 d\tau\right)^{\frac{1}{3}}.$$

Similarly one has

$$
\left| \int_0^t \alpha^2 g'(\tau) u_x(0,\tau) d\tau \right|
$$

$$
\leq \int_0^t \alpha^2 |g'(\tau)| \sqrt{2} \|u_x(\cdot,\tau)\|^{\frac{1}{2}} \|u_{xx}(\cdot,\tau)\|^{\frac{1}{2}} d\tau
$$

$$
\leq \sqrt{2}\alpha^2 \left(\int_0^t |g'(\tau)|^{\frac{3}{2}} d\tau \right)^{\frac{2}{3}} \left(\int_0^t \|u_x(\cdot,\tau)\|^6 d\tau \right)^{\frac{1}{12}} \left(\int_0^t \|u_{xx}(\cdot,\tau)\|^2 d\tau \right)^{\frac{1}{4}} \quad (3.19)
$$

$$
\leq C_T |g|_{1,T} \left[\left(\int_0^t \|u_x(\cdot,\tau)\|^6 d\tau \right)^{\frac{1}{3}} + \left(\int_0^t \|u_{xx}(\cdot,\tau)\|^2 d\tau \right)^{\frac{1}{3}} \right],
$$

and

$$
|2\alpha^2 g(t) u_x(0,t)| \leq \frac{\alpha^2}{2} \|u_x(\cdot,t)\|^2 + \frac{3\alpha^2}{2} |g(t)|^{\frac{4}{3}} \|u_{xx}(\cdot,t)\|^{\frac{2}{3}}. \quad (3.20)
$$

Using (3.18), (3.19) and (3.20) in (3.17) yields

$$
\|u(\cdot,t)\|_1^2 + \nu \int_0^t \|u_x(\cdot,\tau)\|^2 d\tau
$$

$$
\leq C(\|f\|_1, |g|_{C^1(0,T)}) + C|g(t)|^{\frac{4}{3}} \|u_{xx}(\cdot,t)\|^{\frac{2}{3}} \quad (3.21)
$$

$$
+ C_T |g|_{1,T} \left[\left(\int_0^t \|u_x(\cdot,\tau)\|^6 d\tau \right)^{\frac{1}{3}} + \left(\int_0^t \|u_{xx}(\cdot,\tau)\|^2 d\tau \right)^{\frac{1}{3}} \right],
$$

or, what is the same,

$$
\|u(\cdot,t)\|_1^6 + \nu^3 \left(\int_0^t \|u_x(\cdot,\tau)\|^2 d\tau \right)^3
$$

$$
\leq C(\|f\|_1, |g|_{C^1(0,T)}) + C|g(t)|^4 \|u_{xx}(\cdot,t)\|^2 \quad (3.22)
$$

$$
+ C_T |g|_{1,T}^3 \int_0^t \left[\|u_x(\cdot,\tau)\|^6 + \|u_{xx}(\cdot,\tau)\|^2 \right] d\tau.
$$

Now multiply (3.1a) by $2u_{xx}(x,t)$ and integrate the result over $\mathbb{R}^+ \times [0,t]$. After integration by parts and using Lemma 3.3, we reach the equation

$$
\|u_x(\cdot,t)\|^2 + \alpha^2 \|u_{xx}(\cdot,t)\|^2 + \nu \int_0^t \|u_{xx}(\cdot,\tau)\|^2 d\tau
$$

$$
= \|f'\|^2 + \alpha^2 \|f''\|^2 - \int_0^t \int_0^{+\infty} P''(u) u_x^3 \, dx \, d\tau \quad (3.23)
$$

$$
- \int_0^t \left[2g'(\tau) u_x(0,\tau) + u_x^2(0,\tau) + \frac{1}{2} P'(g(\tau)) u_x^2(0,\tau) \right] d\tau.
$$

By using (**), (3.14) and some other elementary inequalities, we deduce from (3.23) that

$$\left| \int_0^t \int_0^{+\infty} P''(u)u_x^3 \, dx \, d\tau \right|$$

$$\leq \int_0^t \int_0^{+\infty} \varepsilon |u^2 u_x^3| \, dx \, d\tau$$

$$\leq \int_0^t \varepsilon \|u(\cdot,\tau)\|_{L_\infty}^2 \|u_x(\cdot,\tau)\|_{L_\infty} \|u_x(\cdot,\tau)\|^2 d\tau \qquad (3.24)$$

$$\leq \int_0^t C(\varepsilon)\|u(\cdot,\tau)\|\|u_x(\cdot,\tau)\|^{7/2}\|u_{xx}(\cdot,\tau)\|^{1/2} d\tau$$

$$\leq \int_0^t C(\varepsilon)\big[\|u(\cdot,\tau)\|_1^6 + \|u_{xx}(\cdot,\tau)\|^2\big] d\tau.$$

Using (3.24) in (3.23), one obtains

$$\|u_{xx}(\cdot,t)\|^2 + \nu \int_0^t \|u_{xx}(\cdot,\tau)\|^2 d\tau$$

$$\leq C(\|f\|_2, |g|_{C^1(0,T)}) + C(\alpha)\|u_x(\cdot,t)\|^2 \qquad (3.25)$$

$$+ \int_0^t C(\varepsilon)\big[\|u(\cdot,\tau)\|_1^6 + \|u_{xx}(\cdot,\tau)\|^2\big] d\tau.$$

If (3.25) is multiplied by a suitable constant and the result added to (3.22), there appears

$$\|u(\cdot,t)\|_1^6 + C(|g|_{C^1(0,T)})\|u_{xx}(\cdot,t)\|^2$$

$$+ C(\nu, |g|_{C^1(0,T)}) \left[\left(\int_0^t \|u_x(\cdot,\tau)\|^2 d\tau \right)^3 + \int_0^t \|u_{xx}(\cdot,\tau)\|^2 d\tau \right]$$

$$\leq C(\|f\|_2, |g|_{C^1(0,T)}) \qquad (3.26)$$

$$+ \int_0^t C(\nu, \varepsilon, |g|_{C^1(0,T)}, T)\big[\|u(\cdot,\tau)\|_1^6 + \|u_{xx}(\cdot,\tau)\|^2\big] d\tau.$$

An appeal to Gronwall's lemma now concludes the proof.   □

REMARK 3.7: From (3.10) and (3.17), one sees immediately that an H^1-bound can be obtained without a growth restriction on P if a homogeneous boundary condition $g \equiv 0$ is posed.

REMARK 3.8: The condition in Lemma 3.6 on the growth of P is not sharp. One may use a more general version of Gronwall's inequality applied to (3.24) to handle cases where the growth of P is a little bit stronger than assumed in (**). While this strengthening of Lemma 3.6 is probably not of practical importance, it is perhaps worth recording.

LEMMA 3.9. *(A particular case of Theorem 3 in [3], Chapter 4, §5) Let $T > 0$ and let G be an increasing positive function defined on $[0, T]$ which is bounded away from 0. Let u be a continuous function on $[0, T]$ such that*

$$u(t) \leq k + \int_0^t G(u(s))ds \qquad \text{for } 0 \leq t \leq T,$$

where k is a positive constant. Then

$$u(t) \leq Q^{-1}(t) \qquad \text{for } 0 \leq t \leq T_1,$$

where $Q(t) = \int_k^t \frac{ds}{G(s)}$, range $G = [0, T^*]$, $T^* \in (0, +\infty]$ and $T_1 = \min\{T, T^*\}$.

COROLLARY 3.10. *Let f and g satisfy the conditions in Lemma 3.6. Suppose the nonlinearity P satisfies the growth condition*

$$\limsup_{|s| \to \infty} \frac{|P''(s)|}{|s|^2 \log(2 + |s^6|)} \leq \varepsilon_1, \qquad (***)$$

for some constant ε_1. Let $u(x, t)$ be a classical solution of (3.1) up to the boundary on $\mathbb{R}^+ \times [0, T']$ where $T' \leq T$. Then for all $t \in [0, T']$ $u(\cdot, t) \in H^2(\mathbb{R}^+)$ and there exists a constant a_2' only depending on $T, \|f\|_2$ and $|g|_{C^1(0,T)}$ such that

$$\|u(\cdot, t)\|_2 \leq a_2'. \tag{3.27}$$

PROOF: The hypothesis $(***)$ implies that

$$\left| \int_0^t \int_0^{+\infty} P''(u)u_x^3 \, dx \, d\tau \right|$$
$$\leq C(\varepsilon_1) \int_0^t \left(\|u(\cdot, \tau)\|_1^6 + \|u_{xx}(\cdot, \tau)\|^2 \right) \log(2 + \|u(\cdot, \tau)\|_1^6) d\tau. \tag{3.28}$$

Using (3.28) in (3.23) yields a new version of (3.24), namely

$$\|u(\cdot, t)\|_1^6 + C_2 \|u_{xx}(\cdot, t)\|^2 + C_3 \left(\int_0^t \|u_x(\cdot, \tau)\|^2 d\tau \right)^3 + C_4 \int_0^t \|u_{xx}(\cdot, \tau)\|^2 d\tau$$
$$\leq C_5(\|f\|_2, |g|_{C^1(0,T)}) + \int_0^t C_6(\varepsilon_1, \nu, |g|_{C^1(0,T)}) \big[\|u(\cdot, \tau)\|_1^6$$
$$+ \|u_{xx}(\cdot, \tau)\|^2 \big] \log(2 + \|u(\cdot, \tau)\|_1^6) d\tau. \tag{3.29}$$

Applying Lemma 3.9 to the above inequality gives (3.27). $\qquad \square$

An immediate consequence of the just derived *a priori* bounds is our main existence theorem.

Theorem 3.11. *Let f and g be given with $f(0) = g(0)$, and suppose that $T > 0$. Let the nonlinearity P in equation (3.1) be specified and assume that Λ is defined as before by $\Lambda'(z) = P(z)$ for $z \in \mathbb{R}$ and $\Lambda(0) = 0$.*

(1) If $f \in C_b^2(\mathbb{R}^+) \cap H^1(\mathbb{R}^+)$, $g \in C^1(0,T)$, Λ satisfies the one-sided growth condition

$$\limsup_{|s| \to \infty} |s|^{-4} \Lambda(s) \leq 0, \tag{*}$$

and $\Lambda \in C^2(\mathbb{R}^+)$, then the system (3.1) with $\nu = 0$ has a unique solution $u \in \mathcal{B}_T^{2,1} \cap C(0,T;H^1(\mathbb{R}^+))$ corresponding to the auxiliary specification of f as initial data and g as boundary data. Moreover, if $g \in C^1(\mathbb{R}^+) \cap H^1(\mathbb{R}^+)$, then $u \in C_b(\mathbb{R}^+;H^1(\mathbb{R}^+))$.

(2) If $f \in C_b^2(\mathbb{R}^+) \cap H^2(\mathbb{R}^+)$, $g \in C^1(0,T)$, P satisfies the growth condition

$$\limsup_{|s| \to \infty} |s|^{-2} |P''(s)| \leq \varepsilon, \tag{**}$$

for some constant ε and $P \in C^2(\mathbb{R}^+)$, then the equation (3.1) has a unique solution $u \in \mathcal{B}_T^{2,1} \cap C(0,T;H^2(\mathbb{R}^+))$ corresponding to the initial and boundary conditions (f,g).

(3) If $f \in C_b^r(\mathbb{R}^+) \cap H^k(\mathbb{R}^+)$ and $g \in C^s(0,T)$ where $k \geq 2$, $r \geq 2$, $s \geq 1$, $r > s$, then the solution u of (3.1) corresponding to initial data f and boundary data g lies in $\mathcal{B}_T^{r,s} \cap C(0,T;H^k(\mathbb{R}^+))$.

(4) In all the above cases, the solution u depends continuously on variations of the auxiliary data. That is, the mapping that assigns to (f,g) the associated solution of (3.1) is continuous from the function class of the initial data to the function class of the solution. In case (2) and (3), the solution also depends continuously on ν.

(5) If the boundary condition g is the zero function, then the above conclusion hold without the growth conditions ($*$) or ($**$) on the nonlinearity P.

By Lemma 3.3 and Proposition 3.2, the initial-boundary-value problem (3.1) has a solution u in $\mathcal{B}_T^{2,1} \cap C(0,T_0;H^k(\mathbb{R}^+))$ for $k = 1,2$, for small T_0. Then Lemma 3.5 and Lemma 3.6 show that on any finite time interval $[0,T]$, $\|u(\cdot,t)\|_1$ and hence $\|u(\cdot,t)\|_{C_b}$, is uniformly bounded. Thus the quantity $\|u(\cdot,t)\|_{C_b} + 2\|g\|_{C(0,T)}$ is uniformly bounded for $0 \leq t \leq T$. This in turn determines a lower bound on how far a solution, defined already on $[0,T_0]$, can be extended by an application of the local existence result in Proposition 3.1 and Proposition 3.2. As this term is bounded above, then the extension length is bounded below by a positive constant. By iteration of the existence proof of Proposition 3.1 and Proposition 3.2 one can extend the solution u from $[0,T_0]$ to $[0,T]$ in a finite number of temporal steps. $\square$

If $\nu \neq 0$, it would be expected that solutions of (3.1) decay in time. The following result will be useful for estimating temporal decay.

COROLLARY 3.12. Let $f \in C_b^2(\mathbb{R}^+) \cap H^2(\mathbb{R}^+)$, $g \in C^1(\mathbb{R}^+) \cap H^1(\mathbb{R}^+)$, $\nu > 0$ and suppose P to satisfy the condition $(**)$ in such a way that if equality occurs in $(**)$, then either ε is sufficiently small, ν is sufficiently big, or the boundary data g is small in the sense that $|g|_{1,+\infty}$ is sufficiently small. The corresponding solution u of (3.1) then has the properties that $u_x, u_{xx} \in L_2(\mathbb{R}^+ \times \mathbb{R}^+)$ and $u \in C_b(\mathbb{R}^+; H^2)$.

PROOF: Since $\nu > 0$, the left-hand side of (3.19) may be estimated as

$$
\begin{aligned}
&\left| \int_0^t \alpha^2 g'(\tau) u_x(0,\tau) d\tau \right| \\
&\leq \int_0^t \sqrt{2} \alpha^2 |g'(\tau)| \cdot \|u_x(\cdot,\tau)\|^{\frac{1}{2}} \|u_{xx}(\cdot,\tau)\|^{\frac{1}{2}} d\tau \\
&\leq \frac{\nu}{4} \int_0^t \|u_x(\cdot,\tau)\|^2 d\tau + \frac{C|g|_{1,+\infty}^{\frac{4}{3}}}{\nu^{\frac{1}{3}}} \left(\int_0^t \|u_{xx}(\cdot,\tau)\|^2 d\tau \right)^{\frac{1}{3}}.
\end{aligned}
\tag{3.30}
$$

Because of this new estimate in regard to (3.19), (3.22) can be rewritten as

$$
\begin{aligned}
&\|u(\cdot,t)\|_1^6 + \nu^3 \left(\int_0^t \|u_x(\cdot,\tau)\|^2 d\tau \right)^3 \\
&\leq C_1(\|f\|_1, |g|_{1,+\infty}) + C_2 |g|_{1,+\infty}^4 \|u_{xx}(\cdot,t)\|^2 \\
&\quad + \frac{C_3 |g|_{1,+\infty}^4}{\nu} \int_0^t \|u_{xx}(\cdot,\tau)\|^2 d\tau.
\end{aligned}
\tag{3.31}
$$

The elementary inequality (3.14) and the hypothesis $(**)$ show that

$$
\left| \int_0^t \int_0^{+\infty} P''(u) u_x^3 \, dx \, d\tau \right| \leq \int_0^t \left[\frac{C_4 \varepsilon^{\frac{4}{3}}}{\nu^{\frac{1}{3}}} \|u(\cdot,\tau)\|^{\frac{4}{3}} \|u_x(\cdot,\tau)\|^{\frac{14}{3}} + \frac{\nu}{4} \|u_{xx}(\cdot,\tau)\|^2 \right] d\tau. \tag{3.32}
$$

Using (3.32) and Young's inequality, the inequality in (3.23) may be revised to read

$$
\begin{aligned}
&\|u_{xx}(\cdot,t)\|^2 + \nu \int_0^t \|u_{xx}(\cdot,\tau)\|^2 d\tau \\
&\leq C_5(\|f\|_2, |g|_{1,+\infty}) + \delta \|u_x(\cdot,t)\|^6 + \int_0^t \frac{C_4 \varepsilon^{\frac{4}{3}}}{\nu^{\frac{1}{3}}} \|u(\cdot,\tau)\|^{\frac{4}{3}} \|u_x(\cdot,\tau)\|^{\frac{14}{3}} d\tau,
\end{aligned}
\tag{3.33}
$$

for any $\delta > 0$. Multiplying (3.33) by the constant C^* determined by the relation

$$
C^* \frac{|g|_{1,+\infty}^4}{\nu^2} = \max\left\{ 1 + C_3 \frac{|g|_{1,+\infty}^4}{\nu^2}, 1 + C_2 |g|_{1,+\infty}^4 \right\},
$$

where C_2 and C_3 are the constants appearing earlier in the proof, and then adding the result to (3.31), it is deduced that

$$
\begin{aligned}
&\|u(\cdot,t)\|_1^6 + C_7(\nu, |g|_{1,+\infty}) \|u_{xx}(\cdot,t)\|^2 + \nu^2 C_8 |g|_{1,+\infty}^4 \left(\int_0^t \|u_x(\cdot,\tau)\|^2 d\tau \right)^3 \\
&\quad + C_9(\nu, |g|_{1,+\infty}) \int_0^t \|u_{xx}(\cdot,\tau)\|^2 d\tau \\
&\leq C_{10}(\|f\|_2, |g|_{1,+\infty}, \nu) + \varepsilon^{\frac{4}{3}} C_{11} \frac{|g|_{1,+\infty}^4}{\nu^{\frac{4}{3}}} \int_0^t \|u(\cdot,\tau)\|^{\frac{4}{3}} \|u_x(\cdot,\tau)\|^{\frac{14}{3}} d\tau.
\end{aligned}
\tag{3.34}
$$

It is then straightforward to show that

$$
\int_0^t C_{11}\varepsilon^{\frac{4}{3}} \frac{|g|_{1,+\infty}^4}{\nu^{\frac{4}{3}}} \|u(\cdot,\tau)\|^{\frac{4}{3}} \|u_x(\cdot,\tau)\|^{\frac{14}{3}} d\tau
$$

$$
\leq C_{11}\varepsilon^{\frac{4}{3}} \frac{|g|_{1,+\infty}^4}{\nu^{\frac{4}{3}}} \|u\|_{C(\mathbb{R}^+;L_2)}^{\frac{4}{3}} \|u_x\|_{C(\mathbb{R}^+;L_2)}^{\frac{8}{3}} \int_0^t \|u_x(\cdot,\tau)\|^2 d\tau
$$

$$
\leq \frac{2C_{11}^{\frac{3}{2}}\varepsilon^2|g|_{1,+\infty}^4}{3C_8^2\nu^3} \|u\|_{C(\mathbb{R}^+;L_2)}^2 \|u_x\|_{C(\mathbb{R}^+;L_2)}^4
$$

$$
+ \frac{C_8\nu^2|g|_{1,+\infty}^4}{3} \Big(\int_0^t \|u_x(\cdot,\tau)\|^2 d\tau\Big)^3.
$$

(3.35)

Putting (3.35) in (3.34) and assuming that $\frac{2C_{11}^{\frac{3}{2}}\varepsilon^2|g|_{1,+\infty}^4}{3C_8^2\nu^3} \leq \frac{1}{2}$, it follows that

$$
\|u\|_{C(\mathbb{R}^+;H^1)}^6 + C_6\|u_{xx}\|_{C(\mathbb{R}^+;L_2)}^2 + C_7\Big(\int_0^{+\infty} \|u_x(\cdot,\tau)\|^2 d\tau\Big)^3
$$

$$
+ C_8 \int_0^{+\infty} \|u_{xx}(\cdot,\tau)\|^2 d\tau \leq C_9(\|f\|_2, |g|_{1,+\infty}, \nu).
$$

(3.36)

Hence the corollary is proved.   □

LEMMA 3.13. *Let u be the solution of (3.1a) corresponding to initial data $f \in H^2(\mathbb{R}^+)$ and boundary data $g \in H^1(\mathbb{R}^+)$. If P satisfies the conditions delineated in Corollary 3.12 and $\nu > 0$, then*

$$
\|u_x(\cdot,t)\|, \ \|u_{xx}(\cdot,t)\| \to 0, \qquad \text{as} \qquad t \to +\infty,
$$

and

$$
\|u(\cdot,t)\|_{L_\infty} \to 0 \qquad \text{as} \qquad t \to +\infty.
$$

PROOF: By Corollary 3.12, it is known that $\|u_x(\cdot,t)\|$, $\|u_{xx}(\cdot,t)\|$ lie in $L_2(\mathbb{R}^+)$. From (3.23), it then follows that both $\|u_x(\cdot,t)\|$ and $\|u_{xx}(\cdot,t)\|$ have limits as t tends to infinity. In consequence, these two limits at infinity must be zero. Noting that

$$
\|u(\cdot,t)\|_{L_\infty}^2 \leq 2\|u(\cdot,t)\|\|u_x(\cdot,t)\| \leq 2C(\|f\|_2, |g|_{1,+\infty}, \nu)\|u_x(\cdot,t)\|
$$

and remarking that the right-hand side of this inequality tends to zero as $t \to +\infty$, the result follows.   □

LEMMA 3.14. *Let u be the solution of (3.1a) with $\nu > 0$ corresponding to initial data $f \in H^2(\mathbb{R}^+)$ and boundary data $g \in H^2(\mathbb{R}^+)$ and suppose P satisfies the condition in Corollary 3.12. It then follows that u_t and u_{xt} lie in $L_2(\mathbb{R}^+ \times \mathbb{R}^+)$.*

PROOF: Multiply (3.1a) by u_t and integrate the result over $\mathbb{R}^+ \times [0, t)$. After integration by parts and using Lemma 3.3, we are reduced to

$$
\begin{aligned}
\frac{\nu}{2} &\|u_x(\cdot, t)\|^2 + \int_0^t \left[\|u_t(\cdot, \tau)\|^2 + \alpha^2 \|u_{xt}(\cdot, \tau)\|^2 \right] d\tau \\
&= \frac{\nu}{2} \|f'\|^2 + \int_0^t \left[\alpha^2 g''(\tau) - \nu g'(\tau) \right] u_x(0, \tau) d\tau - \alpha^2 g'(t) u_x(0, t) \\
&\quad + \alpha^2 g'(0) f(0) - \int_0^t \int_0^{+\infty} u_t (u_x + P(u)_x) \, dx \, d\tau \\
&\leq C(\|f\|_2, |g|_{2,+\infty}) + \frac{1}{2} \int_0^t \|u_t(\cdot, \tau)\|^2 d\tau \\
&\quad + C|g|_{2,+\infty} \left(\int_0^t \|u_x(\cdot, \tau)\|^2 d\tau \right)^{\frac{1}{4}} \left(\int_0^t \|u_{xx}(\cdot, \tau)\|^2 d\tau \right)^{\frac{1}{4}} \\
&\quad + \frac{1}{2}(1 + \|P'(u)\|_{L_\infty}^2) \int_0^t \|u_x(\cdot, \tau)\|^2 d\tau,
\end{aligned}
\tag{3.37}
$$

where (3.14) has been used. The result follows from (3.37) since u_x, $u_{xx} \in L_2(\mathbb{R}^+ \times \mathbb{R}^+)$ and $\|P'(u)\|_{L_\infty} \leq C(\|f\|_1, |g|_{1,\infty})$. $\square$

3.3. Continuous Dependence

Attention is now given to showing that solutions of (3.1) depend continuously on the specified data. That is, small perturbations of the initial and boundary data f and g lead to small perturbations of the corresponding solution. This is a very important aspect of model equations for waves that apply in regimes where breaking and other singularity formation are not countenanced. Indeed, this property is crucial if laboratory measurements are to be successfully compared with numerical approximations of the solutions of the model.

Let (f_1, g_1) and (f_2, g_2) be two sets of data for problem (3.1). Theorem 3.11 shows that if $f_i \in C_b^2(\mathbb{R}^+) \cap H^1(\mathbb{R}^+)$ (respectively, $f_i \in C_b^2(\mathbb{R}^+) \cap H^2(\mathbb{R}^+)$) and $g_i \in C^1(0, T)$, $i = 1, 2$, then with suitable growth restrictions on P, the corresponding solutions u_1 and u_2 of (3.1) lie in $\mathcal{B}_T^{2,1} \cap C(0, T; H^1)$ (respectively, $\mathcal{B}_T^{2,1} \cap C(0, T; H^2)$).

For $k = 1, 2$, let $\mathbb{U}_k$ denote the mapping that takes the auxiliary specifications ν, f and g into the corresponding solutions of (3.1). Thus $\mathbb{U}_k$ maps X_k into Y_k, $k = 1, 2$, where

$$
\begin{aligned}
X_1 &= \{(f, g) : (f, g) \in H^1(\mathbb{R}^+) \cap C_b^2(\bar{\mathbb{R}}^+) \times C^1(0, T)\}, \\
X_2 &= \{(\nu, f, g) : (\nu, f, g) \in \mathbb{R}^+ \times H^2(\mathbb{R}^+) \cap C_b^2(\bar{\mathbb{R}}^+) \times C^1(0, T)\}
\end{aligned}
$$

and

$$
Y_k = \mathcal{B}_T^{2,1} \cap C(0, T; H^k).
$$

THEOREM 3.15. *The mapping* $\mathbb{U}_k$ *defined above is continuous,* $k = 1, 2$.

PROOF: We deal only with showing that $\mathbb{U}_2$ is continuous. Similar arguments suffice to show that $\mathbb{U}_1$ is continuous.

It is first shown that $\mathbb{U}_2$ is continuous for a fixed positive ν. Let $(\nu, f_i, g_i) \in X_2$, and $u_i = \mathbb{U}_2(\nu, f_i, g_i)$ for $i = 1, 2$ and define w to be $w = u_1 - u_2$. Then w satisfies the initial- and boundary-value problem

$$w_t + w_x + \{P(u_1) - P(u_2)\}_x - \nu w_{xx} - \alpha^2 w_{xxt} = 0, \qquad \text{for } x, t \geq 0, \qquad (3.38a)$$

$$w(x, 0) = f(x), \qquad \text{for } x \geq 0, \qquad (3.38b)$$

$$w(0, t) = g(t), \qquad \text{for } t \geq 0. \qquad (3.38c)$$

where

$$f(x) = f_1(x) - f_2(x), \quad g(t) = g_1(t) - g_2(t).$$

Multiply equation (3.38a) by $2w - 2w_{xx}$ and integrate the result over $\mathbb{R}^+ \times [0, t)$. After integration by parts, one has

$$\begin{aligned}
&\|w(\cdot, t)\|^2 + (1 + \alpha^2)\|w_x(\cdot, t)\|^2 + \alpha^2\|w_{xx}(\cdot, t)\|^2 \\
&\quad + \int_0^t \left(w_x^2(0, \tau) + 2\nu\left[\|u_x(\cdot, \tau)\|^2 + \|u_{xx}(\cdot, \tau)\|^2\right]\right) d\tau \\
&= \|f\|^2 + (1 + \alpha^2)\|f'\|^2 + \alpha^2\|f''\|^2 - 2\alpha^2 g(t)w_x(0, t) + 2\alpha^2 g(0)f'(0) \\
&\quad + \int_0^t \left[g^2(\tau) + 2(\alpha^2 - 1)g'(\tau)w_x(0, \tau) - 2\nu g(\tau)w_x(0, \tau)\right] d\tau \qquad (3.39) \\
&\quad - \int_0^t \int_0^{+\infty} 2\left[w(x, \tau) - w_{xx}(x, \tau)\right]\left(P(u_1(x, \tau)) - P(u_2(x, \tau))\right)_x dx d\tau.
\end{aligned}$$

By Lemma 3.6, there exists a constant $\bar{a}_2$ which only depends upon $T, \|f_i\|_2$ and $|g_i|_{1,T}$ such that

$$\|u_i(\cdot, t)\|_2 \leq \bar{a}_2, \qquad (i = 1, 2). \qquad (3.40)$$

From (3.39), (3.40) and some elementary inequalities, including (3.14) applied to w_x, it follows that

$$\|w(\cdot, t)\|_2^2 \leq \|f\|_2^2 + C_1\|g\|_{C^1(0,T)}^2 + C_2(\gamma(\mathbf{B}_T)) \int_0^t \|w(\cdot, \tau)\|_2^2 d\tau, \qquad (3.41)$$

where $\gamma(\mathbf{B}_T)$ is a constant such that $|P(z_1) - P(z_2)| \leq \gamma(\mathbf{B}_T)|z_1 - z_2|$ and $\mathbf{B}_T = \{w \in \mathcal{B}_T : \|w\|_{\mathcal{B}_T} \leq 2\bar{a}_2\}$. By Gronwall's lemma, it follows from (3.41) that

$$\|w(\cdot, t)\|_2^2 \leq C(\|f\|_2^2 + \|g\|_{C^1(0,T)}^2),$$

or what is the same,

$$\|u_1 - u_2\|^2_{C(0,T;H^2)} \leq C(\|f_1 - f_2\|^2_2 + \|g_1 - g_2\|^2_{C^1(0,T)}). \tag{3.42}$$

By a Sobolev embedding theorem, (3.42) implies that

$$\|u_1 - u_2\|_{\mathcal{B}_T} \leq C(\|f_1 - f_2\|_2 + \|g_1 - g_2\|_{C^1(0,T)}). \tag{3.43}$$

In consequence, the mapping $\mathbb{U}_2 \colon X_2 \to \mathcal{B}_T$ is continuous, at least for fixed ν.

To prove that $\mathbb{U}_2 \colon X_2 \to \mathcal{B}_T^{2,1}$ is continuous for fixed ν, write (3.38a) as an integral equation, analogous to (3.8), namely

$$
\begin{aligned}
w(x,t) = {} & \exp(-t\nu/\alpha^2)f(x) + \tilde{g}(t)e^{-x/\alpha} \\
& + \int_0^t \int_0^{+\infty} \exp(-\nu(t-\tau)/\alpha^2)K(x,\xi)\big[P(u_1(\xi,\tau)) - P(u_2(\xi,\tau)) + w(\xi,\tau)\big]d\xi d\tau \\
& - \frac{\nu}{\alpha} \int_0^t \int_0^{+\infty} \exp(-\nu(t-\tau)/\alpha^2)L(x,\xi)w(\xi,\tau)d\xi d\tau,
\end{aligned} \tag{3.44}
$$

where $\tilde{g}(t) = g(t) - \exp(-t\nu/\alpha^2)g(0)$ and K and L are as in (3.5) and (3.6), respectively. By using (3.43) and (3.44), along with the assumption $\tilde{g}(t) \in C^1(0,T)$ and $f(x) \in H^2(\mathbb{R}^+) \cap C_b^2(\mathbb{R}^+)$, one can easily show that

$$\|w\|_{\mathcal{B}_T^{2,1}} \leq C(\|f_1 - f_2\|_2 + \|g_1 - g_2\|_{C^1(0,T)}). \tag{3.45}$$

As an example, by differentiating (3.44) with respect to t, it follows that

$$
\begin{aligned}
w_t(x,t) = {} & -\frac{\nu}{\alpha^2}\exp(-t\nu/\alpha^2)f(x) + \tilde{g}'(t)e^{-x/\alpha} \\
& - \frac{\nu}{\alpha^2} \int_0^t \int_0^{+\infty} \exp(-\nu(t-\tau)/\alpha^2)K(x,\xi)\Big[P(u_1(\xi,\tau)) \\
& \hspace{6cm} - P(u_2(\xi,\tau)) + w(\xi,\tau)\Big]d\xi d\tau \\
& + \int_0^{+\infty} K(x,\xi)\big[P(u_1(\xi,\tau)) - P(u_2(\xi,\tau)) + w(\xi,\tau)\big]d\xi d\tau \\
& + \frac{\nu^2}{\alpha^3} \int_0^t \int_0^{+\infty} \exp(-\nu(t-\tau)/\alpha^2)L(x,\xi)w(\xi,\tau)d\xi d\tau \\
& - \frac{\nu}{\alpha} \int_0^{+\infty} L(x,\xi)w(\xi,\tau)d\xi d\tau.
\end{aligned} \tag{3.46}
$$

Notice that

$$
\begin{aligned}
\int_0^{+\infty} |K(x,\xi)|d\xi & = \frac{1}{2\alpha^2}\int_0^{+\infty} |\exp(-(x+\xi/\alpha)) + sgn(x-\xi)\exp(-|x-\xi|/\alpha)|d\xi \\
& = \frac{1}{2\alpha^2}\int_x^{+\infty} |\exp(-(x+\xi/\alpha)) - \exp((x-\xi)/\alpha)|d\xi \\
& \hspace{2cm} + \frac{1}{2\alpha^2}\int_0^x (\exp(-(x+\xi/\alpha)) + \exp((\xi-x)/\alpha))d\xi \\
& \leq \frac{1}{\alpha},
\end{aligned} \tag{3.47}
$$

and similarly,

$$\int_0^{+\infty} |L(x,\xi)|d\xi \le \frac{2}{\alpha}. \tag{3.48}$$

Using some elementary inequalities, (3.46), together with (3.47) and (3.48) give

$$\|w_t\|_{\mathcal{B}_T} \le \frac{\nu}{\alpha^2}\|f\|_2 + \|\tilde{g}\|_{C^1(0,T)} + C(T\nu + 1 + \gamma(\mathbf{B}_T))\|w\|_{\mathcal{B}_T}. \tag{3.49}$$

By using (3.43), one then obtains

$$\|w_t\|_{\mathcal{B}_T} \le C(\|f_1 - f_2\|_2 + \|g_1 - g_2\|_{C^1(0,T)}). \tag{3.50}$$

Differentiating (3.44) with respect to x and then using (3.43) and the same sort of considerations that led to (3.50), one sees that

$$\|w_x\|_{\mathcal{B}_T} \le C(\|f_1 - f_2\|_2 + \|g_1 - g_2\|_{C^1(0,T)}). \tag{3.51}$$

Differentiating (3.44) with respect to x again and arguing as above yields

$$\|w_{xx}\|_{\mathcal{B}_T} \le C(\|f_1 - f_2\|_2 + \|g_1 - g_2\|_{C^1(0,T)}). \tag{3.52}$$

Similar results may be obtained with regard to u_{xt} and u_{xxt}, and in this manner (3.45) is verified.

Next, we show that $\mathbb{U}_2$ is continuous in the variable ν when the initial and boundary data are fixed. Let $v_i = \mathbb{U}_2(\nu_i, f_2, g_2)$ for $i = 1, 2$, and let $y = v_1 - v_2$. Then y satisfies the initial- and boundary-value problem

$$y_t + y_x + \{P(v_1) - P(v_2)\}_x - (\nu_1 - \nu_2)(v_2)_{xx} - \nu_1 y_{xx} - \alpha^2 y_{xxt} = 0, \tag{3.53a}$$
$$\text{for } x, t \ge 0,$$
$$y(x,0) = 0, \qquad\qquad\qquad\qquad \text{for } x \ge 0, \tag{3.53b}$$
$$y(0,t) = 0, \qquad\qquad\qquad\qquad \text{for } t \ge 0. \tag{3.53c}$$

Multiply (3.53a) by $2y - 2y_{xx}$ and integrate the result over $\mathbb{R}^+$. After integration by parts and using Lemma 3.6 applied to the solutions v_1 and v_2, one has

$$\frac{d}{dt}\left(\|y\|^2 + (1 + \alpha^2)\|y_x\|^2 + \alpha^2\|y_{xx}\|^2\right) + y_x^2(0,t) + 2\nu_1\left(\|y_x\|^2 + \|y_{xx}\|^2\right)$$
$$= \int_0^{+\infty} [2y - 2y_{xx}]\left[(\nu_1 - \nu_2)(v_2)_{xx} - \{P(v_1) - P(v_2)\}_x\right]dx$$
$$\le \left(|\nu_1 - \nu_2|\|v_2\|_{\mathcal{B}_T^{2,1}}\|y\|_2 + \gamma(\mathbf{B}_T)\|y\|_2^2\right) \tag{3.54}$$
$$\le C^*(|\nu_1 - \nu_2| \cdot \|y\|_2 + \|y\|_2^2),$$

where $C^* = max\{\gamma(\mathbf{B}_T), \|v_2\|_{\mathcal{B}_T^{2,1}}\}$. Since $y(x,0) = 0$, it follows from Gronwall's lemma that

$$\|y\|_2 \leq |\nu_1 - \nu_2|(\exp(C^* c(\alpha)t) - 1)$$

for $0 \leq t \leq T$, where $c(\alpha) = \frac{1}{\min\{1,\alpha^2\}}$. By a Sobolev embedding theorem, this in turn shows that

$$\|v_1 - v_2\|_{\mathcal{B}_T} \leq C|\nu_1 - \nu_2|. \tag{3.55}$$

Using of an integral equation version of (3.53), the last inequality may be extended to a bound of the form

$$\|v_1 - v_2\|_{\mathcal{B}_T^{2,1}} \leq C|\nu_1 - \nu_2|, \tag{3.56}$$

thus showing that $\mathbb{U}_2$ is a Lipschitz continuous function of the parameter ν.

The triangle inequality applied thusly,

$$\|\mathbb{U}_2(\nu_1, f_1, g_1) - \mathbb{U}_2(\nu_2, f_2, g_2)\|_{\mathcal{B}_T^{2,1} \cap C(0,T;H^2(\mathbb{R}^+))}$$

$$\leq \|\mathbb{U}_2(\nu_1, f_1, g_1) - \mathbb{U}_2(\nu_1, f_2, g_2)\|_{\mathcal{B}_T^{2,1} \cap C(0,T;H^2(\mathbb{R}^+))}$$

$$+ \|\mathbb{U}_2(\nu_1, f_2, g_2) - \mathbb{U}_2(\nu_2, f_2, g_2)\|_{\mathcal{B}_T^{2,1} \cap C(0,T;H^2(\mathbb{R}^+))},$$

allows one to infer from (3.45) and (3.57) that

$$\mathbb{U}_2 \colon X_2 = \{[H^2(\mathbb{R}^+) \cap C_b^2(\mathbb{R}^+)] \times C^1(0,T)\} \longmapsto Y = \mathcal{B}_T^{2,1} \cap C(0,T;H^2(\mathbb{R}^+))$$

is continuous. This concludes the proof of the theorem. $\qquad\square$

COROLLARY 3.16. *Let $\mathbb{U}$ denote the mapping that associates with the triple (v, f, g) the corresponding solution of (3.1). For any $k \geq 2$ and $l \geq 1$,*

$$\mathbb{U} \colon\; X = \{(v, f, g) : (v, f, g) \in \mathbb{R}^+ \times H^k(\mathbb{R}^+) \cap C_b^k(\bar{\mathbb{R}}^+) \times C^l(0,T)\}$$

$$\longmapsto Y = \mathcal{B}_T^{k,l} \cap C(0,T;H^k(\mathbb{R}^+)),$$

and this correspondence is continuous. $\qquad\square$

4. WELL-POSEDNESS OF TWO-POINT BOUNDARY-VALUE PROBLEM

In this section, interest will be focused on the initial- and two-point boundary-value problem

$$u_t + u_x + P(u)_x - \nu u_{xx} - \alpha^2 u_{xxt} = 0, \qquad \text{for } x \in (0,1),\ t \geq 0, \tag{4.1a}$$

$$u(x,0) = f(x), \qquad \text{for } x \in [0,1], \tag{4.1b}$$

$$u(0,t) = g(t), \qquad \text{for } t \geq 0. \tag{4.1c}$$

$$u(1,t) = h(t), \qquad \text{for } t \geq 0. \tag{4.1d}$$

For consistency, the restrictions

$$u(0,0) = f(0) = g(0), \quad \text{and} \quad u(1,0) = f(1) = h(0) \tag{4.2}$$

will be imposed and it will be supposed that $\nu \geq 0$ and $\alpha \neq 0$. Throughout this section, Ω will stand for the interval $[0,1]$.

4.1. Local Solutions

By converting the differential equation with initial condition (4.1b) and boundary conditions (4.1c) and (4.1d) into an integral equation as in (3.3) and applying the contraction-mapping theorem to this formulation of the problem, a local solution can be established. The argument closely parallels that worked out above for (3.1) (see also [6]), and consequently we content ourselves with a statement of the conclusions that may be derived by this approach.

PROPOSITION 4.1. *Let $T > 0$, $f \in C(\Omega)$, g and $h \in C(0,T)$, and P locally Lipschitz continuous. Then there exists a positive constant $T' = T'(\|f\|_{C_b(\Omega)}, \|g\|_{C(0,T)})$ such that for any t with $0 < t < T_0 = min(T', T)$, there is a unique solution of (4.1) in $\mathcal{B}_{T_0}$. If $f \in C^k(\Omega)$, g and $h \in C^l(0,T)$ where $k \geq 2$, $l \geq 1$ and $k > l$, then the corresponding solution u of (4.1) is an element of $\mathcal{B}_{T_0}^{k,l}$ and is a classical solution of the initial- and boundary-value problem (4.1).*

4.2. Global Solutions

The local solution of (4.1) whose existence was just confirmed can be extended to a global solution by appeal to appropriate *a priori* estimates. The *a priori* bounds which allow one to extend the local solution of (4.1) to arbitrary time intervals is derived by energy estimates just as was done for (3.1). We have been unable to derive an analog of Lemma 3.4 for the two-point boundary-value problem (4.1). Similar problems were noted for nonhomogeneous boundary-value problem for KdV (see [6, 17]). Thus we are not able to obtain the helpful $H^1(\mathbb{R}^+)$-bound that is the outcome of Lemma 3.4 in the present circumstances. However, $H^2(\mathbb{R}^+)$-bounds can be derived as is now demonstrated.

LEMMA 4.2. *Suppose that $f \in C^2(\Omega) \cap H^2(\Omega)$, g, $h \in C^1(0,T)$ with $f(0) = g(0)$ and $f(1) = h(0)$, and that $P \in C^2(\mathbb{R})$ satisfies the growth condition*

$$\limsup_{|s| \to \infty} |s|^{-2} |P''(s)| \leq \varepsilon \tag{**}$$

for some finite constant ε. Let $u(x,t)$ be a classical solution of (4.1) up to the boundary on $\Omega \times [0,T]$. Then for all $t \in [0,T)$ $(T \leq T_0)$, $u(\cdot,t) \in H^2(\Omega)$ and there exists a constant a_2 only depending on T, $\|f\|_2$, $|g|_{C^1(0,T)}$ and $|h|_{C^1(0,T)}$ such that $\|u(\cdot,t)\|_2 \leq a_2$.

Lemma 4.2 for the problem (4.1) may be proved in exactly the same way as we proved Lemma 3.6 for the problem (3.1). The principal difference is that the term $u_x(1,t)$ arises on account of integrations by parts. This quantity can be controlled by use of the analogue

$$\|v\|_{C_b(0,1)} \leq [\|v\|(\|v\| + 2\|v'\|)]^{\frac{1}{2}} \tag{4.3}$$

of (3.14). Applying this relation and an estimate like that appearing in (3.18), a suitable differential inequality may be derived to which Gronwall's lemma applies and yields the desired results.

Using the bound obtained from Lemma 4.2, the local solution obtained in Proposition 4.1 can be extended to arbitrary time intervals. Results of continuous dependence on the data f, g and h and on the parameter ν can be obtained just as for (3.1). Thus the following theorem for (4.1) emerges.

THEOREM 4.3. *Let there be given $T > 0$, $f \in C^k(\Omega) \cap H^k(\Omega)$, g, $h \in C^l(0,T)$, where $k \geq 2$, $l \geq 1$. Suppose that $f(0) = g(0)$, $f(1) = h(0)$, and that nonlinearity P is smooth and satisfies the conditions (**) specified in Lemma 4.2. Then the initial-boundary-value problem (4.1) has a unique solution $u \in \mathcal{B}_T^{k,l}$. Let $\mathbb{U}$ denote the mapping that associates to ν and the triple (f,g,h) of data in (4.1b, c, d) the corresponding solutions of (4.1). Then $\mathbb{U}$ is a continuous mapping of X into Y where*

$$X = \{(\nu,f,g,h):\ (\nu,f,g,h) \in \mathbb{R}^+ \times C^k(\Omega) \times C^l(0,T) \times C^l(0,T)\} \quad \text{and}$$
$$Y = \mathcal{B}_T^{k,l}.$$

The following result is similar to Corollary 3.12. It is useful for estimating decay in the temporal variable in case $\nu > 0$.

COROLLARY 4.4. *Let initial data f defined on $\Omega = (0,1)$ be given and boundary data g, h defined on all of $\mathbb{R}^+$ specified. Suppose that (f,g,h) satisfies the hypotheses in Lemma 4.2 for any $T > 0$, and that in addition g, $h \in H^1(\mathbb{R}^+)$. Suppose $\nu > 0$ and that either ε is small enough, ν is big enough, or that the boundary data g and h are small enough in $H^1(\mathbb{R}^+)$. If u is the solution of (4.1) corresponding to (f,g,h), then u_x, $u_{xx} \in L_2(\Omega \times \mathbb{R}^+)$ and $\|u(\cdot,t)\|_{L_\infty} \to 0$ as $t \to +\infty$.*

4.3. Decay Rates

The asymptotic behaviour as $t \to +\infty$ of solutions of (4.1) will be considered in this subsection. To simplify the presentation, it is assumed that $(P(x))_x = u^p u_x$. It is worth note that the proof of decay obtained for the pure initial-value problem posed on all of $\mathbb{R}$ in [2, 9] can be taken over intact in case g, $h \equiv 0$. However, since Ω is bounded, stronger temporal decay is expected than the algebraic rates obtained in [2, 9] in case $\nu > 0$. Moreover, the imposition of homogeneous boundary conditions at both sides of the spatial domain is artificial. In consequence, we undertake now a direct derivation of temporal decay rates for the initial-boundary-value problem (4.1).

LEMMA 4.5. *Suppose $\nu > 0$ and that the auxiliary data (f, g, h) and P satisfy the conditions specified in Corollary 4.4. Let u be the associated solution of (4.1). Then, $u \in C_b(\mathbb{R}^+; H^2(\Omega)) \cap L_2(\mathbb{R}^+; H^2(\Omega))$ and u_{xt}, $u_{xxt} \in L_2(\Omega \times \mathbb{R}^+)$.*

PROOF: First multiply (4.1a) by the combination $2(b_1 - x)u(x, t)$ for some positive constant b_1 to be specified presently, and then integrate the result over Ω. After simplification, we reach the relation

$$\int_0^1 \left[u(x,t)^2 + 2\nu(b_1 - x)u_x^2(x,t)\right] dx + \frac{d}{dt}\int_0^1 \left[(b_1 - x)[u(x,t)^2 + \alpha^2 u_x(x,t)^2]\right] dx$$

$$= b_1 \left[2\alpha^2 g(t)u_{xt}(0,t) + 2\nu g(t)u_x(0,t) - g^2(t) - \frac{2}{p+2}g^{p+2}(t)\right]$$

$$- (b_1 - 1)\left[2\alpha^2 h(t)u_{xt}(1,t) + 2\nu h(t)u_x(1,t) - h^2(t) - \frac{2}{p+2}h^{p+2}(t)\right]$$

$$+ \nu h(t)^2 - \nu g(t)^2 + 2\alpha^2 h(t)h'(t) - 2\alpha^2 g(t)g'(t)$$

$$- \int_0^1 \left[\frac{2}{p+2}u^{p+2}(x,t) + 2\alpha^2 u_x(x,t)u_t(x,t)\right] dx. \tag{4.4}$$

Equation (4.1a) gives the crude inequality

$$\|u_t(\cdot, t)\|^2 \leq C(|g'(t)| + |h'(t)|)$$

$$+ C\left[\|u_x(\cdot, t)\|^2 + \|u_{xx}(\cdot, t)\|^2 + \|u_{xt}(\cdot, t)\|^2 + \|u_{xxt}(\cdot, t)\|^2\right]. \tag{4.5}$$

Applying (4.3) to u_x and u_{xt}, using (4.5) in (4.4) and making other elementary estimates leads to

$$\int_0^1 \left[u(x,t)^2 + 2\nu(b_1 - x)u_x^2(x,t)\right] dx + \frac{d}{dt}\int_0^1 \left[(b_1 - x)[u(x,t)^2 + \alpha^2 u_x(x,t)^2]\right] dx$$

$$\leq C(\delta)(|g(t)|^2 + |h(t)|^2 + |g'(t)|^2 + |h'(t)|^2) + \frac{2}{p+2}\|u(\cdot, t)\|_{L_\infty}^p \|u(\cdot, t)\|^2$$

$$+ C(\delta)\|u_x(\cdot, t)\|^2 + \delta\left[\|u_{xx}(\cdot, t)\|^2 + \|u_{xt}(\cdot, t)\|^2 + \|u_{xxt}(\cdot, t)\|^2\right], \tag{4.6}$$

which holds for any $\delta > 0$.

Multiply (4.1a) by u_{xx} and integrate the result over Ω. After integration by parts and using Lemma 4.2, we derive

$$
\begin{aligned}
\frac{1}{2}\frac{d}{dt}&\|u_x(\cdot,t)\|^2 + \frac{\alpha^2}{2}\frac{d}{dt}\|u_{xx}(\cdot,t)\|^2 + \nu\|u_{xx}(\cdot,t)\|^2 \\
&= [\frac{1}{2}u_x^2(1,t) + h'(t)u_x(1,t)] - [\frac{1}{2}u_x^2(0,t) + g'(t)u_x(0,t)] - \int_0^1 u^p u_x u_{xx}\,dx \\
&\leq C(|g'(t)|^2 + |h'(t)|^2) + C(\nu)(1 + \|u(\cdot,t)\|_{L_\infty}^{2p})\|u_x(\cdot,t)\|^2 + \frac{\nu}{2}\|u_{xx}(\cdot,t)\|^2,
\end{aligned}
\tag{4.7}
$$

where inequalities (4.3) has been used to control the terms $u_x(0,t)$ and $u_x(1,t)$.

Multiply (4.1a) by u_{xxt} and integrate the result over Ω. After integration by parts and use of (4.3), it is adduced that

$$
\begin{aligned}
\frac{\nu}{2}\frac{d}{dt}&\|u_{xx}(\cdot,t)\|^2 + \alpha^2\|u_{xxt}(\cdot,t)\|^2 + \|u_{xt}(\cdot,t)\|^2 \\
&= h'(t)u_{xt}(1,t) - g'(t)u_{xt}(0,t) + \int_0^1 [u^p u_x u_{xxt} + u_x u_{xxt}]\,dx \\
&\leq C(|g'(t)|^2 + |h'(t)|^2) + C(\alpha)(1 + \|u(\cdot,t)\|_{L_\infty}^{2p})\|u_x(\cdot,t)\|^2 \\
&\qquad + \frac{1}{2}\|u_{xt}(\cdot,t)\|^2 + \frac{\alpha^2}{2}\|u_{xxt}(\cdot,t)\|^2.
\end{aligned}
\tag{4.8}
$$

Adding (4.6), (4.7) and (4.8) together yields the differential inequality

$$
\begin{aligned}
\int_0^1 &[u(x,t)^2 + 2\nu(b_1 - x)u_x^2(x,t)]\,dx + \frac{\nu}{2}\|u_{xx}(\cdot,t)\|^2 + \frac{\alpha^2}{2}\|u_{xxt}(\cdot,t)\|^2 + \frac{1}{2}\|u_{xt}(\cdot,t)\|^2 \\
&+ \frac{d}{dt}\Big[\int_0^1 [(b_1 - x)[u(x,t)^2 + \alpha^2 u_x(x,t)^2]]\,dx + \frac{1}{2}\|u_x(\cdot,t)\|^2 + \Big(\frac{\alpha^2}{2} + \frac{\nu}{2}\Big)\|u_{xx}(\cdot,t)\|^2\Big] \\
&\leq C(|g(t)|^2 + |h(t)|^2 + |g'(t)|^2 + |h'(t)|^2) + \frac{2}{p+2}\|u(\cdot,t)\|_{L_\infty}^p \|u(\cdot,t)\|^2 \\
&\quad + [C(\delta) + C(\nu)(1 + \|u(\cdot,t)\|_{L_\infty}^{2p}) + C(\alpha)(1 + \|u(\cdot,t)\|_{L_\infty}^{2p})]\|u_x(\cdot,t)\|^2 \\
&\quad + \delta[\|u_{xx}(\cdot,t)\|^2 + \|u_{xt}(\cdot,t)\|^2 + \|u_{xxt}(\cdot,t)\|^2].
\end{aligned}
\tag{4.9}
$$

First, choose T large enough so that, simultaneously,

$$
\|u(\cdot,t)\|_{L_\infty}^{2p} \leq 1 \quad\text{and}\quad \frac{2}{p+2}\|u(\cdot,t)\|_{L_\infty}^p \leq \frac{1}{2},
$$

for $t > T$. This is possible since $\|u(\cdot,t)\|_{L_\infty} \to 0$ for $t \to +\infty$. Next, choose the constant b_1 in (4.9) so that

$$
b_1 \geq \max\Big\{2, \frac{C(\delta) + 2C(\nu) + 2C(\alpha)}{\nu}\Big\}.
$$

Finally, choose δ small, say

$$\delta = \frac{1}{4}\min\{\alpha^2, \nu, 1\}.$$

With these choice, (4.9) implies that

$$\begin{aligned}
&\beta\|u(\cdot,t)\|_2^2 + \frac{d}{dt}\|u(\cdot,t)\|_2^2 \\
&+ C_1\|u_{xt}(\cdot,t)\|^2 + C_2\|u_{xxt}(\cdot,t)\|^2 \\
&\leq C(|g(t)|^2 + |h(t)|^2 + |g'(t)|^2 + |h'(t)|^2),
\end{aligned} \tag{4.10}$$

for all $t > T$, where β, C_1 and C_2 are some positive constants depending on α, δ and ν. Integrating (4.10) over $[T, t)$, the result stated in the lemma comes into view upon applying Theorem 4.3 to $\|u(\cdot, T)\|_2$. $\square$

If stronger hypotheses are posited on the boundary conditions, stronger results can be derived. In fact, if one multiplies (4.10) by βt and adds the result to (4.10), there appears

$$\begin{aligned}
&t\beta^2\|u(\cdot,t)\|_2^2 + \frac{d}{dt}\left((1 + t\beta)\|u(\cdot,t)\|_2^2\right) \\
&+ t\beta(C_1\|u_{xt}(\cdot,t)\|^2 + C_2\|u_{xxt}(\cdot,t)\|^2) \\
&\leq C(t\beta + 1)(|g(t)|^2 + |h(t)|^2 + |g'(t)|^2 + |h'(t)|^2),
\end{aligned} \tag{4.11}$$

for $t > T$. If $t[|g(t)|^2 + |h(t)|^2 + |g'(t)|^2 + |h'(t)|^2] \in L_1(\mathbb{R}^+)$, then integrating (4.11) over $[T, t)$ leads to

$$\begin{aligned}
&(1 + t\beta)\|u(\cdot,t)\|_2^2 + \int_T^t \tau\beta^2\|u(\cdot,\tau)\|_2^2 d\tau \\
&\leq (1 + T\beta)\|u(\cdot,T)\|_2^2 + C\int_T^t (\tau\beta + 1)(|g(\tau)|^2 + |h(\tau)|^2 + |g'(\tau)|^2 + |h'(\tau)|^2)d\tau,
\end{aligned} \tag{4.12}$$

for $t > T$. From (4.12), one sees that

$$\|u(\cdot,t)\|_2^2 = 0(t^{-1})$$

as $t \to +\infty$, by using Theorem 4.3 applied to $\|u(\cdot, T)\|_2$.

Repeating this procedure, one may deduce that if $t^n(|g(t)|^2 + |h(t)|^2 + |g'(t)|^2 + |h'(t)|^2) \in L_1(T, +\infty)$ for some $T > 0$, then $\|u(\cdot,t)\|_2^2 = 0(t^{-n})$. In consequence, it is seen that

$$\|u(\cdot,t)\|_{L_\infty} = 0(t^{-\frac{n}{2}}),$$

as $t \to +\infty$, since $\|u(\cdot,t)\|_{L_\infty}^2 \leq \|u(\cdot,t)\|(\|u(\cdot,t)\| + 2\|u_x(\cdot,t)\|)$.

If equation (4.10) is multiplied by $e^{\beta t}$ and the result integrated over $[T, t)$, there is derived

$$e^{\beta t}\|u(\cdot, t)\|_2^2 \leq e^{\beta T}\|u(\cdot, T)\|_2^2$$
$$+ \int_T^{+\infty} C e^{\beta \tau} \left[|g(\tau)|^2 + |h(\tau)|^2 + |g'(\tau)|^2 + |h'(\tau)|^2\right] d\tau. \qquad (4.13)$$

If $e^{\beta t}(|g(t)|^2 + |h(t)|^2 + |g'(t)|^2 + |h'(t)|^2) \in L_1(T, +\infty)$, then (4.13) yields

$$\|u(\cdot, t)\|_2^2 = 0(e^{-\beta t}), \qquad (4.14)$$

for $t > T$, again by using Theorem 4.3 applied to $\|u(\cdot, T)\|_2$. Moreover combining (4.3) and (4.14) shows that

$$\|u(\cdot, t)\|_{L_\infty} = 0(e^{-\frac{\beta t}{2}}),$$

as $t \to +\infty$.

In particular, if (4.1) is endowed with homogeneous boundary condition, the decay of solutions of (4.1) is exponential. The results just explained are summarized in the following corollary.

COROLLARY 4.6. *Let the initial condition f lie in $C_b^2(\Omega)$ and suppose the nonlinearity satisfies condition $(**)$ and the other stipulations in Corollary 4.4.*

(1) If the boundary data g and h are in $H^1(\mathbb{R}^+)$ and $t^{\frac{n}{2}} g$ and $t^{\frac{n}{2}} h$ are also in $H^1(\mathbb{R}^+)$, then $\|u(\cdot, t)\|_2^2 = 0(t^{-n})$ and

$$\|u(\cdot, t)\|_{L_\infty} = 0(t^{-\frac{n}{2}}),$$

as $t \to +\infty$, $n = 0, 1, 2, \cdots$.

(2) If $e^{\frac{\beta t}{2}} g$ and $e^{\frac{\beta t}{2}} h$ are in $H^1(\mathbb{R}^+)$, then $\|u(\cdot, t)\|_2^2 = 0(e^{-\beta t})$ and

$$\|u(\cdot, t)\|_{L_\infty} = 0(e^{-\frac{\beta t}{2}}),$$

as $t \to +\infty$. $\square$

5. CONCLUSION

In this paper, initial- and boundary-value problems for the generalized regularized long-wave equation have been considered. Both the quarter-plane problem in which an initial condition together with boundary data at the end of a semi-infinite stretch of the medium and the two-point boundary-value problem in which initial conditions are coupled with boundary conditions at both ends of a finite extent of the medium of propagation are shown to be well posed subject only to suitable smoothness together with obvious compatibility conditions. Extending earlier work on these problems, results of global existence, uniqueness and continuous dependence of the solutions on the initial and boundary data is established for nonlinearities that grow at infinity no faster than quartically in the case of nonhomogeneous boundary conditions, and for smooth nonlinearities of any growth rate in the case of homogeneous boundary conditions.

When dissipative effects modeled by the term $-\nu u_{xx}$ are present, solutions of the two-point boundary-value problems are found to decay at rates up to exponential. Indeed, the rates of decay are determined by the evanescence of the boundary data for large values of t. This is in contrast to the sharp decay rates that obtain for the pure initial-value problem (see [2] and [9]), which are only algebraic. Preliminary decay results are also obtained for the quarter-plane problem, and these appear similar to those that obtain for the pure initial-value problem, reflecting the difference between bounded and unbounded domains.

This work points to a number of open questions. As far as well posedness is concerned, we have not yet determined if the quartic restriction on the growth rate of the nonlinearity P imposed in our theory is merely an artifact of our proof, or whether singularities in solutions may form in finite time for nonlinearities not respecting this restriction. The decay results for the quarter-plane problem could use some refinement and it could certainly be interesting to treat more general forms of both dispersion and dissipation.

REFERENCES

[1] J. P. ALBERT AND J. L. BONA, *Comparisons between model equations for long waves*, J. Nonlinear Sci., **1** (1991) 345-374.

[2] C. J. AMICK, J. L. BONA AND M. E. SCHONBEK, *Decay of solutions of some nonlinear wave equations*, J. Differential Equations, **81** (1989) 1-49.

[3] E.F. BECKENBACH AND R. BELLMAN, *Inequalities*, Springer-Verlag: Berlin, New York, (1961).

[4] T.B. BENJAMIN, J.L. BONA AND J.J. MAHONY, *Model equations for long waves in nonlinear dispersive systems*, Philos. Trans. Roy. Soc. London Ser. A, **272** (1972) 47-78.

[5] J. L. BONA AND P. J. BRYANT, *A mathematical model for long waves generated by wavemakers in nonlinear dispersive systems*, Proc. Camb. Phil. Soc., **73** (1973) 391-405.

[6] J. L. BONA AND V. A. DOUGALIS, *An initial- and boundary-value problem for a model equation for propagation of long waves*, J. Math. Anal. Appl., **75** (1980) 503-522.

[7] J. L. BONA, V. A. DOUGALIS, O. A. KARAKASHIAN AND W. MCKINNEY, *Computations of blow up and decay for periodic solutions of the generalized Korteweg-de Vries equation*, Applied Numerical Math., **10** (1992) 335-355.

[8] ———, *Conservative, high-order numerical schemes for the generalized Korteweg-de Vries equation*, (submitted).

[9] J. L. BONA AND L. LUO, *Decay of solutions to nonlinear, dispersion, dissipative wave equations*, Differential & Integral Equs., **6** (1993) 961-980.

[10] ———, *Generalized KdV equation in a quarter plane*, (Preprint).

[11] ———, *Nonhomogeneous initial- and boundary-value problems for KdV equation*, (Preprint).

[12] ———, *On the decay of solutions in a quarter-plane for some nonlinear wave equations*, (in preparation).

[13] J. L. BONA, W. G. PRITCHARD AND L. R. SCOTT, *An evaluation of a model equation for water waves*, Phil. Trans. Roy Soc. London A, **302** (1981) 457-510.

[14] ———, *A comparison of solutions of model equations for long waves*, in Lectures in Applied Mathematics, **20** (N. Lebovitz, ed.) American Mathematical Society: Providence, RI (1983).

[15] J. L. BONA AND J. C. SAUT, *Singularités dispersives de solutions d'équations de type Korteweg-de Vries*, C. R. Acad. Sci. Paris, **303** (1986) 101-103.

[16] J. L. BONA AND R. SMITH, *The initial-value problem for the Korteweg-de Vries equation*, Philos. Trans. Roy. Soc. London A, **278** (1975) 555-601.

[17] J. L. BONA AND R. WINTHER, *The Korteweg-de Vries equation, posed in a quarter-plane*, SIAM J. Math. Anal., **14** (1983) 1056-1106.

[18] ———, *The Korteweg-de Vries equation in a quarter-plane, continuous dependence results*, Diff. & Int. Equations, **2** (1989) 228-250.

[19] DANG DINH ANG AND TRAN THANH, *A nonlinear pseudoparabolic equation*, Proc. Roy. Soc. Edinburgh, **114A** (1990) 119-133.

[20] DIX, D. B., *The dissipation of nonlinear dispersive waves: The case of asymptotically weak nonlinearity*, Comm. P.D.E., **17** (1992) 1665-1693.

[21] J. GINIBRE AND Y. TSUTSUMI, *Uniqueness of solutions for the generalized Korteweg-de Vries equation*, SIAM J. Math. Anal., **6** (1989) 1388-1425.

[22] T. KATO, *On the Cauchy problem for the (generalized) Korteweg-de Vries equation*, Studies in Applied Math., Advances in Math. Suppl. Studies, **8** (1983) 93-128.

[23] C. E. KENIG, G. PONCE AND L. VEGA, *On the (generalized) Korteweg-de Vries equation*, Duke. Math. J., **59** (1989) 585-610.

[24] D. J. KORTEWEG AND G. DE VRIES, *On the change of form of long wave advancing in a rectangular canal, and on a new type of long stationary waves*, Philos. Mag., **39** (1895) 422-443.

[25] J.-L. LIONS, *Quelques méthodes de résolution des problèmes aux limites non-linéaires* Dunod - Gauthier—Villars: Paris (1969).

[26] L. A MEDEIROS AND G. PERLA MENZALA, *Existence and uniqueness for periodic solutions of the Benjamin-Bona-Mahony equation*, SIAM J. Math Anal., **8** (1977) 792-799.

[27] L. A MEDEIROS AND M. MILLA MIRANDA, *Weak solutions for a nonlinear dispersive equation*, J. Math. Anal. Appl., **59** (1977) 432-441.

[28] D. H. PEREGRINE, *Calculations of the development of an undular bore*, J. Fluid Mech., **25** (1966) 321-330.

[29] P. J. OLVER, *Euler operators and conservation laws of the BBM equation*, Math. Proc. Camb. Phil. Soc., **85** (1979) 143-160.

[30] F. TREVES, *Topological Vector Spaces, Distributions and Kernels*. Academic Press: New York (1967).

Hyperbolicity and Dissipativity

C. Chicone and Y. Latushkin   University of Missouri, Columbia, Missouri

1. INTRODUCTION

Consider a C_0-semigroup $\{T^t\}_{t\geq 0}$ with infinitesimal generator A in a separable Hilbert space $\mathcal{H}$ with scalar product $\langle .\,,\,. \rangle$. The semigroup is called *hyperbolic* if $\sigma(T^t) \cap \mathbb{T} = \emptyset$, $t \neq 0$, where σ denotes the spectrum and $\mathbb{T} = \{z : |z| = 1\}$. This notion is important because often the hyperbolicity of T^t is equivalent to the existence of an exponential dichotomy for the solutions of the associated differential equation $v' = Av$ in $\mathcal{H}$. If the group has a *bounded* generator $A \in B(\mathcal{H})$, then as is well known (see, e.g. Daleckij and Krein (1974)), the existence of an exponential dichotomy can be characterized in terms of A. In fact, the hyperbolicity of the group $\{e^{tA}\}$ is equivalent to the existence of a Hermitian operator W on $\mathcal{H}$ such that A is uniformly W-dissipative, i. e., for some $\delta > 0$ and all $v \in \mathcal{H}$ we have $\langle (WA + A^*W)v, v \rangle \leq -\delta \|v\|^2$, an inequality we will denote by $WA + A^*W \ll 0$. In this case, the operator W defines a quadratic Lyapunov function $\mathcal{L}$, a perhaps indefinite scalar product on $\mathcal{H}$, given by $\mathcal{L}(u, v) = \langle u, v \rangle_W = \langle Wu, v \rangle$ such that $t \mapsto \mathcal{L}(v(t), v(t))$ decreases monotonically for each solution v of the associated differential equation. In the present paper we generalize these results to the case of a C_0-semigroup, $\{T^t\}_{t\geq 0}$, whose infinitesimal generator may be an *unbounded* operator A in $\mathcal{H}$.

In Section 2 we prove that the hyperbolicity of $\{T^t\}$ is equivalent to the existence of two Hermitian operators W and V, such that A is uniformly W-dissipative and A^* is V-dissipative. Moreover, we show that if the semigroup $\{T^t\}$ is hyperbolic, then the operator

W is a solution of the operator equation $WA + A^*W = -H$ for some $H \gg 0$. This solution can be obtained by an integral formula similar to the one used in the bounded case (see Daleckij and Krein (1974)). The dual fact holds for V. In addition, using Gearhart's spectral mapping theorem (see Nagel (1984), p. 95), we derive the following fact, related to the classical Lumer-Phillips theorem: If a C_0-semigroup is generated by a W-dissipative operator A and for some $\xi_0 \in \mathbb{R}$ the range of the operator $i\xi_0 - A$ is all of $\mathcal{H}$, then this semigroup is hyperbolic. In addition, if A generates a hyperbolic semigroup, then A is W-dissipative and the range of $i\xi - A$ is all of $\mathcal{H}$ for each $\xi \in \mathbb{R}$.

Section 3 deals with hyperbolic evolutionary families $\{U(x,s)\}_{x \geq s}$. These can be viewed as propagators of nonautonomous differential equations of the form $v' = A(x)v$ in $\mathcal{H}$. The evolutionary family is called hyperbolic if there exists a projection-valued function $P : \mathbb{R} \to B(\mathcal{H})$, such that solutions of the differential equation starting in $\operatorname{Im} P(x)$ (resp. in $\operatorname{Ker} P(x)$) exponentially decrease (resp. increase). If the operators in the evolutionary family are all invertible, that is, the family $\{U(x,s)\}_{x \geq s}$ is extendible to all $(x,s) \in \mathbb{R}^2$, then hyperbolicity of the family is equivalent to the existence of an exponential dichotomy for the solutions of the differential equation.

We associate with $\{U(x,s)\}_{x \geq s}$ the so-called *evolutionary semigroup* $\{e^{tC}\}_{t \geq 0}$, acting in the space $L_2(\mathbb{R}; \mathcal{H})$ by the rule $(e^{tC} f)(x) = U(x, x - t)f(x - t)$, $x \in \mathbb{R}$. These semigroups were first introduced by Howland (1974) and recently have become the object of intensive study (see Ben-Artzi and Gohberg (1992), Nagel (1984), Nagel and Rhandi (1992), Rau (1992) and the detailed bibliography in Rau (1992)). A theorem of Rau (1992) shows that the hyperbolicity of $\{U(x,s)\}_{x \geq s}$ is equivalent to the hyperbolicity of the semigroup $\{e^{tC}\}_{t \geq 0}$. To characterize the hyperbolicity of $\{U(x,s)\}_{x \geq s}$, we apply the results from Section 2 to the generator C of the semigroup $\{e^{tC}\}_{t \geq 0}$. In this case the operators W and V are operators of multiplication by Hermitian-valued functions $W(\cdot)$ and $V(\cdot)$ defined on $\mathbb{R}$.

Under additional assumptions the generator C of the evolutionary semigroup has the form $(Cf)(x) = -\dfrac{d}{dx}f(x) + A(x)f(x)$. In this case the condition for W-dissipativity of C and C^* in $L_2(\mathbb{R}; \mathcal{H})$ can be given more explicitly. In fact, if the equation $WC + C^*W = -H$ has a solution for some multiplication operator $H = H(\cdot) \gg 0$, then there is an operator valued function $W(\cdot)$ and $\delta > 0$ such that

$$\frac{d}{dx}\langle W(x)u, u \rangle + 2\operatorname{Re}\langle A(x)u, W(x)u \rangle \leq -\delta \|u\|^2$$

uniformly for all $x \in \mathbb{R}$ and $u \in \mathcal{D}$, where $\mathcal{D}$ is the domain of the unbounded operator $A(x)$ for each $x \in \mathbb{R}$. If, for example, the mapping $\mathbb{R} \ni x \mapsto A(x) \in B(\mathcal{H})$ is norm-continuous and bounded, the weak derivative here can be replaced by the uniform derivative, and the existence of an exponential dichotomy for the equation $v' = A(x)v$ is equivalent to the existence of an invertible Hermitian-valued function $W : \mathbb{R} \to B(\mathcal{H})$, such that

$$\frac{dW}{dx} + A(x)W(x) + W(x)A^*(x) \ll 0$$

in $\mathcal{H}$ uniformly for $x \in \mathbb{R}$.

The results presented here generalize the results of Chicone and Latushkin (1993) where similar facts were obtained for the group case. Also, we mention the paper by Nagel and Rhandi (1992), where similar characterizations for the stability of semigroups and evolutionary families are obtained using different methods. Instead of finding explicit formulas for the solution of the equation $WA + A^*W = -H$, they consider an associated positive semigroup given by $W \mapsto T^{t*}WT^t$ in $B(\mathcal{H})$ and then apply results from the theory of the asymptotic behavior of positive semigroups. This approach can be applied [R.Nagel, private communication] for the case of hyperbolicity as well.

The first author was supported by the Air Force Office of Scientific Research and the National Science Foundation under the grant DMS-9022621.

2   HYPERBOLIC SEMIGROUPS AND DISSIPATIVITY

Let A denote the infinitesimal generator of a C_0-semigroup $\{T^t\}_{t \geq 0}$ in a Hilbert space $\mathcal{H}$ and let $\mathcal{D}(A)$ denote the domain of A. We say A is uniformly W-dissipative for some bounded Hermitian operator $W \in B(\mathcal{H})$ if there exists $\delta > 0$ such that for each $v \in \mathcal{D}(A)$

$$\mathrm{Re}\langle Av, Wv \rangle \leq -\delta\|v\|^2. \tag{1}$$

The following result describes the hyperbolicity of $\{e^{tA}\}_{t \geq 0}$ in terms of the dissipativity of A and A^* (cf. Corollary 1.4.4 in Pazy (1983)).

THEOREM 1   The C_0-semigroup $\{T^t\}_{t \geq 0}$, $T^t = e^{tA}$ is hyperbolic if and only if there are Hermitian operators W and V in $B(\mathcal{H})$, such that A is uniformly W-dissipative and A^* is uniformly V-dissipative.

The following reformulation of this result is related to the classical Lumer-Phillips theorem (Pazy (1983), p. 14).

COROLLARY 1   If for some $\xi_0 \in \mathbb{R}$ the operator $i\xi_0 - A$ is "onto" and if A is uniformly W-dissipative for some Hermitian W, then the C_0-semigroup $\{T^t\}_{t \geq 0}$, $T^t = e^{tA}$ is hyperbolic. Conversely, if the semigroup is hyperbolic, then for all $\xi \in \mathbb{R}$ the operator $i\xi - A$ is "onto" and A is uniformly W-dissipative.

The next result allows one to obtain the operators W and V as the solutions of linear Riccati equations. If $T = e^A$ is hyperbolic, we let P denote the Riesz projection for T corresponding to the part of $\sigma(T)$ lying inside the unit disk $\mathbb{D}$ and set $Q = I - P$. Since $\mathrm{Im}\,P$ and $\mathrm{Im}\,Q$ are T^t-invariant, we have C_0-semigroups $T_P^t := T^t|\,\mathrm{Im}\,P = PT^tP$ and $T_Q^t := T^t|\,\mathrm{Im}\,Q = QT^tQ$, $t > 0$. Since $\sigma(T_Q^1)$ is outside of $\mathbb{D}$, the operator T_Q^1 is invertible and $\{T_Q^t\}_{t \geq 0}$ extends to a C_0-group $\{T_Q^t\}_{t \in \mathbb{R}}$ in $\mathrm{Im}\,Q$. Also, if H is a bounded operator we define $H_{PQ} = P^*HP + Q^*HQ$.

THEOREM 2   The C_0-semigroup $\{T^t\}_{t \geq 0}$, $T^t = e^{tA}$ is hyperbolic if and only if there are bounded operators $R \ll 0$ and $S \ll 0$ such that the operator equations

$$(WA + A^*W)v = Rv, \quad v \in \mathcal{D}(A) \tag{2}$$

and

$$(VA^* + AV)v = Sv, \quad v \in \mathcal{D}(A^*) \tag{3}$$

have, respectively, solutions $W : \mathcal{D}(A) \to \mathcal{D}(A^*)$ and $V : \mathcal{D}(A^*) \to \mathcal{D}(A)$. In addition, if $\{T^t\}_{t \geq 0}$ is hyperbolic, then W and V are invertible. Also, for any bounded operator $H \gg 0$ the operators W and V can be obtained for $R = -H_{PQ}$ in (2) and $S = -H_{P^*Q^*}$ in (3) by the integral formulas

$$W = \int_0^\infty \{T_P^{t*} H T_P^t - T_Q^{-t*} H T_Q^{-t}\} \, dt \tag{4}$$

and

$$V = \int_0^\infty \{T_P^t H T_P^{t*} - T_Q^{-t} H T_Q^{-t*}\} \, dt. \tag{5}$$

With this choice of W and V the subspace $\operatorname{Im} P$ is uniformly W-positive and $\operatorname{Im} Q$ is uniformly W-negative while the subspace $\operatorname{Im} P^*$ is uniformly V-positive and $\operatorname{Im} Q^*$ is uniformly V-negative. Here $\operatorname{Im} P$ and $\operatorname{Im} Q$ are W-orthogonal while $\operatorname{Im} P^*$ and $\operatorname{Im} Q^*$ are V-orthogonal.

It is convenient to give the proofs of Theorems 1-2 and Corollary 1 simultaneously.

PROOF. SUFFICIENCY: If there is an operator $R \ll 0$ such that equation (2) has a solution W with the property $W : \mathcal{D}(A) \to \mathcal{D}(A^*)$, then (1) states $\operatorname{Re}(WA) \ll 0$, that is,

$$2\operatorname{Re}\langle Av, Wv \rangle = \langle (WA + A^*W)v, v \rangle \leq -2\delta \|v\|^2, \quad v \in \mathcal{D}(A). \tag{6}$$

Thus, A is uniformly W-dissipative. Since $\langle Wv, v \rangle \in \mathbb{R}$ for all $v \in \mathcal{H}$, inequality (1) implies for all $\xi \in \mathbb{R}$ and $v \in \mathcal{D}(A)$:

$$\delta \|v\|^2 \leq |\operatorname{Re}\langle Av, Wv \rangle| = |\operatorname{Re}\langle (A - i\xi)v, Wv \rangle|$$
$$\leq |\langle (A - i\xi)v, Wv \rangle| \leq \|(A - i\xi)v\| \cdot \|W\| \cdot \|v\|.$$

Hence $\|(A - i\xi)v\| \geq \delta \|W\|^{-1} \|v\|$, and $A - i\xi$ is uniformly injective for all $\xi \in \mathbb{R}$.

To complete the proof of sufficiency in Theorems 1 and 2, we note that similarly for $v \in \mathcal{D}(A^*)$ and $\xi \in \mathbb{R}$ if A^* is V-dissipative, then $\|(A^* - i\xi)v\| \geq \delta \|V\|^{-1} \|v\|$. Hence, for all $\xi \in \mathbb{R}$ the operator $A - i\xi$ is invertible, and the resolvent $(A - i\xi)^{-1}$ of A is uniformly bounded along $i\mathbb{R}$, that is $\|(A - i\xi)^{-1}\| \leq \delta^{-1} \|W\|$. By Gearhart's spectral mapping theorem (see Nagel (1984), p. 95) we have $\sigma(T^t) \cap \mathbb{T} = \emptyset$, $t \neq 0$.

To complete the proof of sufficiency for Corollary 1 just note that, as we have seen, for all $\xi \in \mathbb{R}$ and for some $\delta > 0$ the inequality $\|(A - i\xi)v\| \geq \delta \|W\|^{-1} \|v\|$ holds provided A is W-dissipative. Hence, for the approximate point spectrum $\sigma_{ap}(\cdot)$ one has $\sigma_{ap}(A) \cap i\mathbb{R} = \emptyset$. Since the boundary of $\sigma(A)$ is in $\sigma_{ap}(A)$, we have either $\sigma(A) \cap i\mathbb{R} = \emptyset$ or for each $\xi \in \mathbb{R}$ the operator $i\xi - A$ is not "onto". Since for $\xi = \xi_0$ the operator $i\xi - A$ is "onto", we conclude that $i\xi - A$ is invertible for all $\xi \in \mathbb{R}$. Moreover, the resolvent of A is bounded, $\|(A - i\xi)^{-1}\| \leq \delta^{-1} \|W\|$, $\xi \in \mathbb{R}$, and $\sigma(T^t) \cap \mathbb{T} = \emptyset$, $t \neq 0$ by Gearhart's spectral mapping theorem.

NECESSITY: Suppose T^t, $t \neq 0$ is hyperbolic. By the spectral inclusion theorem (see Pazy (1983), Theorem 2.3) $\sigma(A) \cap i\mathbb{R} = \emptyset$. Thus, the first condition in Corollary 1 holds.

If $\{T^t\}$ is hyperbolic, we will show that for any bounded operator $H >> 0$ the operator W in (4) is bounded, $W : \mathcal{D}(A) \to \mathcal{D}(A^*)$, and W satisfies (2) with $R = -H_{PQ}$. Then A is W-dissipative by (6) as required in Theorem 1 and Corollary 1.

We show first that W_P and W_Q given by

$$W_P = \int_0^\infty T_P^{t*} H T_P^t \, dt, \quad W_Q = -\int_0^\infty T_Q^{-t*} H T_Q^{-t} \, dt$$

are bounded. Indeed, since the spectral radius of both T_P^1 and $(T_Q^1)^{-1}$ is less than one, the integrals $\int_0^\infty \|T_P^t\|^2 \, dt$ and $\int_0^\infty \|T_Q^{-t}\|^2 \, dt$ both converge. Next, we show that if $u \in \mathcal{D}(A)$, then $W_P u \in \mathcal{D}(A^*)$ and

$$A^* W_P u = -P^* H P u - W_P A u. \tag{7}$$

By definition, $A^* W_P u = \lim_{t \to 0} t^{-1}(T_P^{t*} W_P u - W_P u)$. We will prove that

$$I_t := \frac{1}{t}(T_P^{t*} W_P u - W_P u) + P^* H P u + W_P A u \to 0$$

as $t \to 0$. Since $T_P^{t*} W_P u = \int_t^\infty T_P^{s*} H T_P^{(s-t)} u \, ds$, one has:

$$I_t = P^* H P u - \frac{1}{t} \int_0^t T_P^{s*} H T_P^s u \, ds \tag{8}$$

$$+ \int_0^t T_P^{s*} H T_P^s A u \, ds \tag{9}$$

$$+ \int_t^\infty \left[\frac{1}{t} \{ T_P^{s*} H T_P^{(s-t)} u - T_P^{s*} H T_P^s u \} + T_P^{s*} H T_P^s A u \right] ds. \tag{10}$$

Obviously, (8) and (9) tend to 0 as $t \to 0$. Since (10) is equal to

$$T_P^{t*} \int_0^\infty T_P^{s*} H T_P^s \, ds \left[\frac{1}{t}(u - T_P^t u) + T_P^t A u \right],$$

we have $I_t \to 0$ and (7) holds. Similarly, if $v \in \mathcal{D}(A)$ then $W_Q v \in \mathcal{D}(A^*)$ and $A^* W_Q v = -Q^* H Q v - W_Q A v$. Hence (4) satisfies (2) with $R = -H_{PQ}$.

As $\{T^t\}_{t \geq 0}$ is assumed hyperbolic, $\sigma(A) \cap i\mathbb{R} = \emptyset$ and A is invertible. But, this implies W is invertible. In fact, since $R \gg 0$ is a bounded operator and $(WA + A^*W)v = -Rv$ for $v \in \mathcal{D}(A)$, for each $v \in \mathcal{H}$ and some $\delta > 0$

$$2 \operatorname{Re}\langle A^{-1}v, Wv \rangle = \langle (A^*W + WA)A^{-1}v, A^{-1}v \rangle = \langle RA^{-1}v, A^{-1}v \rangle \leq -2\delta \|A^{-1}\|^2 \|v\|^2.$$

It follows that

$$\delta \|A^{-1}\|^2 \|v\|^2 \leq |\operatorname{Re}\langle A^{-1}v, Wv \rangle| \leq |\langle A^{-1}v, Wv \rangle| \leq \|A^{-1}\| \cdot \|v\| \cdot \|Wv\|.$$

Hence, $\|Wv\| \geq \delta \|A^{-1}\| \|v\|$ and, as a result, W is invertible.

To prove that $\operatorname{Im} P$ is uniformly W-positive we note that for $T = T^1$

$$
\begin{aligned}
T^* W T - W &= \int_1^\infty T_P^{t*} H T_P^t \, dt - \int_0^\infty T_P^{t*} H T_P^t \, dt \\
&\quad - \int_{-1}^\infty T_Q^{-t*} H T_Q^{-t} \, dt + \int_0^\infty T_Q^{-t*} H T_Q^{-t} \, dt \\
&= - \int_0^1 \left(T_P^{t*} H T_P^t + T_Q^{t*} H T_Q^t \right) dt \\
&= - \int_0^1 T^{*s} H_{PQ} T^s \, ds \ll 0
\end{aligned}
$$

and for each $v \in \operatorname{Im} P$ also $\langle W v, v \rangle = \langle (W P + P^* W - W) v, v \rangle$. Then, using the Riesz integral representation (see, e.g., Daleckij and Krein (1974), p. 20) for the spectral projection P of $T = T^1$ we find

$$
W P + P^* W - W = \frac{1}{2\pi} \int_0^{2\pi} (e^{-i\tau} - T^*)^{-1} (W - T^* W T)(e^{i\tau} - T)^{-1} \, d\tau,
$$

and $W P + P^* W - W$ is uniformly positive since $W - T^* W T$ is uniformly positive. Similarly, $\operatorname{Im} Q$ is uniformly W-negative.

Obviously, $\operatorname{Im} P$ and $\operatorname{Im} Q$ are W-orthogonal: $\langle W f, g \rangle = 0$ for $f \in \operatorname{Im} P$, $g \in \operatorname{Im} Q$. Also, since $\{T^{*t}\}_{t \geq 0}$ is hyperbolic provided $\{T^t\}_{t \geq 0}$ is hyperbolic, all statements about the operator V in Theorems 1 and 2 can be obtained by applying the arguments above to the C_0-semigroup $\{T^{*t}\}_{t \geq 0}$. ∎

REMARK If the operator W, that satisfies (2), $W : \mathcal{D}(A) \to \mathcal{D}(A^*)$, is invertible and $W^{-1} : \mathcal{D}(A^*) \to \mathcal{D}(A)$, then $V = W^{-1}$ is a solution of (3) with $S = W^{-1} R W^{-1}$. If $\{T^t\}_{t \in \mathbb{R}}$ is a hyperbolic C_0-group, then (see Chicone and Latushkin (1993)) the operator W, computed by (4), is invertible and for this W also $W^{-1} : \mathcal{D}(A^*) \to \mathcal{D}(A)$. Hence the hyperbolicity of a group $\{T^t\}_{t \in \mathbb{R}}$ is equivalent to the existence of an invertible operator W, such that A is W-dissipative and A^* is W^{-1}-dissipative.

A Lyapunov function $\mathcal{L}(u, v) = \langle W u, v \rangle$ for the differential equation $v' = A v$ in $\mathcal{H}$ is defined by the operator W satisfying Theorems 1 and 2, i. e., for each solution $v(t) = e^{tA} v(0)$ of this differential equation the function $t \mapsto \mathcal{L}(v(t), v(t))$ decreases monotonically. Similarly, V defines a Lyapunov function for the differential equation $v' = A^* v$. We also note that the condition of V-dissipativity of A^* in Theorem 1 cannot be dropped, see Massera and Schäffer (1959), Example 3.2.

3 HYPERBOLIC EVOLUTIONARY FAMILIES AND DISSIPATIVITY

Let $\{U(x, s)\}_{x \geq s}$, $\infty > x \geq s > -\infty$ be a family of bounded operators in $\mathcal{H}$. The family $\{U(x, s)\}_{x \geq s}$ is called an *evolutionary family* if the following conditions are fulfilled:

a) The mapping $(x, s) \mapsto U(x, s)v$, $x \geq s$ is continuous for each $v \in \mathcal{H}$;

b) $U(x, x) = I$, $U(x, s) = U(x, r)U(r, s)$, $x \geq r \geq s$,

c) $\|U(x, s)\| \leq ce^{\beta(x-s)}$ for some $c, \beta > 0$ and all $x \geq s$.

As the following examples show, one can view the evolutionary family $\{U(x, s)\}_{x \geq s}$ as the propagator for the nonautonomous differential equation $v' = A(x)v$, that is the solution is given by $v(x) = U(x, s)v(s)$, $x \geq s$.

EXAMPLE 1   If $A(t) \equiv A$ generates a C_0-semigroup in $\mathcal{H}$, then the propagator for the equation $v' = Av$ is the evolutionary family $U(x, s) = e^{(x-s)A}$, $x \geq s$.

EXAMPLE 2   If the mapping $\mathbb{R} \ni x \mapsto A(x) \in B(\mathcal{H})$ is (norm) continuous and bounded, then the propagator $U(x, s)$ of the equation $v' = A(x)v$ is an evolutionary family (see Daleckij and Krein (1974)). Moreover, $U(x, s)$ is defined for all $(x, s) \in \mathbb{R}^2$ and $(x, s) \mapsto U(x, s)$ is norm-continuous.

EXAMPLE 3   Let $\{A(x)\}_{x \in \mathbb{R}}$ be a stable (see Tanabe (1979), p. 93) family of generators of C_0-semigroups such that their domains $\mathcal{D}(A(x)) = \mathcal{D}$ are independent of $x \in \mathbb{R}$ and for each $v \in \mathcal{D}$ the map $\mathbb{R} \ni x \mapsto A(x)v \in \mathcal{H}$ is a strongly continuously differentiable function. Then (see Tanabe (1979), p. 102) the propagator $U(x, s)$ of the equation $v' = A(x)v$ is an evolutionary family. Moreover, $U(x, s) : \mathcal{D} \to \mathcal{D}$, the functions $x \mapsto U(x, s)v$ and $s \mapsto U(x, s)v$ are strongly continuously differentiable for each $v \in \mathcal{D}$ and, in addition, for each $v \in \mathcal{D}$ the following equations hold:

$$\frac{\partial}{\partial x}U(x, s)v = A(x)U(x, s)v \tag{11}$$

$$\frac{\partial}{\partial s}U(x, s)v = -U(x, s)A(s)v. \tag{12}$$

The following notion of (spectral) hyperbolicity generalizes the notion of exponential dichotomy for differential equations $v' = A(x)v$ (cf. Daleckij and Krein (1974)).

DEFINITION 1   The evolutionary family $\{U(x, s)\}_{x \geq s}$ is called (spectrally) *hyperbolic* if there exists a projection-valued function $P : \mathbb{R} \to L(\mathcal{H})$ and constants $M, \lambda > 0$ such that for $x \geq s$ the following conditions are satisfied:

1) The mapping $\mathbb{R} \ni x \mapsto P(x)v \in \mathcal{H}$ is continuous and bounded for each $v \in \mathcal{H}$;

2) $P(x)U(x, s) = U(x, s)P(s)$;

3) The restriction $U(x, s)|\operatorname{Im}Q(s)$ has dense range as an operator from $\operatorname{Im}Q(s)$ to $\operatorname{Im}Q(x)$ for $Q(x) = I - P(x)$;

4) $\|U(x, s)v\| \leq Me^{-\lambda(x-s)}\|v\|$ for $v \in \operatorname{Im}P(s)$ and

$\|U(x, s)v\| \geq M^{-1}e^{\lambda(x-s)}\|v\|$ for $v \in \operatorname{Im}Q(s)$.

For the evolutionary family of Example 2 condition 1) is usually replaced by the norm-continuity of $P : \mathbb{R} \to L(\mathcal{H})$. In this form Definition 1 is exactly the definition of exponential dichotomy for $v' = A(x)v$ on $\mathbb{R}$.

If $U(x, s)$ is invertible, more precisely, the family $\{U(x, s)\}_{x \geq s}$ extends to the family

$\{U(x,s)\}_{(x,s)\in\mathbb{R}^2}$, then condition 3) automatically follows from 2). If $\{U(x,s)\}$, $x \geq s$ consists of compact operators, then condition 3) also follows from 2) and 4).

For an evolutionary family $\{U(x,s)\}_{x\geq s}$, we define the associated evolutionary semigroup $\{e^{tC}\}_{t\geq 0}$ in $L_2(\mathbb{R};\mathcal{H})$ by the following rule:

$$(e^{tC})f(x) = U(x, x - t)f(x - t), \ f \in L_2(\mathbb{R};\mathcal{H}), \ x \in \mathbb{R}.$$

Since $\{U(x,s)\}$ is an evolutionary family, $\{e^{tC}\}_{t\geq 0}$ is a strongly continuous semigroup. In some important special cases, the generator C can be computed explicitly.

EXAMPLE 4   The generator of the evolutionary semigroup for the evolutionary family $\{U(x,s)\}$ of Example 2 is given by the closure of the operator $(Cf)(x) = -\frac{df}{dx} + A(x)f(x)$ where $\mathcal{D}(C) = \{$ absolutely continuous $f \in L_2(\mathbb{R};\mathcal{H})$ with strong derivative $\frac{d}{dx}f \in L_2(\mathbb{R};\mathcal{H})\}$ (see Nagel and Rhandi (1993)).

EXAMPLE 5   The evolutionary semigroup for the evolutionary family $\{U(x,s)\}$ of Example 1 is given by $\left(e^{tC}f\right)(x) = e^{tA}f(x - t)$. Here the generator of the semigroup $\{e^{tC}\}$ is the closure of the operator $(Cf)(x) = -\frac{df}{dx} + Af(x)$, with $\mathcal{D}(C) = \{$ absolutely continuous $f : \mathbb{R} \to \mathcal{D}(A)$ with strong derivative $\frac{d}{dx}f \in L_2(\mathbb{R};\mathcal{H})$ and $Af \in L_2(\mathbb{R};\mathcal{H})\}$. Here the function $Af : \mathbb{R} \to \mathcal{H}$ is defined by $(Af)(x) = Af(x)$.

We mention two important facts about evolutionary semigroups: $\sigma(C)$ is invariant under translations along the imaginary axis and the spectral mapping theorem $\sigma(e^{tC})\backslash\{0\} = \exp t\sigma(C)$, $t \neq 0$ is valid (Rau (1992), see also Latushkin and Stepin (1991)).

From now on we will assume in addition to the properties a)–c) that our evolutionary family satisfies the following property:

d) The mapping $\mathbb{R} \ni x \mapsto U(x, x - 1) \in B(\mathcal{H})$ is (norm) continuous.

If d) is fulfilled, then e^C belongs to the C^*-algebra of operators in $L_2(\mathbb{R};\mathcal{H})$ generated by the translation operator $f(\cdot) \mapsto f(\cdot - 1)$ and the operators of multiplication by bounded norm-continuous operator valued functions. Hence, the technique from Antonevich (1985) can be applied (cf. also Latushkin and Stepin (1991)) and the following result (see Rau (1992)) can be derived.

THEOREM 3   The evolutionary family $\{U(x,s)\}_{x\geq s}$ is hyperbolic in $\mathcal{H}$ in the sense of Definition 1 if and only if the evolutionary semigroup $\{e^{tC}\}_{t\geq 0}$ is hyperbolic in $L_2(\mathbb{R};\mathcal{H})$. Moreover, the Riesz projection $\mathcal{P}$ of the hyperbolic operator e^C is the operator of multiplication $(\mathcal{P}f)(x) = P(x)f(x)$ by the projection-valued function $P : \mathbb{R} \to B(\mathcal{H})$ from Definition 1.

Under assumption d) Theorem 3 remains valid even if condition 1) in Definition 1 is replaced by the condition of the norm-continuity of the function $\mathbb{R} \ni x \mapsto P(x) \in B(\mathcal{H})$.

Theorem 3 remains valid (see Rau (1992)) without assumption d) for the case of an evolutionary group $\{e^{tC}\}_{t\in\mathbb{R}}$ (invertible evolutionary family $\{U(x,s)\}_{(x,s)\in\mathbb{R}^2}$).

CONJECTURE   Theorem 3 is valid for an arbitrary evolutionary family, that is, a family of opertators $\{U(x,s)\}_{x\geq s}$ that satisfies the conditions a), b) and c) but not necessarily d).

In view of Theorem 3, one can apply Theorems 1 and 2 to describe the hyperbolicity of the evolutionary family $\{U(x,s)\}_{x\geq s}$. To state this application we need a few definitions. For the hyperbolic family $\{U(x,s)\}_{x\geq s}$ and the projection $P(x)$ we define $Q(x) := I - P(x)$ as well as the operator $U_P(x,s) := P(x)U(x,s)P(s)$, as an operator from $\operatorname{Im}P(s)$ to $\operatorname{Im}P(x)$, and $U_Q(x,s) := Q(x)U(x,s)Q(s)$ as an operator from $\operatorname{Im}Q(s)$ to $\operatorname{Im}Q(x)$, $x \geq s$. The operator $U_Q(x,s)$ is invertible by 3) and 4). Finally, if we define $U_Q(s,x) := [U_Q(x,s)]^{-1}$ for $x \geq s$, then $\{U_Q(x,s)\}$ is defined for all $(x,s) \in \mathbb{R}^2$.

THEOREM 4   The evolutionary family $\{U(x,s)\}_{x\geq s}$ is hyperbolic if and only if there exist Hermitian operators W, $V \in B(L_2(\mathbb{R};\mathcal{H}))$ such that the generator C of the evolutionary semigroup $\{e^{tC}\}_{t\geq 0}$ is uniformly W-dissipative and the operator C^* is uniformly V-dissipative in $L_2(\mathbb{R};\mathcal{H})$:

$$\operatorname{Re}\langle Cf, Wf\rangle_{L_2} \leq -\delta\|f\|_{L_2}^2, \ f \in \mathcal{D}(C); \tag{13}$$

$$\operatorname{Re}\langle C^*f, Vf\rangle_{L_2} \leq -\delta\|f\|_{L_2}^2, \ f \in \mathcal{D}(C^*). \tag{14}$$

The operators W and V also enjoy all properties listed in Theorem 2. Moreover, if $\{U(x,s)\}_{x\geq s}$ is hyperbolic and $H : \mathbb{R} \to B(\mathcal{H})$ is any bounded norm-continuous operator valued function such that $H(x) \ll 0$ uniformly for $x \in \mathbb{R}$, then the solutions W and V of the operator equations

$$(WC + C^*W)f = -H_{PQ}f, \quad f \in \mathcal{D}(C) \tag{15}$$

$$(VC^* + CV)f = -H_{P^*Q^*}f, \quad f \in \mathcal{D}(C^*) \tag{16}$$

can be chosen as the operators of multiplication by the bounded functions $W(\cdot) : \mathbb{R} \to B(\mathcal{H})$ and $V(\cdot) : \mathbb{R} \to B(\mathcal{H})$, given by the formulas

$$W(x) = \int_0^\infty \{U_P^*(x+t,x)H(x+t)U_P(x+t,x) - U_Q^*(x-t,x)H(x-t)U_Q(x-t,x)\}\, dt \tag{17}$$

$$V(x) = \int_0^\infty \{U_P(x,x-t)H(x-t)U_P^*(x,x-t) - U_Q(x,x+t)H(x+t)U_Q^*(x,x+t)\}\, dt. \tag{18}$$

PROOF: The theorem follows from the fact that the Riesz projection $\mathcal{P}$ for the hyperbolic operator e^C is the operator of multiplication by $P(\cdot)$.   ∎

Under some additional assumptions on the evolutionary family $\{U_{(x,s)}\}_{x\geq s}$ the generator C of the semigroup $\{e^{tC}\}_{t\geq 0}$ can be computed explicitly. Therefore, equations (13)–(16) can be written explicitly as well. We will carry this out for Example 3.

From now on we will assume that the evolutionary family $\{U(x,s)\}_{x\geq s}$ with properties (a)-(d) is a propagator for the equation $v' = A(x)v$, where the operators $A(x)$ and $A^*(x)$ are generators of C_0-semigroups that satisfy the properties described in Example 3. In particular, $\mathcal{D}(A(x)) = \mathcal{D}$, $\mathcal{D}(A^*(x)) = \mathcal{D}^*$, $U(x,s) : \mathcal{D} \to \mathcal{D}$, $U^*(x,s) : \mathcal{D}^* \to \mathcal{D}^*$, the functions $x \mapsto A(x)v, v \in \mathcal{D}$ and $x \mapsto A^*(x)v, v \in \mathcal{D}^*$ are continuously differentiable and equations (11) and (12) are valid. Also we assume that the generators of the semigroup $\{e^{tC}\}$ and its adjoint are the closure of the operators given by the following formulas:

$$(Cf)(x) = -(\frac{d}{dx}f)(x) + A(x)f(x), \quad (C^*f)(x) = (\frac{d}{dx}f)(x) + A^*(x)f(x),$$

where $\mathcal{D}(C) = \{f : \mathbb{R} \to \mathcal{D} | f$ is absolutely continuous, $\frac{d}{dx}f \in L_2(\mathbb{R};\mathcal{H})$ and $Af \in L_2(\mathbb{R};\mathcal{H})\}$; and $\mathcal{D}(C^*) = \{f : \mathbb{R} \to \mathcal{D}^* | f$ is absolutely continuous, $\frac{d}{dx}f \in L_2(\mathbb{R};\mathcal{H})$ and $A^*f \in L_2(\mathbb{R};\mathcal{H})\}$. The derivative $\frac{d}{dx}$ is understood in the strong sense in $\mathcal{H}$ and the operators A and A^* are the multiplication operators $(Af)(x) = A(x)f(x)$, $(A^*f)(x) = A^*(x)f(x)$.

THEOREM 5   Under the assumptions above the evolutionary family $\{U(x,s)\}_{x \geq s}$ is hyperbolic if and only if there exist (norm) bounded Hermitian-valued functions $W : \mathbb{R} \to B(\mathcal{H})$ and $V : \mathbb{R} \to B(\mathcal{H})$ that satisfy the following conditions:
(i)   For each $u, v \in \mathcal{H}$ the functions $x \mapsto \langle W(x)v, u \rangle$ and $x \mapsto \langle V(x)v, u \rangle$ are continuous;
(ii) For each $u \in \mathcal{D}$ and $v \in \mathcal{D}^*$ the functions $x \mapsto \langle W(x)u, u \rangle$ and $x \mapsto \langle V(x)v, v \rangle$ are continuously differentiable;
(iii) There exists $\delta > 0$ such that for all $x \in \mathbb{R}$ the following inequalities hold:

$$\frac{d}{dx}\langle W(x)u, u \rangle + 2\operatorname{Re}\langle A(x)u, W(x)u \rangle \leq -2\delta\|u\|^2, \quad u \in \mathcal{D}; \tag{19}$$

$$\frac{d}{dx}\langle V(x)v, v \rangle + 2\operatorname{Re}\langle A(x)v, V(x)v \rangle \leq -2\delta\|v\|^2, \quad v \in \mathcal{D}^*. \tag{20}$$

PROOF. NECESSITY: Suppose $W(\cdot)$ and $V(\cdot)$ satisfy (i)–(iii). Consider in $L_2(\mathbb{R};\mathcal{H})$ the bounded Hermitian multiplication operators defined by W and V. We claim W satisfies condition (13) of Theorem 4 (condition (14) for V can be proved similarly). To show this, fix $f \in \mathcal{D}(C)$. Since the function $x \mapsto \|W(x)\|$ is bounded, conditions (i) and (ii) yield the following "Leibniz rule":

$$\frac{d}{dx}\langle W(x)f(x), f(x) \rangle_{\mathcal{H}} = \frac{d}{dt}\langle W(x+t)f(x), f(x) \rangle_{\mathcal{H}}\Big|_{t=0} + 2\operatorname{Re}\langle \frac{d}{dx}f(x), W(x)f(x) \rangle_{\mathcal{H}}.$$

Since $f \in L_2(\mathbb{R};\mathcal{H})$, we find

$$2\operatorname{Re}\langle \frac{d}{dx}f, Wf \rangle_{L_2} = -\int_{\mathbb{R}} \frac{d}{dt}\langle W(x+t)f(x), f(x) \rangle_{\mathcal{H}}\Big|_{t=0} dx.$$

Now since $C = -\frac{d}{dx} + A$, we have from (19) that

$$2\operatorname{Re}\langle Cf, Wf \rangle_{L_2} = -2\operatorname{Re}\langle \frac{d}{dx}f, Wf \rangle_{L_2} + 2\operatorname{Re}\langle Af, Wf \rangle_{L_2}$$

$$= \int_{\mathbb{R}} \{\frac{d}{dt}\langle W(x+t)f(x), f(x) \rangle_{\mathcal{H}}\Big|_{t=0} + 2\operatorname{Re}\langle A(x)f(x), W(x)f(x) \rangle_{\mathcal{H}}\}dx$$

$$\leq -2\delta\|f\|_{L_2}^2,$$

and (13) is proved.

SUFFICIENCY: Suppose $\{U(x,s)\}_{x \geq s}$ is hyperbolic. We will show the function $W(\cdot)$, given by (17) satisfies (i)–(iii) (similarly, $V(\cdot)$, given by (18), satisfies (i)-(iii)).

Split W from (17) as $W(x) = W_P(x) + W_Q(x)$, where

$$W_P(x) = \int_0^\infty U_P^*(x+t,x)H(x+t)U_P(x+t,x)\,dt = \int_x^\infty U_P^*(t,x)H(t)U_P(t,x)\,dt.$$

Recall, that the function $x \mapsto \|H(x)\|_{B(\mathcal{H})}$ is bounded. Condition 4) of Definition 1 implies $\int_x^\infty \|U_P(t,x)\|^2\,dt < M(2\lambda)^{-1}$ for all $x \in \mathbb{R}$. For each $u,v \in \mathcal{H}$ the function

$$\langle W_P(x)u, v\rangle = \int_x^\infty \langle H(t)U_P(t,x)u, U_P(t,x)v\rangle\,dt$$

is continuous provided $x \mapsto U_P(t,x)u$ is continuous and the last integral converges absolutely and uniformly for $x \in \mathbb{R}$. For $u \in \mathcal{D}$ one also has from (12):

$$\frac{d}{dx}\langle W_P(x)u, u\rangle = -\langle H(x)U_P(x,x)u, U_P(x,x)u\rangle$$
$$- \int_x^\infty \{\langle H(t)U_P(t,x)A(x)u, U_P(t,x)u\rangle + \langle H(t)U_P(t,x)u, U_P(t,x)A(x)u\rangle\}\,dt$$
$$= -\langle P^*(x)H(x)P(x)u, u\rangle - 2\operatorname{Re}\langle A(x)u, W_P(x)u\rangle.$$

The same arguments can be applied for $W_Q(x)$. As a result,

$$\frac{d}{dx}\langle W(x)u, u\rangle = -\langle H_{PQ}(x)u, u\rangle - 2\operatorname{Re}\langle A(x)u, W(x)u\rangle,$$

and (ii)-(iii) hold for W. ∎

Theorem 5 can be simplified for the evolutionary family $\{U(x,s)\}_{(x,s)\in\mathbb{R}^2}$ of Example 2. Note that C for this case is described in Example 4. Also (see Daleckij and Krein (1974)) for the derivatives taken in the uniform sense the following hold:

$$\frac{\partial}{\partial x}U(x,s) = A(x)U(x,s), \quad \frac{\partial}{\partial s}U(x,s) = -U(x,s)A(s), \quad (x,s) \in \mathbb{R}^2.$$

COROLLARY 2 The exponential dichotomy of the equation $v' = A(x)v$ with bounded norm-continuous $A : \mathbb{R} \to B(\mathcal{H})$ is equivalent to the existence of a bounded differentiable (in the uniform sense) function $W : \mathbb{R} \to B(\mathcal{H})$ with Hermitian and invertible values such that $W^{-1} : \mathbb{R} \to B(\mathcal{H}) : x \mapsto [W(x)]^{-1}$ is bounded and

$$\hat{W}(x) = \frac{dW}{dx}(x) + A^*(x)W(x) + W(x)A(x) \tag{21}$$

is negative in $\mathcal{H}$ uniformly for $x \in \mathbb{R}$.

PROOF: Since $\{U(x,s)\}$ is defined for all $(x,s) \in \mathbb{R}^2$, the exponential dichotomy of the equation $v' = A(x)v$ is just the hyperbolicity of $\{U(x,s)\}$. By Theorem 4 the hyperbolicity of $\{U(x,s)\}$ is equivalent to the existence of the operators of multiplication $W = W(\cdot)$ and $V = V(\cdot)$ that satisfy (15) and (16). By the Remark we can take $V = W^{-1}$, since $\{e^{tC}\}$ is a C_0-group. Since $C = -\dfrac{d}{dx} + A$, condition (15) means $\hat{W} \ll 0$ for $\hat{W}$ from (21). Indeed, if

W is an operator of multiplication by $W(\cdot)$ in $L_2(\mathbb{R};\mathcal{H})$, then $\dfrac{d}{dx} \cdot W = \frac{dW}{dx} + W\dfrac{d}{dx}$, where $\frac{dW}{dx}$ is the operator of multiplication by function $\frac{dW}{dx}(\cdot)$ in $L_2(\mathbb{R};\mathcal{H})$. Also, we have

$$\frac{d}{dx}W^{-1} = -W^{-1}\frac{dW}{dx}W^{-1} + W^{-1}\frac{d}{dx}.$$

Thus, with $V = W^{-1}$ equation (16) becomes

$$W^{-1}\left(\frac{d}{dx} + A^*\right) + \left(-\frac{d}{dx} + A\right)W^{-1}$$

$$= W^{-1}\left(\frac{dW}{dx} + A^*W + WA\right)W^{-1} = W^{-1}\hat{W}W^{-1} = -H_{P^*Q^*} \ll 0. \qquad \blacksquare$$

Added to proof. After this paper was submitted for publication, the Conjecture stated above was proved (see Y. Latushkin and S. Montgomery-Smith, Lyapunov theorems for Banach spaces, Bull. Amer. Math. Soc., to appear). Thus, Theorems 3, 4, and 5 are valid for an arbitrary evolutionary family, that is, for a family of operators that satisfies the conditions a), b), and c) but not necessarily d).

REFERENCES

Antonevich, A. B. (1985). On two methods of studying the invertibility of operators in C^*-algebras induced by dynamical systems, *Math USSR Sbornik*, **52**: 1–20.

Ben-Artzi, A., and Gohberg I. (1992). Dichotomy of systems and invertibility of linear ordinary differential operators, *Oper. Th. Advances and Appl.*, **56**: 90–119.

Chicone, C., and Latushkin, Y. (1993). Quadratic Lyapunov functions for linear skew-product flows and weighted composition operators, *Diff. and Int. Eqns.*, to appear.

Daleckij Y., and Krein M. (1974). *Stability of Solutions of Differential Equations in Banach Space*, AMS Transl., Providence, RI.

Howland, J. S. (1974). Stationary scattering theory for time-dependent Hamiltonians, *Math. Ann.*, **207**: 315–335.

Latushkin Y., and Stepin A. (1991). Weighted translation operators and linear extensions of dynamical systems, *Russian Math. Surveys* , **46**, no 2: 95–165.

Massera, J. L., and Schäffer, J. J. (1959). Linear differential equations and functional analysis, III. Lyapunov's second method in the case of conditional stability, *Annals of Math.*, no 3: 535–574.

Nagel, R., ed. (1984). *One-parameter Semigroups of Positive Operators*, Lect. Notes Math. 1184.

Nagel, R., and Rhandi A. (1992). Positivity and Lyapunov stability conditions for linear systems, *Adv. Math. Sci. Appl.*, to appear.

Rau, R. (1992). Hyperbolic evolution semigroups, *Dissertation, Universität Tübingen*, 1–84.

Pazy, A. (1983). *Semigroups of Linear Operators and Applications to Partial Differential Equations*, Springer-Verlag.

Tanabe H. (1979). *Equations of Evolution*, Pitnam, London.

Local Convoluted Semigroups

Ioana Cioranescu   University of Puerto Rico, Rio Piedras, Puerto Rico

Abstract. This work is concerned with the concept of local convoluted semigroup which permits an approach to the Abstract Cauchy Problem for linear closed operators including the ultradistributional one.

§ 1. Preliminaries

Recently W. Arendt, O. El-Mennaoui and V. Keyantuo [2] developed the concept of local n-times integrated semigroup which can be briefly described as follows.

Let X be a Banach space and A a closed linear operator on X with domain D(A) endowed with the graph norm; let $0 < \tau \le \infty$ and $n \in N_0 = N \cup \{0\}$. If there exists a strongly continuous operator family $(S_t)_{0 \le t < \tau}$ such that

i) for $x \in D(A)$, $S_t x \in D(A)$ and $AS_t x = S_t Ax$, $0 \le t < \tau$

ii) for $x \in X, \int_0^t S_s x\, ds \in D(A)$ and $S_t x = A\int_0^t S_s x\, ds + \dfrac{t^n}{n!} x$ for $0 \le t < \tau$,

then $(S_t)_{0 \le t < \tau}$ is called the local n-times integrated semigroup generated by A.

We note that if $\tau = \infty$, "a global" n-times integrated semigroup of exponential growth is just an n-times integrated semigroup in the sense of W. Arendt [1] and F. Neubrander [8].

Some of the most important results of the theory are given by the following equivalent statements:

1) A generates an n-times integrated semigroup on $[0, \tau)$;

2) the following Cauchy Problem is well posed for all $x \in X$ (existence and uniqueness of the solution)

$$PC_{n+1}(\tau) \begin{cases} v \in C([0,\tau); \ D(A)) \cap C^1([0,\tau); \ X) \\ v'(t) = Av(t) + \dfrac{t^n}{n!} x, \qquad 0 \le t < \tau, \\ v(0) = 0 \end{cases}$$

3) $\rho(A) \neq \varnothing$ and for all $x \in D\left(A^{n+1}\right)$ the following Cauchy Problem is well posed:

$$PC_0(\tau) \begin{cases} u \in C([0,\tau); \ D(A)) \cap C^1([0,\tau); \ X) \\ u'(t) = Au(t), \qquad 0 \le t < \tau, \\ u(0) = x \end{cases}$$

In the case $\overline{D(A)} = X$ one can add the following equivalent assertion:

4) A generates a distribution semigroup.

We remark that the core of this theory is the spectral characterization of the generators of local n-times integrated semigroups which allows Ljubich's uniqueness criterium to be applied.

We intend to develop analogously a theory of local "convoluted semigroups" which includes both Chazarain's [4] ultradistributional and Beal's [3] constructive approach to the $PC_0(\infty)$. Although the results may be presented in a more general frame, we give them here, for simplicity, only for the Gevrey case.

Thus let be $0 < a < 1$, ℓ, $L > 0$ and P an ultradifferential polynomial of Gevrey class $n!^{1/a}$, this is an entire function such that

$$e^{|\ell z|^a} \le |P(z)| \le e^{|L z|^a} \quad \text{for Re} \, z \ge 0. \tag{1.1}$$

Then we have $P(z) = \displaystyle\sum_{n=0}^{\infty} a_n z^n$ where

$$|a_n| = O\left(L^n \cdot n!^{-\frac{1}{a}}\right), \quad n = 0, 1, 2, \ldots \tag{1.2}$$

and $P\left(\frac{d}{dt}\right)$ is called an ultradifferential operator of Gevrey class $n!^{\frac{1}{a}}$.

A typical example for an ultradifferential polynomial of Gevrey class $n!^{\frac{1}{a}}$ is

$$P(z) = \prod_{n=1}^{\infty}\left(1 + \frac{\ell z}{n^{\frac{1}{a}}}\right), \text{ where (1.1) is satisfied for } L = 2\ell.$$

Let us define the function

$$E(t) = \frac{1}{2\pi} \int_{-\infty}^{+\infty} \frac{e^{ist}}{P(is)} ds, \quad t \in R. \tag{1.3}$$

Then E is a C^{∞}-bounded function such that $P\left(\frac{d}{dt}\right) E = \delta$; moreover

$$P(z)^{-1} = \int_{0}^{\infty} e^{-zt} E(t) dt \quad \text{for Re } z > 0. \tag{1.4}$$

For all these facts one can consult [7].

Throughout this work P will be an ultradifferential polynomial of Gevrey class $n!^{\frac{1}{a}}$, E the fundamental solution given by (1.3) and

$$F(t) = \int_{0}^{t} E(s) ds.$$

We give now the

Definition 1.1. *Let A be closed and $0 < \tau \leq \infty$; if there exists a strongly continuous operator family $(S_t)_{0 \leq t < \tau}$ such that:*

i) for $x \in D(A)$, $S_t x \in D(A)$ and $A S_t x = S_t A x$, $\qquad 0 \leq t < \tau$

ii) for $x \in X$, $\int_{0}^{t} S_s x \, ds \in D(A)$ and

$$S_t x = A \int_{0}^{t} S_s x \, ds + F(t)x \text{ for } 0 \leq t < \tau \tag{1.5}$$

then we call $(S_t)_{0 \leq t < \tau}$ the local E-convoluted semigroup generated by A.

Motivation for the terminology is the fact that if $(T_t)_{t \geq 0}$ is a C_0-semigroup, then a simple computation shows that

$$S_t = \int_0^t E(t-s)T_s \, ds$$

is an E-convoluted semigroup on every $[0, \tau)$ for $0 < \tau \leq \infty$.

§ 2. The spectral characterization of the resolvent

In what follows A will be a closed linear operator on X.

Lemma 2.1. *Assume that A generates the E-convoluted semigroup* $(S_t)_{0 \leq t < \tau}$ *and define*

$$L_z(t) = \int_0^t e^{-zs} S_s \, ds, \quad z \in \mathbb{C}.$$

Then $L_z(t) \in \mathcal{L}(X; D(A))$ *and*

$$(z - A)L_z(t)x = e^{-zt}(\varphi_z(t) - S_t)x, \quad x \in X, \ 0 \leq t < \tau, \ z \in C$$

where $\varphi_z(t) = \int_0^t e^{(t-s)z} E(s) \, ds.$

Proof. Let $z \in C$, $0 \leq t < \tau$ and $x \in X$; we have

$$L_z(t)x = e^{-zt} \int_0^t S_u \, x \, du + z \int_0^t e^{-zs}\left[\int_0^s S_u \, x \, du\right] ds.$$

Since $\int_0^t S_u \, x \, du \in D(A)$ and A is closed, it follows that $L_z(t)x \in D(A)$. Moreover, using (1.5) we obtain:

$$(z - A)L_z(t)x = zL_z(t)x - e^{-zt} A\int_0^t S_u \, x \, du - z\int_0^t e^{-zs} A\left[\int_0^s S_u \, x \, du\right] ds$$

$$= zL_z(t)x - e^{-zt}(S_t x - F(t)x) - z\int_0^t e^{-zs}[S_s x - F(s)x] ds$$

$$= -e^{-zt}S_t x + e^{-zs}F(t) + z\int_0^t e^{-zs}F(s)x\,ds$$

$$= -e^{-zt}S_t x + \int_0^t e^{-zs}E(s)x\,ds = e^{-zt}\big(\varphi_z(t) - S_t\big)x. \qquad\blacksquare$$

Theorem 2.2. *Assume that A generates the E-convoluted semigroup* (S_t) *on* $[0,\tau)$; *then for all* $\alpha > \dfrac{L^a}{\tau}$ *there exists* $\beta \geq 1$ *such that*

$$\Lambda_{\alpha,\beta} = \big\{z \in \mathbb{C};\ \operatorname{Re}z \geq \alpha|z|^a + \beta\big\} \subset \rho(A)$$

and

$$\|R(z;\,A)\| = 0\!\left(e^{(|Lz|)^a}\right),\ \textit{for } z \in \Lambda_{\alpha,\beta}. \qquad (2.1)$$

Proof. Let $\alpha_0 > \dfrac{1}{\tau}$, $t_0 = \dfrac{1}{\alpha_0}$ and

$$\beta_0 = \frac{1}{t_0}\ln\!\left(2\|S_{t_0}\| + \int_{t_0}^{\infty}e^{t_0-s}|E(s)|ds\right).$$

We shall first prove that

$$\Omega_{\alpha_0,\beta_0} = \big\{z \in \mathbb{C};\ \operatorname{Re}z \geq 1;\ \operatorname{Re}z \geq \alpha_0\ln|P(z)| + \beta_0\big\} \subset \rho(A).$$

Let $z \in \Omega_{\alpha_0,\beta_0}$ and φ_z defined in Lemma 2.1; then

$$|\varphi_z(t_0)| = \left|\int_0^{t_0} e^{(t_0-s)z}E(s)ds\right|$$

$$\geq e^{\operatorname{Re}zt_0}\left|\int_0^{\infty}e^{-zs}E(s)ds\right| - \left|\int_{t_0}^{\infty}e^{z(t_0-s)}E(s)ds\right|$$

$$\geq e^{\beta_0 t_0}e^{\ln|P(z)|}|P(z)|^{-1} - \int_{t_0}^{\infty}e^{\operatorname{Re}z(t_0-s)}|E(s)|ds$$

$$= 2\|S_{t_0}\| + \int_{t_0}^{\infty}e^{t_0-s}|E(s)|ds - \int_{t_0}^{\infty}e^{t_0-s}|E(s)|ds$$

$$= 2\|S_{t_0}\|.$$

Thus we obtained that

$$\|S_{t_0}\| \leq \frac{1}{2}|\varphi_z(t_0)|. \qquad (2.2)$$

Consequently the operator $\left(\varphi_z(t_0) - S_{t_0}\right)$ is invertible and

$$\left\|\left(\varphi_z(t_0) - S_{t_0}\right)^{-1}\right\| \le 2\left|\varphi_z(t_0)\right|^{-1}. \tag{2.3}$$

Let us denote $R = e^{zt_0} L_z(t_0)\left(\varphi_z(t_0) - S_{t_0}\right)^{-1}$.

By Lemma 2.1 we have that
$$(z - A)Rx = x, \text{ for all } x \in X.$$
By the Definition 1.1 (i) we see that R and A commute on $D(A)$ so that $z \in \rho(A)$ and

$$R(z; A) = R = e^{zt_0} L_z(t_0)\left(\varphi_z(t_0) - S_{t_0}\right)^{-1} \tag{2.4}$$

Let $\alpha = L^a \alpha_0$ and $\beta = \max\{1, \beta_0\}$.

Since $P(z)$ satisfies (1.1), then $\Lambda_{\alpha,\beta} \subset \Omega_{\alpha_0,\beta_0} \subset \rho(A)$.

In order to prove the estimate (2.1) for $z \in \Lambda_{\alpha, \beta}$ we use (2.3) and (2.4) together with the fact that

$$\left\|L_z(t_0)x\right\| \le \int_0^{t_0} e^{-\mathrm{Re}\, z \cdot s}\left\|S_s x\right\| ds \le \int_0^{t_0} e^{-s}\left\|S_s x\right\| ds, \ x \in X$$

to obtain that

$$\left\|R(z;\ A)\right\| \le \mathrm{const.}\left|e^{zt_0}\varphi_z(t_0)^{-1}\right| = \mathrm{const.}\left|\int_0^{t_0} e^{-zs} E(s)ds\right|^{-1}. \tag{2.5}$$

We further note that for $z \in \Lambda_{\alpha, \beta} \subset \Omega_{a_0, \beta_0}$, we have

$$\left|P(z)\int_{t_0}^{\infty} e^{-zs} E(s)ds\right| \le \mathrm{const.}\left|P(z)\right|\int_{t_0}^{\infty} e^{-\mathrm{Re}\, z \cdot s}ds =$$

$$= \frac{\mathrm{const.}}{\mathrm{Re}\, z} \cdot \left|P(z)\right| \cdot e^{-\mathrm{Re}\, z \cdot t_0} \le \frac{\mathrm{const.}}{\mathrm{Re}\, z}\left|P(z)\right|\left(e^{-\alpha_0 t_0 \ln|P(z)| - \beta_0 t_0}\right)$$

$$\le \frac{\mathrm{const.}}{\mathrm{Re}\, z} \cdot \left|P(z)\right| \cdot \left|P(z)\right|^{-1} = \frac{\mathrm{const.}}{\mathrm{Re}\, z} \xrightarrow[\mathrm{Re}\, z \to \infty]{} 0.$$

Consequently, for $z \in \Lambda_{\alpha, \beta}$

$$\left| P(z) \int_0^{t_0} e^{-zs} E(s)ds \right| =$$

$$\left| P(z) \cdot \int_0^{\infty} e^{-zs} E(s)ds - P(z) \int_{t_0}^{\infty} e^{-zs} E(s)ds \right| \xrightarrow[\operatorname{Re} z \to \infty]{} 1.$$

Now it is clear that (1.1) and (2.5) yield

$$\|R(z; A)\| = O(|P(z)|) = O\left(e^{|Lz|^a} \right). \qquad \blacksquare$$

Using the above result we prove the

Theorem 2.3. *A is the generator of an E-convoluted semigroup on* $[o, \tau)$ *if and only if the following Cauchy Problem is well posed:*

$$PC_E(\tau) \begin{cases} v \in C([0,\tau); D(A)) \cap C^1([0,\tau); X) \\ v'(t) = Av(t) + F(t)x \qquad 0 \le t < \tau. \\ v(0) = 0 \end{cases}$$

Proof. Suppose A is the generator of an E-convoluted semigroup on $[o, \tau)$ and let $v(t) = \int_0^t S_s x ds$; then v is a solution of the $PC_E(\tau)$. For the uniqueness we use the estimate (2.1) to show that Ljubich's criterion is fulfilled; namely we have:

$$\lim_{z \to \infty} \frac{\ln \|R(z; A)\|}{z} \le \lim_{z \to \infty} \frac{C_1 + C_2 |z|^a}{z} = 0$$

(for some constants $C_1, C_2 > 0$).
Assume that $PC_E(\tau)$ is well-posed and define for $x \in X$ and $t \in [0, \tau)$, $S_t x = v'(t)$, where v is the solution associated to x; then $S_t x : [0, \tau) \to X$ is continuous and ii) from the Definition 1.1. is satisfied.
We show further that $S_t \in L(X)$, for $t \in [0, \tau)$. Consider the mapping $V: X \to C([0, \tau); D(A))$ defined by $V(x) = v(\cdot)x$. Then by a standard argument one can prove that V is linear and closed hence continuous.

Taking into account the topology of $C([0,\tau); D(A))$ we have that for all $\varepsilon > 0$ there exists $c_\varepsilon > 0$ such that

$$\|Av(t)x\| \le c_\varepsilon \|x\|, \quad x \in X, \ t \in [0, \tau - \varepsilon].$$

Since $S_t x = Av(t)x + F(t)x, \ x \in X$ it follows that $S_t \in L(X)$, for all $t \in [0,\tau)$.

At this point we note that by the first part of the proof of Theorem 2.2, for all $z \in \Lambda_{\alpha,\beta}$ the operator $z - A$ is surjective. We shall prove that it is also injective.

Indeed, let $x \in D(A)$ be such that $Ax = zx$. Then the solution of $PC_E(\tau)$ is given by $v(t) = \int_0^t \varphi_z(s)x\,ds$. Then by the uniqueness we have that $S_t x = \varphi_z(t)x, \ t \in [0,\tau)$.

By (2.3), for $t = t_0$ we have $\|S_{t_0}\| \le \dfrac{1}{2}|\varphi_z(t_0)|$ so that $\|x\| \le \dfrac{1}{2}\|x\|$ which yields $x = 0$.

Finally, consider $z \in \rho(A), \ x \in D(A)$ and

$$w(t) = R(z;A) \int_0^t S_s(z - A)x\,ds, \quad t \in [0,\tau).$$

Then $w(t) \in D(A)$ and

$$Aw(t) = R(z;A)A\int_0^t S_s(z - A)x\,ds =$$

$$= R(z;A)S_t(z - A)x - F(t)x$$

$$= w'(t) - F(t)x.$$

Hence, by the uniqueness, $w'(t) = S_t x, \ t \in [0,\tau)$ so that

$$R(z;A)S_t(z - A)x = S_t x.$$

It follows that

$$S_t x \in D(A) \quad \text{and} \quad AS_t x = S_t Ax, \ t \in [0,\tau). \qquad \blacksquare$$

Corollary 2.4. *Assume that A generates the E-convoluted semigroup $(S_t)_{0 \le t < \tau}$; then $S_t S_s = S_s S_t$ for all $0 \le s, \ t < \tau$.*

Proof. Let $s \in [0,\tau)$ be fixed, $x \in X$ and $v(t) = S_s S_t x$; since $\int_0^t S_u x\,du \in D(A)$, we have

$$\int_0^t v(u)du = \int_0^t S_s S_u x\,du = S_s \int_0^t S_u x\,du \in D(A)$$

and

$$A \int_0^t v(u)du = A \int_0^t S_s S_u x\, du = S_s A \int_0^t S_u x\, du$$

$$= S_s\left[S_t x - F(t)x\right] = v(t) - F(t)S_s x.$$

Hence v is the solution to the $PC_E(0,\tau)$ corresponding to $S_s x$ so that by the uniqueness we have $v(t) = S_t S_s x$. ∎

Theorem 2.5. *Suppose that there are $0 < a < 1$ and $\alpha,\ \beta,\ \gamma > 0$ such that*

$$\Lambda_{\alpha,\beta} \subset \rho(A) \text{ and } \|R(z;A)\| = 0\left(e^{\gamma|z|^a}\right), \text{ for } z \in \Lambda_{\alpha,\beta}. \tag{2.6}$$

Let $\tau > 0$, $\ell^a > \alpha\tau + \gamma, P(z) = \prod_{n=0}^{\infty}\left(1 + \dfrac{\ell z}{n^{\frac{1}{a}}}\right)$ and E defined by (1.3); then A generates an E-convoluted semigroup on $[0,\tau)$.

Proof. Let $\Gamma = \left\{z \in \mathbb{C}; \operatorname{Re}z = \alpha|z|^a + \beta\right\}$ and define $S_t x = \dfrac{1}{2\pi i}\displaystyle\int_\Gamma \dfrac{e^{zt}R(z;A)}{P(z)} x\, dz$,

$x \in X,\ t \in [0,\tau)$.

We see that for $z \in \Gamma$

$$\left\|e^{zt}P^{-1}(z)R(z;A)\right\| \le \text{const.}e^{\operatorname{Re}z.t + \gamma|z|^a - \ell^a|z|^a} \le Me^{\beta\tau} \cdot e^{\left(\alpha\tau + \gamma - \ell^a\right)|z|^a}$$

Since $\alpha\tau + \gamma - \ell^a < 0$ it follows that $S_t \in L(X)$ and $t \to S_t x$ is continuous for $x \in X$ and $0 \le t < \tau$. Moreover:

$$A\int_0^t S_s\, x\, ds = \dfrac{1}{2\pi i}\int_\Gamma \dfrac{e^{zt}AR(z;A)x}{zP(z)}dz = \dfrac{1}{2\pi i}\int_\Gamma \dfrac{e^{zt}R(z;A)x}{P(z)}dz - \dfrac{1}{2\pi i}\int_\Gamma \dfrac{e^{zt}x}{zP(z)}dz$$

$$= S_t x - F(t)x.$$

(Note that $E(t) = \dfrac{1}{2\pi}\displaystyle\int_{-\infty}^{+\infty}\dfrac{e^{ist}}{P(is)}ds = \dfrac{1}{2\pi i}\int_\Gamma \dfrac{e^{zt}}{P(z)}dz$ and $F(t) = \dfrac{1}{2\pi i}\int_\Gamma \dfrac{e^{zt}}{zP(z)}dz$).

Hence all the properties of the Definition 1.1. are fulfilled. ∎

Proposition 2.6. *Let $\tau > 0$ $h > L^{-1}$ and assume that A generates an E-convoluted semigroup $(S_t)_{0 \le t < \tau}$; define*

$$X_h = \left\{x \in X; t \to S_t x\ C^\infty \text{ on } [0,\tau) \text{ and } \|S_t^{(n)}x\| = 0\left(h^n n!^{\frac{1}{a}}\right) \text{ for all } n \in \mathbb{N}\right\}.$$

Then if the $PC_1(\tau)$ is well posed for all $x \in X_h$.

Proof. We note first that for $x \in X_h$ we have by (1.2)

$$\left\| P(d/dt)S_t x \right\| = \left\| \sum_{n=1}^{\infty} a_n S_t^{(n)} x \right\| \le M.\text{const}(x) \sum_{n=0}^{\infty} L^n . h^n < \infty.$$

Let $x \in X_h$ and define $u(t) = P(d/dt)S_t x$, $0 \le t < \tau$. Then

$$P(d/dt)A \int_0^t S_s x \, ds = P(d/dt)(S_t x - F(t)x) = u(t) - x, \quad 0 \le t < \tau.$$

By the closedness of A it follows

$$A \int_0^t u(s)ds = AP(d/dt)\int_0^t S_s x \, ds = P(d/dt)A \int_0^t S_s x \, ds$$

$$= u(t) - x, \quad 0 \le t < \tau. \qquad \blacksquare$$

§ 3.   A functional equation and the extension of the solution

Let $\tau > 0$ and suppose that the linear closed operator A generates the E-convoluted semigroup $(S_t)_{0 \le t < \tau}$. If $H:[0,\tau) \to C$ is continuous, then we denote

$$(H * S)(t,x) = \int_0^t H(s)S_{t-s}x \, ds.$$

Lemma 3.1. *For* $x \in X$ *and* $0 \le t < \tau$ *we have*

$$A \int_0^t (H * S)(t,x)ds = (H * S)(t,x) - (H * F)(t)x. \tag{3.1}$$

Moreover, if $H \in C^1[0,\tau)$, *then*

$$A(H * S)(t,x) = (H' * S)(t,x) - (H' * F)(t)x + H(0)(S_t x - F(t)x). \tag{3.2}$$

Proof. Since A is closed, $\int_0^t (H * S)(s,x)ds \in D(A)$, for all $x \in X$, $0 \le t < \tau$.

We have:

$$A \int_0^t (H * S)(s,x)ds = A \int_0^t \left(\int_0^s H(r)S_{s-r}x \, dr \right)ds$$

$$= \int_0^t H(r)\left(A \int_r^t S_{s-r}x \, ds \right)dr = \int_0^t H(r)\left(A \int_0^{t-r} S_s x \, ds \right)dr$$

$$= \int_0^t H(r)\big[S_{t-r}x - F(t-r)x\big]dr = (H*S)(t,x) - (H*F)(t)x.$$

We note now that

$$(H*S)(t,x) = \int_0^t H(s)S_{t-s}x\,ds = \int_0^t H(t-s)S_s x\,ds$$

and consequently (3.2) can be derived from (3.1) by differentiation. ∎

Theorem 3.2. *For $x \in X$ and $0 \le t,\ s < \tau$, we have*

$$S_t S_s x = \int_s^{t+s} E(t+s-r)S_r x\,dr - \int_0^t E(t+s-r)S_r x\,dr. \qquad (3.3)$$

Proof. For $h > 0$, we shall denote E_h the function
$$E_h(t) = E(t+h).$$

Consider $0 \le s < \tau$ and x fixed and denote

$$w(t) = \int_s^{t+s} E(t+s-r)S_r x\,dr - \int_0^t E(t+s-r)S_r x\,dr, \ \ 0 \le t < \tau.$$

Then we can write

$$w(t) = \int_s^{t+s} E(t+s-r)S_r x\,dr - \int_0^t E(t+s-r)S_r x\,dr - \int_0^s E(t+s-r)S_r x\,dr$$

$$= (E*S)(t+s,x) - (E_s*S)(t,x) - (E_t*S)(s,x).$$

We use the formulas (3.1) and (3.2) to calculate

$$A\int_0^t w(r)dr = A\int_0^t (E*S)(s+r,x)dr - A\int_0^t (E_s*S)(r,x)dr - A\int_0^t (E_r*S)(s,x)dr$$

$$= A\int_0^{t+s} (E*S)(r,x)dr - A\int_0^s (E*S)(r,x)dr$$

$$- A\int_0^t (E_s*S)(r,x)dr - A\big[(F_t*S)(s) - (F*S)(s)\big]$$

$$= (E*S)(t+s,x) - (E*F)(t+s)x - (E*S)(s,x) + (E*F)(s)x$$

$$-(E_s*S)(t,x) + (E_s*F)(t)x - (E_t*S)(s,x) + (E_t*F)(s)x$$

$$-F(t)S_s x + F(t)F(s)x + (E*S)(s,x) - (E*F)(s)$$

$$= (E*S)(t+s,x) - (E_s*S)(t,x) - (E_t*S)(s,x) - F(t)S_s x$$

$$+\big[F(t)F(s) - (E*F)(t+s) + (E_s*F)(t) + (E_t*F)(s)\big]x$$

$$= w(t) - F(t)S_s x.$$

We used the following equality:

$$F(t)F(s) = (E*F)(t+s) - (E_s*F)(t) - (E_t*F)(s) \qquad (3.4)$$

$$= \int_s^{t+s} E(t+s-r)F(r)dr - \int_0^t E(t+s-r)F(r)dr$$

which can be proved by part integration.

By Theorem 2.3 if follows that $w(t) = S_t S_s x$ so that the functional equation (3.3) is proved. ∎

Remark 3.3. Consider the function

$$(E*E)(t) = \int_0^t E(t-r)E(r)dr.$$

A simple computation shows that

$$P^{-2}(z) = \int_0^\infty e^{-zt}(E*E)(t)dt, \quad \text{Re } z > 0$$

and by (1.1) we also have

$$e^{2\ell|z|^a} \le \left|P^2(z)\right| \le e^{2|Lz|^a}, \quad \text{Re } z \ge 0.$$

Hence P^2 is an ultradifferential polynomial of Gevrey class $n!^{1/a}$. $P^2(d/dt)(E*E) = \delta$ and $(F*E)' = E*E$.

Inspired by the functional equation (3.3) we can give now the

Theorem 3.4. *Let $\tau_0 > 0$ and assume that the $PC_E(\tau_0)$ is well-posed; then the $PC_{E*E}(2\tau_0)$ is well posed.*

Proof. By Theorem 2.2 and 2.3 it is sufficient to show that the $PC_{E*E}(2\tau_0)$ has a solution. Let $\tau < \tau_0$; we define:

$$S_t x = \begin{cases} (E*S)(t,x) & \text{if } 0 \le t \le \tau \\[2mm] S_\tau S_{t-\tau} + \int_0^{t-\tau} E(t-r)S_r x dr + \int_0^\tau E(t-r)S_r x dr & \text{if } \tau < t \le 2\tau \end{cases}$$

For $0 \le t \le \tau$ we use the formula (3.1) to prove that $\int_0^t S_t x ds$ is a solution of the $PC_{E*E}(\tau)$.

Suppose $\tau < t \le 2\tau$; we have

$$S_t x = S_\tau S_{t-\tau} + (E_\tau * S)(t-\tau, x) + (E_{t-\tau} * S)(\tau, x).$$

Hence:

$$A\int_0^t S_s x\,ds = A\int_0^\tau S_s x\,ds + A\int_\tau^t S_s x\,ds$$

$$= (E*S)(\tau,x) - (E*F)(\tau)x + A\int_0^{t-\tau} S_{s+\tau}x\,ds$$

$$= (E*S)(\tau,x) - (E*F)(\tau)x + A\int_0^{t-\tau} S_\tau S_s x\,ds$$

$$+A\int_0^{t-\tau}(E_\tau *S)(s,x)\,ds + A\int_0^{t-\tau}(E_s *S)(\tau,x)\,ds$$

$$= (E*S)(\tau,x) - (E*F)(\tau)x + S_\tau S_{t-\tau}x - F(t-\tau)S_\tau x$$

$$+(E_\tau *S)(t-\tau,x) - (E_\tau *F)(t-\tau)x + (E_{t-\tau}*S)(\tau,x)$$

$$-(E_{t-\tau}*F)(\tau)x + F(t-\tau)S_\tau x - F(\tau)F(t-\tau)x$$

$$-(E*S)(\tau,x) + (E*F)(\tau)x$$

$$= S_\tau S_{t-\tau}x + (E_\tau *S)(t-\tau,x) + (E_{t-\tau}*S)(\tau,x)$$

$$-\big[(E_\tau *F)(t-\tau) + F(\tau)F(t-\tau) + (E_{t-\tau}*F)(\tau)\big]$$

$$= S_t x - (E*F)(t)x.$$

We used for the last equality the formula (3.4). ∎

§ 4. Example and final comments

a). The notion of well posedness of the Cauchy Problem in the sense of Gevrey's ultradistributions for the equation $u'-Au = T$ is due to J. Chazarain [4]. Comparing his results with our spectral characterizations in Theorems 2.2 and 2.5 we can give the

Proposition 4.1. *Let A be a closed densely defined operator on X and* $0 < a < 1$: *the following are equivalent*

i) *there is an - L(X) valued ultradistribution $\mathcal{E}$ of Gevrey class $n!^{1/a}$ such that*
  $$\mathcal{E}'-A\mathcal{E} = \delta \otimes I \quad and \quad \mathcal{E}'-\mathcal{E}A = \delta \otimes I_{D(A)}$$

ii) *there is $\tau > 0$ and an ultradifferential polynomial P of Gevrey class $n!^{1/a}$ such that A generates an E-convoluted semigroup on $[0,\tau)$, where E is the fundamental solution to P*

iii) *there are $\alpha,\beta,\gamma > 0$ such that*

$$\Lambda_{\alpha,\beta} = \left\{ z \in \mathbb{C};\ \operatorname{Re} z \geq \alpha |z|^a + \beta \right\} \subset \rho(A) \tag{4.1}$$

and

$$\|R(z;A)\| = O\!\left(e^{\gamma |z|^a} \right) \text{ for } z \in \Lambda_{\alpha,\beta}.$$

We also recall that densely defined closed operators satisfying the spectral condition iii) were studied by R. Beals [3].

Examples of such operators can be found in [4]; we shortly mention here the following:

Let $X = L^2(\mathbb{R})$ and $A = i\,\dfrac{d^4}{dt^4} - \dfrac{d^2}{dt^2}$; then $\sigma(A) = \left\{ z \in \mathbb{C};\ z = \xi^2 - i\xi^4,\ \xi \in \mathbb{R} \right\}$ and $\|R(z;A)\| = O\!\left(|z| / \operatorname{Re} z \right)$ for $z \notin \sigma(A)$.

Thus the condition iii) of the above Proposition is fulfilled with $a = \dfrac{1}{2}$ and it is an easy matter to see that for each $\tau > 0$, there is $\ell > 0$ (for example $\ell > (\tau+1)^2$) such that A generates an E-convoluted semigroup on $[0,\tau)$, where E is the fundamental solution of the operator associated to $P(z) = \displaystyle\prod_{n=1}^{\infty}\left(1 + \frac{\ell z}{n^2}\right)$.

b) Our next application is to multiplication operators on $L^p(\Omega)$, where Ω is a σ-finite measure space and $1 \leq p \leq \infty$; we recall that such operators were studied in [2] and [6] in the context of the local integrated semigroups.

Let $m:\Omega \to \mathbb{C}$ be a measurable function and define A by $A\mathcal{F} = m\mathcal{F}$ with $D(A) = \left\{ \mathcal{F} \in L^p(\Omega);\ m\mathcal{F} \in L^p(\Omega) \right\}$.

Suppose that for some $0 < a < 1$, $\alpha, \beta > 0$ we have that $\Lambda_{\alpha,\beta} \subset \rho(A)$, where $\Lambda_{\alpha,\beta}$ is given by (4.1). Define further:

$$(S_t F)(\xi) = \left(\int_0^t E_\ell(t-s) e^{m(\xi)s}\,ds \right)(F(\xi)),\ \xi \in \mathbb{R}$$

where E_ℓ is the fundamental solution of the ultradifferential operator associated with

$$P_\ell(z) = \prod_{n=1}^{\infty}\left(1 + \frac{\ell z}{n^{1/a}}\right).$$

We shall prove that for every $\tau > 0$ there is $\ell > 0$ such that A generates an E_1-convoluted semigroup on $[0,\tau)$.

Since the spectrum of A is the essential image of m, we must have

$$\sup_{z \in \sigma(A)} \left| \int_0^t E(t-s) e^{zs}\,ds \right| < \infty. \tag{4.2}$$

By a carefull examination of the proof of the Theorem 2.2 we see that (4.2) is equivalent to

$$\sup_{z\in\sigma(A)} \frac{e^{\mathrm{Re}\, zt}}{|P_l(z)|} < \infty. \tag{4.3}$$

But in order to have (4.3) it is sufficient that

$$\sup_{|z|\to\infty} e^{\alpha|z|^a} \Big/ |P_l(z)| < \infty.$$

Taking now into account (1.1) we find that $l^a > \alpha\tau$.

c) In [5] Emami-Rad studied systems of pseudo-differential operators with constant coefficients using ultradistributional techniques. In a forthcoming work we shall approach this problem in the context of the local convoluted semigroups.

d) We remark that in our considerations we were strongly motivated by the local structure of the ultradistributions (see Komatsu [7]). In this sense Theorem 3.4 shows us exactly how we have to change the ultradifferential operator when we extend the solution, namely, convoluting sufficiently times the fundamental solution we can obtain a solution on every interval. These aspects are not at all reflected in the theory of ultradistribution semigroups.

e) We mention finally that all these results can be presented in a much more general context, including not only all the ultradistributional theories but also the theory of the C-semigroups; this was remarked by G. Lumer and we shall discussed these topics in a subsequent joint paper.

The author is indebted to the referee for many useful remarks.

References

1. Arendt W.: Vector valued Laplace Transforms and Cauchy Problems, Israel J. of Math. 59 (1987) 327-352.

2. Arendt W., El-Mennaoui O. and Keyantuo V.: Local integrated semigoups: Evolution with jumps of regularity. (1992), preprint.

3. Beals R.: On the abstract Cauchy Problem. J. of Funct. Anal. 10 (1972) 281-299.

4. Chazarain J.: Problemes de Cauchy abstraits et applications a quelques problemes mixtes, J. of Funct. Analysis, 7 (1971) 386-445.

5. Emami-Rad H.A.: Systemes pseudo-differentiels d'evolution bien posés au sens des distributions de Beurling, Boletino U.M.I. Anal. Func. e Appl., serie VI, Vol. I (1982) 303-322.

6. Keyantuo V.:   Semi-groupes distributions, semi-groups intégrés et problemes d'évolution, These, Univ de Franche-Comté, (1992), Besancon.

7. Komatsu H.:   Ultradistributions, I: Structure Theorems and characterization, J. Fac. Sci. Univ. of Tokyo, Vol. 20, No. 1 (1973) 25-105.

8. Neubrander F.: Integrated Semigroups and their Applications to the Abstract Cauchy Problem, Pacific J. of Math. 135 (1989) 233-251.

9. Tanaka N., Okazawa N.:   Local C-Semigroups and Local Integrated Semigroups, Proc. of the London Math. Soc. (3) 61 (1990) 63-90.

Empathy, C-Semigroups, and Integrated Semigroups

Wanda L. Conradie and Niko Sauer* Faculty of Sciences, University of Pretoria, Pretoria, South Africa

1. INTRODUCTION

This paper is concerned with the abstract Cauchy problem

$$\frac{d}{dt}[Bu(t)] = Au(t)$$
$$Bu(t)|_{t=0} = y \tag{1.1}$$

where A and B are (generally) unbounded linear operators with a common domain $\mathfrak{D} \subset X$ and range in Y with X and Y complex Banach spaces. In applications to so-called *dynamical boundary conditions* the operators A and B are often uncloseable [10, 7]. The problem (1.1) has been studied in [6] with the aid of the concept of *B-evolution*. These are bounded evolution operators $S(t) : Y \to \mathfrak{D}$ for which it is assumed that

$$S(t + s) = S(t)BS(s). \tag{1.2}$$

From the *B-evolution property* (1.2) it is concluded that the *associated evolution operators* $E(t) := BS(t)$ are semigroups and that the *empathy property*

$$S(t + s) = S(t)E(s) \tag{1.3}$$

---

* Supported by the Foundation for Research Development

holds. The relationship between the concepts of empathy and B-evolution has been studied in [9] where it was shown that empathy is a more general notion than B-evolution.

The notion of *C-semigroup* and the related notion of *integrated semigroup* have in recent years been studied by several authors e.g. [1, 2, 3, 5, 11]. The setting concerns strongly continuous, bounded, linear evolution operators $\{S(t)\}$ in a single Banach space X and a bounded injective operator $C : X \to X$ for which the following holds:

$$CS(t + s) = S(t)S(s)$$
$$\lim_{t \to 0} S(t)x = Cx \quad \text{for every } x \in X. \tag{1.4}$$

From (1.4) follows that an empathy relationship may be associated with the C-semigroup $S(t)$ by letting $E(t) := C^{-1}S(t)$. Thus, for the case $X = Y$, the notions of C-semigroup and B-evolution are related. Moreover, in the case of a C-semigroup $S(t)$ it may be proved that

$$S(t)E(s) = E(t)S(s) \quad \text{for every } s, t > 0. \tag{1.5}$$

In this paper it will be shown that the notion of empathy, under suitable assumptions on the operators $S(t)$ and $E(t)$, generates operators A, B and C such that the Cauchy problem (1.1) may be treated and such that a 'natural' generalization of the notion of C-semigroup is obtained. Thus, as in the analytic theory of semigroups, the (unbounded) operators in the Cauchy problem are generated by the associated evolution operators.

2. EMPATHIES OF CLASS (C_0)

Let X and Y be complex Banach spaces and let $\{S(t) : t > 0\}$ be a family of bounded linear operators from Y to X. Let $E(t)$ be a semigroup defined on Y. The subspace Y_0 of Y is defined as

$$Y_0 = \bigcup \{E(t)[Y] : t > 0\}.$$

The family of pairs $\{\langle S(t), E(t)\rangle : t > 0\}$ is called an *empathy* of class (C_0) if the following holds:

$$S(t + s) = S(t)E(s) = S(s)E(t); \quad s, t > 0, \tag{2.1}$$
$$E(t) \quad \text{is of class } (C_0). \tag{2.2}$$

If (2.1) holds we also say that $S(t)$ *is in empathy with* $E(t)$, or that $S(t)$ and $E(t)$ *are in empathy*.

In what follows we shall assume that the operators $S(t)$ and $E(t)$ are in (C_0) empathy. In addition it will be assumed that there exists a family $\{P(\lambda) : \text{Re } \lambda > 0\}$ of bounded linear operators from Y to X such that

$$P(\lambda)y = \int_0^\infty e^{-\lambda t}S(t)y\,dt \quad \text{for } y \in Y_0 \tag{2.3}$$

$$\|\lambda P(\lambda)\| = O(1) \quad \text{as } \lambda \to \infty. \tag{2.4}$$

These assumptions are in line with assumptions made in the study of B-evolutions in a general setting [8]. See also [4].

PROPOSITION 2.1. *$S(t)$ is strongly right continuous on Y in the interval $0 < t < \infty$. For $y \in Y_0$ the mappings $t \to S(t)y$ are also continuous at $t = 0$.*

These are direct consequences of the boundedness of the operators $S(t)$, (2.1) and (2.2). We shall see later on that, under an additional assumption, $S(t)$ is strongly continuous on $[0, \infty)$ (Corollary 3.7).

3. THE GENERATOR OF AN EMPATHY

It is assumed that on Y_0 the Laplace transform of $S(t)$ exists and is equal to $P(\lambda)$. Let A_E be the infinitesimal generator of $E(t)$ and let

$$R(\lambda)y := \int_0^\infty e^{-\lambda t}E(t)y\,dt = (\lambda - A_E)^{-1}y$$

for all $y \in Y$.

LEMMA 3.1. *For λ and μ with positive real parts the following holds:*

$$P(\mu)R(\lambda) = P(\lambda)R(\mu) \tag{3.1}$$

$$P(\lambda) - P(\mu) = (\mu - \lambda)P(\lambda)R(\mu). \tag{3.2}$$

PROOF: From the 'commutation rule' $S(t)E(s) = S(s)E(t)$ and (2.3), it follows by taking the Laplace transform, that for any $y \in Y$,

$$S(t)R(\lambda)y = P(\lambda)E(t)y. \tag{3.3}$$

If $y \in Y_0$, it is easily shown that $R(\lambda)y \in Y_0$. Thus, by taking the Laplace transform in (3.3), it follows that (3.1) is valid on Y_0. Since Y_0 is dense in Y, the identity extends by continuity. The following calculations are likewise valid for $y \in Y_0$:

$$(\mu - \lambda)P(\lambda)R(\mu)y = (\mu - \lambda)\int_0^\infty e^{-\lambda t}S(t)\int_0^\infty e^{-\mu s}E(s)y\,ds\,dt$$

$$= (\mu - \lambda)\int_0^\infty \int_0^\infty e^{-(\lambda t + \mu s)}S(t + s)y\,ds\,dt.$$

Upon the change of variable $s \to \tau := t + s$ the right hand side becomes

$$\int_0^\infty \frac{d}{dt} e^{(\mu-\lambda)t} \int_t^\infty e^{-\mu\tau} S(\tau)\, d\tau\, dt = e^{(\mu-\lambda)t} \int_t^\infty e^{-\mu\tau} S(\tau) y\, d\tau\, dt \Big|_{t=0}$$

$$+ \int_0^\infty e^{-\lambda t} S(t) y\, dt$$

$$= P(\lambda)y - P(\mu)y + \lim_{t\to\infty} F(t)y.$$

with

$$F(t)y := e^{(\mu-\lambda)t} \int_t^\infty e^{-\mu\tau} S(\tau) y\, d\tau.$$

Clearly, if $\operatorname{Re}\lambda \geq \operatorname{Re}\mu$, $F(t) \to 0$ as $t \to \infty$. By interchanging the roles of λ and μ in the calculation, it is seen that the limit is also zero when $\operatorname{Re}\lambda \leq \operatorname{Re}\mu$. After extension by continuity the proof is complete. $\qquad\square$

LEMMA 3.2. *The domain* $\mathfrak{D}_\lambda := P(\lambda)[Y]$ *is independent of* λ.

PROOF: By (3.2), $P(\lambda)y = P(\mu)[I_Y + (\mu - \lambda)R(\lambda)]y \in P(\mu)[Y]$. $\qquad\square$

We shall, by the provided justification, write $\mathfrak{D} := \mathfrak{D}_\lambda \subset X$.

LEMMA 3.3. *If for some* μ *with positive real part,* $P^{-1}(\mu)$ *exists,* $P(\lambda)$ *is invertible for all* λ.

PROOF: Suppose $P(\lambda)y = 0$. Then, by (3.2),

$$[I_Y + (\mu - \lambda)R(\lambda)]y = 0.$$

Since $R(\lambda)$ is a resolvent operator, it is now a simple matter to prove that $y = 0$. $\qquad\square$

We shall assume that the operators $P(\lambda)$ are invertible, and define the operators $A(\lambda), B(\lambda) : \mathfrak{D} \to Y$ as

$$A(\lambda) = [\lambda R(\lambda) - I_Y]P^{-1}(\lambda)$$
$$B(\lambda) = R(\lambda)P^{-1}(\lambda).$$

LEMMA 3.4. *The operators* $A(\lambda)$ *and* $B(\lambda)$ *do not depend on* λ.

PROOF: Equation (3.2) may be rewritten in the form

$$P^{-1}(\mu) - P^{-1}(\lambda) = (\mu - \lambda)R(\mu)P^{-1}(\mu) = (\mu - \lambda)R(\lambda)P^{-1}(\lambda)$$

and it follows that $B(\mu) = B(\lambda)$. Now

$$
\begin{aligned}
A(\lambda) - A(\mu) &= \lambda R(\lambda)P^{-1}(\lambda) - \mu R(\mu)P^{-1}(\mu) + P^{-1}(\mu) - P^{-1}(\lambda) \\
&= \lambda B(\lambda) - \mu B(\mu) + (\mu - \lambda)R(\mu)P^{-1}(\mu) \\
&= (\lambda - \mu)B(\mu) + (\mu - \lambda)B(\mu) = 0.
\end{aligned}
$$

$\square$

By virtue of Lemma 3.4, we may set $A(\lambda) = A$ and $B(\lambda) = B$.

LEMMA 3.5. *The operator B is invertible. $B[\mathfrak{D}] = \mathfrak{D}(A_E)$ and $A = A_E B$. Moreover, $P(\lambda) = (\lambda B - A)^{-1}$, and the operator $\langle A, B \rangle : u \in \mathfrak{D} \to \langle Au, Bu \rangle \in Y \times Y$ is closed.*

PROOF: Since $P(\lambda)$ and $R(\lambda)$ are invertible, so is B. Furthermore, $B[\mathfrak{D}] = R(\lambda)[Y] = \mathfrak{D}(A_E)$. The fact that $P(\lambda) = (\lambda B - A)^{-1}$ follows from the definition of A and B. The assertion that $A = A_E B$ follows from the fact that $P^{-1}(\lambda) = R^{-1}(\lambda)B$.

Since $(\lambda B - A)^{-1} : Y \to \mathfrak{D}$ is bounded, it follows that $\lambda B - A$ is closed for at least two different values of λ and the closedness statement follows [6]. $\square$

The closedness of the operator pair $\langle A, B \rangle$ means that A and B act jointly to let convergences work. It means that if $\{x_n\} \subset \mathfrak{D}, x_n \to x$ in $X, Ax_n \to y$ in Y and $Bx_n \to z$ in Y, then $x \in \mathfrak{D}$ and $Ax = y$ as well as $Bx = z$. This property was first identified in [6] in conjunction with the so-called *generating pair* of a B-evolution. We shall also refer to the closedness of $\langle A, B \rangle$ by saying that A and B are *jointly closed*.

LEMMA 3.6. *There exists a bounded linear operator $C : Y \to X$ such that $B^{-1} = C_0$, the restriction of C to $\mathfrak{D}(A_E)$, and*

$$
S(t) = CE(t).
$$

If B is closed, it maps $\mathfrak{D}$ onto Y and $C = B^{-1}$.

PROOF: Let $C_\lambda = \lambda P(\lambda)$. By Lemma 3.5,

$$
x = \lambda^{-1}C_\lambda(\lambda B - A)x = C_\lambda Bx - \lambda^{-1}C_\lambda Ax
$$

for any $x \in \mathfrak{D}$. Since $\|C_\lambda Ax\| = O(1)$ as $\lambda \to \infty$, it follows that

$$
\lim_{\lambda \to \infty} C_\lambda Bx = x \quad \text{for all } x \in \mathfrak{D}. \tag{3.4}
$$

Thus, for all $y \in \mathfrak{D}(A_E)$ the limit of $C_\lambda y$ exists. Since $\mathfrak{D}(A_E)$ is dense in Y, the Uniform Boundedness Theorem yields the result that the limit exists for all $y \in Y$ and that the linear operator $Cy := \lim C_\lambda y$ is bounded. From (3.4) it follows that $CBx = x$.

From (3.3) we also see that $S(t)\lambda R(\lambda) = C_\lambda E(t)$. The required identity follows by taking the limit $\lambda \to \infty$.

The statement on the invertibility of C when B is closed, is proved in a direct manner, taking into account that $B[\mathfrak{D}]$ is dense in Y. $\square$

COROLLARY 3.7. *$S(t)$ is strongly continuous on $[0, \infty)$. For all $y \in Y$, $\lim_{t \to 0} S(t)y$ $= Cy$.*

Our next assignment is to show that $S(t)$ is almost a B-evolution.

LEMMA 3.8. *For $y \in \mathfrak{D}(A_E) = B[\mathfrak{D}]$, $S(t)y \in \mathfrak{D}$ and*

$$S(t + s)y = S(t)BS(s)y \quad \text{for all } s, t > 0.$$

PROOF: The first statement is obvious from (3.3). The second statement follows from the empathy relationship and Lemma 3.6 in the following way:

$$S(t + s)y = S(t)E(s)y = S(t)BC_0 E(s)y = S(t)BS(s)y.$$

$\square$

In B-evolution theory it is required that those y for which the result holds, should be unrestricted.

The operator pair $\langle A, B \rangle$ will be called the *generator* of the empathy $\langle S(t), E(t) \rangle$. The following result states the relationship of the generator with an abstract Cauchy problem of the form (1.1).

THEOREM 3.9. *Let $\langle S(t), E(t) \rangle$ be an empathy of class (C_0) which satisfies (2.3) and (2.4) and for which the Laplace transform $P(\lambda)$ is invertible for some λ. Let $\langle A, B \rangle$ be the generator and let A_E be the infinitesimal generator of the semigroup $E(t)$. If $y \in \mathfrak{D}(A_E) = B[\mathfrak{D}]$, then $u(t) := S(t)y$ solves the Cauchy problem*

$$\frac{d}{dt}[Bu(t)] = Au(t)$$
$$Bu(t)|_{t=0} = y.$$

There is no other solution in $\mathfrak{D}$.

PROOF: Suppose $y \in B[\mathfrak{D}]$. In fact, let $y = R(\lambda)z$, for some $z \in Y$. Then, by (3.3)

$$\begin{aligned}
BS(t)y &= R(\lambda)P^{-1}(\lambda)S(t)R(\lambda)z \\
&= R(\lambda)P^{-1}(\lambda)P(\lambda)E(t)z \\
&= R(\lambda)E(t)z = E(t)R(\lambda)z = E(t)y.
\end{aligned}$$

The differentiability of $u(t)$ is therefore proved and

$$\frac{d}{dt}[Bu(t)] = \frac{d}{dt}[E(t)y] = A_E E(t)y = A_E Bu(t) = Au(t).$$

The initial condition is satisfied for a simpler reason.

To prove uniqueness, let $v(t) \in \mathfrak{D}$ be any solution of the abstract Cauchy problem and, for $0 < s < t$, set $w(s,t) = S(t-s)Bv(s)$. Then

$$\frac{\partial}{\partial s}[Bw(s,t)] = \frac{\partial}{\partial s}[E(t-s)Bv(s)]$$

$$= E(t-s)\{-A_E Bv(s) + \frac{\partial}{\partial s}[Bv(s)]\} = 0$$

Therefore there exists $g(t) \in \mathfrak{D}$ such that $Bw(s,t) = BS(t-s)Bv(s) = Bg(t)$. By the invertibility of B,

$$g(t) = \lim_{s \to 0} S(t-s)Bv(s) = S(t)y = u(t).$$

Again,

$$u(t) = \lim_{s \to t} S(t-s)Bv(s) = C_0 Bv(t) = v(t). \qquad \square$$

The results achieved so far may be turned around to obtain the following test for a given operator pair to be the generator of an empathy of class (C_0):

THEOREM 3.10. *Let $A, B : \mathfrak{D} \subset X \to Y$ be jointly closed and suppose that B has a bounded inverse C_0. If AC_0 generates a semigroup of class (C_0) in Y, then $\langle A, B \rangle$ is the generator of an empathy of class (C_0) which satisfies Conditions (2.3) and (2.4).*

PROOF: Let $E(t)$ be the semigroup generated by $AC_0 = AB^{-1}$. Since $E(t)$ is of class (C_0), $B[\mathfrak{D}] = \mathfrak{D}(AB^{-1})$ is dense in Y. Let C be the extension by continuity of C_0 to Y and set $S(t) := CE(t)$. It is not difficult to verify that $\langle S(t), E(t) \rangle$ is an empathy of class (C_0). To show that Conditions (2.3) and (2.4) are met, is easy because of the representation $S(t) = CE(t)$.

It remains to be shown that $\langle A, B \rangle$ corresponds to the generator of the constructed empathy. Let this generator be denoted by $\langle A_S, B_S \rangle$. It has to be shown that $\mathfrak{D} = P(\lambda)[Y]$ and that $P(\lambda) = (\lambda B - A)^{-1}$. From the definition of $S(t)$ it follows that $P(\lambda) = CR(\lambda) = C_0 R(\lambda)$. But $R(\lambda) = (\lambda I_Y - AB^{-1})^{-1} = B(\lambda B - A)^{-1}$. Hence $(\lambda B - A)$ is invertible and maps $\mathfrak{D}$ onto Y. Finally,

$$(\lambda B_S - A_S)^{-1} = P(\lambda) = C_0 R(\lambda) = C_0 B(\lambda B - A)^{-1} = (\lambda B - A)^{-1}. \qquad \square$$

In the theory of B-evolutions the generating pair has been characterized by the hypothesis of Theorem 3.10 augmented by an additional condition [6]. It is therefore easier to verify that a given operator pair is the generator of an empathy. If the operator B is closed, as is often assumed in work on the Cauchy problem (1.1), it is not difficult to show that the range of B is Y, which means that $C = C_0$. It therefore appears that the class (C_0) restriction on $E(t)$ could under certain circumstances be alleviated.

We conclude this section with a result on jointly closed operators which will be needed in the final section. This observation was made by Christiaan Le Roux:

LEMMA 3.11. *Let X and Y be Banach spaces, and let $A, B : \mathfrak{D} \subset X \to Y$ be linear operators. If B has a bounded inverse $B^{-1} : B[\mathfrak{D}] \to \mathfrak{D}$, then A and B are jointly closed if and only if AB^{-1} is closed.*

Once the statement is known, the proof is direct.

4. C-SEMIGROUPS

Under assumptions additional to (1.4), the theory of C-semigroups may be brought within the scope of the theory of empathy as described in the preceeding sections. These assumptions are:

$$S(t)[X] \subset C[X] \tag{4.1}$$

$$\lim_{t \to 0} C^{-1}S(t)x \quad \text{exists for all } x \in X. \tag{4.2}$$

Under these assumptions, the linear operators $E(t) := C^{-1}S(t) : X \to X$ are semigroups. From the closedness of the operator C^{-1}, the boundedness of $S(t)$ and the closed graph theorem, we deduce that the $E(t)$ are bounded. Again from the closedness of C^{-1} and (4.2) it follows that $E(t)$ is of class (C_0).

THEOREM 4.1. *Let $S(t)$ be a C-semigroup which satisfies (4.1) and (4.2). Then $\langle S(t), E(t) \rangle$ is an empathy of class (C_0) satisfying (2.3) and (2.4). The generator of the C-semigroup is A_E and $C = s - \lim_{\lambda \to \infty} \lambda P(\lambda)$. Moreover, the commutation property*

$$S(t)E(s) = E(t)S(s) \quad \text{for all } s, t > 0 \tag{4.3}$$

holds.

PROOF: The empathy property follows from the fact that C and therefore C^{-1} commutes with $S(t)$ (see e.g. [2]), as does (4.3). Indeed,

$$S(t + s) = C^{-1}S(t)S(s) = E(t)S(s)$$
$$= S(t)C^{-1}S(s) = S(t)E(s).$$

Conditions (2.3) and (2.4) are fulfilled because of the identity $S(t) = CE(t)$. In fact this implies that $P(\lambda) = CR(\lambda)$ from which the statement about the operator C folows.

If $\mathcal{A}$ denotes the infinitesimal generator of the C-semigroup, it follows that for $x \in \mathfrak{D}(\mathcal{A})$,

$$\mathcal{A}x = C^{-1} \lim_{h \to 0} h^{-1}[S(h) - C]x$$
$$= C^{-1} \lim_{h \to 0} C[E(h) - I]x = A_E x.$$

A moment of reflection will complete the proof. $\square$

Empathy theory also yields a version of C-semigroup theory in the setting discussed above. For this the spaces X and Y are identical, and the operator C is constructed, in much the same way as the generator of a semigroup is constructed. The constructed operator is, however, not injective.

THEOREM 4.2. *Let $Y = X$ and let $\langle S(t), E(t) \rangle$ be an empathy of class C_0 satifying (2.3) and (2.4). Suppose that the commutation rule (4.3) holds for $S(t)$ and $E(t)$. Then, if $C = s - \lim_{\lambda \to \infty} \lambda P(\lambda)$, $S(t)$ satisfies the conditions (1.4) and (4.1).*

PROOF: By Lemma 3.6, $S(t) = CE(t)$. Therefore, $CS(t + s) = CS(t)E(s) = CE(t)S(s) = S(t)S(s)$. The rest of the proof is even more elementary. $\square$

5. INTEGRATED SEMIGROUPS

Let $A : \mathfrak{D}(A) \subset X \to X$ be a closed linear operator with bounded inverse A^{-1}, and suppose that A generates a semigroup $E(t)$ of class (C_o). We wish to study the Cauchy-problem

$$\frac{d}{dt}[A^n u(t)] = A^{n+1} u(t); \quad u(t) \in \mathfrak{D}(A^{n+1})$$
$$A^n u(t)|_{t=0} = x \in X.$$

According to Theorem 3.10 we need to view the operator pair $\langle A^n, A^{n+1} \rangle$ as the generator of an empathy. By Lemma 3.11, the pair is closed, and hence all the conditions of Theorem 3.10 are satisfied. The operator $C = A^{-n}$, and it is therefore clear that (4.3) is satisfied. Therefore $S(t)$ is a $C = A^{-n}$-semigroup. A direct calculation shows that

$$(\lambda - A)^{-1} y = \lambda^n \int_0^\infty e^{-\lambda t} S(t) y \, dt - \sum_{j=0}^{n-1} \lambda^{n-1-j} A^{j-n} y$$
$$= \lambda^n \int_0^\infty e^{-\lambda t} [S(t) - \sum_{j=0}^{n-1} \frac{t^j}{j!} A^{j-n}] y \, dt,$$

which means that A is the generator of the n-times integrated semigroup

$$U(t) = S(t) - \sum_{j=0}^{n-1} \frac{t^j}{j!} A^{j-n}$$

[3, 5].

REFERENCES

[1] W. Arendt, Vector valued Laplace transforms and Cauchy problems, *Israel J. Math.*, 59(1987), 327–352.

[2] E.B. Davies & M.M. Pang, The Cauchy problem and a generalization of the Hille-Yosida Theorem, *Proc. London Math. Soc.* 55(1987), 181–208.

[3] R. De Laubenfels, Integrated semigroups, C-semigroups and the abstract Cauchy problem, *Semigroup Forum*, 41(1990), 83–95.

[4] E. Hille, & R.S. Phillips, *Functional Analysis and Semigroups*, Amer. Math. Soc. Colloq. Publ., Vol. 31, Providence, R.I., 1957.

[5] F. Neubrander, Integrated semigroups and their applications to the abstract Cauchy problem, *Pacific J. Math.*, 135(1988), 111–155.

[6] N. Sauer, Linear evolution equations in two Banach spaces, *Proc. Royal Soc. Edinburgh*, 113A(1982), 287–303.

[7] N. Sauer, *Dynamical Boundary Conditions and related Evolution Processes*, Elsevier Science Publishers, Amsterdam (to appear).

[8] N. Sauer & J.E. Singleton, Evolution operators related to semi-groups of class (A), *Semigroup Forum*, 35(1987), 317–335.

[9] N. Sauer & J.E. Singleton, Evolution operators in empathy with a semi-group, *Semigroup Forum*, 39(1989), 85–94.

[10] N. Sauer & A.J. van der Merwe, Eigenvalue problems with the spectral parameter also in the boundary condition, *Quaestiones Mathematicae*, 5(1982), 1–27.

[11] I. Tanaka, On exponentially bounded C-semigroups, *Tokyo J. Math.*, 10(1987), 107–117.

Is Boundary Control a Realistic
Approach to the Stabilization of
Vibrating Elastic Systems?

Richard Datko   Georgetown University, Washington, D.C.

0 INTRODUCTION

We have no precise answer to the question posed in the title of this note, only some observations. The type of problem we consider here is the stabilization of a linear hyperbolic partial differential equation with nonhomogeneous boundary values. The stabilization is achieved by selecting the boundary value to be a velocity derivative (see [6] for an extensive discussion of this problem). The resulting dynamical system then has the following abstract structure

$$(0.1) \qquad \ddot{x} + B\dot{x} + Ax = 0,$$

where A is an unbounded positive definite operator on a Hilbert space H and B is a symmetric operator on Banach space $V \supset H$ and satisfies the property that $A^{-\frac{1}{2}}BA^{-\frac{1}{2}}$ is a compact operator on H.

In [3] it has been shown that many systems of the type (0.1) cannot tolerate small time delays in their damping. That is, if $r > 0$ the system

$$(0.2) \qquad \ddot{x}(t) + B\dot{x}(t - r) + Ax(t) = 0$$

is unstable provided the eigenvalue problem

$$(0.3) \qquad (w^2 I - wB - A)x = 0, w - \text{real},$$

has an infinite number of simple eigenvalues.

Since the stabilizers in the elastic systems in this note are of the feedback variety it is entirely natural to consider small time delays in the feedback, i.e. the term $B\dot{x}(t-r)$ in (0.2). This poses a dilemma, since linear control systems governed by ordinary- and functional-differential equations are able to tolerate small time delays in their stabilizers. Does this invalidate the existing theory for elastic systems or is there some hidden unknown feature which will resolve the problem? Our view is that there is no undiscovered aspect of the theory that will correct the problem of nonrobustness with respect to time delays but that the general theory is viable provided one recognizes its intrinsic limitations. One of these is that linear models for elastic systems have a very limited range of applicability. Even in the uncontrolled case they reflect physical reality only for small displacements, relatively low frequencies and, if conservative, finite but unspecified intervals of time. In the case of robustness with respect to time delays we believe one can offer a rough quantification of this applicability. For example, as we shall indicate in Section 2, one can construct obvious Galerkin approximations to these systems each of which is robust with respect to small time delays. However as the dimension of approximation increases the tolerance to delays decreases in a precisely defined manner. Thus, if one accepts the possibility of time delays and can estimate their size, there is also an estimate on the dimension of the Galerkin approximation which will tolerate these delays. In effect there is a Heisenberg-type relationship between the size of the delay and the dimension of the approximating system.

In Section 1 we analyze a simple collection of one-dimensional elastic models in which delays are allowed to occur in the feedback. Even though two of these models contain intrinsic damping, small delays can cause catastrophic behavior in the higher frequencies. In Section 2 we present a Galerkin approximation scheme for elastic boundary control systems which has the property that each approximation is completely controllable and robust with respect to small time delays. These approximations also possess another attribute. Since they are finite-dimensional, they are completely controllable for any closed bounded control set which contains the origin in its interior. This property does not hold for hyperbolic partial differential equations with boundary controls.

1 SOME SYSTEMS WHICH ARE NOT ROBUST WITH RE-SPECT TO DELAYS

Consider the following three models of the vibrating string. Let $0 < x < 1$ and $t > 0$ we consider:

Model 1.

$$(1.1) \qquad w_{tt} = w_{xx}, 0 < x < 1, t > 0;$$

$$(1.2) \qquad w(0,t) = 0, w_x(1,t) = -Kw_t(1,t-r),$$

where $0 \leq K$ and $0 \leq r$ are fixed.

Model 2.

$$(1.3) \qquad w_{tt} = w_{xx} - 2\epsilon w_t - \epsilon^2 w, 0 < x < 1, t > 0,$$

where $\epsilon > 0$ is fixed. The boundary conditions for this model are (1.2).

Model 3.

$$(1.4) \qquad w_{tt} = w_{xx} + \epsilon w_{txx}, 0 < x < 1, t > 0,$$

$\epsilon > 0$ fixed, plus the boundary conditions (1.2).

When $K > 0$ and $r = 0$ each of the above models is uniformly exponentially stable. The nonhomogeneous boundary condition in (1.2) is termed a velocity feedback. It stabilizes Model 1 in the sense that when $K = 0$ in (1.2) the system (1.1)-(1.2) is conservative. When $K = 0$ in (1.2) Models 2 and 3 are uniformly exponentially stable. When $K \neq 0$ in (1.2), the stability of Models 1 and 2 is enhanced, although only marginally in the case of Model 3 (see e.g. [2]).

Simple calculations which are carried out in [2] show that for $K \neq 1$ the eigenvalues of Model 1 are the solutions to the scalar equation

$$(1.5) \qquad e^{-\lambda} = \frac{K+1}{K-1}, \quad \text{i.e.}$$

$$(1.6) \qquad Re\,\lambda = -\frac{1}{2}Log\left|\frac{K+1}{K-1}\right| < 0.$$

If $K = 1$, Model 1 has no eigenvalues and, in fact, all solutions become zero after time $T = 2$, see e.g. [1]. In this case feedback stabilization also results in complete controllability.

For $K > 0$ the eigenvalue equations of Models 2 and 3 are respectively given by

$$(1.7) \qquad (\lambda + \epsilon)\cosh(\lambda + \epsilon) + K\lambda\sinh(\lambda + \epsilon) = 0$$

and

$$(1.8) \qquad \frac{1}{\sqrt{1+\epsilon\lambda}}\cosh\frac{\lambda}{\sqrt{1+\epsilon\lambda}} + K\sinh\frac{\lambda}{\sqrt{1+\epsilon\lambda}} = 0$$

Notice that as $|\lambda| \to \infty$ the solutions of (1.7) have their real parts asymptotic to (1.6) when $K \neq 1, r = 0$. When $K = 1$, (1.7) has solutions in $Re\,\lambda < 0$ and hence a finite decay rate exists for Model 2, whereas for $K = 1$ the decay rate of Model 1 is minus infinity. Thus for $K = 1$ the decay rate of Model 1 is significantly better than that of Models 2 and 3.

We now assume $r > 0$ and $K > 0$ in (1.2). This produces a time delay in the feedback. The resulting eigenvalue equations are

$$(1.9) \qquad \cosh\lambda = -Ke^{-r\lambda}\sinh\lambda$$

for Model 1,

$$(1.10) \qquad (\lambda + \epsilon)\cosh(\lambda + \epsilon) = -\lambda K e^{-r\lambda}\sinh(\lambda + \epsilon)$$

for Model 2 and

$$(1.11) \qquad \frac{1}{\sqrt{\epsilon\lambda + 1}}\cosh\frac{\lambda}{\sqrt{\epsilon\lambda + 1}} + K e^{-r\lambda}\sinh\frac{\lambda}{\sqrt{\epsilon\lambda + 1}} = 0$$

for Model 3.

It is shown in [1] that there exists $r_n \to 0^+$ for which equations (1.9) and (1.10) have solutions with $Re\,\lambda > 0$. More dramatically if $K > 1$ in (1.2) Model 1 has solutions of the form

$$(1.12) \qquad \lambda = R^{\frac{1}{2}} + iR^{\frac{1}{2}}\sqrt{R - 1}, \quad R > 1,$$

where $R^{\frac{1}{2}}$ and the delay, r, satisfy equations which are respectively asymptotic in terms of the negative integer k to

$$(1.13) \qquad R^{\frac{1}{2}} \doteq \frac{-(2k + 1)\pi}{Log\,K}$$

and

$$(1.14) \qquad r \doteq \frac{(Log\,K)^2}{-2(k + 1)\pi} \ (\text{see e.g. } [4]).$$

A similar result probably also holds for Model 2.

In the case of Model 3 we may assume without loss of generality that $K = 1$ in (1.2) and $\epsilon = 1$ in (1.4). We then attempt to find for small values of r in (1.2) solutions of (1.11) of the form

$$(1.15) \qquad \lambda = Re^{\frac{\pi}{4}i}, \quad R > 0.$$

The result given in [2] is that R and r satisfy equations which are respectively asymptotic in the positive integer k to

$$(1.16) \qquad Log\,R \doteq \frac{\pi}{4} + 2(2k + 1)\pi$$

and

$$(1.17) \qquad r \doteq \frac{R^{-1}}{\sqrt{2}}Log\,R.$$

Thus as $r \to 0^+$ there exist solutions of Model 3 with eigenvalues having arbitrarily large real parts.

Comments

1. Feedback stabilization of the type (1.2) when $r = 0$ and $K > 0$ leads to uniform exponential stability for Model 1, enhances the stability of Model 2 and modestly enhances the stability of Model 3. Model 3 is an internally stable model of a vibrating string. For small $\epsilon > 0$ its nonoscillatory eigenvalues "approximate" those of Model 1, but there are only a finite number of them (This may be seen by examining equation (1.7) when $K = 0$.). Thus in a sense for small values of ϵ Model 3 qualitatively represents observed physical phenomena better than either Models 1 or 2. For such systems there is a more reasonable method of boundary control and one which is robust with respect to small time delays (see e.g. [8]).

2. Small time delays in the feedback (1.2) can result in catastrophic behavior in all three models. However does this occur in actual physical situations? We do not know. If it does the current theory of elastic control is invalid. However what is more probable is that the frequencies where small time delays create havoc in the mathematical models do not represent physical reality. One thing is clear. If linear models are used to represent elastic control systems, one must be aware that they have severe limitations. For example in Model 1 when $K = 1$ and $r = 0$ in (1.2) the velocity feedback system drives all initial values to zero in time $T = 2$, an unheard of event for finite dimensional systems.

2 A PROJECTION SCHEME

In this section we describe a Galerkin projection scheme for control of a wave equation in $R^n, n \geq 2$. The method is quite general. It can also be used for systems of the type represented by Models 1,2, and 3 in Section 1 and for vibrating plates and beams so long as the models are linear and autonomous.

Preliminaries

1. Ω is a bounded simply connected domain in $R^n, n \geq 2$. The boundary of Ω, Γ, is assumed to be smooth enough to satisfy Green's Theorem, i.e.

$$\int_\Omega [v\Delta w = w\Delta v]dx = \int_\Gamma [v\frac{\partial v}{\partial \nu} - w\frac{\partial v}{\partial \nu}]d\sigma,$$

where $\frac{\partial w}{\partial \nu}$-represents the outward normal derivative of a function w on Γ.

2. We assume $P : \Omega \cup \Gamma \to R^+ = (0, \infty)$ is continuous.

3. $H_1^0(\Omega), L_2(\Omega)$ and $L_2(\Gamma)$ will respectively represent the Sobolev space and Hilbert spaces associated with the following control problem (see e.g. [5])

$$(2.1) \qquad w_{tt} = \Delta w - Pw, \text{ on } \Omega$$

$$(2.2) \qquad \frac{\partial w}{\partial \nu} = \mu \text{ on } \Gamma,$$

where $\mu \in L_2(\Gamma) \times L_2(R^+)$, and

$$(2.3) \qquad \begin{bmatrix} w(x,0) = \phi(x) \in H_0^1(\Omega) \\[2mm] w_t(x,0) = \psi(x) \in L_2(\Omega) \end{bmatrix}.$$

DEFINITION 2.1. Let $w(x,t)$ be defined from $(\Omega \cup \Gamma) \times R^+$ into R, then its Laplace transform with respect to the t-variable is

$$(2.4) \qquad \hat{w}(x,\lambda) = \int_0^\infty e^{-\lambda t} w(x,t)dt,$$

provided the integral exists for $Re\,\lambda > \delta$ in the complex plane and all x in $\Omega \cup \Gamma$. The inverse transform of $\hat{w}$ will be denoted by

$$(2.5) \qquad \mathcal{L}^{-1}(\hat{w}(x,\lambda))(t) = w(x,t).$$

ASSUMPTION 2.1. We assume Ω has a geometric structure such that the eigenvalue problem

$$(2.6) \qquad \Delta v_j - Pv_j = -\lambda_j^2 v_j, \text{ on } \Omega;$$

$$(2.7) \qquad \frac{\partial v_j}{\partial \nu} = 0, \text{ on } \Gamma$$

has only simple eigenvalues.

REMARK 2.1. Assumption 2.1 is made for convenience. Given any Ω of the type described in Preliminaries 1 it can be "bent" an arbitrary small amount so that Assumption 2.1 is satisfied (see e.g. [7]). Assumption 2.1 is used below to construct a sequence of Galerkin approximate control problems each of which is of rank-one.

The formal Laplace transform of (2.1)-(2.3) results in the following equations

$$(2.8) \qquad \lambda^2 \hat{w} + P\hat{w} - \lambda\phi - \psi = \Delta \hat{w}$$

$$(2.9) \qquad \frac{\partial \hat{w}}{\partial \nu} = \hat{\mu}.$$

We express the solution of (2.8) in terms of the orthonormal set $\{v_j\}$ by the Fourier expansions

$$(2.10) \qquad \hat{w} = \sum \hat{a}_j v_j, \phi = \sum \phi_j v_j, \psi = \sum \psi_j v_j.$$

For each v_j we apply Green's Formula to (2.8)-(2.9), making use of (2.10), to obtain

$$\int_\Omega [v_j \Delta \hat{w} - \hat{w} \Delta v_j] dj$$

$$(2.11) \qquad = \int_\Omega v_j [\lambda^2 \hat{w} - \lambda\phi - \psi + \lambda_j^2 \hat{w}]dx$$

$$= \lambda^2 \hat{a}_j - \lambda\phi_j - \psi_j + \lambda_j^2 \hat{a}_j = \int_\Gamma v_j \hat{\mu} d\sigma.$$

Thus

(2.12)
$$\hat{a}_j = \frac{1}{\lambda^2 + \lambda_j^2}[\lambda\phi_j + \psi_j + \int_\Gamma v_j\hat{\mu}d\sigma],$$
$$j = 1,2,\cdots.$$

Let

(2.13)
$$a_j(t) = \mathcal{L}^{-1}(\hat{a}_j(\lambda))(t).$$

Then inversion of (2.12) leads to the representations

$$a_j(t) = \phi_j(\cos\lambda_j t) + \frac{\psi_j}{\lambda_j}\sin(\lambda_j t)+$$

(2.14)
$$\frac{1}{\lambda_j}\int_0^t \sin\lambda_j(t - \beta)[\int_\Gamma v_j(\sigma)\mu(\sigma,\beta)d\sigma]d\beta,$$

$$j = 1,2,\cdots.$$

In [5] we show that given any positive integer N we can find a function q on Γ such that the sequence $\{b_j\}$ defined by

(2.15)
$$b_j = \int_\Gamma v_j(\sigma)q(\sigma)dv \neq 0, \quad 1 \leq j \leq N.$$

Thus if we let

(2.16)
$$\mu(\sigma,t) = q(\sigma)\mu_0(t), t \in R^+, \sigma \in \Gamma,$$

the system of equations (2.12) or (2.14) reduces for $1 \leq j \leq N$ to

(2.17)
$$\left[\begin{array}{l} \ddot{x}_j(t) + \lambda_j^2 x_j(t) = b_j\mu_0(t), \\ x_j(0) = \phi_j, \quad \dot{x}_j(0) = \psi_j, \quad 1 \leq j \leq N. \end{array}\right]$$

We also show in [5] that the system (2.17) is completely controllable. It has the structure

(2.18)
$$\ddot{x} + Ax = b\mu_0,$$

where A is a diagonal matrix

(2.19)
$$A = diag[\lambda_1^2,\ldots,\lambda_N^2]$$

and b is the N-vector

(2.20)
$$b = \begin{pmatrix} b_1 \\ \vdots \\ b_N \end{pmatrix}.$$

Moreover b is not an eigenvalue of A in (2.19) and the feedback

$$(2.21) \qquad \mu_0 = -K(\dot{x}, b), K > 0$$

stabilizes the system (2.18), where $(\cdot, \cdot)$ denotes the usual inner product in R^N.

As mentioned in the introduction the system (2.18) with the feedback (2.21) in robust with respect to small time delays. However it is well known that as the dimension of the system increases the margin of robustness decreases. In fact one can quantify this phenomenon rather precisely by finding the smallest $r > 0$ for which there exists a solution of the equation

$$(2.22) \qquad -1 = \lambda e^{\lambda r} \sum_{j=1}^{N} \frac{b_j^2}{\lambda^2 + \lambda_j^2}$$

where $\lambda = iw, w$-real (see e.g. [2]).

REFERENCES

1. R. Datko, J. Lagnese and M.P. Polis, An example of the effect ot time delays in boundary stabilization of wave equations,*SIAM J. Control Optim.*, *24* (1986), 152-156.

2. R. Datko, Two questions concerning the boundary control of elastic systems,*J. Differential Equations, 92* (July 1991), 27-44.

3. R. Datko and Y.C. You, Some second-order vibrating systems cannot tolerate small time delays in their damping, *J. Optim. Theory and Applic., 70* (Sept. 1991), 521-536.

4. R. Datko, Two examples of ill-posedness with respect to small time delays in stabilized elastic systems, *IEEE Trans. on Automatic Control, 38* (Jan. 1993), 163-166.

5. R. Datko, Galerkin projections for some wave and plate boundary control problems, to appear in *Applicable Analysis*.

6. J.L. Lions, Exact controllability, stabilization and perturbations for distributed systems, *SIAM Review, 30* (1988), 1-68.

7. A.M. Micheletti, Perturbazione dello spettro de un opperatore ellitico tips variazionale in relazione ad una variazione del campo, *Richerche de Mathematica, Vol. XXV*, Fasc. II (1976).

8. Y. Sakawa, Feedback control of second-order evolution equations with damping, *SIAM J. Control Optim., 22* (1984), 343-361.

Regularized Functional Calculi
and Evolution Equations

Ralph deLaubenfels   Ohio University, Athens, Ohio

I. INTRODUCTION.

Suppose A is a linear operator on a Banach space and C is a bounded linear operator. For any Banach algebra, $\mathcal{F}$, of complex-valued functions, I will introduce a *C-regularized $\mathcal{F}$ functional calculus* for A; this is the same as a $\mathcal{F}$ functional calculus for A, except that I only require $f(A)C$ to be bounded, for $f \in \mathcal{F}$.

I give two simple examples of where regularized functional calculi arise and how they may be applied to evolution equations.

The applications to evolution equations that appear in Sections II–V have appeared elsewhere (see [13], [14], [15] and [27]); what is new in this presentation is that they are unified and simplified, as being simple consequences of regularized functional calculi constructions.

Terminology 1.1.   All operators are linear, on a Banach space, X. I will write $\mathcal{D}(A)$ for the domain of the operator A, $Im(A)$ for the image, $\rho(A)$ for the resolvent set, $\sigma(A)$ for the spectrum. The Banach space $B(X)$ will be all bounded operators from X to itself.

I will write RHP for the open right half-plane $\{z \in \mathbf{C}|\, Re(z) > 0\}$, LHP for the open left half-plane, $S_\phi \equiv \{re^{i\psi}|\, r > 0, |\psi| < \phi\}(0 < \phi \leq \pi)$, $H_\epsilon \equiv \{z \in \mathbf{C}|\, |Im(z)| < \epsilon\}$.

First, I should define a functional calculus.

Definition 1.2. Suppose $\mathcal{F}$ is a Banach algebra of complex-valued functions on a subset of the complex plane containing $f_0(z) \equiv 1$ and $g_\lambda(z) \equiv (\lambda - z)^{-1}$, for some complex λ. By a *$\mathcal{F}$-functional calculus for A* I will mean a continuous linear map, $f \mapsto f(A)$, from $\mathcal{F}$ into $B(X)$, such that

(1) $f_0(A) = I$;

(2) If $g_\lambda \in \mathcal{F}$, then $\lambda \in \rho(A)$ and $g_\lambda(A) = (\lambda - A)^{-1}$; and

(3) $f(A)g(A) = (fg)(A)$, for all $f, g \in \mathcal{F}$.

For *bounded* operators (see Example 1.5 and Section II) a well-known example is the Riesz-Dunford functional calculus

$$f(A) \equiv \frac{1}{2\pi i} \int_{\partial O} f(w)(w - A)^{-1}\, dw, \tag{1.3}$$

where O is an open subset of the complex plane containing $\sigma(A)$ whose boundary ∂O is a positively oriented countable system of piecewise smooth, mutually nonintersecting arcs.

It is well known that (1.3) defines an $H^\infty(O)$ functional calculus for A, where $H^\infty(O)$ is defined to be the set of all bounded, holomorphic complex-valued functions on O, with the supremum norm.

A famous example of a functional calculus for unbounded operators is the spectral theorem, which may be stated as follows, where I write $BC(\mathbf{R})$ for the set of all bounded continuous $f : \mathbf{R} \to \mathbf{C}$.

Spectral Theorem 1.4. *Suppose A is a self-adjoint operator on a Hilbert space. Then A has a $BC(\mathbf{R})$ functional calculus.*

Using the representation of the dual space of $C_0(\mathbf{R})$, and the fact that $\mathcal{D}(A)$ is dense, it can be shown that this implies that the functional calculus is then given by

$$f(A)x = \int_{\mathbf{R}} f(s)\, dE(s)x \quad (f \in BC(\mathbf{R}), x \in X),$$

for some projection-valued measure E. In particular, the strongly continuous group generated by $-iA$ is given by the Fourier transform

$$e^{-itA}x = \int_{\mathbf{R}} e^{-its}\, dE(s)x \quad (t \in \mathbf{R}, x \in X).$$

If A has nonnegative spectrum, then $-A$ generates a cosine family, given by

$$C(t)x = \int_0^\infty \cos(t\sqrt{s})\, dE(s)x \quad (t \in \mathbf{R}, x \in X).$$

Thus we obtain not only existence of a desired family of operators (and the corresponding well-posedness of Cauchy problems), but a nice representation of the operators (or the solutions of a Cauchy problem). Questions about qualitative behaviour, such as asymptotic behaviour, can be easily answered, given such a representation.

Example 1.5. Once one leaves a Hilbert space, one cannot expect such good behaviour, even for generators of strongly continuous bounded groups.

Let D be the unit disc in the complex plane, $\Gamma \equiv \partial D$, the unit circle, and let $A \equiv \frac{d}{d\theta}$, the generator of rotation $(e^{tA}f)(e^{i\theta}) \equiv f(e^{i(\theta+t)})$, on $L^p(\Gamma), 1 \le p < \infty$, $V \equiv V_1 \cup V_2$, where $V_1 \equiv (-\frac{1}{3} + S_{\frac{\pi}{4}})$, $V_2 \equiv (-\frac{2}{3} - S_{\frac{\pi}{4}})$. Then it is well known that $-iA$ has a $BC(\mathbf{R})$ functional calculus if and only if $p = 2$ (see [19]), and $-iA$ has an $H^\infty(V)$ functional calculus if and only if $1 < p < \infty$ (see [21] for $1 < p < \infty$; for $p = 1$, see the next paragraph).

For $p = 1$, it is clear how to define $f(-iA)$, at least on the dense set of trigonometric polynomials,

$$f(-iA)\left(\sum_{k=-N}^{N} \alpha_k h_k\right) \equiv \sum_{k=-N}^{N} \alpha_k f(k) h_k \quad (h_k(z) \equiv z^k),$$

but $f(-iA)$ may be unbounded. Let $f \equiv 1_{V_1}$. Then $f \in H^\infty(V)$, but $f(-iA)$ is the projection of $L^1(\Gamma)$ onto $H^1(D) \equiv \{f \in L^1(\Gamma) \mid f \text{ has an analytic extension to } D\}$, which is well known to be unbounded.

We shall see, however (Example 6.1), that $f(-iA)(1 - A^2)^{-r} \in B(X)$, whenever f is a bounded Borel function and $r > \frac{1}{4}$ or $f \in H^\infty(V)$ and $r > 0$. The map $f \mapsto f(-iA)(1 - A^2)^{-r}$ defines what I will call a regularized functional calculus (Definition 1.7.)

Definitions 1.6. Suppose $C \in B(X)$.

(a) The complex number λ is in $\rho_C(A)$, the *C-resolvent set of* A, if $(\lambda - A)$ is injective and $Im(C) \subseteq Im(\lambda - A)$.

  We remark that the map $\lambda \mapsto (\lambda - A)^{-1}C$ may not be analytic on $\rho_C(A)$.

(b) Denote by $B_C(X)$ the set of all operators A such that $AC \in B(X)$, with the norm $\|A\|_{B_C(X)} \equiv \|AC\|_{B(X)}$.

(c) Denote by $[Im(C)]$ the Banach space $Im(C)$ with norm

$$\|y\|_{[Im(C)]} \equiv \inf\{\|x\| \mid Cx = y\}.$$

When C is injective and commutes with $f(A)$, for all $f \in \mathcal{F}$, the following may be shown to be equivalent to Definition 3.11 in [13], by defining $\Lambda f \equiv f(A)C$ or $f(A) \equiv C^{-1}(\Lambda f)C$.

Definition 1.7. Suppose $\mathcal{F}$ is as in Definition 1.2 and $C \in B(X)$. By a *C-regularized $\mathcal{F}$-functional calculus for* A I will mean a continuous linear map, $f \mapsto f(A)$, from $\mathcal{F}$ into $B_C(X)$ such that

(1) $f_0(A) = I$;

(2) If $g_\lambda \in \mathcal{F}$, then $\lambda \in \rho_C(A)$ and $g_\lambda(A)C = (\lambda - A)^{-1}C$; and

(3) $f(A)g(A)C = (fg)(A)C$, for all $f, g \in \mathcal{F}$.

Note that, when $C = I$, this is the definition of a $\mathcal{F}$-functional calculus for A. This is analogous to a C-existence family, $\{W(t)\}$, for A (see [12]); the operator $W(t)$ may be thought of as $e^{tA}C$.

As with C-existence families, the idea is that one may have unbounded operators, but the regularizing operator, C, provides a uniform control over the unboundedness.

Proposition 1.8. *Suppose $C \in B(X)$ is injective and there exists a C-regularized $\mathcal{F}$ functional calculus for A such that $Cf(A) \subseteq f(A)C$, for all $f \in \mathcal{F}$. Then*

(a) *for all $f \in \mathcal{F}$, $f(A)$ is closable, with its closure contained in $C^{-1}f(A)C$; and*

(b) *there exists a Banach space Z such that*

$$[Im(C)] \hookrightarrow Z \hookrightarrow X$$

and $A|_Z$ has a $\mathcal{F}$ functional calculus.

Outline of Proof: For (a), it is clear that $C^{-1}f(A)C$ is a closed extension of $f(A)$. For (b), define Z to be the set of all x in $\bigcap_{f,g \in \mathcal{F}} \mathcal{D}(C^{-1}f(A)g(A)C)$ such that

$$\|x\|_Z \equiv \sup\{\|f(A)x\| \mid f \in \mathcal{F}, \|f\|_{\mathcal{F}} \le 1\}$$

is finite. ∎

Remark 1.9. It is clear that exponential and cosine functions, in Definition 1.7, are of particular interest to evolution equations. Through a much more indirect route, imaginary powers are also of great interest (see [2, 8, 17, 18, 28, 29, 33, 34, 36]). As I commented after stating the spectral theorem, we may obtain, with Definition 1.7, much more than existence of the desired families of operators, we obtain a nice representation, merely by using representations of the dual space of $\mathcal{F}$. We gave one example after the spectral theorem. Another example is the following. If Ω is an open region in the complex plane and A has a C-regularized $H^\infty(\Omega)$ functional calculus, then it may be shown that, for any $x \in X, x^* \in X^*$, there exists a complex-valued measure E_{x,x^*} such that

$$< f(A)Cx, x^* > = \int_{\overline{\Omega}} f(w)\, dE_{x,x^*}(w),$$

for all $f \in C(\overline{\Omega}) \cap H^\infty(\Omega)$, and there exists a constant M so that

$$var\, E_{x,x^*} \leq M\|x\|\|x^*\|,$$

for all $x \in X, x^* \in X^*$.

For appropriate Ω, we may choose f to be an exponential, cosine, or imaginary power, to obtain representations of regularized semigroups, cosine families or imaginary powers. Or one may remove the regularizing, by passing to the continuously embedded Banach space Z, of Proposition 1.8.

II. AN ANALOGUE OF THE RIESZ-DUNFORD FUNCTIONAL CALCULUS FOR UNBOUNDED OPERATORS.

Throughout this section, V and O will be open subsets of the complex plane whose complements contain a half-line and whose boundaries, $\partial V, \partial O$, are positively oriented countable systems of piecewise smooth, mutually nonintersecting (possibly unbounded) arcs.

Definition 2.1. Suppose $\alpha \geq -1$. We will say that the operator A is *of α-type* V if $\sigma(A) \subseteq V$ and $\|(w - A)^{-1}\|$ is $O((1 + |w|)^\alpha)$, for w outside V.

Examples 2.2. (1) If $A \in B(X)$, then A is of (-1)-type V, whenever V is an open set containing $\sigma(A)$.

Thus operators of (-1)-type V may be thought of as those that behave like bounded operators, in some sense.

(2) Let $(Af)(z) \equiv zf(z)$, on $L^2([1,\infty))$, with maximal domain. Then, for all $\theta > 0$, A is of (-1)-type S_θ. For all $\epsilon > 0$, A is of 0-type H_ϵ, but is not of (-1)-type H_ϵ.

This illustrates the fact that, although many operators are of (-1)-type V, for *some* V, many natural choices of V require that we consider operators of α-type V, for $\alpha > -1$.

Another example of this is $A \equiv -\Delta$, the Laplacian, on $L^p(\mathbf{R}^n), 1 \leq p \leq \infty$, with $\mathcal{D}(A) = \{f \in L^p(\mathbf{R}^n) \,|\, \Delta f \in L^p(\mathbf{R}^n)\}$, where Δ is taken in the sense of distribution. In particular, if $1 < p < \infty$, then $\mathcal{D}(A) = W^{2,p}(\mathbf{R}^n)$ (the Sobolev space) (see [26]). For all $\theta > 0$, A is of (-1)-type $(S_\theta - 1)$, while for all $\epsilon > 0$, there exists $\alpha > -1$, depending on both p and n, such that A is of α-type H_ϵ, but A is not of (-1)-type H_ϵ. (This may be seen by using the functional calculus for commuting generators of bounded strongly continuous groups; see Section IV.)

Consider also an operator, A, that generates an exponentially decaying strongly continuous semigroup, $\{e^{tA}\}_{t\geq0}$. The operator A is of 0-type LHP and is of (-1)-type LHP if and only if $\{e^{tA}\}$ extends to a bounded holomorphic strongly continuous semigroup.

(3) In many places ([2, 7, 8, 9, 18, 19, 20, 27, 28, 29, 33, 34, 36], for example), operators of *type ω* are considered. These are densely defined operators, A, such that $\sigma(A) \subseteq \overline{S_\omega}$ and for all $\psi > \omega$, there exists $M_\psi < \infty$ such that $\|w(w - A)^{-1}\| \leq M_\psi$, for all w outside S_ψ. Note that a densely defined operator with 0 in its resolvent set is of type ω if and only if it is of (-1)-type S_ϕ, for all $\phi > \omega$.

Even on a Hilbert space, when $f \in H^\infty(S_\phi)$ and A is of type ϕ, $f(A)$ may not be bounded (see [30, 2]).

The proof of the following lemma may be found in [13, Lemma 3.4].

Lemma 2.3. *Suppose A is of α-type V. Then there exists O such that $\overline{O} \subseteq V$ and A is of α-type O.*

CONSTRUCTION I. *Suppose A is of α-type V. Let O be as in Lemma 2.3, $m \equiv [\alpha]+2$. Then, for all λ outside V, the map*

$$f \mapsto (\lambda - A)^m \, f(A)(\lambda - A)^{-m} \, ,$$

where

$$f(A)(\lambda - A)^{-m} \equiv \frac{1}{2\pi i} \int_{\partial O} f(w)(w - A)^{-1} \frac{dw}{(\lambda - w)^m}.$$

defines a $(\lambda - A)^{-m}$-regularized $H^\infty(V)$-functional calculus for A.

Note that by the residue theorem, the definition of $f(A)$ is independent of O.

The proof that this construction defines a $(\lambda - A)^{-m}$-regularized $H^\infty(V)$-functional calculus for A may be found in [13]; one uses the residue theorem and the resolvent identity.

Similar results, for operators of (-1)-type $S_{\frac{\pi}{2}}$, may be found in [7].

III. APPLICATION: SPECTRAL CONDITIONS GUARANTEEING SOLUTIONS OF THE ABSTRACT CAUCHY PROBLEM.

It is natural and very common to ask what conditions on the resolvent of A, $(w - A)^{-1}$, will guarantee that the abstract Cauchy problem

$$\frac{d}{dt}u(t,x) = A(u(t,x))\,(t \geq 0), \quad u(0,x) = x \tag{3.1}$$

has a solution, for all initial data, x, in a dense set. By "solution," I will mean a classical solution, that is, $t \mapsto u(t,x) \in C([0,\infty),[\mathcal{D}(A)]) \cap C^1([0,\infty),X)$.

As an application of Construction I, I will give a simple sufficient condition on the spectrum of A that guarantees that (3.1) will have a solution for all initial data in a dense set, whenever A is densely defined. It is also possible to use Construction I to give a complete characterization of subsets, V, of the complex plane, with the property that, whenever A is of α-type V (Definition 2.1) and densely defined, then (3.1) has a solution, for all initial data, x, in a dense set (see [13, Section VI]).

This theorem reduces the operator theoretic question, of when solutions of (3.1) exist, to the complex analytic question of when there exist sufficiently many holomorphic g, on an appropriate neighborhood of the spectrum of A, such that $z \mapsto e^{tz}g(z)$ is bounded, for all $t \geq 0$.

It is clear how the same techniques, with the exponential function replaced by the cosine function, may be applied to the second-order abstract Cauchy problem (see [13, Section VI]).

Definition 3.2. If V is as in Section II, we will say that V is an *α-spectral dense solution set* if, whenever A is of α-type V and densely defined, then (3.1) has a solution, for all initial data in a dense set.

The simplest sufficient condition for being an α-spectral dense solution set is independent of α. In general, this property is not independent of α (see Example 3.12).

Definition 3.3. We will denote by E_V the set of holomorphic functions g on V that decay more rapidly than any exponential, $E_V \equiv \{g \mid z \mapsto e^{tz}g(z) \in H^\infty(V), \text{ for all } t \geq 0\}$.

Theorem 3.4. *Suppose the constant function $f_0(z) \equiv 1$ is the pointwise limit of a bounded sequence from E_V. Then, for any $\alpha \geq -1$, V is an α-spectral dense solution set.*

Outline of Proof: Let $\{g_k\}_{k=1}^\infty$ be the bounded sequence in E_V converging pointwise to f_0. Fix $\lambda \notin V$. For $t \geq 0, x \in X, k \in \mathbf{N}$, let $h_{t,k}(z) \equiv (\lambda - z)^{-1}e^{tz}g_k(z)$ and let $u_k(t,x) \equiv h_{t,k}(A)(\lambda - A)^{-m}x$, as defined by Construction I. Then the properties of a regularized functional calculus imply that, for any k, u_k is a solution of (3.1), with x replaced by $u_k(0,x)$. Dominated convergence implies that $(\lambda - A)^{-(m+1)}x = \lim_{k \to \infty} u_k(0,x)$. Since $\mathcal{D}(A^{m+1})$ is dense whenever $\mathcal{D}(A)$ is dense, this proves that V is an α-spectral dense solution set. ∎

Example 3.5. For any real $c, \alpha \geq -1$, the left half-plane $V \equiv (c + LHP)$ is an α spectral dense solution set, since $f_0(z) \equiv 1 \in E_V$. In fact, it is known that an operator of α-type V generates an $([\alpha] + 2)$-times integrated semigroup and a $(\lambda - A)^{-([\alpha]+2)}$-semigroup (see [1, 12]).

Example 3.6. For any real c, $\alpha \geq -1$, $0 < \theta < \frac{\pi}{2}$, the sector $V \equiv c + S_\theta$ is an α-spectral dense solution set.

To see this, choose r such that $1 < r < \frac{\pi}{2\theta}$. For any $\epsilon > 0$, let $g_\epsilon(z) \equiv e^{-\epsilon(z-c)^r}$. Then it may be shown that, for all $\epsilon > 0$, g_ϵ is in E_V. Since $\lim_{\epsilon \to 0} g_\epsilon(z) = 1$, for all $z \in V$, we may apply Theorem 3.4.

We shall see that S_θ behaves much differently when $\theta \geq \frac{\pi}{2}$ (Example 3.12).

Example 3.6 may be considered a result about reversibility of solutions.

REVERSIBILITY OF PARABOLIC PROBLEMS 3.7. Consider

$$\frac{d}{dt} u(t,x) = A(u(t,x)) \, (t \in \mathbf{R}), \ u(0,x) = x, \tag{3.8}$$

where A generates a strongly continuous holomorphic semigroup (this is what I mean by parabolic; the reversibility is the fact that t runs over all of the real line rather than just $[0, \infty)$).

We've just shown how Construction I produces solutions of (3.8) for all initial data, x, in a dense set.

Thus time can be "run backwards" on a dense set, for such Cauchy problems as the diffusion equation.

Example 3.9. For any $c > 0$, $\alpha \geq -1$, the horizontal strip H_c is an α-spectral dense solution set.

As in Example 3.6, this may be seen by showing that $g_\epsilon(z) \equiv e^{-\epsilon z^2}$ is in E_{H_c}, for all $\epsilon > 0$.

Example 3.10. For $\theta < \frac{\pi}{4}$, $\alpha \geq -1$, $c \in \mathbf{R}$, the same choice of g_ϵ as in the previous example shows that the double sector $(c + S_\theta) \cup (c - S_\theta)$ is an α-spectral dense solution set.

Example 3.11. In [5, 6], operators of α-type $V_a \equiv \{x + iy \,|\, x < |y|^a\}$ are considered. In the language of this section, it is shown there that V_a is an α-spectral dense solution set, whenever $0 < a < 1$. This follows from Theorem 3.4, by considering $g_\epsilon(z) \equiv e^{-\epsilon(-z)^b}$, for $a < b < 1$, and showing that this is in E_{V_a}, for all $\epsilon > 0$; in fact, this is the key to the construction in [5].

Example 3.12. Let V be the open right half-plane RHP. Then V is a (-1)-spectral dense solution set, but is not a 0-spectral dense solution set.

To see this, note first that, if A is of (-1)-type V, then there exists $\theta < \frac{\pi}{2}$ such that A is of (-1)-type S_θ (see [4]). In Example 3.6, I showed that S_θ is a (-1)-spectral dense solution set, when $\theta < \frac{\pi}{2}$. Thus (3.1) has a solution, for all initial data in a dense set, so that V is a (-1)-spectral dense solution set.

To see that V is not a 0-spectral dense solution set, consider $A \equiv 1 + \frac{d}{dx}$, on $X \equiv \{f \in C[0,1] \,|\, f(0) = 0\}$, $\mathcal{D}(A) \equiv \{f \in X \,|\, f' \in X\}$. Then it is not hard to show that A is of 0-type V, since $1 - A$ generates a bounded strongly continuous semigroup, but (3.1) has no nontrivial solutions.

IV. GENERATORS OF BOUNDED STRONGLY CONTINUOUS GROUPS.

Terminology 4.1. Throughout this section, n is a fixed natural number. Define $k \equiv \left[\frac{n}{2}\right] + 1$.

We will need some standard multivariable terminology. We will write $s = (s_1, \ldots s_n)$, for vectors in $\mathbf{R}^n$, $\alpha = (\alpha_1, \ldots \alpha_n)$ for vectors in $\mathbf{N}^n$. We will write $s^\alpha \equiv s_1^{\alpha_1} \ldots s_n^{\alpha_n}$, $|\alpha| \equiv \sum_{j=1}^n \alpha_j$, $\|s\| \equiv \left(\sum_{j=1}^n |s_j|^2\right)^{\frac{1}{2}}$.

Operator Terminology 4.2. Throughout this section, $iB_1, \cdots, iB_n$ will be commuting generators of bounded strongly continuous groups of operators, $\{e^{itB_j}\}_{t \in \mathbf{R}}$. We will write B for $(B_1, \ldots, B_n)$.

In particular, the operator D_j will be $i\frac{\partial}{\partial s_j}$, on $L^p(\mathbf{R}^n)$, $C_0(\mathbf{R}^n)$, $BUC(\mathbf{R}^n)$, or any Banach space of functions in n variables where translation is uniformly bounded and strongly continuous, and $D \equiv (D_1, \ldots, D_n)$.

We will write $e^{i(s \cdot B)}$ for $\prod_{j=1}^{n} e^{is_j B_j}$. We will use the following well-known functional calculus (see [10, chapter 8] for the case $n = 1$). Define a bounded operator $f(B)$ by

$$f(B)x \equiv \int_{\mathbf{R}^n} e^{i(s \cdot B)} x \, \mathcal{F}f(s) \, ds \quad (x \in X), \tag{4.3}$$

where $\mathcal{F}$ is the Fourier transform, whenever $\mathcal{F}f \in L^1(\mathbf{R}^n)$.

The map $f \mapsto f(B)$ has the following properties.

(1) It is an algebra homomorphism.

(2) There exists $M < \infty$ such that

$$\|f(B)\| \leq M \|\mathcal{F}f\|_1,$$

whenever $\mathcal{F}f \in L^1(\mathbf{R}^n)$.

(3)

$$\prod_{j=1}^{n} (\lambda_j - B_j)^{-\alpha_j} = \left(\prod_{j=1}^{n} (\lambda_j - s_j)^{-\alpha_j} \right) (B),$$

for any $\alpha \in \mathbf{N}^n, Im(\lambda_j) \neq 0$.

The operator $-|B|^2$ is defined to be $-\sum_{j=1}^{n} B_j^2$, the generator of the strongly continuous semigroup

$$e^{-t|B|^2} \equiv f_t(B), \quad f_t(s) \equiv e^{-t|s|^2}.$$

There are many equivalent ways to define fractional powers of $(1 + |B|^2)^{-1}$ (see, for example, [22, 24]); here, let us put the Laplace transform definition,

$$(1 + |B|^2)^{-r}x \equiv \frac{1}{\Gamma(r)} \int_0^\infty e^{-t} t^{r-1} e^{-t|B|^2} x \, dt \quad (x \in X),$$

where $r > 0$. This defines a bounded, injective operator, so that $(1 + |B|^2)^r \equiv \left((1 + |B|^2)^{-r} \right)$ is defined as a closed operator.

The following may be derived from results in [33]; we will show here how it follows when $n = 1$. Note that $H^k(\mathbf{R}^n)$ is a Sobolev space.

Lemma 4.4. *There exists a constant K such that*

$$\|\mathcal{F}f\|_1 \leq K \sum_{|\alpha| \leq k} \|D^\alpha f\|_2,$$

for all $f \in H^k(\mathbf{R}^n)$.

Proof when $n = 1$:

$$\|\mathcal{F}f(s)\|_1 = \| \left(\mathcal{F}(f + f')(s) \right) (1 + is)^{-1} \|_1$$
$$\leq \|\mathcal{F}(f + f')\|_2 \|(1 + is)^{-1}\|_2 = \|(f + f')\|_2 \sqrt{\pi}.$$

∎

Using Lemma 4.4, I obtain the following (see [15] for details of the proof), where $BC^k(\mathbf{R}^n)$ is the space of complex-valued functions on $\mathbf{R}^n$ with k bounded continuous derivatives, with $\|f\|_{BC^k(\mathbf{R}^n)} \equiv \Sigma_{|\alpha| \leq k} \|D^\alpha f\|_\infty$.

CONSTRUCTION II. *For any $r > \frac{n}{4}$, the map*

$$f \mapsto (1 + |B|^2)^r \left(\left(f(s)(1 + |s|^2)^{-r} \right)(B) \right),$$

as defined by (4.3), defines a $(1 + |B|^2)^{-r}$-regularized $BC^k(\mathbf{R}^n)$ functional calculus for B.

Remark 4.5. For $n = 1$, a similar construction, in [16], is done for a more general class of operators, generators of m-times integrated groups that are $O(|t|^m)$ (see [3, 27]).

V. APPLICATION: PETROWSKY CORRECT SYSTEMS OF CONSTANT COEF-FICIENT PARTIAL DIFFERENTIAL EQUATIONS.

We will apply Construction II to systems of partial differential equations. Throughout this section, m and n are fixed natural numbers, and I use the same terminology as in Section IV.

More Operator Terminology 5.1. Also, throughout this section, $\mathcal{M} = (p_{i,j})_{i,j=1}^m$ will be an $m \times m$ matrix of polynomials in n variables. We will denote by N the maximum, over i, j between 1 and m, of the degree of $p_{i,j}\}$.

We define $\mathcal{M}(D) \equiv (p_{i,j}(D))_{i,j=1}^m, \mathcal{D}(\mathcal{M}(D)) \equiv \left(\mathcal{D}(\Delta^{\frac{N}{2}}) \right)^m$, where Δ is the Laplacian, and both D and Δ are acting on $X \equiv L^p(\mathbf{R}^n)(1 \leq p < \infty), C_0(\mathbf{R}^n)$ or $BUC(\mathbf{R}^n)$ and for $r > 0$, $\mathcal{D}(\Delta^r) \equiv \{f \in X \mid \Delta^r f \in X\}$, where Δ^r is taken in the sense of distributions.

When A is an operator on X, I will also write A for AI_m, the operator on X^m with domain $(\mathcal{D}(A))^m$.

Analogues of Theorem 5.3 may be found in [27, 14]; Theorem 5.4 may be found in [14].

A system of constant coefficient partial differential equations can be written as

$$\frac{d}{dt}\vec{u}(t, \vec{f}) = \mathcal{M}(D)\vec{u}(t, \vec{f}) \ (t \geq 0), \ \vec{u}(0, \vec{f}) = \vec{f}, \tag{5.2}$$

where $\vec{f} \in X^m$.

By *Petrowsky correct* (see [23]) we mean that there exists $\omega \in \mathbf{R}$ such that for all $s \in \mathbf{R}^n$,

$$\sigma(\mathcal{M}(s)) \subseteq \{z \in \mathbf{C} \mid Re(z) \leq \omega\}.$$

Theorem 5.3. *Suppose $\mathcal{M}$ is Petrowsky correct and $r > \frac{1}{2}\left(k(N-1) + \frac{n}{2} + Nm)\right)$. Then for all $\epsilon > 0, \vec{f} \in (\mathcal{D}(\Delta^r))^m$, there exists a $O(e^{(\omega+\epsilon)t})$ solution of (5.2).*

If, in the definition of a Petrowsky correct matrix, we replace spectrum with numerical range, we can weaken the regularity requirements.

If B is an $m \times m$ matrix, I will write n.r.(B) for the *numerical range of B,*

$$n.r.(B) \equiv \{< s, Bs > \mid s \in \mathbf{C}^m, \|s\| = 1\},$$

where $< \cdot >$ is the inner product in $\mathbf{C}^m, < s, y > \equiv \sum_{k=1}^m s_k \overline{y_k}$.

Theorem 5.4. *Suppose $\omega \in \mathbf{R}$ and for all $s \in \mathbf{R}^n$,*

$$n.r.(\mathcal{M}(s)) \subseteq \{z \in \mathbf{C} \mid Re(z) \leq \omega\},$$

and $r > \frac{1}{2}\left(k(N-1) + \frac{n}{2} + N\right)$. Then for all $\epsilon > 0, \vec{f} \in (\mathcal{D}(\Delta^r))$, there exists a $O(e^{(\omega+\epsilon)t})$ solution of (5.2).

OUTLINE OF PROOF OF THEOREMS 5.3 and 5.4: The following is primarily a calculus and linear algebra calculation.

Lemma 5.5. *Suppose $\mathcal{M}_1 \equiv (p_{i,j})_{i,j=1}^m$ and $\mathcal{M}_2 \equiv (q_{i,j})_{i,j=1}^m$ are $m \times m$ matrices of polynomials in n variables, $\mathcal{M}_1$ is Petrowsky correct with ω less than zero and $\mathcal{M}_2$ has numerical range contained in a half-plane $\{z \mid Re(z) \leq \omega\}$, for some ω less than zero. Then, for $1 \leq i, j \leq m$, we have the following.*

(a) *The map $t \mapsto (e^{t\mathcal{M}_1(s)}(1 + |s|^2)^{-r})_{(i,j)}$ defines a uniformly bounded continuous map from $[0, \infty)$ into $BC^k(\mathbf{R}^n)$, for all $r > \frac{1}{2}(k(N-1) + N(m-1))$.*

(b) *The map $t \mapsto (e^{t\mathcal{M}_2(s)}(1 + |s|^2)^{-r})_{(i,j)}$ defines a uniformly bounded continuous map from $[0, \infty)$ into $BC^k(\mathbf{R}^n)$, for all $r > \frac{1}{2}(k(N-1))$.*

Without loss of generality, we may assume, in Theorems 5.3 and 5.4, that $(\omega + \epsilon) = 0$. Combining the "smoothing" of the matrix exponentials in Lemma 5.5 with the regularizing of Construction II gives us uniformly continuous families of operators

$$W_1(t) \equiv (e^{t\mathcal{M}_1(D)}(1 + \triangle)^{-r_1}),$$

for $r_1 > \frac{1}{2}\left(k(N-1) + \frac{n}{2} + N(m-1)\right)$, and

$$W_2(t) \equiv (e^{t\mathcal{M}_2(D)}(1 + \triangle)^{-r_2}),$$

for $r_2 > \frac{1}{2}\left(k(N-1) + \frac{n}{2}\right)$. Using the properties of the map in Construction II, this then implies that, for $j = 1, 2, \vec{g} \in X^m$, $W_j(t)(1+\triangle)^{-\frac{N}{2}}\vec{g}$ is a classical solution of (5.2), yielding both Theorems 5.3 and 5.4.

Example 5.6. The equation describing sound propagation in a viscous gas is (see [23], Example 3, page 134)

$$\frac{\partial^2 u}{\partial t^2} = 2\frac{\partial^3 u}{\partial t \partial s^2} + \frac{\partial^2 u}{\partial s^2}.$$

After the usual matrix reduction, this becomes

$$\frac{d}{dt}\vec{u} = \mathcal{M}(D)\vec{u}, \tag{5.7}$$

where $D \equiv i\frac{d}{ds}, \mathcal{M}(s) \equiv \begin{bmatrix} 0 & 1 \\ (-s^2) & (-2s^2) \end{bmatrix}$. Since the spectrum of $\mathcal{M}(s)$ is $\{y \mid y = -s^2 \pm s\sqrt{s^2 - 1}\}$, which has real part bounded above, for s real, this is a Petrowsky correct system, so that Theorem 5.3 implies that we have exponentially bounded solutions of (5.7), for all initial data in $\mathcal{D}(D^{2r})$, for all $r > \frac{11}{4}$.

Example 5.8. Note that Theorem 5.4 includes all symmetric hyperbolic systems, that is, operators A of the form

$$A = \sum_{j=1}^n \mathcal{M}_j D_j,$$

for some constant $m \times m$ complex-valued symmetric matrices $\mathcal{M}_j$, on $(L^p(\mathbf{R}^n))^m (1 \leq p < \infty), C_0(\mathbf{R}^n)^m$, etc. More generally, it includes operators of the form

$$A = \sum_{j=1}^n \mathcal{M}_j D_j + \mathcal{M}_0,$$

for some constant $m \times m$ complex-valued symmetric matrices $\mathcal{M}_j (0 \leq j \leq n)$, by choosing

$$\mathcal{M}(s) \equiv \sum_{j=1}^n \mathcal{M}_j s_j + \mathcal{M}_0,$$

for $s \in \mathbf{R}^n$.

This includes the wave equation, Maxwell's equations, the abstract Cauchy problem corresponding to the Dirac operator, and many other famous partial differential equations; see [25, Section II.9] and [24] for more examples.

VI. MORE GENERATORS OF BOUNDED STRONGLY CONTINUOUS GROUPS.

We will write $B(\mathbf{R}^n)$ for the space of bounded Borel measurable functions on $\mathbf{R}^n$, with the supremum norm.

CONSTRUCTION III. *Suppose $B_1, \ldots B_n$ are as in Section IV and, in addition, there exists $\epsilon > 0$ such that $|x - y| > \epsilon$, for all $x \neq y \in \sigma(B_k), 1 \leq k \leq n$. Then for any $r > \frac{n}{4}$, the map*

$$f \mapsto (1 + |B|^2)^r \left(\left(f(s)(1 + |s|^2)^{-r} \right) (B) \right),$$

as defined by (4.3), defines a $(1 + |B|^2)^{-r}$-regularized $B(\mathbf{R}^n)$ functional calculus for B.

Outline of Proof: In (4.3), $f(B) \equiv 0$ whenever $f \equiv 0$ in $\times_{j=1}^n \sigma(B_j)$.

Our hypotheses on $\sigma(B_j)$ imply that there exists a constant M, so that, for any $f \in B(\mathbf{R}^n)$, there exists $g \in BC^k(\mathbf{R}^n)$, such that $(g - f) \equiv 0$ in $\times_{j=1}^n \sigma(B_j)$ and

$$\|g\|_{BC^k(\mathbf{R}^n)} \leq M \|f\|_\infty.$$

Thus, by Construction II, for $f \in B(\mathbf{R}^n)$, we may define $f(B) \equiv g(B)$, where g is any function in $BC^k(\mathbf{R}^n)$ that equals f on $\times_{j=1}^n \sigma(B_j)$, and obtain a $(1 + |B|^r)^{-r}$-regularized $B(\mathbf{R}^n)$ functional calculus for B. ∎

In fact, on a subspace Z as in Proposition 1.8, B_k is a scalar-type spectral operator, for $1 \leq k \leq n$ (see [16], for $n = 1$).

Example 6.1. As in Example 1.5, let Γ be the unit circle in the complex plane and let $A \equiv \frac{d}{d\theta}$, the generator of rotation $(e^{tA}f)(e^{i\theta}) \equiv f(e^{i(\theta+t)})$, on $L^p(\Gamma), 1 \leq p < \infty$, $V \equiv V_1 \cup V_2$, where $V_1 \equiv (-\frac{1}{3} + S_{\frac{\pi}{4}})$, $V_2 \equiv (-\frac{2}{3} - S_{\frac{\pi}{4}})$.

By Construction III, with $B \equiv -iA$, the map $f \mapsto f(-iA)(1 - A^2)^{-r}$ defines, for $r > \frac{1}{4}$, a $(1 - A^2)^{-r}$-regularized $B(\mathbf{R})$ functional calculus, while, for $r > 0$, an unbounded version of the Riesz-Dunford functional calculus, as in Construction I, produces a $(1 - A^2)^{-r}$-regularized $H^\infty(V)$ functional calculus, with

$$f(-iA)(1 - A^2)^{-r} \equiv \frac{1}{2\pi i} \int_{\partial O} f(w)(w + iA)^{-1} \frac{dw}{(1 + w^2)^r},$$

where O is as in Lemma 2.3.

REFERENCES

[1] W. Arendt and H. Kellermann, *Integrated solutions of Volterra integro-differential equations and applications*, Integro-differential equations in Banach spaces and applications, Proc. Conf. Trento (1987), 21–51, G. Da Prato and M. Ianelli(eds.), Pitman Res. Notes Math. Ser. 190, Longman Sci. Tech., Harlow, 1989.

[2] J. B. Baillon and Ph. Clement, *Examples of unbounded imaginary powers of operators*, J. Func. An. 100(1991), 419–434.

[3] M. Balabane, H. Emamirad and M. Jazar, *Spectral distributions and generalization of Stone's theorem*, Act. Appl. Math., to appear.

[4] A.V. Balakrishnan, *Fractional powers of closed operators and the semigroups generated by them*, Pac. J. Math. 10(1960), 419–437.

[5] R. Beals, *On the abstract Cauchy problem*, J. Func. An. 10(1972), 281–299.

[6] R. Beals, *Semigroups and abstract Gevrey spaces*, J. Func. An. 10(1972), 300–308.

[7] K. Boyadzhiev and R. deLaubenfels, H^∞-*functional calculus for perturbations of generators of holomorphic semigroups*, Houston J. Math. 17(1991), 131–147.

[8] K. Boyadzhiev and R. deLaubenfels, *Semigroups and resolvents of bounded variation, imaginary powers and H^∞ functional calculus*, Semigroup Forum 45(1992), 372–384.

[9] M. Cowling, I. Doust, A. McIntosh and A. Yagi, *Banach space operators with an H^∞ functional calculus*, preprint.

[10] E.B. Davies, "One-Parameter Semigroups," Academic Press, London, 1980.

[11] R. deLaubenfels, *Entire solutions of the abstract Cauchy problem*, Semigroup Forum 42(1991), 83–105.

[12] R. deLaubenfels, *Existence and uniqueness families for the abstract Cauchy problem*, J. London Math. Soc. 44 (1991), 310–338.

[13] R. deLaubenfels, *Unbounded holomorphic functional calculus and abstract Cauchy problems for operators with polynomially bounded resolvents*, J. Func. An., to appear.

[14] R. deLaubenfels, *Matrices of operators and regularized semigroups*, Math. Z., to appear.

[15] R. deLaubenfels, *Simultaneous well-posedness*, Evolution Equations, Control Theory and Biomathematics: the Third International Conference, Hans-sur-Lesse 1991, Lecture Notes in Pure and Applied Mathematics, Marcel-Dekker (1993), 101–115.

[16] R. deLaubenfels, H. Emamirad and M. Jazar, *Spectral distributions and regularized scalar operators*, submitted.

[17] G. Dore and A. Venni, *On the closedness of the sum of two closed operators*, Math. Z. 196(1987), 189–201.

[18] G. Dore and A. Venni, *Some results about complex powers of closed operators*, J. Math. Anal. Appl. 149(1990), 124–136.

[19] H. R. Dowson, "Spectral Theory of Linear Operators," Academic Press, 1978.

[20] X. T. Duong, H_∞ *functional calculus of second order elliptic partial differential operators on L^p spaces*, Miniconference on Operators in Analysis, Proc. of the Center for Math. Analysis, ANU, Canberra, 24 (1989), 91–102.

[21] X. T. Duong, H^∞ *functional calculus of elliptic operators with C^∞ coefficients on L^p spaces of smooth domains*, J. Austral. Math. Soc. (Ser. A) 48(1990), 113–123.

[22] H.O. Fattorini, "The Cauchy Problem", Addison Wesley, Reading, Mass., 1983

[23] I.M. Gelfand and G.E. Shilov, "Generalized Functions," Vol.3, Academic Press, New York, 1967.

[24] D. Gilliam and J.R. Schulenberger, *A class of symmetric hyperbolic systems with special properties*, Comm. Part. Diff. Equations 4 (1979), 509–536.

[25] J.A. Goldstein, "Semigroups of Linear Operators and Applications," Oxford, New York, 1985.

[26] R. Hempel and J. Voigt, *On the L_p-spectrum of Schrödinger operators*, J. Math. Anal. Appl. 121(1987), 138–159.

[27] M. Hieber, A. Holderrieth and F. Neubrander, *Regularized semigroups and systems of linear partial differential equations*, Ann. Scuola Norm. di Pisa 19 (1992), 363–379.

[28] M. Jazar, *Sur la théorie de la distribution spectrale et applications aux problèmes de Cauchy*, Thèse de l'Université de Poitiers 1991.

[29] A. McIntosh, *Operators which have an H^∞ functional calculus*, Miniconference on Operator Theory and PDE, Proc. of the Center for Math. Analysis, ANU, Canberra, 14 (1986), 210–231.

[30] A. McIntosh and A. Yagi, *Operators of type ω without a bounded H^∞- functional calculus*, Miniconference on Operators in Analysis 1989, Proc. of the Center for Math. Analysis, ANU, Canberra, 24 (1989).

[31] J. Prüss and H. Sohr, *On operators with bounded imaginary powers in Banach spaces*, Math. Z. 203(1990), 429–452.

[32] W. Ricker, *Spectral properties of the Laplace operator in $L^p(\mathbf{R})$*, Osaka J. Math., 25(1988), 399–410.

[33] E. M. Stein, "Singular Integrals and Differentiability Properties of Functions", Princeton University Press, New Jersey, 1970.

[34] A. Venni, *Some instances of the use of complex powers of linear operators*, Semesterbericht Funktionalanalysis, Tübingen, Sommersemester 1988, 235–246.

[35] A. Yagi, *Coincidence entre des espaces d'interpolation et des domaines de puissances fractionnaires d'operateurs*, C. R. Acad. Sci. Paris (Ser. I), 299(1984), 173–176.

[36] A. Yagi, *Applications of the purely imaginary powers of operators in Hilbert spaces*, J. Func. An. 73(1987), 216–231.

Dynamic Identification of Implicit Parabolic Systems

M. A. Demetriou   North Carolina State University, Raleigh, North Carolina

I. G. Rosen   University of Southern California, Los Angeles, California

1. INTRODUCTION

We consider a dynamic, or adaptive, identification scheme for implicit, and possibly degenerate, abstract parabolic systems. A system is said to be implicit if the highest order time derivative does not appear explicitly. Such a system is said to be degenerate when the operator coefficient of the highest order time deriviative is non-coercive. An extensive study of the theory and application of degenerate distributed parameter systems can be found in the monograph [3] by Carroll and Showalter.

System identification is concerned with the determination or estimation of unknown parameters in a mathematical model for a given physical system, or plant. The parameters may be scalar or functional and the identification is carried out based upon knowledge of the input signal and measurements of the system output. The identification can be carried out either off-line, or staticly, by first exciting the system and taking output measurements, and then fitting an appropriate model (i.e. determining the unknown parameters) via the minimization of a functional which measures the difference between the output of the model and the output of the system. Least squares is one possible approach. The identification can also be carried out on-line, adaptively, or dynamicly. In this case the unknown parameters are estimated in real time. The estimator takes both the system input and output as input, and produces a time varying estimate of the parameters. The dynamics governing the evolution of the parameter estimator are known as an adaptation rule. This latter approach is attractive in that it compensates for modeling errors, changing environmental conditions, and noise in the data. It is also useful in the design of adaptive control schemes. In analyzing on-line identification schemes, the primary objective is to establish parameter convergence as time tends to infinity. A survey of the literature dealing with adaptive identification of distributed parameter systems is given in [4].

In this short paper we consider linear systems in which the dependence on the unknown parameters is linear as well. Our treatment here is an extension of earlier results for the adaptive identification of non-implicit parabolic systems described in [4] and [12]. In [6] we developed adaptive schemes for the identification of abstract hyperbolic systems. The estimator we develop here takes the form of an initial value problem for an infinite dimensional evolution equation. The state of the estimator consists of an estimator for the state of the plant and an estimator for the parameters. It is worth noting that if the operator coefficient of the time derivative term is unknown and is to be estimated, then measurements of both the state and the time derivative of the plant are required. It is also worth noting that even if the system to be identified is degenerate, the estimator is not.

We are able to establish convergence of the state estimator directly via a Lyapunov-like argument. Parameter convergence requires an additional assumption. This assumption, which is a *richness* condition on the output data of the plant, is referred to as *persistence of excitation*. This assumption is in the spirit of similar conditions used to establish parameter convergence in [6] and [4]. It is also an extension of the notion of persistence of excitation as it was originally defined in the context of the adaptive identification of finite dimensional systems (see, for example, [10] and [11]).

The proof of parameter convergence given here, which is indirect, is based upon a similar result for finite dimensional systems in [2]. A direct and far longer and more technical argument which provides additional insight into the optimal tuning of the scheme, is given in [7]. In [7] we have also developed a rather complete finite dimensional approximation theory and established corresponding convergence results. In that paper we also report on the results of our numerical studies.

We provide a brief outline of the remainder of the paper. In Section 2 we define the plant and the estimator and establish that they are well posed. In Section 3 we argue convergence of the state estimator, define persistence of excitation, and establish parameter convergence. In Section 4 we discuss a simple example.

2. THE PLANT AND THE ESTIMATOR

Let H be a real Hilbert space with inner product $\langle \cdot, \cdot \rangle$ and corresponding norm $|\cdot|$. Let V be a real reflexive Banach space with norm denoted by $\|\cdot\|$, and assume that V is embedded densely and continuously in H. Let V^* denote the dual of V and let $\|\cdot\|_*$ denote the usual norm on V^*. It then follows (see, for example, [13]) that

$$V \hookrightarrow H = H^* \hookrightarrow V^*, \tag{2.1}$$

where H^* and V^* denote the continuous duals of H and V, respectively. All of the embeddings in (2.1) are dense and continuous. In particular we assume that

$$|\varphi| \leq K\|\varphi\|, \qquad \varphi \in V, \tag{2.2}$$

for some positive constant K. Let Q be a real Hilbert space with inner product $\langle \cdot, \cdot \rangle_Q$ and corresponding norm $|\cdot|_Q$. For each $q \in Q$, let $a(q, \cdot, \cdot) : V \times V \to R^1$ be a bilinear form on V satisfying the following assumptions

$(A1)$ (Boundedness) There exists a subset $\tilde{Q} \subset Q$ such that for each $q^* \in \tilde{Q}$ there exists $\alpha^0(q^*) > 0$ for which

$$|a(q^*; \varphi, \psi)| \leq \alpha^0(q^*)\|\varphi\|\|\psi\|, \qquad \varphi, \psi \in V.$$

$(A2)$ (V-H-Coercivity) For each $q^* \in \tilde{Q}$ there exists $\lambda_a(q^*) \in R^1$ and $\alpha_0(q^*) > 0$ for which

$$a(q^*; \varphi, \varphi) + \lambda_a(q^*)|\varphi|^2 \geq \alpha_0(q^*)\|\varphi\|^2, \qquad \varphi \in V.$$

$(A3)$ (q-Linearity) The mapping $q \to a(q; \cdot, \cdot)$ is linear from Q into the space of bilinear forms on $V \times V$.

For each $q \in Q$, let $m(q;, \cdot, \cdot)$ be a bilinear form on $H \times V$ satisfying the following assumptions.

$(M0)$ (Symmetry) For each $q^* \in \tilde{Q}$, $m(q^*; \cdot, \cdot)$ is a symmetric bilinear form on $H \times H$.

$(M1)$ (Boundedness) For each $q^* \in \tilde{Q}$ there exists $m^0(q^*) > 0$ such that

$$|m(q^*; \varphi, \psi)| \leq m^0(q^*)|\varphi||\psi|, \quad \varphi, \psi \in H.$$

$(M2)$ (H-Coercivity) For each $q^* \in \tilde{Q}$ there exists $m_0(q^*) > 0$ such that

$$m(q^*; \varphi, \varphi) \geq m_0(q^*)|\varphi|^2, \quad \varphi \in H.$$

$(M3)$ (q-Linearity) The mapping $q \to m(q; \cdot, \cdot)$ is linear from Q into the space of bilinear forms on $H \times V$.

For $q \in Q$ let $A(q)$ be the linear operator from V into V', the algebraic dual of V, determined by the bilinear form $a(q; \cdot, \cdot)$ via

$$\langle A(q)\varphi, \psi \rangle_{V',V} = a(q; \varphi, \psi), \qquad \varphi, \psi \in V,$$

and let $M(q)$ be the operator from H into V' induced by the form $m(q; \cdot, \cdot)$. That is,

$$\langle M(q)\varphi, \psi \rangle = m(q; \varphi, \psi), \qquad \varphi \in H \text{ and } \psi \in V.$$

For $q \in Q$, $u_0 \in H$, and $f \in L_2^{loc}(0, \infty; V^*)$, we consider the first order linear implicit and possibly degenerate initial value problem given in weak form by

$$m(q; D_t u(t), \varphi) + a(q; u(t), \varphi) = \langle f(t), \varphi \rangle, \qquad \varphi \in V, t > 0, \tag{2.3}$$

$$(M(q)u)(0) = M(q)u_0. \tag{2.4}$$

The well posedness of the initial value problem (2.3), (2.4) was treated in [3] and [13]. If $q \in Q$ is such that $m(q; \cdot, \cdot)$ is a symmetric, nonnegative bilinear form on $H \times H$ and there exist constants, $\alpha, \beta \in R$, with $\alpha > 0$ such that

$$a(q; \varphi, \varphi) + \beta m(q; \varphi, \varphi) \geq \alpha \|\varphi\|^2, \qquad \varphi \in V,$$

then for each $T > 0$, there exists a weak solution to the initial value problem (2.3), (2.4). By a weak solution we mean a function $u \in L_2(0, T; V)$ such that

$$\int_0^T a(q; u(t), \varphi(t)) dt - \int_0^T m(q; u(t), D_t\varphi(t)) dt$$
$$= \int_0^T \langle f(t), \varphi(t) \rangle dt + m(q; u_0, \varphi(0)),$$

for all $\varphi \in L_2(0, T; V) \cap H^1(0, T; H)$ with $\varphi(T) = 0$. If in addition, the form $a(q; \cdot, \cdot)$, is symmetric, then this solution is unique. If for some $q \in Q$, $m(q; \cdot, \cdot)$ is a symmetric, nonnegative bilinear form on $H \times H$, there exists $\alpha > 0$ such that $a(q; \varphi, \varphi) \geq \alpha \|\varphi\|^2$, $\varphi \in V$, and f is such that $f = M(q)g$ for some $g \in C(0, \infty; H)$ which is Hölder continuous, then the initial value problem (2.3), (2.4) admits a unique strong solution. Let $H^*_{m(q)}$ denote the dual of the semi-normed space $H_{m(q)}$ consisting of the space H endowed with the semi-norm induced by the nonnegative symmetric form $m(q; \cdot, \cdot)$. By a strong solution to the initial value problem (2.3), (2.4) we mean a function $u : [0, \infty) \to V$ with $M(q)u \in C([0, \infty); H^*_{m(q)}) \cap C^1((0, \infty); H^*_{m(q)})$ satisfying (2.3), (2.4).

DEFINITION 2.1 *A plant is a pair $(\overline{q}, \overline{u})$ for which $\overline{q} \in Q$, $\overline{u}$ is a strong solution to (2.3), (2.4) with $q = \overline{q}$ and there exists a constant $\lambda = \lambda(\overline{u}) > 0$ such that*

$$|\langle M(q) D_t\overline{u}(t), \varphi \rangle + \langle A(q)\overline{u}(t), \varphi \rangle| \leq \lambda |q|_Q \|\varphi\|,$$

for all $t > 0, q \in Q$ and $\varphi \in V$.

Let $(\overline{q}, \overline{u})$ be a plant and assume that $\overline{u}$ and $D_t\overline{u}$ are available and that $\overline{q}$ is unknown. The objective is to use both $\overline{u}$ and $D_t\overline{u}$ to asymptotically identify $\overline{q}$. As a direct extension of the finite dimensional results in [10], and similar results for parabolic and hyperbolic systems in [6] and [4], we define our estimator for $\overline{q}$ and $\overline{u}$ in the form of the initial value problem

$$m(q^*; D_t u(t), \varphi) + a(q^*; u(t), \varphi) + \lambda_a(q^*)\langle u(t), \varphi \rangle \tag{2.5}$$
$$+m(q(t); D_t\overline{u}(t), \varphi) + a(q(t); \overline{u}(t), \varphi) = \langle f(t), \varphi \rangle$$
$$+a(q^*; \overline{u}(t), \varphi) + \lambda_a(q^*)\langle \overline{u}(t), \varphi \rangle + m(q^*; D_t\overline{u}(t), \varphi), \qquad \varphi \in V, t > 0,$$

$$\langle D_t q(t), p \rangle_Q - m(p; D_t\overline{u}(t), u(t)) - a(p; \overline{u}(t), u(t)) \tag{2.6}$$
$$= -m(p; D_t\overline{u}(t), \overline{u}(t)) - a(p; \overline{u}(t), \overline{u}(t)), \qquad p \in Q, t > 0,$$

$$u(0) \in H, \qquad q(0) \in Q. \tag{2.7}$$

The parameter q^*, known as the tuning parameter, is chosen so that $q^* \in \tilde{Q}$. That is, it is chosen so that Assumptions $(A1)$, $(A2)$, $(M0)$, $(M1)$, and $(M2)$ are satisfied. Note that $q^* \in \tilde{Q}$ (in particular, that Assumptions $(M0)$, $(M1)$ and $(M2)$ are satisfied) implies that the estimator, (2.5) - (2.7), is non-degenerate. Note also that Assumptions $(A3)$ and $(M3)$ imply that the initial value problem (2.5) - (2.7) is linear. Since $(\overline{q}, \overline{u})$ is assumed to be a plant, and in particular that $\overline{u}$ is a strong solution to the initial value problem (2.3), (2.4) with $q = \overline{q}$, the well-posedness of the initial value problem (2.5) - (2.7) can be argued as follows. Let X be the Hilbert space $H \times Q$ endowed with the inner product

$$\langle (\varphi, q), (\psi, p) \rangle_X = m(q^*; \varphi, \psi) + \langle q, p \rangle_Q, \qquad (\varphi, q), (\psi, p) \in X.$$

That $\langle \cdot, \cdot \rangle_X$ as defined above is in fact an inner product on X is implied by Assumptions $(M0)$ - $(M2)$. Let Y be the reflexive Banach space $V \times Q$ endowed with the norm

$$\|(\varphi, q)\|_Y = \left\{ \|\varphi\|^2 + |q|_Q^2 \right\}^{\frac{1}{2}}, \qquad (\varphi, q) \in Y.$$

Then $Y \hookrightarrow X \hookrightarrow Y^*$ with the embeddings dense and continuous. For $t > 0$, define the operator $\mathcal{A}(t) : Y \to Y^*$ by

$$\begin{aligned}
\langle \mathcal{A}(t)(\varphi, q), (\psi, p) \rangle_{Y^*, Y} &= \langle \{A(q^*) + \lambda_a(q^*)\}\varphi + M(q)D_t\overline{u}(t) + A(q)\overline{u}(t), \psi \rangle \\
&\quad - \langle M(p)D_t\overline{u}(t) + A(p)\overline{u}(t), \varphi \rangle, \quad (\varphi, q), (\psi, p) \in Y.
\end{aligned}$$

It follows from Assumption $(A1)$ and Definition 2.1 that for $(\varphi, q), (\psi, p) \in Y$ we have

$$\begin{aligned}
&|\langle \mathcal{A}(t)(\varphi, q), (\psi, p) \rangle_{Y^*, Y}| \\
&= |\langle \{A(q^*) + \lambda_a(q^*)\}\varphi, \psi \rangle + \langle M(q)D_t\overline{u}(t) + A(q)\overline{u}(t), \psi \rangle - \langle M(p)D_t\overline{u}(t) + A(p)\overline{u}(t), \varphi \rangle| \\
&\leq \{\alpha^0(q^*) + |\lambda_a(q^*)|K^2\}\|\varphi\|\|\psi\| + \lambda|q|_Q\|\psi\| + \lambda|p|_Q\|\varphi\| \\
&\leq c\|(\varphi, q)\|_Y\|(\psi, p)\|_Y,
\end{aligned}$$

for some $c > 0$, and therefore that $\mathcal{A}(t) \in \mathcal{L}(Y, Y^*)$. Moreover, from Assumption $(A2)$ it follows that

$$\begin{aligned}
\langle \mathcal{A}(t)(\varphi, q), (\varphi, q) \rangle_{Y^*, Y} &= \langle \{A(q^*) + \lambda_a(q^*)\}\varphi, \varphi \rangle \\
&\geq \alpha_0(q^*)\|\varphi\|^2 \geq \alpha_0(q^*)\|(\varphi, q)\|_Y^2 - \alpha_0(q^*)|(\varphi, q)|_X^2,
\end{aligned}$$

for $(\varphi, q) \in Y$.

For almost every $t > 0$, define $F(t) \in Y^*$ by

$$\begin{aligned}
\langle F(t), (\varphi, q) \rangle_{Y^*, Y} &= \langle f(t) + \{A(q^*) + \lambda_a(q^*)\}\overline{u}(t) + M(q^*)D_t\overline{u}(t), \varphi \rangle \\
&\quad - \langle M(q)D_t\overline{u}(t) + A(q)\overline{u}(t), \overline{u}(t) \rangle, \qquad (\varphi, q) \in Y.
\end{aligned}$$

The fact that $(\overline{q}, \overline{u})$ is a plant implies that $F \in L_2^{loc}(0, \infty; Y^*)$. If we rewrite (2.5) - (2.7) as

$$D_t x(t) + \mathcal{A}(t)x(t) = F(t), \qquad a.e.\ t > 0,$$

$$x(0) = (u(0), q(0)),$$

with $x = (u, q)$, then standard results from the theory of linear evolution systems (see, for example, [9] and [14]) imply the existence of a unique solution, (u, q), to the initial value problem (2.5) - (2.7) with $u \in L_2(0, T; V) \cap C([0, T]; H) \cap H^1(0, T; V^*)$ and $q \in H^1(0, T; Q)$, for all $T > 0$.

Setting $e(t) = u(t) - \overline{u}(t)$ and $r(t) = q(t) - \overline{q}$, where $(\overline{q}, \overline{u})$ is a plant and (q, u) is a solution to the initial value problem (2.5) - (2.7), it is desired that $\lim_{t \to \infty} |e(t)| = 0$, and $\lim_{t \to \infty} |r(t)|_Q = 0$. The functions e and r are solutions to the *error equations* given by

$$m(q^*; D_t e(t), \varphi) + a(q^*; e(t), \varphi) + \lambda_a \langle e(t), \varphi \rangle$$
$$+ m(r(t); D_t \overline{u}(t), \varphi) + a(r(t); \overline{u}(t), \varphi) = 0, \quad \varphi \in V, \, t > 0, \tag{2.8}$$

$$\langle D_t r(t), p \rangle_Q - m(p; D_t \overline{u}(t), e(t)) - a(p; \overline{u}(t), e(t)) = 0, \quad p \in Q, \, t > 0, \tag{2.9}$$

$$e(0) \in H, \qquad r(0) \in Q. \tag{2.10}$$

In the next section we show that under no further assumptions, the state error, e, goes to zero asymptotically with time, and that under the additional assumption of *persistence of excitation*, parameter convergence is achieved.

3. CONVERGENCE

In this section we establish convergence of the state estimate (i.e $\lim_{t \to \infty} |e(t)| = 0$) and, with the additional assumption of *persistence of excitation*, parameter convergence. That is, $\lim_{t \to \infty} |r(t)|_Q = \lim_{t \to \infty} |q(t) - \overline{q}|_Q = 0$. We assume throughout this section that $(\overline{q}, \overline{u})$ is a plant (see Definition 2.1).

We begin by establishing a Lyapunov-like estimate for the system (2.8) - (2.10).

LEMMA 3.1 *There exist constants ρ, $\sigma > 0$ such that for all $t > 0$*

$$|e(t)|^2 + |r(t)|_Q^2 + \rho \int_0^t \|e(s)\|^2 ds \leq \xi,$$

where $\xi = \sigma \left\{ |e(0)|^2 + |r(0)|_Q^2 \right\}$.

Proof. We define an energy functional, $E : [0, \infty) \to R^1$, for the system (2.8), (2.9) by

$$E(t) = m(q^*; e(t), e(t)) + |r(t)|_Q^2. \tag{3.1}$$

Then, using (2.8), (2.9), and Assumption $(A2)$, we obtain

$$\begin{aligned}
D_t E(t) &= 2m(q^*; D_t e(t), e(t)) + 2\langle D_t r(t), r(t) \rangle_Q \tag{3.2} \\
&= -2a(q^*; e(t), e(t)) - 2\lambda_a(q^*)|e(t)|^2 \\
&\leq -2\alpha_0(q^*)\|e(t)\|^2.
\end{aligned}$$

Consequently,

$$E(t) + 2\alpha_0(q^*) \int_0^t \|e(s)\|^2 ds \le E(0). \tag{3.3}$$

From Assumptions $(M1)$ and $(M2)$ we find that

$$\sigma_0 \left\{ |e(t)|^2 + |r(t)|^2 \right\} \le E(t) \le \sigma_1 \left\{ |e(t)|^2 + |r(t)|^2 \right\}, \tag{3.4}$$

for some positive constants σ_0 and σ_1. The desired result then follows from (3.3) and (3.4).
$\square$

The next theorem establishes the convergence of the state estimate. The proof is in the spirit of the arguments used to verify an analogous result for elliptic systems in [1] and for non-implicit parabolic systems in [4] and [12].

THEOREM 3.1 *The energy functional E is non-increasing and $\lim_{t\to\infty} |e(t)| = 0$.*

Proof. That E is nonincreasing follows immediately from the estimate (3.3). For $t > 0$, let

$$\theta(t) = m(q^*; e(t), e(t)). \tag{3.5}$$

Assumptions $(M1)$ and $(M2)$ imply that

$$m_0(q^*)|e(t)|^2 \le \theta(t) \le m^0(q^*)|e(t)|^2. \tag{3.6}$$

If we can show that $\lim_{t\to\infty} \theta(t) = 0$, then the theorem will be proved. Now, for $t_2 > t_1$, (2.2) and (2.8) yield

$$|\theta(t_2) - \theta(t_1)| = |\int_{t_1}^{t_2} D_s\theta(s)ds|$$

$$= 2 \left| \int_{t_1}^{t_2} -a(q^*; e(s), e(s)) - \lambda_a(q^*)|e(s)|^2 - m(r(s); D_t\overline{u}(s), e(s)) - a(r(s); \overline{u}(s), e(s))ds \right|$$

$$\le 2 \left\{ a^0(q^*) + |\lambda_a(q^*)|K^2) \right\} \int_{t_1}^{t_2} \|e(s)\|^2 ds + 2\lambda \int_{t_1}^{t_2} \|e(s)\| |r(s)|_Q ds, \tag{3.7}$$

since the assumption that $(\overline{q}, \overline{u})$ is a plant implies that

$$|m(r(t); D_t\overline{u}(t), e(t)) + a(r(t); \overline{u}(t), e(t))| \le \lambda |r(t)|_Q \|e(t)\|.$$

Using the above estimate (3.7), Lemma 3.1, and the Cauchy-Schwarz inequality, we find that

$$|\theta(t_2) - \theta(t_1)| \le 2 \left\{ a^0(q^*) + |\lambda_a(q^*)|K^2 \right\} \int_{t_1}^{t_2} \|e(s)\|^2 ds + 2\lambda(t_2 - t_1)^{\frac{1}{2}} \xi^{\frac{1}{2}} \left(\int_{t_1}^{t_2} \|e(s)\|^2 ds \right)^{1/2}.$$

Now, (3.3) implies that

$$\lim_{t\to\infty} \int_t^{t+L} \|e(s)\|^2 ds = 0, \tag{3.8}$$

for all $L > 0$. It follows from (3.8) that for every $\delta > 0$ and $\epsilon > 0$, there exists $t_0 > 0$ such that for all $t_2 > t_1 > t_0$ with $t_2 - t_1 \le \delta$ we have

$$|\theta(t_2) - \theta(t_1)| < \epsilon. \tag{3.9}$$

Now, suppose that $\lim_{t\to\infty}\theta(t)\neq 0$. Then, there exists $\mu > 0$ and $\{t_n\}_{n=1}^{\infty} \subset R^+$ with $\lim_{n\to\infty} t_n = \infty$ and $t_{n+1} - t_n \geq 2$ such that

$$\theta(t_n) \geq \mu. \tag{3.10}$$

Setting $\delta = 1$ and $\epsilon = \frac{\mu}{2}$ in (3.9), it follows that there exists $t_0 > 0$ such that if $t_2 > t_1 > t_0$ with $t_2 - t_1 < 1$, then

$$|\theta(t_2) - \theta(t_1)| < \frac{\mu}{2}. \tag{3.11}$$

Let n_0 be such that $t_{n_0} > t_0 + 1$. Then, from (3.10), and (3.11) we have that for all $n \geq n_0$ and $t \in [t_n - 1, t_n + 1]$

$$\theta(t) = \theta(t) - \theta(t_n) + \theta(t_n) > -\frac{\mu}{2} + \mu = \frac{\mu}{2}.$$

Therefore it follows that

$$\int_0^{\infty} \theta(s)ds \;\geq\; \sum_{n=1}^{\infty}\int_{t_n}^{t_n+1}\theta(s)ds \geq \sum_{n=n_0}^{\infty}\int_{t_n-1}^{t_n+1}\theta(s)ds \geq \sum_{n=n_0}^{\infty}\int_{t_n-1}^{t_n+1}\frac{\mu}{2}ds$$

$$=\; \frac{\mu}{2}\sum_{n=n_0}^{\infty}\{(t_n+1) - (t_n-1)\} = \infty,$$

which, together with (3.6), contradicts (3.3). Hence

$$\lim_{t\to\infty}|e(t)|^2 = 0,$$

and the proof is complete. $\square$

In order to establish parameter convergence we require the notion of *Persistence of Excitation*. We extend the definition of persistence of excitation to implicit systems and argue parameter convergence as was done in [1] for elliptic systems and in [8] for parabolic systems.

DEFINITION 3.1 A plant $(\overline{q},\overline{u})$ is said to be *persistently excited*, or an input f is called *persistently exciting* for the plant $(\overline{q},\overline{u})$, if there exists $T_0, \delta_0, \epsilon_0 > 0$ such that for each $p \in Q$ with $|p|_Q = 1$ and each $t > 0$ sufficiently large, there exists a $\tilde{t} \in [t, t+T_0]$ such that

$$\left\|\int_{\tilde{t}}^{\tilde{t}+\delta_0} M(p)D_\tau\overline{u}(\tau) + A(p)\overline{u}(\tau)d\tau\right\|_* \geq \epsilon_0.$$

THEOREM 3.2 *If the plant $(\overline{q},\overline{u})$ is persistently excited then*

$$\lim_{t\to\infty}|r(t)|_Q = 0.$$

Proof. Since E defined by (3.1) is bounded below by zero, and since by Theorem 3.1 it is nonincreasing, we must have that $\lim_{t\to\infty} E(t)$ exists. Moreover, Theorem 3.1 together with Assumption $(M1)$ implies that $\lim_{t\to\infty} m(q^*; e(t), e(t)) = 0$. Consequently, we must have that $\lim_{t\to\infty}|r(t)|_Q$ exists. Suppose that

$$\lim_{t\to\infty}|r(t)|_Q \neq 0. \tag{3.12}$$

It follows from (3.12) that there exists an increasing sequence of positive real numbers, $\{t_k\}_{k=1}^{\infty}$, with $t_{k+1} - t_k > T_0$ (T_0 as it was defined in Definition 3.1) such that

$$|r(t_k)| \geq \delta, \tag{3.13}$$

for some $\delta > 0$. Integrating (2.8) over the interval $[\tau_1, \tau_2]$ where $0 \leq \tau_1 \leq \tau_2$, we obtain

$$\int_{\tau_1}^{\tau_2} m(r(t); D_t\overline{u}(t), \varphi) + a(r(t); \overline{u}(t), \varphi)dt$$
$$= m(q^*; e(\tau_1), \varphi) - m(q^*; e(\tau_2), \varphi) - \int_{\tau_1}^{\tau_2} a(q^*; e(t), \varphi) + \lambda_a(q^*)\langle e(t), \varphi \rangle dt, \quad \varphi \in V.$$

Taking norms in V^* and applying Assumptions $(M1)$ and $(A1)$ together with the Cauchy-Schwarz inequality, we find that

$$\left\| \int_{\tau_1}^{\tau_2} M(r(t)) + A(r(t))dt \right\|_* \tag{3.14}$$
$$\leq \|M(q^*)e(\tau_1)\|_* + M(q^*)e(\tau_2)\|_* + \int_{\tau_1}^{\tau_2} \|(A(q^*) + \lambda_a(q^*))e(t)\|_* \, dt$$
$$\leq m^0(q^*)K|e(\tau_1)| + m^0(q^*)K|e(\tau_2)| + \int_{\tau_1}^{\tau_2} (\alpha^0(q^*) + |\lambda_a(q^*)|K^2)\|e(t)\|dt$$
$$\leq m^0(q^*)K|e(\tau_1)| + m^0(q^*)K|e(\tau_2)| + \mu(\tau_2 - \tau_1)^{\frac{1}{2}} \left\{ \int_{\tau_1}^{\tau_2} \|e(t)\|^2 dt \right\}^{\frac{1}{2}},$$

where $\mu = \alpha^0(q^*) + |\lambda_a(q^*)|K^2$. Integrating (2.9) over the interval $[\tau_1, \tau_2]$, taking norms in Q^*, and applying Definition 2.1, we obtain

$$|r(\tau_2) - r(\tau_1)|_Q = |r(\tau_2) - r(\tau_1)|_{Q^*} \tag{3.15}$$
$$= \left\| \int_{\tau_1}^{\tau_2} \langle M(\cdot)D_t\overline{u}(t) + A(\cdot)\overline{u}(t), e(t) \rangle dt \right\|_{Q^*}$$
$$\leq \int_{\tau_1}^{\tau_2} \lambda\|e(t)\| \leq \lambda(\tau_2 - \tau_1)^{\frac{1}{2}} \left\{ \int_{\tau_1}^{\tau_2} \|e(t)\|^2 dt \right\}^{\frac{1}{2}}.$$

Since $(\overline{q}, \overline{u})$ is assumed to be persistently excited, for all k sufficietly large, there exists $t_k^* \in [t_k, t_k + T_0]$ such that

$$\left\| \int_{t_k^*}^{t_k^*+\delta_0} M(p_k)D_t\overline{u}(t) + A(p_k)\overline{u}(t)dt \right\|_* \geq \epsilon_0, \tag{3.16}$$

where $p_k = r(t_k)/|r(t_k)|_Q$. Integrating the V^*-valued function $t \to M(r(t_k))D_t\overline{u}(t) + A(r(t_k))\overline{u}(t)$ from t_k^* to $t_k^* + \delta_0$, taking norms in V^*, applying the triangle inequality, the fact that $(\overline{q}, \overline{u})$ is a plant, and the estimates (3.14) and (3.15) above, we obtain

$$\left\| \int_{t_k^*}^{t_k^*+\delta_0} M(r(t_k))D_t\overline{u}(t) + A(r(t_k))\overline{u}(t)dt \right\|_* \tag{3.17}$$
$$\leq \left\| \int_{t_k^*}^{t_k^*+\delta_0} M(r(t))D_t\overline{u}(t) + A(r(t))\overline{u}(t)dt \right\|_*$$

$$
+ \left\| \int_{t_k^*}^{t_k^*+\delta_0} M(r(t_k) - r(t)) D_t \overline{u}(t) + A(r(t_k) - r(t)) \overline{u}(t) dt \right\|_*
$$

$$
\leq m^0(q^*) K |e(t_k^*)| + m^0(q^*) K |e(t_k^* + \delta_0)|
$$

$$
+ \mu \sqrt{\delta_0} \left\{ \int_{t_k^*}^{t_k^*+\delta_0} \|e(t)\|^2 dt \right\}^{\frac{1}{2}} + \lambda \int_{t_k^*}^{t_k^*+\delta_0} |r(t_k) - r(t)|_Q dt
$$

$$
\leq m^0(q^*) K |e(t_k^*)| + m^0(q^*) K |e(t_k^* + \delta_0)|
$$

$$
+ \mu \sqrt{\delta_0} \left\{ \int_{t_k^*}^{t_k^*+\delta_0} \|e(t)\|^2 dt \right\}^{\frac{1}{2}} + \lambda^2 \delta_0 \sqrt{T_0 + \delta_0} \left\{ \int_{t_k}^{t_k+T_0+\delta_0} \|e(t)\|^2 dt \right\}^{\frac{1}{2}}
$$

$$
\leq m^0(q^*) K \left\{ |e(t_k^*)| + |e(t_k^* + \delta_0)| \right\}
$$

$$
+ \left\{ \mu \sqrt{\delta_0} + \lambda^2 \delta_0 \sqrt{T_0 + \delta_0} \right\} \int_{t_k}^{t_k+T_0+\delta_0} \|e(t)\|^2 dt.
$$

Combining (3.13), (3.16), and (3.17) it follows that

$$
\begin{aligned}
0 < \epsilon_0 \delta \;=\; & |r(t_k)|_Q \left\| \int_{t_k^*}^{t_k^*+\delta_0} M(p_k) D_t \overline{u}(t) + A(p_k) \overline{u}(t) dt \right\|_* \\
=\; & \left\| \int_{t_k^*}^{t_k^*+\delta_0} M(r(t_k)) D_t \overline{u}(t) + A(r(t_k)) \overline{u}(t) dt \right\|_* \\
\leq\; & m^0(q^*) K \left\{ |e(t_k^*)| + |e(t_k^* + \delta_0)| \right\} \\
& + \left\{ \mu \sqrt{\delta_0} + \lambda^2 \delta_0 \sqrt{T_0 + \delta_0} \right\} \int_{t_k}^{t_k+T_0+\delta_0} \|e(t)\|^2 dt.
\end{aligned}
$$

Taking limits as $k \to \infty$ and recalling Theorem 3.1 and (3.5) we obtain a contradiction, and the theorem is proved.$\square$

Finally, it is also worth noting that persistence of excitation yields an identifiability result.

THEOREM 3.3 *If the plant* $(\overline{q}, \overline{u})$ *is persistently excited then the parameter* $\overline{q}$ *is identifiable.*

Proof. Assume the contrary. That is, there exists $\overline{q}_1$ and $\overline{q}_2$ such that the initial value problem (2.3), (2.4) is satisfied with $u = \overline{u}$ and $\overline{q} = \overline{q}_1$ or $\overline{q} = \overline{q}_2$. A simple subtraction yields

$$
m(\overline{q}_1 - \overline{q}_2; D_t \overline{u}(t), \varphi) + a(\overline{q}_1 - \overline{q}_2; \overline{u}(t), \varphi) = 0, \qquad \varphi \in V, \; a.e. \; t > 0. \tag{3.18}
$$

Now Definition 3.1 implies that there exist a $\tilde{t}$ such that

$$
0 \leq \epsilon_0 |\overline{q}_1 - \overline{q}_2|_Q \leq \left\| \int_{\tilde{t}}^{\tilde{t}+\delta_0} M(\overline{q}_1 - \overline{q}_2) D_t \overline{u}(\tau) + A(\overline{q}_1 - \overline{q}_2) \overline{u}(\tau) d\tau \right\|_*.
$$

But from (3.18) it then must follow that

$$
|\overline{q}_1 - \overline{q}_2|_Q = 0,
$$

and the result is obtained. $\square$

4. AN EXAMPLE

We consider the identification of the one dimensional degenerate heat equation given by

$$\chi_{[\alpha,\beta]}(x)\frac{\partial u}{\partial t}(t,x) = \frac{\partial}{\partial x}\gamma(x)\frac{\partial u}{\partial x}(t,x) + f(t,x), \quad t > 0, \quad 0 < x < 1,$$

where $\gamma \in L_\infty(0,1)$, $\gamma(x) > 0$, a.e. $x \in (0,1)$, and $f \in L_2((0,T) \times (0,1))$ for all $T > 0$. We require that u satisfies the Dirichlet boundary conditions

$$u(t,0) = 0 = u(t,1), \qquad t > 0,$$

and the initial conditions

$$\chi_{[\alpha,\beta]}(x)u(0,x) = \chi_{[\alpha,\beta]}(x)u_0(x), \qquad 0 < x < 1,$$

where $u_0 \in L_2(0,1)$. In this case we have $H = L_2(0,1)$ and $V = H_0^1(0,1)$ endowed with the usual inner products and corresponding norms. The objective here is to identify adaptively the constant scalar parameters α and β with $0 \le \alpha \le \beta \le 1$, and the functional parameter γ, having the state u and its time derivative $D_t u$ available for measurement. We note that it is actually the function $\chi_{[\alpha,\beta]}$ rather than the constant parameters α and β that must be identified. Otherwise the assumption that the form $m(\cdot;\cdot,\cdot)$ be trilinear on $Q \times H \times V$ would be violated. It is also interesting to note that while the system being identified may be degenerate, the estimator is not. And finally we also point out that unlike the non-degenerate case where only $\overline{u}$ is required, the degenerate case requires knowlwdge of both $\overline{u}$ and $D_t\overline{u}$.

Let Q be the Hilbert space $H^1(0,1) \times L_2(0,1)$ endowed with the inner product

$$\langle q, p \rangle_Q = \langle (q_1, q_2), (p_1, p_2) \rangle_Q \tag{4.1}$$
$$= \omega_1 \int_0^1 q_1(x)p_1(x)dx + \omega_2 \int_0^1 Dq_1(x)Dp_1(x)dx + \omega_3 \int_0^1 q_2(x)p_2(x)dx,$$

for $q, p \in Q$, where $\omega_j > 0$, $j = 1,2,3$. For each $q = (q_1, q_2) \in Q$, the forms $a(q;\cdot,\cdot) : V \times V \to R$ and $m(q;\cdot,\cdot) : H \times V \to R$ are given by

$$a(q,\varphi,\psi) = \int_0^1 q_1(x)D\varphi(x)D\psi(x)dx, \qquad \varphi, \psi \in H_0^1(0,1), \tag{4.2}$$

and

$$m(q;\varphi,\psi) = \int_0^1 q_2(x)\varphi(x)\psi(x)dx, \qquad \varphi \in L_2(0,1), \ \psi \in H_0^1(0,1), \tag{4.3}$$

respectively. It is not difficult to verify that Assumptions $(A1)$ - $(A2)$ and $(M0)$ - $(M2)$ are satisfied so long as the gain $q^* = (q_1^*, q_2^*) \in Q$ is chosen so that $q_1^*(x)$, $x \in [0,1]$, and $q_2^* \in L_\infty(0,1)$ with $q_2^*(x) > 0$ a.e. $x \in [0,1]$. In general one would most likely choose $q_1^*(x) = q_1^* > 0$ and $q_2^*(x) = q_2^* > 0$.

The condition that $(\overline{q}, \overline{u})$ is a plant requires that $\overline{u}$ is a strong solution to the initial-boundary value problem

$$\overline{q}_2(x)\frac{\partial \overline{u}}{\partial t}(t,x) = \frac{\partial}{\partial x}\overline{q}_1(x)\frac{\partial \overline{u}}{\partial x}(t,x) + f(t,x), \ t > 0, \ 0 < x < 1,$$

$$\overline{u}(t,0) = 0 = \overline{u}(t,1), \ t > 0,$$

$$\overline{q}_2(x)\overline{u}(0,x) = \overline{q}_2(x)u_0(x), \ 0 < x < 1,$$

where $\overline{q} = (\overline{q}_1, \overline{q}_2) \in H^1(0,1) \times L_2(0,1)$, and that there exists a $\lambda > 0$ such that

$$\left| \int_0^1 q_2(x)\frac{\partial \overline{u}}{\partial t}(t,x)\varphi(x) + q_1(x)\frac{\partial \overline{u}}{\partial x}(t,x)D\varphi(x)dx \right|$$

$$\leq \lambda \left\{ \omega_1 \int_0^1 q_1(x)^2 dx + \omega_2 \int_0^1 (Dq_1(x))^2 dx + \omega_3 \int_0^1 q_2(x)^2 dx \right\}^{\frac{1}{2}} \left\{ \int_0^1 (D\varphi(x))^2 dx \right\}^{\frac{1}{2}},$$

for all $t > 0$, $q = (q_1, q_2) \in H^1(0,1) \times L_2(0,1)$ and $\varphi \in H_0^1(0,1)$. The condition that $(\overline{q}, \overline{u})$ be persistently excited requires that there exist positive constants T_0, δ_0, and ϵ_0, such that for each $q = (q_1, q_2)) \in H^1(0,1) \times L_2(0,1)$ and each $t > 0$ sufficiently large, there exists $\tilde{t} \in [t, t + T_0]$ for which

$$\sup_{\|\varphi\| \leq 1} \left| \int_{\tilde{t}}^{\tilde{t}+\delta_0} \int_0^1 q_1(x)\frac{\partial \overline{u}}{\partial t}(t,x)\varphi(x) + q_2(x)\frac{\partial \overline{u}}{\partial x}(t,x)D\varphi(x)dxdt \right| \geq \epsilon_0 |(q_1, q_2)|_Q,$$

where $\| \cdot \|$ denotes the standard norm in $H_0^1(0,1)$ and $| \cdot |_Q$ is the norm on Q induced by the inner product defined in (4.1).

We note that in general, it is difficult, if not impossible, to ensure that a plant is persistently excited. However, based upon a somewhat longer and far more technical proof of Theorem 3.2, one that is given in [7], it is possible to gain insight into how the values of the parameters δ_0, T_0, and ϵ_0 along with the value of the tuning parameter, q^*, affect the rate of parameter convergence. (Note that for δ_0 fixed, the value of ϵ_0 can be increased by increasing the gain on $\overline{u}$.) In addition, in [5] it was observed (in the case of non-implicit parabolic systems) that choosing the values of these parameters to effect optimal performance is related to tuning the relative levels of damping and stiffness in a damped linear harmonic oscillator. Indeed, the value of q^* determines the level of damping, while the values of δ_0 and ϵ_0 are related to the degree of stiffness. If q^* is too large, the estimator is overdamped and convergence is sluggish. If δ_0 is too small, or ϵ_0 is too large, the estimator is underdamped (or too stiff) and oscillations result.

Finally, we note that the estimator given by (2.5) - (2.7) is infinite dimensional. Thus its implementation requires some form of finite dimensional approximation. A complete approximation theory, corresponding convergence results, and the results of some of our numerical studies are presented in [7].

References

[1] H. T. BANKS AND K. KUNISCH, *Estimation Techniques for Distributed Parameter Systems*, Birkhauser, Boston, 1989.

[2] J. BAUMEISTER AND W. SCONDO, *Adaptive methods for parameter identification*, Methoden und Verfahren der Mathematischen Physik, (1987), pp. 87–116.

[3] R. W. CARROLL AND R. E. SHOWALTER, *Singular and Degenerate Cauchy Problems*, Academic Press, New York, 1976.

[4] M. A. DEMETRIOU AND I. G. ROSEN, *On-line estimation for infinite dimensional dynamical systems*, CAMS Report 93-2, Center for Applied Mathematical Sciences, Department of Mathematics, University of Southern California, Los Angeles, CA 90089-1113, February 1993.

[5] ———, *On the persistence of excitation in the adaptive estimation of distributed parameter systems*, CAMS Report 93-4, Center for Applied Mathematical Sciences, Department of Mathematics, University of Southern California, Los Angeles, CA 90089-1113, March 1993 also *IEEE Trans. Automatic Control*, to appear.

[6] ———, *Adaptive identification of second order distributed parameter systems*, CAMS Report 93-3, Center for Applied Mathematical Sciences, Department of Mathematics, University of Southern California, Los Angeles, CA 90089-1113, March, 1993, also *Inverse Problems*, to appear.

[7] ———, *Adaptive parameter estimation for degenerate parabolic systems*, CAMS Report 93-5, Center for Applied Mathematical Sciences, Department of Mathematics, University of Southern California, Los Angeles, CA 90089-1113, April 1993 also *Journal of Mathematical Analaysis and Applications*, to appear.

[8] M. A. DEMETRIOU, I. G. ROSEN, AND P. A. IOANNOU, *Adaptive parameter estimations for a class of distributed parameter systems*, in Proceedings of the 31st Conference on Decision and Control, Tuscon, Arizona, December 14-18 1992, pp. 1741–1742.

[9] J. L. LIONS, *Optimal Control of Systems Governed by Partial Differential Equations*, Springer-Verlag, New York, 1971.

[10] A. P. MORGAN AND K. S. NARENDRA, *On the stability of nonautonomous differential equations $\dot{x} = [A + B(t)]x$, with skew symmetric matrix $B(t)^*$*, SIAM J. Control and Optimization, 15 (1977), pp. 163–176.

[11] K. S. NARENDRA AND A. M. ANNASWAMY, *Stable Adaptive Systems*, Prentice Hall, Englewood Cliffs, NJ, 1989.

[12] W. SCONDO, *Ein Modellabgleichsverfahren zur adaptiven Parameteridentifikation in Evolutionsgleichungen*, PhD thesis, Johann Wolfgang Goethe-Universitat zu Frankfurt am Main, Frankfurt am Main, Germany, 1987.

[13] R. E. SHOWALTER, *Hilbert Space Methods for Partial Differential Equations*, Pitman, London, 1977.

[14] H. TANABE, *Equations of Evolution*, Pitman, London, 1979.

Elliptic–Parabolic Problems in the Space of Continuous Functions

Joachim Escher   Mathematisches Institut, University of Basel, Basel, Switzerland

1. INTRODUCTION

There are problems in the theory of diffusion, see [5, 16], which can be described as elliptic equations in an open domain subject to so-called dynamical boundary conditions of parabolic type. More precisely, suppose that Ω is a bounded domain in $\mathbb{R}^n$, $n \geq 2$, having a smooth boundary $\Gamma = \partial\Omega$. We are looking for a function $u : \overline{\Omega} \times [0, \infty) \to \mathbb{R}$ satisfying the following equations:

$$
\begin{aligned}
\mathcal{A}u &= 0 & in &\quad \Omega \times (0, \infty)\,, \\
\partial_t(\gamma u) + \mathcal{B}u &= g(\cdot, \gamma u) & on &\quad \Gamma \times (0, \infty)\,, \\
\gamma u(\cdot, 0) &= \gamma u_0 & on &\quad \Gamma\,.
\end{aligned}
\qquad (P)
$$

Here, $\mathcal{A}$ denotes a second order uniformly elliptic operator with smooth coefficients, i.e.,

$$
\mathcal{A}u := \sum_{j,k=1}^{n} -a_{jk}(x)\partial_j u \partial_k u + a_j(x)\partial_j u + a_0(x)u
$$

where

$$
a_{jk},\ a_j,\ a_0 \in C^2(\overline{\Omega}) \quad \text{for} \quad 1 \leq j,\, k \leq n
\qquad (1)
$$

---

supported by Schweizerischer Nationalfonds

with

$$\sum_{j,k=1}^{n} a_{jk}(x)\xi^j\xi^k > 0 \quad \text{for} \quad x \in \overline{\Omega}, \ \xi \in \mathbb{R}^n \setminus \{0\} \tag{2}$$

and

$$a_0(x) \geq 0 \quad \text{for} \quad x \in \overline{\Omega}. \tag{3}$$

The second equation in (P) describes a so-called **dynamical boundary condition** of parabolic type. Here, γ denotes the trace (or restriction) operator with respect to Γ. $\mathcal{B}$ stands for a boundary operator of Neumann type with respect to $\mathcal{A}$, i.e.,

$$\mathcal{B}u := \sum_{j,k=1}^{n} a_{jk}(x)\nu^j(x)\gamma\partial_k u + b_0(x)\gamma u, \tag{4}$$

where $\nu := (\nu^1, \ldots, \nu^n)$ denotes the outer normal unit vector and where $b_0 \in C^1(\Gamma)$. Finally, g is a given smooth nonlinear function and u_0 represents an initial value which is assumed to be contiuous on $\overline{\Omega}$.

Recall that $C^{k+\alpha}(\overline{\Omega})$ and $C^{k+\alpha}(\Gamma)$, $k \in \mathbb{N}$, $\alpha \in (0,1)$, denote the Banach spaces of all functions in $C^k(\overline{\Omega})$ and $C^k(\Gamma)$, respectively, having uniformly α–Hölder continuous derivatives (of order k) over the compact manifolds $\overline{\Omega}$ and Γ, respectively. The Fréchet space of all functions in $C^k(\Omega)$ having locally α–Hölder continuous functions over Ω is denoted by $C^{k+\alpha}(\Omega)$. Finally, $C^{\alpha,2-}(\Gamma \times \mathbb{R})$ stands for the Fréchet space of all functions f in $C(\Gamma \times \mathbb{R})$ such that $f(\cdot,\xi) \in C^\alpha(\Gamma)$ for $\xi \in \mathbb{R}$ and such that $f(x,\cdot) \in C^1(\mathbb{R})$, $x \in \Gamma$, has locally Lipschitz continuous derivatives.

Let us now state our main result.

THEOREM 1: *Suppose that $g \in C^{\alpha,2-}(\Gamma \times \mathbb{R})$ and that $u_0 \in C(\overline{\Omega})$. Then there exists a unique maximal classical solution*

$$u(\cdot,u_0) \in C([0,t^+),C^{2+\alpha}(\Omega)) \cap C^1((0,t^+),C^{1+\alpha}(\overline{\Omega})) \cap C([0,t^+),C(\overline{\Omega})) \tag{5}$$

of problem (P), where $t^+ := t^+(u_0)$ denotes the positive escape time of u_0.

It should be mentioned that we do not require any growth condition for the nonlinear function g appearing in the dynamical boundary condition. Also it is not necessary in our appraoch to assume any structural conditions for g like monotonicity, sign conditions, positivity, or aymptotic behaviour at infinity.

Observe that the solution $u(\cdot,u_0)$ of (P) possesses the same spatial regularity on the whole interval of existence $[0,t^+(u_0))$ as we assumed for the initial condition u_0. Thus we may study the semiflow properties of the mapping $\Phi := [(t,u_0) \mapsto u(t,u_0)]$. It turns out that in general Φ is not a semiflow on the whole of $C(\overline{\Omega})$. However, there exists a nontrivial closed subspace X of $C(\overline{\Omega})$ such that $\Phi|X$ is a semiflow on X. Before stating presicely this result, let us recall the concept of local semiflows. Suppose that Y is a metric space and that $t^+ : Y \to (0, \infty]$. We define

$$\mathcal{Y} := \bigcup_{x \in Y} [0, t^+(x)) \times \{x\}.$$

A mapping $\varphi : \mathcal{Y} \to Y$ is called a **local semiflow on Y** if

    (1) $\mathcal{Y}$ is open in $[0, \infty) \times Y$

    (2) $\varphi \in C(\mathcal{Y}, Y)$

    (3) $\varphi(0, \cdot) = 1_Y$

    (4) Given $x \in Y$, $s \in [0, t^+(x))$, and $t \in [0, t^+(\varphi(s, x)))$ it

        holds that $s + t \in [0, t^+(x))$ and $\varphi(t + s, x) = \varphi(t, \varphi(s, x)))$.

Our next result now reads as follows:

THEOREM 2: $X := \{u \in C(\overline{\Omega}) \cap C^{2+\alpha}(\Omega) \,;\, \mathcal{A}u = 0\}$ *is a closed subspace of* $C(\overline{\Omega})$ *and* $[(t, u_0) \mapsto u(t, u_0)]$ *defines a local semiflow on* X.

2. THE DIRICHLET-NEUMANN OPERATOR

Concern of this section is to associate to problem (P) an abstract evolution equation on suitable function spaces. The linear operator appearing in this evolution equation is called Dirichlet-Neumann operator. We show that this operator has a natural realization in the space of continuous functions over the boundary Γ and that it is the negative generator of an analytic semigroup on this space.

Let us begin with the elliptic problem induced by (P). It is well-known that our hypotheses (1)-(3) guarantee that the Dirichlet problem

$$\mathcal{A}u = 0 \quad \text{in} \quad \Omega, \qquad \gamma u = z \quad \text{on} \quad \Gamma \tag{6}$$

possesses for each $z \in C(\Gamma)$ a unique solution $u \in C(\overline{\Omega}) \cap C^{2+\alpha}(\Omega)$. We refer to [10, Theorem 6.13] for a proof of this classical result. Let $\mathcal{T}$ denote the corresponding solution operator, i.e., $\mathcal{T}z := u$, where u is the solution of (6). It follows from the weak maximum principle [10, Theorem 3.1] and the so-called Schauder interior estimates [10, Theorem 6.2] that

$$\mathcal{T} \in \mathcal{L}(C(\Gamma), C(\overline{\Omega})) \cap \mathcal{L}(C(\Gamma), C^{2+\alpha}(\Omega)). \tag{7}$$

On the other hand the L_p−estimates of Agmon-Douglis-Nirenberg, see [1, 2] imply that

$$\mathcal{T} \in \mathcal{L}(B_{pp}^{2-1/p}(\Gamma), W_p^2(\Omega)), \quad 1 < p < \infty. \tag{8}$$

Here, $B_{pp}^s(\Gamma)$, $s \in \mathbb{R}$, stand for the Besov spaces over Γ, see [3, 18], and $W_p^2(\Omega)$ denotes the usual Sobolev space over Ω.

It follows from the trace theorem [18, Theorem 4.4.2] and known pointwise multiplication results in Besov spaces [18, Theorem 4.2.2] that

$$\mathcal{B} \in \mathcal{L}(W_p^2(\Omega), B_{pp}^{1-1/p}(\Gamma)),$$

where $\mathcal{B}$ denotes the boundary operator from (4). Next, observe that due to Sobolev's embedding theorem we have that

$$B_{pp}^{1-1/p}(\Gamma) \hookrightarrow C(\Gamma) \quad \text{for} \quad p > n.$$

Thus we see that

$$\mathcal{B} \in \mathcal{L}(W_p^2(\Omega), C(\Gamma)) \quad \text{for} \quad p > n.$$

Given $p > n$, we now define the following unbounded linear operator in $C(\Gamma)$:

$$\Lambda_p : \ \operatorname{dom}(\Lambda_p) := B_{pp}^{2-1/p}(\Gamma) \subset C(\Gamma) \to C(\Gamma) \, , \quad \Lambda_p z := \mathcal{B}\mathcal{T}z.$$

In general, Λ_p is not a closed operator in $C(\Gamma)$. However, it can be shown that Λ_p is closable in $C(\Gamma)$, see (1.5), (1.6) in [9]. The closure of Λ_p in $C(\Gamma)$ is denoted by Λ and we call Λ the **generalized Dirichlet-Neumann operator** on $C(\Gamma)$. It follows from Lemma 6.4 in [9] that Λ is well-defined, i.e., the definition is independent of $p > n$. Further properties of the Dirichlet-Neumann operator which are important for our purposes are collected in the following result:

THEOREM 3: $-\Lambda$ *generates a strongly continuous, positive, compact, and analytic semigroup on* $C(\Gamma)$. *Moreover,* $C^{1+\alpha}(\Gamma) \hookrightarrow \operatorname{dom}(\Lambda) \hookrightarrow B_{pp}^{1-1/p}(\Gamma)$ *for* $\alpha \in (0,1)$ *and* $p > n$.

A proof of Theorem 3 can be found in [9]. It uses precise resolvent estimates for the Dirichlet-Neumann operator on the Besov spaces $B_{pp}^s(\mathbb{R}^{n-1})$, Sobolev's embedding theorem, and a careful localization procedure.

The Dirichlet-Neumann operator is well-studied in the Hilbert space setting, see [4, 11, 13, 14, 17], i.e., in the case $B_{22}^s(\Gamma) = H^s(\Gamma)$, where $H^s(\Gamma)$ denote the usual Sobolev spaces over Γ. Moreover, in these papers the authors consider only self-adjoint operators and mostly the situation where $\mathcal{A} = \Delta$ and $\mathcal{B} = \partial_\nu$. It should be mentioned that the generalized Dirichlet-Neumann operator, defined as above, is generally not self-adjoint. Nonsymmetric operators in general Besov spaces are studied in [7, 12].

Finally, there are some investigation of the Dirichlet-Neumann operator on the space of continuous functions due to Ph. Bénilan in [6]. It is proved, among other properties, that the resolvent set of the negative Dirichlet-Neumann operator (in the case $\mathcal{A} = \Delta$, $\mathcal{B} = \partial_\nu$) contains $\mathbb{R}_+$ and that $(\lambda + \Lambda)^{-1}$ is a positive and compact operator on $C(\Gamma)$. However, no resolvent estimates or generation properties are shown there. These results are, at least to our knowledge, the only ones studying the Dirichlet-Neumann operator on $C(\Gamma)$.

3. PROOF OF THEOREM 1 AND THEOREM 2

a) Let G denote the Nemytskii operator induced by g, i.e.,

$$G(z)(x) := g(x, z(x)) \quad \text{for} \quad z \in C(\Gamma) \quad \text{and} \quad x \in \Gamma.$$

Using the fact that Lipschitz continuous functions are uniformly Lipschitz continuous on compact sets and the mean value theorem (in integral form) it is not difficult to see that $G \in C^{2-}(C(\Gamma))$ and that G is bounded on bounded sets.

b) Given $z_0 \in C(\Gamma)$, we consider the abstract Cauchy problem

$$\dot{z} + \Lambda z = G(z), \quad 0 < t < \infty, \quad z(0) = z_0 \tag{9}$$

in the space $C(\Gamma)$. It follows from Theorem 3 and well-known results for semilinear Cauchy problems, cf. [15, Theorem 6.1.4], that there exists a unique maximal mild solution

$$z(\cdot, z_0) \in C([0, s^+(z_0)), C(\Gamma)) \tag{10}$$

of (9). Additionally,

$$\mathcal{Z} := \bigcup_{z_0 \in C(\Gamma)} [0, s^+(z_0)) \times \{z_0\} \tag{11}$$

is open in $[0, \infty) \times C(\Gamma)$ and

$$[(t, z_0) \mapsto z(t, z_0)] : \mathcal{Z} \to C(\Gamma)$$

defines a local semiflow on $C(\Gamma)$.

c) Recall that $-\Lambda$ generates an analytic semigroup on $C(\Gamma)$ such that $\mathrm{dom}(\Lambda) \subset B_{pp}^{1-1/p}(\Gamma)$ if $p > n$. Moreover, it follows from Lemma 6.4 in [9] that the $B_{pp}^{1-1/p}(\Gamma)$-realization of Λ is given by $\mathcal{BT}$ and that $-\mathcal{BT}$ generates an analytic semigroup on $B_{pp}^{1-1/p}(\Gamma)$. Hence using the smoothing effect of analytic semigroups, the regularity of G, and the fact that $\mathrm{dom}(\mathcal{BT}) = B_{pp}^{2-1/p}(\Gamma)$, a bootstrap argument similar as in [8, section 3] shows that the mild solution of (9) possesses the additional regularity

$$z(\cdot, z_0) \in C^1((0, s^+(z_0)), B_{pp}^{2-1/p}(\Gamma)). \tag{12}$$

d) Regularity. Let $u_0 \in C(\overline{\Omega})$ be given and let $z(\cdot, \gamma u_0)$ denote the solution of (9) to the initial value $z_0 := \gamma u_0$. The corresponding maximal interval of existence is denoted by $[0, s^+(\gamma u_0))$. We now define $t^+ := t^+(u_0) := s^+(\gamma u_0)$ and

$$u(t, u_0) := \mathcal{T} z(t, \gamma u_0) \quad \text{for} \quad t \in [0, t^+(u_0)). \tag{14}$$

Here, $\mathcal{T}$ stands for the solution operator induced by (6). It follows from (10) and (7) that

$$u(\cdot, u_0) \in C([0, t^+), C(\overline{\Omega})) \cap C([0, t^+), C^{2+\alpha}(\Omega))$$

and from (12) and (8) that $u(\cdot, u_0) \in C^1((0, t^+), W_p^2(\Omega))$. Choosing $p > \frac{n}{1-\alpha}$, Sobolev's embedding theorem implies that $W_p^2(\Omega) \hookrightarrow C^{1+\alpha}(\overline{\Omega})$. Consequently, we see that $u(\cdot, u_0) \in C^1((0, t^+), C^{1+\alpha}(\overline{\Omega}))$. This shows the regularity of $u(\cdot, u_0)$ asserted in (5).

e) Existence. Next we verify that $u(\cdot, u_0)$ is in fact a classical solution of (P). By definition of $\mathcal{T}$ we have $\mathcal{A}\mathcal{T} = 0$ and $\gamma\mathcal{T} = \mathbf{1}_{C(\Gamma)}$. Therefore we find that

$$\mathcal{A}u(t, u_0) = \mathcal{A}\mathcal{T} z(t, \gamma u_0) = 0, \quad t \in [0, t^+) \tag{15}$$

and that

$$\gamma u(t, u_0) = \gamma \mathcal{T} z(t, \gamma u_0) = z(t, \gamma u_0), \quad t \in [0, t^+). \tag{16}$$

In particular we have

$$\gamma u(0, u_0) = z(0, \gamma u_0) = \gamma u_0. \tag{17}$$

Finally, $z(t, \gamma u_0) \in B_{pp}^{2-1/p}(\Gamma)$ and $\mathcal{B}\mathcal{T} \subset \Lambda$ imply that

$$\mathcal{B}u(t, u_0) = \mathcal{B}\mathcal{T} z(t, \gamma u_0) = \Lambda z(t, \gamma u_0), \quad t \in (0, t^+).$$

Consequently, it follows that

$$\partial_t(\gamma u(t, u_0)) + \mathcal{B}u(t, u_0) = \partial_t z(t, \gamma u_0) + \Lambda z(t, \gamma u_0) = g(\cdot, \gamma u(t, u_0)) \tag{18}$$

for $t \in (0, t^+)$. Summarizing, we infer from (15), (17), and (18) that $u(\cdot, u_0)$ is a classical solution of (P).

f) Given $\alpha \in (0, 1)$, we have the following representation for Λ:

$$\Lambda w = \mathcal{B}\mathcal{T} w \quad \text{for} \quad w \in C^{1+\alpha}(\Gamma). \tag{19}$$

Indeed, note that $\mathcal{T} \in \mathcal{L}(C^{1+\alpha}(\Gamma), C^1(\overline{\Omega}))$, $\alpha \in (0, 1)$. This follows from the Notes to section I.6 in [10] (see also the proof of Theorem 3.4 in [19]). Consequently, $\mathcal{B}\mathcal{T} \in \mathcal{L}(C^{1+\alpha}(\Gamma), C(\Gamma))$ for $\alpha \in (0, 1)$. Now, let $w \in C^{1+\alpha}(\Gamma)$ be given and fix $\alpha' \in (0, \alpha)$ Then there is a sequence $(w_n)_{n \in \mathbb{N}}$ in $C^\infty(\Gamma)$ such that $w_n \to w$ in $C^{1+\alpha'}(\Gamma)$ as $n \to \infty$.

Observe that $\Lambda w_n = \Lambda_p w_n = \mathcal{B}\mathcal{T} w_n$ for $n \in \mathbb{N}$. Hence, letting $n \to \infty$ we obtain (19), since $C^{1+\alpha'}(\Gamma) \hookrightarrow \mathrm{dom}(\Lambda)$, see Theorem 3.

g) Uniqueness. Suppose that

$$v \in C([0, t^+), C(\overline{\Omega}) \cap C^{2+\alpha}(\Omega)) \cap C^1((0, t^+), C^{1+\alpha}(\overline{\Omega}))$$

is another classical solution of (P). Using the the indentity $v = \mathcal{T}\gamma v$ and (19), a calculation similar to (18) shows that γv is a solution of the Cauchy problem (9). Thus by the unique solvability of (9) and by (16) we find that

$$\gamma u(t, u_0) = \gamma v(t), \quad t \in [0, t^+).$$

Now we set $w(t) := u(t, u_0) - v(t)$ for $t \in [0, t^+)$. Then we see that

$$\mathcal{A}w(t) = 0 \quad \text{in} \quad \Omega, \qquad \gamma w(t) = 0 \quad \text{on} \quad \Gamma.$$

Consequently, the maximum principle implies that $w = 0$. This completes the proof of Theorem 1. $\quad\square$

Proof of Theorem 2: a) First we show that

$$X := \{u \in C(\overline{\Omega}) \cap C^{2+\alpha}(\Omega) \,;\, \mathcal{A}u = 0\}$$

is a closed subspace of $C(\overline{\Omega})$. Indeed, using the definition of $\mathcal{T}$ and (7), we see that $P := \mathcal{T}\gamma$ is a continuous projection in $C(\overline{\Omega})$ such that $X = Im(P)$. Hence X is a closed subspace of $C(\overline{\Omega})$.

b) Let

$$\mathcal{X} := \bigcup_{u \in X} [0, t^+(u)) \times \{u\}$$

and fix $(t_0, u_0) \in \mathcal{X}$, i.e., we have $t_0 \in [0, t^+(u_0))$. Since $\mathcal{Z}$ is open in $[0, \infty) \times C(\Gamma)$, see (11), there is an $\varepsilon > 0$ such that

$$\{t \in [0, \infty) \,;\, |t_0 - t| < \varepsilon\} \times \{z \in C(\Gamma) \,;\, \|\gamma u_0 - z\|_{C(\Gamma)} < \varepsilon\} \subset \mathcal{Z}. \tag{20}$$

Now define $U := \{t \in [0, \infty) \,;\, |t_0 - t| < \varepsilon\} \times \{u \in X \,;\, \|u - u_0\|_{C(\overline{\Omega})} < \varepsilon\}$ and let $(t, u) \in U$ be given. Because of $\|\gamma u_0 - \gamma u\|_{C(\Gamma)} \le \|u_0 - u\|_{C(\overline{\Omega})} < \varepsilon$ it follows from (20) that $(t, \gamma u) \in \mathcal{Z}$, i.e., $t \in [0, s^+(\gamma u))$. Since, by definition, we have $s^+(\gamma u) = t^+(u)$ this shows that $(t, u) \in \mathcal{X}$. Therefore, $\mathcal{X}$ is open in $[0, \infty) \times X$.

c) We now define

$$\Phi : \mathcal{X} \to X, \quad \Phi(t, u_0) := u(t, u_0),$$

where $u(\cdot, \cdot)$ denotes the solution of (P). By Theorem 1, Φ is well-defined. Moreover, because of $u(t, u_0) = \mathcal{T}z(t, \gamma u_0)$, cf. (14), and since $[(t, z_0) \mapsto z(t, z_0)]$ is a local semiflow on $C(\Gamma)$, it follows that $\Phi \in C(\mathcal{X}, X)$.

Next let $u_0 \in X$, $s \in [0, t^+(u_0))$, and $t \in [0, t^+(u(s, u_0)))$ be given. Recall that $t^+(u_0) = s^+(\gamma u_0)$ and that $t^+(u(s, u_0)) = s^+(z(s, \gamma u_0))$. Thus we have $s + t \in [0, t^+(u_0))$ and $z(t, z(s, \gamma u_0)) = z(t + s, \gamma u_0)$ since $z(\cdot, \cdot)$ has the semiflow property. Together with (16) we now obtain:

$$u(t, u(s, u_0)) = \mathcal{T}z(t, \gamma u(s, u_0)) = \mathcal{T}z(t, z(s, \gamma u_0))$$
$$= \mathcal{T}z(t + s, \gamma u_0) = u(t + s, u_0).$$

Finally, the identity $X = Im(P)$ implies that

$$u(0, u_0) = \mathcal{T}z(0, \gamma u_0) = \mathcal{T}\gamma u_0 = Pu_0 = u_0$$

for $u_0 \in X$. This completes the proof of Theorem 2. $\square$

Ackowledgement: I am grateful to Gieri Simonett for carefully reading the manuscript to this paper.

REFERENCES

[1] **S. Agmon, A. Douglis, L. Nirenberg:** Estimates near the boundary for solutions of elliptic partial differential equations satisfying general boundary conditions, II. Comm. Pure Appl. Math. **17** (1964), 35-92.

[2] **H. Amann:** Nonhomogeneous Linear and Quasilinear Elliptic and Parabolic Boundary Value Problems. Preprint, 1993.

[3] **J. Bergh & J. Löfström:** "Interpolation Spaces. An Introduction", Springer, Berlin, 1976.

[4] **A. P. Calderón:** On an inverse boundary value problem. Seminar on Numerical Analysis and its Application to Continuum Physics, Soc. Brasileira de Matemática, Rio de Janeiro (1980), 65-73.

[5] **J. Cannon, G. Meyer:** On diffusion in a fractured medium, S.I.A.M. J. Math. Anal., **20** (1971), 434-448.

[6] **R. Daurtray & J.-L. Lions:** "Analyse Mathématique et Calcul Numérique pour les Sciences et les Techniques", Masson, Paris, 1984.

[7] **J. Escher:** Nonlinear elliptic systems with dynamic boundary conditions. Math. Z. **210** (1992), 413-439.

[8] **J. Escher:** Smooth solutions of nonlinear elliptic systems with dynamic boundary conditions. To appear in Proc. "Semigroup Theory & Evolution Equations", M. Dekker, New York, 1993.

[9] **J. Escher:** The Dirichlet-Neumann operator on continuous functions. Preprint, 1993.

[10] **D. Gilbarg & N.S. Trudinger:** "Elliptic Partial Differential Equations of Second Order", Springer, Berlin, 1977.

[11] **K. Gröger:** Personal communication.

[12] **T. Hintermann:** Evolution equations with dynamic boundary conditions. Proc. Roy. Soc. Edinburgh Sect. A **113** (1989), 43-60.

[13] **J.-L. Lions:** "Quelques Méthodes de Résolution des Problèmes aux Limites Non Linéaires", Dunod, Paris, 1969.

[14] **A. Nachmann:** Reconstruction from boundary measurements. Ann. Math. **128** (1988), 531-576.

[15] **A. Pazy:** "Semigroups of Linear Operators and Application to Partial Differential Equations". Springer, Berlin, New York, 1983.

[16] **R. Showalter:** Degenerate evolution equations. Indiana Univ. Math. J. **23** (1974), 655-677.

[17] **J. Sylvester & G. Uhlmann:** A global uniqueness theorem for an inverse boundary value problem. Ann. Math. **125** (1987), 153-169.

[18] **H. Triebel:** "Theory of Function Spaces II". Birkhäuser, Basel, 1992.

[19] **K. O. Widman:** Inequalities for the Green function and boundary continuity of the gradient of solutions of elliptic differential equations. Math. Scand. **21**, 17-37, (1967).

Diffractive Diffusion System with Locally Defined Reaction

W. E. Fitzgibbon   University of Houston, Houston, Texas

J. J. Morgan   Texas A&M University, College Station, Texas

1. INTRODUCTION

We envision the dispersion of m-chemical species through a bounded spatial domain Ω by means of a diffusion process. We assume that the region contains a finite number of reacting cells and that chemical species diffuse through the boundary into the reacting cells where chemical reaction transpires.

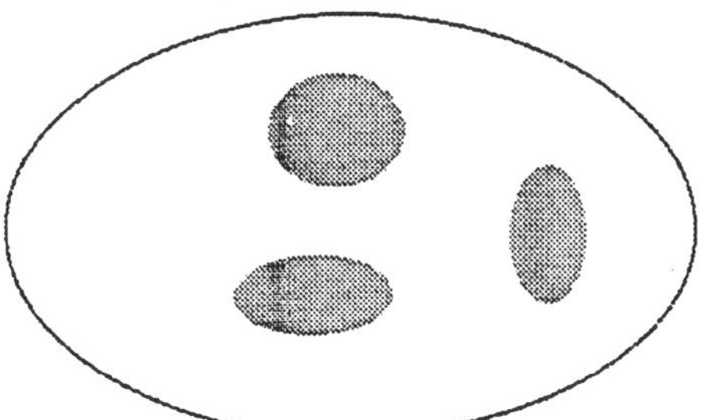

Figure 1 (Reacting Cells Shaded)

These authors gratefully acknowledge support from NSF Grants DMS 9207064 and DMS 9208046 respectively.

We assume reaction and diffusion interior to the reacting cells and diffusion in the ambient region. We should expect distinct diffusion processes interior and exterior to the cells. Furthermore we assume that no material is absorbed by the interface between the reacting cells and the surrounding region, and we make the assumptions that the species concentrations and their fluxes are continuous across the interface between the reacting cells and the surrounding region. If the diffusitiveties differ inside and outside of the reacting cells and the species concentrations are not constant then the assumption of flux continuity implies a discontinuity in the normal derivative across the interface.

Mathematically we expect this process to be described by a system of reaction diffusion equations on the interior of the reacting cells which are coupled across an interface to a system of diffusion equations modelling dispersion in the ambient region. We will not obtain the smooth classical solutions normally associated with reaction diffusion, since we expect normal derivatives to be discontinuous across the interface of the reacting cells and the surrounding region. The reaction kinetics are confined to the reacting cells and thus will appear as discontinuous source terms. Finally, we are prescribing discontinuities in the diffusion operators. Thus we should expect to at best be able to obtain some notion of a weak or generalized solution and consequently an approach which combines variational and semigroup techniques seems both natural and appropriate.

2. THE MATHEMATICAL PROBLEM

For reasons of notational convenience we confine our attention to the case of one reacting cell. If the reacting cells are not in contact with one another and none of the reacting cells meet the boundary of Ω the general case described in the introduction presents no additional analytic challenge and requires only a complicated tracking of indices listing reaction components and reacting cells.

We shall assume that Ω is a bounded region contained in $\mathbf{R}^n$ whose boundary is an $(n-1)$ dimensional $C^{2+\varepsilon}$ $(0 < \varepsilon < 1)$ manifold with $\partial\Omega$ lying locally on one side of Ω. We let Ω_1 be a bounded subdomain of Ω with $\overline{\Omega}_1 \subseteq \text{Int}(\Omega)$. We further assume that $\partial\Omega_1$ has the same properties as $\partial\Omega$ and set

$$\Omega_2 = \Omega - \overline{\Omega}_1$$

we refer to the following diagram

Figure 2

We shall be concerned with the spatio-temporal evolution of an m-component temporal dependent variable $u = (u_1 \ldots u_m)^T$ whose components represent the concentrations of the individual chemical species. The chemical kinetics transpiring in Ω_1 are represented by a vector field $f(\ ) = \{f_i(\ ,\ )\}_{i=1}^m$. Because we want to consider reaction and diffusion in Ω_1 and diffusion in Ω_2 we should expect systems of the form

$$\partial u/\partial t - D_1 \Delta u = f(u) \qquad x \in \Omega_1, t > 0 \qquad (2.2a)$$

and

$$\partial u/\partial t - D_2 \Delta u = 0 \qquad x \in \Omega_2, t > 0, \qquad (2.2b)$$

where D_1 and D_2 are diagonal $m \times m$ matrices of diffusion coefficients with distinct positive entries $d_{ij} > 0$, $i = 1$ or 2, $j = 1$ to m. We do not assume equality of d_{1j} and d_{2j}. We place the following conditions at $\partial \Omega$

$$B \partial u/\partial n + Cu = 0 \qquad (2.3)$$

where B and C are diagonal $m \times m$ matrices with nonnegative entries $b_i \geq 0$, $c_i > 0$ along the diagonal. The continuity of the state variables across the interface of Ω_1 and Ω_2 is prescribed by

$$[u]_{\partial \Omega_1} = 0 \qquad (2.4)$$

where $[w]_{\partial \Omega_1}$ denotes the vector valued saltus of w on $\partial \Omega_1$. If we define

$$(\tilde{D}v)(x) = \begin{array}{ll} (D_1 v)(x) & x \in \Omega_1 \\ (D_2 v)(x) & x \in \Omega_2 \end{array} \qquad (2.5)$$

for $v(\ ) = (v_1(\ ) \ldots v_m(\ ))^T$ we can insure continuity of flux across $\partial \Omega_1$ with

$$[\tilde{D} \partial u/\partial n]_{\partial \Omega_1} = 0 \qquad (2.6)$$

where $\partial u/\partial n$ is the outward normal derivative on $\partial \Omega_1$. We shall collectively refer to (2.4) and (2.6) as the interface compatibility condition.

We shall place the following conditions on the vector field $f(\ ) = (f_i(\ ))_{i=1}^m$

$$f_i(u) = u^T E^i u / \alpha(u) \quad \text{where } E^i \tag{2.7a}$$

is an $m \times m$ matrix and

$$\alpha(u) = 1 + \sum_{j=1}^m \varepsilon_j u^{\delta_j} \quad \text{with each}$$

$$\varepsilon > 0 \quad \text{and} \quad \delta_j \geq 1$$

$$f_i(u) \geq 0 \quad \text{if} \quad u_i = 0 \quad \text{with} \quad u \in \mathbf{R}_+^m. \tag{2.7b}$$

We point out that Condition (2.7a) will guarantee the existence of a Lipschitz constant, $K(\varepsilon)$, so that,

$$|f(u) - f(v)| \leq K(\varepsilon)|u - v| \quad \text{for} \quad u, v \in \mathbf{R}_+^m. \tag{2.8}$$

The second condition will be used to establish the nonnegativity of solutions. Although the hypotheses on the vector field do not allow us to consider general mass action kinetics, we are able to treat Michaelis-Menten kinetics. Moreover it is reasonable to assume that catalysts confined to the reacting cells permit the occurrence of chemical reaction. Physical arguments [2] support the assertion that catalytic reaction kinetics produce vector fields which satisfy (2.7a-b).

3. THE VARIATIONAL SEMIGROUP SETTING

Throughout this section we shall be considering scalar problems. We begin with the scalar eigenvalue problem;

$$\lambda u - d_1 \Delta u = w \qquad x \in \Omega_1, \tag{3.1a}$$

and

$$\lambda u - d_2 \Delta u = w \qquad x \in \Omega_2, \tag{3.1b}$$

subject to the exterior boundary condition,

$$b \partial u / \partial n + c u = 0 \qquad x \in \partial \Omega \tag{3.1c}$$

(with $b \geq 0, c > 0$) and the interface compatibility conditions,

$$[u]_{\partial \Omega_1} = 0, \tag{3.1d}$$

and

$$[\tilde{d}u]_{\partial \Omega_1} = 0, \tag{3.1e}$$

for $w \in L_2(\Omega)$ and $\lambda \geq 0$. We note that (3.1a,b) can be reformulated as

$$\lambda u - \tilde{d}\Delta u = w \qquad x \in \Omega - \partial \Omega_1. \tag{3.2}$$

Problems of this type are called diffractive problems. Diffraction problems have been of interest to those involved in nuclear reactor physics and are treated in both the American and Russian literature [4], [5], [6], [7], [9], [10], [11]. Typically these papers place homogeneous Dirichlet boundary conditions on the exterior boundary $\partial\Omega$, however their analysis carries over to the case at hand and we shall cite the forementioned results as being applicable with a laborious recapitulational of the arguments.

Assume $b > 0$. If we multiply (3.1a-b) by $\varphi \in H^1(\Omega)$ and formally compute the $L_2(\Omega)$ inner product we obtain the weak formulation

$$\lambda \int_\Omega u\varphi dx + d_1 \int_{\Omega_1} \langle \nabla u, \nabla \varphi \rangle dx \tag{3.3}$$
$$+ d_2 \int_{\Omega_2} \langle \nabla u, \nabla \varphi \rangle dx + d_2 \frac{c}{b} \int_{\partial\Omega} u\varphi d\sigma = \int_\Omega w\varphi dx.$$

Following [10] we rewrite (3.3) in the form

$$\langle A_\lambda u, \varphi \rangle = \langle w, \varphi \rangle \tag{3.4}$$

with

$$A_\lambda = A + \lambda I \tag{3.5}$$

and A (mapping $H^1(\Omega)$ onto $L_2(\Omega)$) is a positive, self-adjoint linear operator. As a result, for every $\lambda \geq 0$ we obtain a unique generalized solution of (3.1a-e). Moreover by virtue of [5] we see that the generalized solution belongs to class $C_{0,\alpha}(\overline{\Omega}) \cap H^2(\Omega_i)$ with $i = 1$ or 2. Its derivatives are Hölder continuous up to $\partial\Omega_1$ and $\partial\Omega$. We summarize this information and more with the following proposition.

PROPOSITION 3.6. *For each $\lambda \geq 0$ and $w \in L_2(\Omega)$, (3.1a-e) has a unique generalized solution, u, satisfying (3.3) so that the following are true*

(i) *$u \in C_{0,\alpha}(\overline{\Omega}) \cap H^1(\Omega), (0 < \alpha < 1)$,*
(ii) *u has first order derivatives that are continuous up to $\partial\Omega_1$ and $\partial\Omega$,*
(iii) *the boundary and the interface compatibility conditions hold,*
(iv) *and $u|_{\partial\Omega_i} \in H^2(\Omega_i)$.*

Finally if A is defined via (3.5) then A is self-adjoint $D(A) \subseteq H^1(\Omega)$ and $-A$ is the infinitesimal generator of an analytic semigroup of contractions, $\{T(t) \mid t \geq 0\}$.

We now turn our attention to semilinear parabolic equations with diffractive diffusion. Namely we consider equations of the form:

$$\partial u/\partial t - d_1 \Delta u = f(u) \qquad x \in \Omega_1, t > 0 \tag{3.7a}$$
$$\partial u/\partial t - d_2 \Delta u = 0 \qquad x \in \Omega_2, t > 0 \tag{3.7b}$$

with

$$b\partial u/\partial n + cw = 0 \qquad x \in \partial\Omega, \tag{3.7c}$$

and

$$[u]_{\partial\Omega_1} = 0 \qquad t > 0, \tag{3.7d}$$

$$[\tilde{d}\partial u/\partial n]_{\partial\Omega_1} = 0 \qquad t > 0, \tag{3.7e}$$

and

$$u(\,,0) = u_0(\,) \in L_2(\Omega). \tag{3.7f}$$

We further require that $f(\,)$: $\mathbf{R} \to \mathbf{R}$ is uniformly Lipschitz continuous function. If $\chi(x)$ denotes the characteristic function of Ω_1 we can define a local nonlinear function on $L_2(\Omega)$, $F(\,)$, by

$$(F(u))(x) = \chi(x)f(u(x))) \qquad x \in \Omega. \tag{3.8}$$

It should be clear that $F(\,)$ so defined is uniformly Lipschitz from $L_2(\Omega)$ to itself. We may observe that (3.7a-e) may be reformulated as an abstract Cauchy initial value problem in $L_2(\Omega)$ of the form:

$$du/dt + Au = F(u) \qquad t > 0 \tag{3.9a}$$

$$u(0) = u_0 \tag{3.9b}$$

where A is a operator from the previous proposition. We have the following result.

PROPOSITION 3.10. *For each $u_0 \in L_2(\Omega)$ there exists an unique $u(\,) \in C([0,\infty), L_2(\Omega))$ satisfying*

$$u(t) = T(t)u_0 + \int_0^t T(t-s)F(u(s))ds. \tag{3.10}$$

The solution of the integral equation is continuously differentiable for $t > 0$ and satisfies equations (3.9a,b). Moreover if $u(\,,\,)$ is the pointwise evaluation, i.e.

$$u(t))(x) = u(x,t) \tag{3.11}$$

then the following are true:

   (i) *$u(\,) \in C_{\alpha,\alpha/2}(\overline{Q}_{\varepsilon,T})$ where $\varepsilon > 0$, $0 < \alpha < 1$, $T > 0$ and $Q_{\varepsilon,T} = \Omega \times [\varepsilon, T]$;*
   (ii) *the derivatives $u_{x_i}(\,,\,)$ are Hölder continuous in $Q_{\varepsilon,T}$ up to $\partial\Omega$ and the interface boundary of Ω_1 and Ω_2, and the boundary and interface compatibility conditions hold;*
   (iii) *$u(\,,t) \in H^2(\Omega_i), i = 1$ and 2.*

INDICATION OF PROOF: The existence and uniqueness of a solution to the integral equation and its continuous differentiability follow from a concatenation of existence theory for abstract semilinear theory and regularity theory for mild solutions for analytic semigroups, cf. [8, Sec. 4.3 and Sec. 6.3]. The deeper regularity results follow from work of Ladyshenkaya, Rivkind and Ural'ceva [5].

4. SYSTEMS – MAIN RESULTS

We return to the systems described in the second section. For $i = 1$ to m we have,

$$\partial u_i/\partial t - d_{1,i}\Delta u_i = f_i(u) \qquad x \in \Omega_1, t > 0, \tag{4.1a}$$

$$\partial u_i/\partial t - d_{2,i}\Delta u_i = 0 \qquad x \in \Omega_2, t \geq 0, \tag{4.1b}$$

subject to interface compatibility conditions,

$$[u]_{\partial\Omega_1} = 0, \qquad t > 0 \tag{4.1c}$$

$$[\tilde{d}_i\partial u/\partial n]_{\partial\Omega_1} = 0, \qquad t > 0 \tag{4.1d}$$

with boundary conditions

$$b_i\partial u/\partial n + c_i u = 0 \qquad x \in \partial\Omega, t > 0 \tag{4.1e}$$

and initial condition

$$u_i(x,0) = u_{0_i}(x) \qquad x \in \Omega. \tag{4.1f}$$

Here $d_{1,i}, d_{2,i} > 0, c_i > 0$ and $b_i \geq 0$. We introduce an m-component state space

$$X = (L_2(\Omega))^m. \tag{4.2}$$

Our dependent variables will be of the form

$$u(\ ) = (u_1(\ )\ldots u_m(\ ))^T \in X. \tag{4.3}$$

Proposition 3.6 may be used to define an abstract diffusion operator A_i in each of the m-components. Moreover, it is straightforward to verify that if

$$\hat{A} = (A_i)_{i=1}^m \tag{4.4}$$

then $-\hat{A}$ is the infinitesimal generator of an analytic semigroup of contractions.

$$\{\hat{T}(t) \mid t \geq 0\} = \{T_i(t) \mid t \geq 0\}_{i=1}^m \tag{4.5}$$

on X. The diffractive diffusive process represented by $\{T(t) \mid t \geq 0\}$ is not coupled and the assertions of Proposition 3.6 hold in each component.

The process we wish to describe is weakly and locally coupled through the reaction kinetics. Proceeding in a manner analogous to that used in the scalar case we let $\chi(\ )$

be the characteristic function of Ω_1 and define $\widehat{F}(\ ) = (\widehat{F}_i(\ ))_{i=1}^m : X \to X$ pointwise in each component by

$$(\widehat{F}_i(u))(x) = \chi(x) f_i(u(x)) \qquad x \in \Omega, i = 1 \text{ to } m. \tag{4.6}$$

By virtue of (2.8) it is not difficult to observe that $\widehat{F}(\ )$ satisfies a uniform Lipschitz condition.

Within an $L_2(\Omega)$ framework our system may be written as

$$du/dt + \hat{A}u = \widehat{F}(u) \quad t > 0 \tag{4.7a}$$

$$u(0) = u_0 \in X \tag{4.7b}$$

which has variation of parameter representation

$$u(t) = \widehat{T}(t)u_0 + \int_0^t \widehat{T}(t - s)\widehat{F}(u(s))ds. \tag{4.8}$$

We have the following result.

THEOREM 4.9. *Assume that all hypotheses of the section are in effect and let $u_0(\ ) \in X$ be such that $u_{0_i}(x) \geq 0$ a.e. $x \in \Omega$. Then, there exists a unique continuous function $u(\ ): [0, \infty) \to X$ satisfying (4.8) which is continuously differentiable and satisfies (4.7a) for $t > 0$. Moreover if $u(\ , \ ) = (u_i(\ , \ ))_{i=1}^m$ is the pointwise evaluation, i.e.*

$$((u(t))(x))_i = u_i(x, t) \quad \text{for} \quad i = 1 \text{ to } m \tag{4.10}$$

then the following are true:

    (i) $u_i(x, t) \geq 0$ for $x \in \Omega, t > 0, i = 1$ to m

    (ii) $u_i(\ , \ ) \in C_{\alpha, \alpha/2}(\overline{Q}_{\varepsilon, T})$ where $\varepsilon > 0$,

    (iii) the derivatives $\partial u_i(\ ,)/\partial x_j$ are Hölder continuous in $Q_{\varepsilon, T}$ up to $\partial\Omega$ and the interface boundary of Ω_1 and Ω_2 and the boundary and interface compatibility conditions hold

    (iv) $u_i(\ , t) \in H^2(\Omega_j)$ $j = 1$ or 2, $i = 1$ to m.

INDICATION OF PROOF: The existence and regularity assertions follow from arguments analogous to those establishing Proposition 3.10. To establish positive invariance of solution components we introduce the positive and negative parts of real valued function $z(\ )$ by

$$z^+(x) = (|z(x)| + z(x))/2 \qquad x \in \Omega \tag{4.11}$$

$$z^-(x) = (|z + (x)| - z(x))/2.$$

For $v(\ ) = (v_i(\ ))_{i=1}^m$ we let $v^+(\ ) = (v_i^+(\ ))_{i=1}^m$ and return to the local kinetics to define an auxiliary nonlinear function $\widehat{G}(\ ) = (\widehat{G}_i(\ ))_{i=1}^m$ as

$$(\widehat{G}_i(u))(x) = \chi(x)f_i(u^+(x)). \tag{4.12}$$

For $u(\ ) \in X$ and $x \in \Omega$. It is evident from (4.11) and (2.8) that $\widehat{G}(\ )$ is uniform Lipschitz continuous. Therefore there exists a unique $v(\ )\colon [0,\infty) \to X^+$ satisfying

$$v(t) = \widehat{T}(t)u_0 + \int_0^t \widehat{T}(t-s)\widehat{G}(v(s))ds \tag{4.13}$$

so that $v(\ ) \in C^1((0,\infty), X)$ and

$$dv/dt + \hat{A}v = \widehat{G}(v) \tag{4.14a}$$

$$v(0) = u_0. \tag{4.14b}$$

Moreover all regularity assertions also hold for $v(\ )$. If $\theta(\ ,\ ) = \theta_i(\ )_{i=1}^m \in H^1(Q_{\varepsilon,T})$ then the ith component satisfies

$$\int_\Omega \partial v_i/\partial t\; \theta_i dx + d_{1_i}\int_{\Omega_i} \langle \nabla v_i, \nabla \theta_1\rangle dx \tag{4.15}$$

$$+ d_{2_i}\int_{\Omega_2} \langle \nabla u, \nabla \theta_i\rangle dx + d_{2_i}b_i \int_{\partial\Omega} v_i\theta_i d\sigma = \int_\Omega \widehat{G}_i(v)\theta_i dx.$$

We now argue that the selection of $\theta_i = v_i^-$ leads to the conclusion that $v_i^- \equiv 0$ and hence $v_i^+ = v$. Therefore $\widehat{G}_i(v) = \widehat{F}_i(v)$ and uniqueness guarantee that the solution to (4.13) is the solution to (4.8).

We remark that the absence of a classical maximum principle forced us to develop a more elaborate argument for positivity of solution components. We now introduce an additional hypotheses for our reaction vector field $f(\ ) = (f_i(\ ))_{i=1}^m$

$$\sum_{i=1}^m f_i(u) \le 0 \quad \text{for} \quad u \in \mathbf{R}_+^m. \tag{4.16}$$

We remark that a condition such as (4.16) is physically reasonable and frequently arises from considerations pertaining to conservation of material. We have the following result.

THEOREM 4.17. *Assume the hypotheses of section 2 are in effect. If for each i ($i = 1$ to m) $\|u_{0_i}(\ )\|_{\infty,\Omega} < \infty$ and in addition (4.16) holds then there exists an $M > 0$ so that for $t \ge 0$*

$$\sum_{i=1}^m \|u_i(\ ,t)\|_{\infty,\Omega} \le M.$$

INDICATION OF PROOF: Roughly speaking we integrate each component over the space time cylinder, sum the components and use the positivity of each component to observe that there exists $M > 0$ so that $\sum_{i=1}^{m} \|u_i(\, ,t)\|_{1,\Omega} \leq M$. We then adopt the iterative arguments of Alikakos [1] to bootstrap this a priori $L_1(\Omega)$ estimate through successive spaces $L_{2^k}(\Omega)$ and compute an a priori $L_\infty(\Omega)$ bound.

We remark that uniform a priori $L_\infty(\Omega)$ bounds can be used to guarantee that trajectories are precompact, and that this compactness may be used to describe the asymptotic behavior of solutions. This topic shall be taken up in forthcoming work of the authors.

REFERENCES

1. N. Alikakos, "An application of the invariance principle to reaction diffusion equations", *J.D.E.*, 33 (1979), 201-225.

2. R. Aris, *The Mathematical Theory of Diffusion and Reaction in Permeable Catalysts*, Vol. 1 and 2, Oxford University Press, London, 1975.

3. J. Ball, "Strongly continuous semi-groups, weak solutions and the variation of constants formula", *Proc. Amer. Math. Soc.*, 63 (1977) 370-373.

4. O. Ladyshenkaya, *The Boundary Vale Problems of Mathematical Physics*, Springer-Verlag, Berlin, 1985.

5. O. Ladyshenkaya, V. Rivkind and N. Ural'ceva, "On the classical solvability of diffraction problems for equations of elliptic and parabolic type", *Soviet Math. Dokl.*, 5 1964, 1249-1252.

6. O. Ladyshenkaya, N. Solonnokov and N. Ural'ceva, *Linear and Quasilinear Equations of Parabolic Type*, Trans. AMS, 23, Providence, R.I., 1968.

7. Z. Šeftel, "Estimates in L_p of solutions of elliptic equations with discontinuous coefficients and satisfying general boundary conditions and conjugacy conditions", *Soviet Math.*, 4 (1963), 321-324.

8. A. Pazy, *Semigroups of Linear Operators and Applications in Partial Differential Equations*, Springer-Verlag, Berlin, 1983.

9. H. Stewart, "Generation of analytic semigroups by strongly elliptic operators", *Trans. Amer. Math. Soc.*, 1974, 141-162.

10. H. Stewart, "Spectral theory of heterogeneous diffusion systems", *JMAA*, 1976, 59-78.

11. H. Stewart, "Generation of analytic semigroups by strongly elliptic operators under general boundary conditions", *Trans. Amer. Math. Soc.*, 1980, 299-310.

Global Dynamics of Coupled Systems
Modeling Nonplanar Beam Motion

W. E. Fitzgibbon   University of Houston, Houston, Texas

M. Parrott and Y. You   University of South Florida, Tampa, Florida

1. INTRODUCTION

In the past two decades, considerable effort has been directed toward a description of
the asymptotic behavior of the nonlinear beam equation with external damping

$$\frac{\partial^2 u}{\partial t^2} + \alpha \frac{\partial^4 u}{\partial x^4} + \delta \frac{\partial u}{\partial t} - \left(\beta + \int_0^1 \left(\frac{\partial u(\xi,\,t)}{\partial \xi} \right)^2 d\xi \right) \frac{\partial^2 u}{\partial x^2} = 0. \qquad (1.1a)$$

Here $u(x,\,t)$ represents the transverse deflection of a beam whose ends are held a fixed
distance apart. The constants α and δ are assumed to be positive but no assumptions are

---

1 Supported by NSF Grant DMS 9207064.

2 Supported by USF Research & Creative Grant 1249-931-RO.

made regarding the sign of the constant β. We refer the reader to [12], [4], [1], [5], [2], [3], [8], [11], [15] and the references contained therein. In much of the aforementioned work hinged (or simply supported) boundary conditions are imposed of the form

$$u\left(0,\ t\right) = u\left(1,\ t\right) = u_{xx}\left(0,\ t\right) = u_{xx}\left(1,\ t\right) = 0,\ t \geq 0. \tag{1.1b}$$

Clamped (or built-in) boundary conditions are also commonly considered; they have the form

$$u\left(0,\ t\right) = u\left(1,\ t\right) = u_{x}\left(0,\ t\right) = u_{x}\left(1,\ t\right) = 0,\ t \geq 0. \tag{1.1c}$$

Cantilever beams, where the left endpoint at $x = 0$ is clamped at rest and the right endpoint at $x = 1$ is free of transversal force and of bending torque, are also important in applications, but the mathematical analysis of such beams is more difficult. The boundary conditions of cantilever beams have the form

$$u\left(0,\ t\right) = u_{x}\left(0,\ t\right) = u_{xx}\left(1,\ t\right) = u_{xxx}\left(1,\ t\right) = 0,\ t \geq 0. \tag{1.1d}$$

The nonlinearity in equation (1.1a) appears in the coefficient of the second order spatial derivative and is included to model a change in the tension of the beam due to its extensibility. Much of the qualitative work [4], [2], [3], [11] focuses on this coefficient of the second derivative. In particular, if $\beta > -\lambda_1$, where λ_1 is the first eigenvalue of the eigenvalue problem

$$\alpha \frac{\partial^4 \phi}{\partial x^4} = \lambda \frac{\partial^2 \phi}{\partial x^2} \tag{1.2}$$

with appropriate homogeneous boundary conditions, it can be shown that solutions attenuate exponentially toward the zero equilibrium state. Nontrivial solutions to

$$\alpha u'''' - \left(\beta + \int_0^1 \left(u'\right)^2 dx \right) u'' = 0 \tag{1.3}$$

correspond to buckled states of the beam. The time convergence of solutions to buckled equilibrium states is the subject of [4], [1]. The global dynamics for solutions for general β have been considered in [8] and [15]. In [8], the existence of a global attractor for (1.1a) and (1.1c) was established by the construction of an appropriate Lyapunov function. More recently, the existence of a global attractor and an inertial manifold for (1.1a) and (1.1d) are established in [15].

Experiments conducted by Haight and King [7] on cantilever beams subject to external periodic forcing indicate that at times a more complicated model could be in order. They observed that for a certain range of frequency values of the external force, the planar motion parametrically induced a nonplanar motion and they proposed a coupled system of the form

$$\frac{\partial^2 u}{\partial t^2} + \alpha \frac{\partial^4 u}{\partial x^4} + \delta \frac{\partial u}{\partial t} \qquad (1.4a)$$

$$- \left[\beta + \int_0^1 \left\{ \left(\frac{\partial u\,(\xi,\, t)}{\partial \xi} \right)^2 + \left(\frac{\partial v\,(\xi,\, t)}{\partial \xi} \right)^2 \right\} d\xi \right] \frac{\partial^2 u}{\partial x^2}$$

$$= p\,(x,\, t)$$

$$\frac{\partial^2 v}{\partial t^2} + \alpha \frac{\partial^4 v}{\partial x^4} + \delta \frac{\partial v}{\partial t} \qquad (1.4b)$$

$$- \left[\beta + \int_0^1 \left\{ \left(\frac{\partial u\,(\xi, t)}{\partial \xi} \right)^2 + \left(\frac{\partial v\,(\xi, t)}{\partial \xi} \right)^2 \right\} d\xi \right] \frac{\partial^2 v}{\partial x^2}$$

$$= q\,(x,\, t)$$

as a model. Here $p\,(x,\, t)$ and $q\,(x,\, t)$ represent periodic forcing terms and $u\,(x,\, t)$ and $v\,(x,\, t)$ represent perpendicular components of motion. Ho, Scott and Eisley [9] used a model similar to (1.4a–b) to study the whirling motion of simply-supported beams. Fix and Kannan [6] utilized a Caesari-type method to establish the existence of periodic solutions to (1.4a–b) subject to hinged boundary conditions.

The previously described experimental work has also indicated the desirability of studying free vibrations for the nonplanar motion of the simply-supported beam which occur when $p\,(x,\, t)$ and $q\,(x,\, t)$ are identically zero. Heretofore, analytical work on such systems has not been forthcoming. In the present work, we shall be concerned with the initial boundary value problem for (1.4a–b) where the forcing functions are identically zero and the dependent variables u and v satisfy homogeneous hinged boundary conditions. We shall show that by defining a vector function $w\,(x,\, t) = (u\,(x,\, t),\, v\,(x,\, t))^T$, we can use the semigroup and functional analytic methods of [15] for the scalar damped beam equation to show the existence of a global attractor and an inertial manifold for the coupled system.

2. THE ABSTRACT SETTING AND LOCAL EXISTENCE

The specific system we consider is the following:

$$\frac{\partial^2 u}{\partial t^2} + \alpha \frac{\partial^4 u}{\partial x^4} + \delta \frac{\partial u}{\partial t} \tag{2.1a}$$

$$= \left[\beta + \int_0^1 \left\{ \left(\frac{\partial u\,(\xi,\ t)}{\partial \xi} \right)^2 + \left(\frac{\partial v\,(\xi,\ t)}{\partial \xi} \right)^2 \right\} d\xi \right] \frac{\partial^2 u}{\partial x^2}$$

$$\frac{\partial^2 v}{\partial t^2} + \alpha \frac{\partial^4 v}{\partial x^4} + \delta \frac{\partial v}{\partial t} \tag{2.1b}$$

$$= \left[\beta + \int_0^1 \left\{ \left(\frac{\partial u\,(\xi,\ t)}{\partial \xi} \right)^2 + \left(\frac{\partial v\,(\xi,\ t)}{\partial \xi} \right)^2 \right\} d\xi \right] \frac{\partial^2 v}{\partial x^2}$$

for $x \in (0,\ 1)$, $t \geq 0$. We stipulate homogeneous hinged (simply-supported) boundary conditions:

$$u\,(0,\ t) = u\,(1,\ t) = u_{xx}\,(0,\ t) = u_{xx}\,(1,\ t) = 0, \tag{2.1c}$$

$$v\,(0,\ t) = v\,(1,\ t) = v_{xx}\,(0,\ t) = v_{xx}\,(1,\ t) = 0, \tag{2.1d}$$

for $t \geq 0$, and initial conditions:

$$u\,(x,\ 0) = u_0\,(x),\ u_t\,(x,\ 0) = u_1\,(x), \tag{2.1e}$$

$$v\,(x,\ 0) = v_0\,(x),\ v_t\,(x,\ 0) = v_1\,(x), \tag{2.1f}$$

for $x \in (0,1)$. As before, the constants α and δ are positive and the sign of β is nonspecified. We define a vector function:

$$w\,(x,\ t) = \begin{pmatrix} u\,(x,\ t) \\ v\,(x,\ t) \end{pmatrix}. \tag{2.2}$$

Then from (2.1a–2.1f), w satisfies the equation:

$$\frac{\partial^2 w}{\partial t^2} + \alpha \frac{\partial^4 w}{\partial x^4} + \delta \frac{\partial w}{\partial t} = \left[\beta + \int_0^1 \left| \frac{\partial w}{\partial y}\,(y,\ t) \right|_{R^2}^2 dy \right] \frac{\partial^2 w}{\partial x^2}, \tag{2.3a}$$

and the vector-valued boundary and initial conditions:

$$w\,(0,\ t) = w\,(1,\ t) = w_{xx}\,(0,\ t) = w_{xx}\,(1,\ t) = 0 = \begin{pmatrix} 0 \\ 0 \end{pmatrix}, \tag{2.3b}$$

$$w(x, 0) = w_0(x) = \begin{pmatrix} u_0(x) \\ v_0(x) \end{pmatrix}; \quad w_t(x, 0) = w_1(x) = \begin{pmatrix} u_1(x) \\ v_1(x) \end{pmatrix}. \tag{2.3c}$$

We let $H = L^2(0, 1; R^2)$ and use $< ., . >$ and $|\cdot|$ to denote the standard inner product and norm in H respectively. The notation $H^k(0, 1; R^2)$ denotes the standard Hilbert-Sobolev space of order $k \geq 1$. We define $A : \mathcal{D}(A) \subset H \to H$ by the pointwise evaluation

$$(A\phi)(x) = \frac{d^4\phi}{dx^4} \tag{2.4a}$$

with

$$\mathcal{D}(A) = \left\{ \phi \in H^4(0, 1; R^2) : \phi(0) = \phi(1) = \phi''(0) = \phi''(1) = 0 \right\}. \tag{2.4b}$$

It is well-known that A is a closed, densely defined, positive definite, self-adjoint operator on H which has compact resolvent A^{-1} (c.f. for example [15], [5]). The spectrum of A consists of strictly positive eigenvalues $\{\lambda_j\}_{j=1}^{\infty}$ where $\lambda_j = (j\pi)^4$. Straightforward applications of the spectral calculus allow the computation of fractional powers of A and we may observe that

$$\left(A^{1/2}\phi\right)(x) = -\frac{d^2\phi}{dx^2} \tag{2.5a}$$

with

$$\mathcal{D}\left(A^{1/2}\right) = \left\{ \phi : \phi \in \overline{\mathcal{D}(A)} \text{ in } H^2(0, 1; R^2) \right\}. \tag{2.5b}$$

Moreover, we may show that

$$D\left(A^{1/4}\right) = \left\{ \phi : \phi \in \overline{\mathcal{D}(A)} \text{ in } H^1(0, 1; R^2) \right\} \tag{2.6}$$

and that for $u \in D\left(A^{1/4}\right)$,

$$\left|A^{1/4}\phi\right|^2 = \int_0^1 |\phi'(x)|_{R^2}^2 \, dx. \tag{2.7}$$

In light of the foregoing discussion, it becomes apparent that equation (2.3a) with conditions (2.3b) and (2.3c) may be written as the second-order evolution equation:

$$\frac{d^2w}{dt^2} + \alpha Aw + \delta\frac{dw}{dt} + \left(\beta + \left|A^{1/4}w\right|^2\right)A^{1/2}w = 0, \; t \geq 0 \tag{2.8}$$

$$w(0) = w_0, \; w_t(0) = w_1,$$

where we denote by $w(t)$ the time-dependent abstract-valued function $w(\cdot, t)$.

It is our intention to write (2.8) as a first order equation. Toward this end we let V be the Banach space obtained by imposing the equivalent graph norm on $\mathcal{D}\left(A^{1/2}\right)$, that is,

$$\|\psi\| = \left|A^{1/2}\psi\right| \text{ for } \psi \in \mathcal{D}\left(A^{1/2}\right). \tag{2.9}$$

Let

$$X = V \times H. \tag{2.10}$$

We assume that

$$\begin{bmatrix} w_0 \\ w_1 \end{bmatrix} \in X. \tag{2.11}$$

Define a linear operator

$$\widehat{G} = \begin{bmatrix} 0 & I_V \\ -\alpha A & -\delta I_V \end{bmatrix} : \mathcal{D}\left(\widehat{G}\right) \to X, \tag{2.12}$$

with

$$\mathcal{D}\left(\widehat{G}\right) = \mathcal{D}(A) \times V,$$

where I_V is the identity operator on V. Using the properties of A we can readily establish the following:

PROPOSITION 2.1. *The operator $\widehat{G}$, as defined by (2.12), is the infinitesimal generator of a strongly continuous semigroup of contractions $\{T(t)\}_{t\geq 0}$ on X.*

If we define a nonlinear mapping $f : X \to X$ by

$$f(\phi, \psi) = \begin{bmatrix} 0 \\ -\left(\beta + \left|A^{1/4}\phi\right|^2\right) A^{1/2}\phi \end{bmatrix}, \tag{2.13}$$

then equation (2.8) can be written as the abstract first order equation

$$\frac{d}{dt}\begin{bmatrix} w \\ w_t \end{bmatrix} = \widehat{G}\begin{bmatrix} w \\ w_t \end{bmatrix} + f(w(t), w_t(t)), \; t \geq 0 \tag{2.14}$$

$$\begin{bmatrix} w(0) \\ w_t(0) \end{bmatrix} = \begin{bmatrix} w_0 \\ w_1 \end{bmatrix} \in X. \tag{2.15}$$

Because of Proposition 2.1, the mild solution of (2.14)-(2.15) can be considered as the solution of the original system (2.1a-f). Furthermore, the mild solution of the evolution equation (2.14) with initial data (2.15) is given by

$$\begin{bmatrix} w(t) \\ w_t(t) \end{bmatrix} = T(t) \begin{bmatrix} w_0 \\ w_1 \end{bmatrix} + \int_0^t T(t-s) f(w(s), w_s(s)) \, ds, \quad t \geq 0, \qquad (2.16)$$

for as long as it exists. Since the nonlinear term f defined by (2.13) is locally Lipschitz continuous, we can apply [10, Theorems 6.1.4 and 6.1.6] to obtain the following:

LEMMA 2.2. *For every* $\begin{bmatrix} w_0 \\ w_1 \end{bmatrix} \in X$, *there is a* $\tau = \tau(w_0, w_1) > 0$ *such that the mild solution of the equation (2.14) with initial condition (2.15) exists uniquely on the interval* $[0, \tau)$. *Moreover, if* $\begin{bmatrix} w_0 \\ w_1 \end{bmatrix} \in \mathcal{D}(\widehat{G})$, *then the mild solution* $\begin{bmatrix} w \\ w_t \end{bmatrix} \in C([0, \tau); X)$ *of (2.14)-(2.15) is a strong solution of this initial value problem; that is,* $\begin{bmatrix} w \\ w_t \end{bmatrix}$ *is differentiable a.e. and satisfies (2.14)-(2.15) for* $t \in [0, \tau)$.

3. GLOBAL EXISTENCE AND THE ABSORBING PROPERTY

We recall that a strongly continuous semigroup $\{S(t)\}_{t \geq 0}$ of nonlinear transformations on a Banach space X is said to have the *absorbing property* if there exists a bounded subset $B_0 \subset X$ (called the absorbing set) so that for every bounded subset B there exists a t_1, so that $S(t) B \subset B_0$ for $t > t_1$. We have the following result.

THEOREM 3.1. *For any* $\Phi_0 = \begin{bmatrix} w_0 \\ w_1 \end{bmatrix} \in X$ *there exists a unique global mild solution* $\Phi = \begin{bmatrix} w(\cdot) \\ w_t(\cdot) \end{bmatrix} \in C([0, \infty), X)$ *to (2.14)-(2.15). If we define* $S(t) \Phi_0 = \Phi(t)$, *we obtain a strongly continuous semigroup of nonlinear operators* $\{S(t)\}_{t \geq 0}$. *Moreover, the semigroup* $\{S(t)\}_{t \geq 0}$ *has the absorbing property.*

Discussion of Proof. If we begin with initial data $\Phi_0 \in \mathcal{D}(\widehat{G})$ the regularity results of the last section allow us to work with the abstract evolution equation (2.8) on the maximal interval of existence. We remark that this is the most convenient form of the system for energy estimates. Following the energy arguments of [15, Lemma 3.1], we let $\epsilon > 0$ and take the inner product of (2.8) with $2w_t + w$ in H. Subsequent computations demonstrate that if we define a nonlinear functional $\Gamma(\cdot)$ for sufficiently small $\epsilon > 0$ by

$$\Gamma(t) = |w_t|^2 + \alpha \left| A^{1/2} w \right|^2 + \epsilon \langle w_t, w \rangle + \frac{1}{2} \left(\left| A^{1/4} w \right|^2 + \beta \right)^2, \qquad (3.1)$$

then $\Gamma(t)$ can be shown to satisfy the inequality

$$\frac{d\Gamma(t)}{dt} + \frac{\epsilon}{2}\Gamma(t) \leq \frac{\epsilon\beta^2}{2}, \ t \in [0, t_{\max}). \tag{3.2}$$

(We refer the reader to [15, Lemma 3.1 and Theorem 3.2] for more details). From (3.2) it can be shown that

$$\frac{\min\{1, \ \alpha\}}{2} \left\| S(t) \begin{bmatrix} w_0 \\ w_1 \end{bmatrix} \right\|_X^2 \leq e^{-\epsilon t/2}\Gamma(0) + \beta^2, \ t \in [0, \ t_{\max}). \tag{3.3}$$

Since (3.3) produces a uniform a priori estimate which depends only on the parameters of the equation and the norm of $\Phi_0 = \begin{bmatrix} w_0 \\ w_1 \end{bmatrix}$ in X, we can use the density of $\mathcal{D}\left(\widehat{G}\right)$ in X to extend the estimate (3.3) to mild solutions with any initial data $\Phi_0 \in X$. By [10, Theorem 6.1.4] this estimate also shows that the mild solution $S(t)\Phi_0$ of (2.14)-(2.15) exists uniquely on $[0, \infty)$.

It follows from (3.3) that

$$\limsup_{t \to \infty} \left\| S(t) \begin{bmatrix} w_0 \\ w_1 \end{bmatrix} \right\|_X^2 \leq \frac{2\beta^2}{\min\{1, \ \alpha\}}. \tag{3.4}$$

Letting

$$\rho_0^2 = \frac{2\beta^2}{\min\{1, \ \alpha\}}, \tag{3.5}$$

it is not difficult to show that the bounded ball $B_0 = \{x \in X : \|x\|_X \leq \sqrt{2}\rho_0\}$ is an absorbing set.

4. GLOBAL ATTRACTOR

A set $\mathcal{A}$ is called a *global attractor* for a nonlinear semigroup $\{S(t)\}_{t\geq 0}$ on a Banach space X if:

(i) $\mathcal{A}$ is invariant under $\{S(t)\}_{t\geq 0}$, i.e. $S(t)\mathcal{A} = \mathcal{A}$ for all $t \geq 0$;

(ii) $\mathcal{A}$ is compact and uniformly attracts every bounded subset of X, i.e. $\lim_{t \to 0} [dist_X(S(t)u_0, \mathcal{A})] = 0$ for any $u_0 \in X$ and the convergence is uniform for u_0 in any given bounded subset.

In order to prove the existence of a global attractor in this case, we use the following result in [13, Chapt. 2, Thm. 2.3.9 or Existence III], as stated in Lemma 4.1.

LEMMA 4.1. *Let σ be a dissipative and κ-contracting semiflow on a banach space $\overline{W}$, associated with a strongly continuous semigroup $\{S(t)\}_{t\geq 0}$. If for each compact set $\Delta \subset \overline{W}$, the positive trajectory bundle $\gamma^+(\Delta)$ is bounded, then there is a global attractor $\mathcal{A}$ in $\overline{W}$ for σ or $\{S(t)\}_{t\geq 0}$.*

Note that here κ is the Kuratowski measure of noncompactness in $\overline{W}$, and dissipativity means the existence of absorbing sets. For the solution semigroup $S(t)$ defined in Theorem 3.1, we have already shown that it is dissipative and, by (3.3), the property for each compact set $\Delta \subset X$ mentioned as a condition in Lemma 4.1 is also possessed by our $S(t)$. Therefore, we need only to verify the κ-contracting property for the solution semigroup.

A pseudometric ρ on a Banach space $\overline{W}$ is called precompact if any bounded sequence in $\overline{W}$ has a subsequence which is a Cauchy sequence with respect to ρ. This concept is in [8, p. 16]. The following lemma is very useful.

LEMMA 4.2. *Suppose that a strongly continuous semigroup $\{S(t)\}_{t\geq 0}$ on a Banach space $\overline{W}$ satisfies*

$$\|S(t)w_1 - S(t)w_2\| \leq Ce^{-\theta t}\|w_1 - w_2\| + \rho_t(w_1, w_2), \ t > \tau,$$

$$for \ any \ w_1, \ w_2 \in \overline{W}, \tag{4.1}$$

where $C \geq 0$, $\tau \geq 0$, and $\theta > 0$ are constants, and ρ_t is a precompact pseudometric on $\overline{W}$ for each $t > \tau$. Then $\{S(t)\}_{t\geq 0}$ and the corresponding semiflow are κ-contracting.

The proof of this lemma is omitted, because it is a direct corollary of Lemma 2.3.6 in [8], by the definition of κ-contracting semigroups. Now we want to show that our solution semigroup satisfies (4.1) as above.

Define a functional $E(w, g) : X \to R$ by

$$E(w, g) = \int_0^1 [\alpha|w_{xx}|_{R^2}^2 + b\langle w, g\rangle_{R^2} + |g|_{R^2}^2]\, dx, \tag{4.2}$$

where the parameter $b > 0$ is chosen to be small enough and fixed, so that $E(w, g) \geq 0$ always holds and $E(w, g)^{1/2}$ is an equivalent norm in the space X. For any two trajectories $\begin{pmatrix} w_i(x, t) \\ w_{it}(x, t) \end{pmatrix}$ in X, $i = 1, 2$, of the evolution equation (2.14), denote their difference by

$$\widetilde{w}(x, t) = w_1(x, t) - w_2(x, t). \tag{4.3}$$

LEMMA 4.3. *There is a constant $\theta > 0$ and for any bounded set $B \subset X$, there is a constant $L = L(B)$ such that for each $t > 0$, and for any two mild solutions $\begin{pmatrix} w_i \\ w_{it} \end{pmatrix}$, $i = 1, 2$, of the equation (2.14), with the initial data in the bounded set B, it holds that*

$$E(\widetilde{w}(t), \widetilde{w}_t(t)) \leq e^{-\theta t}E(\widetilde{w}(0), \widetilde{w}_t(0)) + L \sup_{0\leq s\leq t} \|\widetilde{w}(s)\|_H; \tag{4.4}$$

here $\widetilde{w}(t) = \widetilde{w}(\cdot,\, t)$ as defined by (4.3) and $\widetilde{w}_t$ is in the distribution sense.

Proof. $\widetilde{w}(x,\, t)$ satisfies the equation

$$
\begin{aligned}
\widetilde{w}_{tt} + \alpha\widetilde{w}_{xxxx} + \delta\widetilde{w}_t &= \left(\beta + \int_0^1 |w_{1x}|^2_{R^2}\, dx\right) w_{1xx} - \left(\beta + \int_0^1 |w_{2x}|^2_{R^2}\, dx\right) w_{2xx} \\
&= \left(\beta + \int_0^1 |w_{1x}|^2\, dx\right) \widetilde{w}_{xx} + \left(\int_0^1 [|w_{1x}|^2 - |w_{2x}|^2]\, dx\right) w_{2xx} \\
&= \left(\beta + \int_0^1 |w_{1x}|^2\, dx\right) \widetilde{w}_{xx} - \left(\int_0^1 \langle w_1 - w_2,\, w_{1xx} + w_{2xx}\rangle_{R^2}\, dx\right) w_{2xx},
\end{aligned} \tag{4.5}
$$

in which the integration by parts is used to get the last equality. Thanks to (3.3), we have

$$
\left\| \left(\int_0^1 \langle w_1 - w_2,\, w_{1xx} + w_{2xx}\rangle_{R^2}\, dx\right) w_{2xx}\right\|_H \le L_1(B)\|w_1 - w_2\|_H, \tag{4.6}
$$

for each $t > 0$ not explicitly written and for some constant $L_1(B)$. Define a perturbation of the E-functional by

$$
M\left(\widetilde{w},\, \widetilde{w}_t\right) = E\left(\widetilde{w},\, \widetilde{w}_t\right) - \int_0^1 \left(\beta + \int_0^1 |w_{1x}|^2_{R^2}\, dx\right) |\widetilde{w}_x|^2_{R^2}\, dx. \tag{4.7}
$$

By using the multiplier method to conduct *a priori* estimates similar to the proof of [15, Lemma 3.1] on the brief proof of Theorem 3.1, we can show that there is a constant $\theta > 0$ such that, for $t > 0$,

$$
\frac{d}{dt} M\left(\widetilde{w}(t),\, \widetilde{w}_t(t)\right) \le -\theta M\left(\widetilde{w}(t),\, \widetilde{w}_t(t)\right)
$$

$$
+ \left| \int_0^1 \left(\beta + \int_0^1 |w_{1x}|^2_{R^2}\, dx\right) |\widetilde{w}_x|^2_{R^2}\, dx\right| + L_2(B)\|w_1(t) - w_2(t)\|_H
$$

$$
\le -\theta M\left(\widetilde{w}(t),\, \widetilde{w}_t(t)\right) + L_3(B)\|\widetilde{w}(t)\|_H, \tag{4.8}
$$

where $L_2(B)$ and $L_3(B)$ are constants, and the last step follows from

$$
\left| \int_0^1 \left(\beta + \int_0^1 |w_{1x}|^2\, dx\right) |\widetilde{w}_x|^2\, dx\right| = \left| \int_0^1 \left(\beta + \int_0^1 |w_{1x}|^2\, dx\right) \langle \widetilde{w},\, \widetilde{w}_{xx}\rangle\, dx\right|
$$

$$
\le const(B)\, \|\widetilde{w}(t)\|_H.
$$

Integrating the differential inequality (4.8), we obtain

$$
M\left(\widetilde{w}(t),\, \widetilde{w}_t(t)\right) \le e^{-\theta t} M\left(\widetilde{w}(0),\, \widetilde{w}_t(0)\right) + L_4(B) \sup_{0 \le s \le t} \|\widetilde{w}(s)\|_H, \tag{4.9}
$$

which together with (4.7) implies that

$$
E\left(\widetilde{w}(t),\, \widetilde{w}_t(t)\right) \le e^{-\theta t} E\left(\widetilde{w}(0),\, \widetilde{w}_t(0)\right) + L_5(B) \sup_{0 \le s \le t} \|\widetilde{w}(s)\|_H, \tag{4.10}
$$

for each $t > 0$. Letting $L = L_5(B)$, the lemma is proved.

LEMMA 4.4. *The solution semigroup $S(t)$ defined in Theorem 3.1 satisfies (4.1). Therefore, it is a κ-contracting semigroup in X.*

Proof. Since the κ-contracting property is an asymptotic property, without loss of generality, we can only consider all trajectories within the absorbing set $B_0 \subset X$ of $S(t)$, as found in the proof of Theorem 3.1. Regarding these trajectories as started from somewhere in B_0, we can use Lemma 4.3 to assert that

$$\left\| S(t) \begin{pmatrix} w_1(0) \\ w_{1t}(0) \end{pmatrix} - S(t) \begin{pmatrix} w_2(0) \\ w_{2t}(0) \end{pmatrix} \right\|_X \leq C_0 e^{-\theta t} \left\| \begin{pmatrix} w_1(0) \\ w_{1t}(0) \end{pmatrix} - \begin{pmatrix} w_2(0) \\ w_{2t}(0) \end{pmatrix} \right\|_X$$
$$+ C_1 L(B_0) \sup_{0 \leq s \leq t} \|w_1(s) - w_2(s)\|_H , \tag{4.11}$$

where C_0, $C_1 > 0$ come from the norm equivalence, and $t = 0$ is indeed a shifted starting time that is uniform to initial data belonging to any given bounded set in X.

Now define a pseudometric ρ_t by

$$\rho_t \left(\begin{pmatrix} w_1(0) \\ w_{1t}(0) \end{pmatrix}, \begin{pmatrix} w_2(0) \\ w_{2t}(0) \end{pmatrix} \right) = C_1 L(B_0) \sup_{0 \leq s \leq t} \|w_1(s) - w_2(s)\|_H . \tag{4.12}$$

Note that $V \subset H$ is a compact imbedding and that the first component $w(t)$ of any trajectory within B_0 is in a bounded subset $P_V(B_0)$ of V, with $P_V : X \to V$ the orthogonal projection. It follows that the pseudometric ρ_t given by (4.12) is precompact in X. Then by Lemma 4.2, the solution semigroup $S(t)$ is κ-contracting in X.

Finally, by assembling the above four lemmas, we have actually proved the following theorem.

THEOREM 4.5. *The semigroup of nonlinear transformations $\{S(t)\}_{t \geq 0}$ on X possesses a global attractor in X which is compact and maximal in X.*

5. INERTIAL MANIFOLD

An *inertial manifold* for a dynamical system $\{S(t)\}_{t \geq 0}$ in a Banach space X is a set M which has the following properties:

(i) M is a finite dimensional Lipschitz manifold;

(ii) M is positively invariant for $S(t)$, that is, $S(t)M \subset M$, $t \geq 0$;

(iii) M exponentially attracts all the trajectories of the system, that is, for any bounded subset B of X there is a constant $C(B)$ such that

$$dist_X(S(t)\Phi_0, M) \leq C(B) e^{-\nu t}, \ t \geq 0,$$

for any $\Phi_0 \in B$, where ν is a uniform constant.

If an inertial manifold exists, it necessarily contains the global attractor $\mathcal{A}$.

Because of the vector formulation (2.3a-2.3c) of our coupled beam system we can use the results of [15, Section 5] to show the existence of an inertial manifold for the dynamical system associated with the abstract evolution equation (2.14)-(2.15). We shall state the result and refer the reader to [15, Theorem 5.1] for the proof. Let $H_m = span\{x_1, \ldots, x_m\}$ where $\{x_1, \ldots, x_m\}$ are the first m eigenvectors of the operator $A : \mathcal{D}(A) \to H$ defined by (2.4a-b). Let $d = |\beta| + \frac{2\rho_0^2}{n^2\pi^2}$, where ρ_0^2 is defined by (3.5).

THEOREM 5.1. *For a suitably large integer m such that*

$$m^2\pi^2 \geq \max\left\{\frac{4d}{\alpha}, \ \frac{8\rho_0^2}{\epsilon\alpha} \ \frac{2|\beta|+1}{\alpha}\right\},$$

where

$$\epsilon = \min\left\{\frac{1}{2}, \ 2\delta\left[3 + \frac{2\delta^2}{\alpha\pi^4}\right]^{-1}\right\},$$

the flat manifold

$$M_m = H_m \times H_m$$

is an inertial manifold in X for the dynamical system $\{S(t)\}_{t\geq 0}$ associated with (2.14)-(2.15).

We note also that as a consequence of obtaining the inertial manifold M_m, we also obtain a so-called *inertial form* for (2.14)-(2.15), which is the restriction of the initial value problem to M_m, and takes the form of an initial value problem for an ordinary differential system.

References

[1] J.M. Ball, Stability theory for an extensible beam, *J. Diff. Equations* **14** (1973), 399-418.

[2] P. Biler, Remark on the decay for damped string and beam equations, *Nonlinear Anal., TMA* **10** (1986), 839-842.

[3] E.H. De Brito, Nonlinear initial-boundary value problems, *Nonlinear Anal., TMA* **11** (1987), 125-137.

[4] R.W. Dickey, Dynamic stability of equilibrium states of the extensible beam, *Proceedings, AMS* **41** (1973), 94-102.

[5] W.E. Fitzgibbon, Strongly damped quasilinear evolution equations, *J. Math. Anal. Appl.* **79** (1981), 536-550.

[6] G.J. Fix and R. Kannan, Nonplanar oscillation of beams under periodic forcing, *J. Diff. Equations* **96** (1992), 419-429.

[7] E.C. Haight and W.W. King, Stability of nonlinear oscillations of an elastic rod, *J. Acoustical Soc. Am.* **52** (1972), 899-911.

[8] J. Hale, "Asymptotic Behavior of Dissipative Systems", Math. Surveys and Monographs, Vol. 25, AMS, Providence, RI, 1988.

[9] C.-H. Ho, R.A. Scott and J.G. Eisley, Non-planar, non-linear oscillations of a beam — I. forced motions, *Int. J. Non-Linear Mechanics* **10** (1975), 113-127.

[10] A. Pazy, "Semigroups of Linear Operators and Applications to Partial Differential Equations", Springer-Verlag, New York, 1983.

[11] D.C. Pereira, Existence, uniqueness and asymptotic behavior for solutions of the nonlinear beam equation, *Nonlinear Anal., TMA* **14** (1990), 613-623.

[12] E.L. Reiss and B.J. Matkowsky, Nonlinear dynamic buckling of a compressed elastic column, *Quart. Appl. Math.* **29** (1971), 245-260.

[13] G.R. Sell and Y. You, "Infinite Dimensional Dynamical Systems", book manuscript, 1993.

[14] R. Temam, "Infinite Dimensional Dynamical Systems in Mechanics and Physics", Springer-Verlag, New York, 1988.

[15] M. Taboada and Y. You, Global attractor and inertial manifolds of nonlinear damped beam equations, to appear in Comm. Partial Diff. Equs.

New Trace Formulas for Schrödinger Operators

F. Gesztesy   University of Missouri, Columbia, Missouri

1   INTRODUCTION

The purpose of this article is to give an overview of some of our recent work on new trace formulas for one-dimensional Schrödinger operators obtained in various joint collaborations with H. Holden, B. Simon, and Z. Zhao [15], [16], [17], [21].

While [15] is devoted to a detailed study of trace formulas and their relation to the Korteweg-de Vries (KdV) hierarchy in connection with KdV solutions which have sufficiently fast decay at spatial infinity, the main results in [16], [17], and [21] are trace formulas for Schrödinger operators $H = -\frac{d^2}{dx^2} + V$ in $L^2(\mathbb{R}; dx)$ for a large class of potentials which satisfy a minimal set of apriori restrictions. In particular, we were interested in finding a general trace formula for any continuous potential V bounded from below.

As shown in great detail in [7] in connection with short-range potentials, the trace formula (2.36) for $V(x)$ is a central object in inverse scattering theory. Similarly, the trace formula (3.43) studied in [6], [8]–[11], [13], [22], [23], [26]–[30] in connection with periodic potentials, algebro-geometric quasi-periodic finite-gap potentials, and certain classes of almost-periodic potentials has been vital in solving the associated inverse spectral problem and the corresponding Cauchy problem for the KdV equation. Moreover, the description of isospectral manifolds of quasi-periodic finite-gap potentials (as well as solutions of the KdV hierarchy), obtained by algebro-geometric methods, is based on trace formulas. Hence it is natural to look for extensions of these special trace formulas to larger classes of potentials in order to be able to study inverse spectral problems and the KdV initial value problem in a more general setting. Our main strategy in this endeavor is to compare $H = -\frac{d^2}{dx^2} + V$ and H_y^D, the operator H plus an additional Dirichlet boundary condition at the point $y \in \mathbb{R}$. The spectral characteristics of H and H_x^D, especially the corresponding Krein spectral shift function $\xi(\lambda, x)$ associated with the pair (H_x^D, H) [24] then enables one to recover the potential $V(x)$. As will be demonstrated in the course of this article, the function $\xi(\lambda, x)$ provides a powerful new tool in the spectral and inverse spectral analysis of H.

In Section 2 we review some of the results in [15] on short-range trace formulas and KdV invariants (see Theorem 2.1, Corollary 2.2) and also prove a new set of such trace formulas in Theorem 2.3 and Corollary 2.4. We then use these results to show the connection between the trace formula (2.36) for $V(x)$ and an alternative one obtained earlier in [31] (see, in particular, (2.51)–(2.53)). The short-range trace formula (2.51) (and its analogs for higher KdV invariants) is then extended in Section 3 to any smooth potential $V(x)$ bounded from below. Our main results are summarized in (3.32) and Theorem 3.3 following the original treatment in [21]. An alternative proof of the celebrated trace formulas for periodic potentials is then recovered in Theorem 3.4. At the end of Section 3 we indicate how to replace the Dirichlet boundary condition $g(x) = 0$ by any other self-adjoint boundary condition of the type $g'(x) + \beta g(x) = 0$, $\beta \in \mathbb{R}$. Our final Section 4 then provides a brief outlook on work in progress to be published in [16] and [17]. We briefly discuss potentials with different spatial asymptotics, impurity scattering in one-dimensional crystals, the case $V(x) \underset{|x| \to \infty}{\longrightarrow} \infty$ such that H has purely discrete spectrum, second-order finite-difference operators, a powerful new criterion for the absolutely continuous spectrum of H, and a new Borg-type theorem concerning inverse spectral theory for operators H with purely discrete spectrum (see Theorem 4.2).

Pertinent references to the literature are given at the end of each section. Some of the results in [16], [17], and [21] have been announced in [20].

2   TRACE FORMULAS AND KDV INVARIANTS. I. THE SHORT-RANGE CASE

Throughout this section we shall impose the short-range conditions

$$V \in L^1(\mathbb{R}; (1 + |x|)\, dx), \ V^{(m)} \in L^1(\mathbb{R}; dx), \quad m \in \mathbb{N} \tag{2.1}$$

for the real-valued potential V in the self-adjoint one-dimensional Schrödinger operator

$$H = -\frac{d^2}{dx^2} + V, \ \mathcal{D}(H) = H^{2,2}(\mathbb{R}) \tag{2.2}$$

in $L^2(\mathbb{R}; dx)$ (with $H^{m,n}(\mathbb{R})$, $m, n \in \mathbb{N}$ the usual Sobolev spaces). The Jost solutions $f_\pm(z, x)$ of H, given by

$$f_\pm(z, x) = e^{\pm i z^{1/2} x} - \int_x^{\pm\infty} dx'\, z^{-1/2} \sin[z^{1/2}(x - x')] V(x') f_\pm(z, x'),$$

$$z \in \mathbb{C} \setminus [0, \infty), \ \text{Im}(z^{1/2}) > 0, \ x \in \mathbb{R}, \quad (2.3)$$

$$H f_\pm(z) = z f_\pm(z) \quad \text{(distributional sense)}, \tag{2.4}$$

then yield the (unitary) scattering matrix $S(\lambda)$, $\lambda > 0$ in $\mathbb{C}^2$ associated with the pair (H, H_0) $(H_0 = -\frac{d^2}{dx^2}, \mathcal{D}(H_0) = H^{2,2}(\mathbb{R}))$. One obtains explicitly

$$S(\lambda) = \begin{pmatrix} T(\lambda) & R^r(\lambda) \\ R^\ell(\lambda) & T(\lambda) \end{pmatrix}, \quad \lambda > 0, \tag{2.5}$$

$$T(\lambda) = \frac{2i\lambda^{1/2}}{W(f_-(\lambda), f_+(\lambda))} = \left[1 - \frac{1}{2i\lambda^{1/2}} \int_{\mathbb{R}} dx V(x) e^{\pm i\lambda^{1/2}x} f_{\mp}(\lambda, x)\right]^{-1}, \tag{2.6}$$

$$R^\ell(\lambda) = -\frac{W(\overline{f_-(\lambda)}, f_+(\lambda))}{W(f_-(\lambda), f_+(\lambda))} = \frac{T(\lambda)}{2i\lambda^{1/2}} \int_{\mathbb{R}} dx V(x) e^{i\lambda^{1/2}x} f_+(\lambda, x), \tag{2.7}$$

$$R^r(\lambda) = -\frac{W(f_-(\lambda), \overline{f_+(\lambda)})}{W(f_-(\lambda), f_+(\lambda))} = \frac{T(\lambda)}{2i\lambda^{1/2}} \int_{\mathbb{R}} dx V(x) e^{-i\lambda^{1/2}x} f_-(\lambda, x), \tag{2.8}$$

where

$$f_\pm(\lambda, x) := \lim_{\epsilon \downarrow 0} f_\pm(\lambda + i\epsilon, x), \quad \lambda > 0, \tag{2.9}$$

$W(f, g)(x) = f(x)g'(x) - f'(x)g(x)$ denotes the Wronskian of f and g, and

$$T(\lambda) f_\pm(\lambda, x) = R^r(\lambda) f_\mp(\lambda, x) + \overline{f_\mp(\lambda, x)}, \quad \lambda > 0. \tag{2.10}$$

Introducing $\phi_\pm(\lambda, x)$ by

$$f_\pm(z, x) = \exp\left[\pm i z^{1/2}x + \int_{\pm\infty}^x dx' \phi_\pm(z, x')\right], \quad z \in \mathbb{C}\backslash [0, \infty),\ \mathrm{Im}(z^{1/2}) > 0, \tag{2.11}$$

then (2.4) yields the Riccati-type equation

$$\phi'_\pm(z, x) \pm 2i z^{1/2}\phi_\pm(z, x) + \phi_\pm(z, x)^2 - V(x) = 0. \tag{2.12}$$

$\phi_\pm(z, x)$ admits the asymptotic expansion

$$\phi_\pm(z, x) \underset{\substack{|z|\to\infty \\ \mathrm{Im}(z^{1/2})\geq 0}}{\sim} \sum_{n=1}^\infty \phi_{\pm,n}(x)(2i z^{1/2})^{-n}, \tag{2.13}$$

where $\phi_{\pm,n}(x)$ satisfy the recursion relation

$$\phi_{\pm,1}(x) = \pm V(x),\ \phi_{\pm,2}(x) = -V'(x),$$
$$\phi_{\pm,n+1}(x) = \mp\phi'_{\pm,n}(x) \mp \sum_{m=1}^{n-1} \phi_{\pm,m}(x)\phi_{\pm,n-m}(x), \quad n \geq 2 \tag{2.14}$$

and one verifies that

$$\phi_{-,m}(x) = (-1)^m \phi_{+,m}(x), \quad m \in \mathbb{N} \tag{2.15}$$

and that $\phi_{\pm,2m}(x)$ are total derivatives, i.e.,

$$\phi_{\pm,2m}(x) = \eta'_m(x), \quad m \in \mathbb{N} \tag{2.16}$$

for appropriate differential polynomials η_m with respect to $V(x)$. Explicitly, one computes

$$\phi_{\pm,1}(x) = \pm V(x),\ \phi_{\pm,2}(x) = -V'(x),\ \phi_{\pm,3}(x) = \pm[V''(x) - V(x)^2], \text{ etc.} \tag{2.17}$$

Moreover, we have

$$T(z) = \exp\left[\pm \int_{\mathbb{R}} dx \phi_\pm(z, x)\right], \quad z \in \mathbb{C}\backslash [0, \infty),\ \mathrm{Im}(z^{1/2}) > 0, \tag{2.18}$$
$$T(\lambda) = \lim_{\epsilon \downarrow 0} T(\lambda + i\epsilon), \quad \lambda > 0 \tag{2.19}$$

and

$$G(z, x, x') = -\frac{T(z)}{2i z^{1/2}} f_+(z, x) f_-(z, x'), \quad x' \leq x,\ z \in \mathbb{C}\backslash [0, \infty),\ \mathrm{Im}(z^{1/2}) > 0, \tag{2.20}$$

where $G(z, x, x')$ denotes the Green's function of H (i.e., the integral kernel of $(H - z)^{-1}$). One can prove the asymptotic expansion

$$T(z)f_+(z, x)f_-(z, x) \underset{\substack{|z| \to \infty \\ \mathrm{Im}(z^{1/2}) \geq 0}}{\sim} \sum_{n=0}^{\infty} \omega_n(x)(2iz^{1/2})^{-n}, \tag{2.21}$$

where

$$\omega_0(x) = 1, \tag{2.22}$$

$$\omega_{2\ell}(x) = -2\phi_{+,2\ell-1}(x) - 2\sum_{j=1}^{\ell-1} \phi_{+,2(\ell-j)-1}(x)\omega_{2j}(x), \quad \ell \in \mathbb{N},$$

$$\omega_{2\ell+1}(x) = 0, \quad \ell \in \mathbb{N}_0 \tag{2.23}$$

and the fact that

$$\omega_{2\ell+2}(x) = -2(2\ell + 1)\phi_{+,2\ell+1}(x) + \zeta_\ell'(x), \quad \ell \in \mathbb{N}_0 \tag{2.24}$$

for appropriate ζ_ℓ with $\zeta_\ell^{(m)} \in L^1(\mathbb{R}; dx)$, $m \in \mathbb{N}_0$ ($\mathbb{N}_0 := \mathbb{N} \cup \{0\}$). Explicitly, one obtains

$$\omega_0(x) = 1, \ \omega_2(x) = -2V(x), \ \omega_4(x) = 6V(x)^2 - 2V''(x), \ \text{etc.} \tag{2.25}$$

Equations (2.20)–(2.23) then yield a high-energy asymptotic expansion for the diagonal Green's function $G(z, x, x)$ of H. Next, we need a bit more notation associated with the eigenvalues of H. Let

$$\sigma_p(H) = \begin{cases} \{-\kappa_j^2\}_{j\in J}, & J \neq \emptyset \\ \emptyset, & J = \emptyset \end{cases} \tag{2.26}$$

be the (possibly empty) set of eigenvalues of H where

$$\kappa_j > 0 \tag{2.27}$$

and J is a finite (possibly empty) index set. The corresponding normalized real-valued eigenfunctions ψ_j of H associated with the (necessarily simple) eigenvalue $\lambda_j = -\kappa_j^2$ of H then reads

$$\begin{aligned} \psi_j(x) &= c_{\pm,j} f_\pm(\lambda_j, x), \\ H\psi_j &= \lambda_j\psi_j, \ \psi_j \in H^{2,2}(\mathbb{R}), \ \|\psi_j\|_2 = 1, \quad j \in J. \end{aligned} \tag{2.28}$$

$T(z)$ is meromorphic in $\mathbb{C}\backslash[0, \infty)$ and has simple poles at λ_j with residues

$$\operatorname*{Res}_{z=\lambda_j}[T(z)] = -2\kappa_j\gamma_j^{\mp 1}c_{\pm,j}^2, \quad j \in J, \tag{2.29}$$

where

$$f_-(\lambda_j, x) = \gamma_j f_+(\lambda_j, x), \ c_{+,j}^2 = \gamma_j^2 c_{-,j}^2, \ c_{\pm,j}^2 = \|f_\pm(\lambda_j)\|_2^{-2}, \quad c_{+,j} > 0, \ j \in J. \tag{2.30}$$

In addition, $\lim_{\epsilon\downarrow 0} T(\lambda \pm i\epsilon)$ are continuous on $[0, \infty)$. Similarly, $f_\pm(z, x)$, for fixed $x \in \mathbb{R}$, are analytic in $\mathbb{C}\backslash[0, \infty)$ and the boundary values $\lim_{\epsilon\downarrow 0} f_\pm(\lambda \pm i\epsilon, x)$ are continuous for $\lambda \in (0, \infty)$.

Given these preliminaries we can state the following trace formulas.

Theorem 2.1.

$$\omega_{2\ell+2}(x) = 2^{2\ell+3}\sum_{j\in J}\kappa_j^{2\ell+1}\psi_j(x)^2$$

$$+(-1)^\ell 2^{2\ell+2}\pi^{-1}\int_0^\infty d\lambda\lambda^\ell\,\mathrm{Im}[T(\lambda)f_+(\lambda,x)f_-(\lambda,x)], \quad \ell\in\mathbb{N}_0,\ x\in\mathbb{R} \qquad (2.31)$$

and

$$\int_0^\infty d\lambda\lambda^{\ell-1/2}\,\mathrm{Re}\left[T(\lambda)f_+(\lambda,x)f_-(\lambda,x) - \sum_{m=0}^\ell (-1)^m 2^{-2m}\omega_{2m}(x)\lambda^{-m}\right]$$

$$= (-1)^{\ell+1}2\pi\sum_{j\in J}\kappa_j^{2\ell}\psi_j(x)^2, \quad \ell\in\mathbb{N}_0,\ x\in\mathbb{R}. \qquad (2.32)$$

For a proof of Theorem 2.1 one integrates

$$(2iz^{1/2})^{n-1}\left[T(z)f_+(z,x)f_-(z,x) - \sum_{m=0}^n \omega_m(x)(2iz^{1/2})^{-m}\right], \quad n\in\mathbb{N}_0 \qquad (2.33)$$

counter clockwise over the following contour $\Gamma_{R,\epsilon}$, $\epsilon>0$, $R>0$

$$\Gamma_{R,\epsilon} = \left\{z=\epsilon e^{i\varphi}\,\Big|\,\frac{3\pi}{2}\geq\varphi\geq\frac{\pi}{2}\right\}\cup\{z=\lambda+i\epsilon\,|\,0\leq\lambda\leq R\} \qquad (2.34)$$

$$\cup\{z=Re^{i\varphi}\,|\,\arctan(\epsilon/R)\leq\varphi\leq 2\pi-\arctan(\epsilon/R)\}\cup\{z=\lambda-i\epsilon\,|\,0\leq\lambda\leq R\}$$

applying (2.29) and the residue theorem. Here R, ϵ^{-1} are chosen large enough to encircle all eigenvalues $\lambda_j = -\kappa_j^2$, $j\in J$ of H. Formulas (2.31) and (2.32) then result in the limit $R\uparrow\infty$, $\epsilon\downarrow 0$.

Using (2.10), (2.31) can be rewritten as follows.

Corollary 2.2.

$$\omega_{2\ell+2}(x) = 2^{2\ell+3}\sum_{j\in J}\kappa_j^{2\ell+1}\psi_j(x)^2$$

$$+(-1)^\ell 2^{2\ell+2}\pi^{-1}\int_0^\infty d\lambda\lambda^\ell\,\mathrm{Im}[R^\ell(\lambda)f_\pm(\lambda,x)^2], \quad \ell\in\mathbb{N}_0,\ x\in\mathbb{R}. \qquad (2.35)$$

In particular, for $\ell=0$ one obtains from (2.25) the well-known trace formula for $V(x)$ (see [7])

$$V(x) = -4\sum_{j\in J}\kappa_j\psi_j(x)^2 - 2\pi^{-1}\int_0^\infty d\lambda\,\mathrm{Im}[\overset{r}{R^\ell}(\lambda)f_\pm(\lambda,x)^2]. \qquad (2.36)$$

It should be stressed in connection with (2.35) and (2.36) that

$$R^{r(\ell)}(.) \in \mathcal{S}(\mathbb{R}) \qquad (2.37)$$

($\mathcal{S}(\mathbb{R})$ the usual Schwartz space) by combining (2.1), (2.7), (2.8), and the Riemann-Lebesgue lemma.

In order to derive a different set of trace formulas we replace $T(z)f_+(z,x)f_-(z,x)$ by $T(z)f_\pm(z,x)e^{\mp iz^{1/2}x}$ and note the asymptotic expansion

$$T(z)f_\pm(z,x)e^{\mp iz^{1/2}x} = \exp\left[\int_{\mp\infty}^x dx'\phi_\pm(z,x')\right]$$

$$\underset{\substack{|z|\to\infty\\ \mathrm{Im}(z^{1/2})\geq 0}}{\sim}\exp\left[\sum_{n=1}^\infty \mu_{\pm,n}(x)(2iz^{1/2})^{-n}\right]\underset{\substack{|z|\to\infty\\ \mathrm{Im}(z^{1/2})\geq 0}}{\sim}\sum_{n=0}^\infty \nu_{\pm,n}(x)(2iz^{1/2})^{-n}, \qquad (2.38)$$

where

$$\mu_{\pm,n}(x) = \int_{\mp\infty}^{x} dx' \phi_{\pm,n}(x), \quad n \in \mathbb{N}, \tag{2.39}$$

$$\nu_{\pm,0}(x) = 1,$$

$$\nu_{\pm,n}(x) = \sum_{m=1}^{n} (m!)^{-1} \sum_{\substack{\tau:\{1,\dots,m\}\to\mathbb{N} \\ |\tau|=\tau(1)+\cdots+\tau(m)=n}} \mu_{\pm,\tau(1)}(x)\mu_{\pm\tau(2)}(x)\cdots\mu_{\pm\tau(m)}(x). \tag{2.40}$$

Explicitly, we have

$$\nu_{\pm,0}(x) = 1, \ \nu_{\pm,1}(x) = \mu_{\pm,1}(x) = \pm\int_{\mp\infty}^{x} dx' V(x'),$$

$$\nu_{\pm,2}(x) = \mu_{\pm,2}(x) + \frac{1}{2}\mu_{\pm,1}(x)^2 = -V(x) + \frac{1}{2}\left[\int_{\mp\infty}^{x} dx' V(x')\right]^2, \ \text{etc.} \tag{2.41}$$

In analogy to Theorem 2.1 one then derives

Theorem 2.3.

$$\nu_{\pm,n+1}(x) = (-1)^{n+1} 2^{n+2} \sum_{j\in J} \kappa_j^n c_{\mp,j} \psi_j(x) e^{\pm\kappa_j x}$$

$$+ 4\pi^{-1} \int_0^{\infty} d\lambda \, \mathrm{Im}\left\{ (2i\lambda^{1/2})^{n-1}\left[T(\lambda) f_{\pm}(\lambda, x) e^{\mp i\lambda^{1/2}x} - \sum_{\substack{m=0 \\ (n-m)\text{even}}}^{n} \nu_{\pm,m}(x)(2i\lambda^{1/2})^{-m}\right]\right\},$$

$$n \in \mathbb{N}_0, \ x \in \mathbb{R}. \tag{2.42}$$

In order to prove (2.42) one integrates

$$(2iz^{1/2})^{n-1}\left[T(z) f_{\pm}(z, x) e^{\mp iz^{1/2}x} - \sum_{m=0}^{n} \nu_{\pm,m}(x)(2iz^{1/2})^{-m}\right] \tag{2.43}$$

again counter clockwise over the contour $\Gamma_{R,\epsilon}$ in (2.34) and computes the limit $R \uparrow \infty$, $\epsilon \downarrow 0$.

Using (2.10), (2.42) can be rewritten in the following form.

Corollary 2.4.

$$\nu_{\pm,n+1}(x) = (-2)^{n+1} \sum_{j\in J} \kappa_j^n c_{\mp,j} \psi_j(x) e^{\pm\kappa_j x}$$

$$+ 2\pi^{-1} \int_0^{\infty} d\lambda \, \mathrm{Im}[(2i\lambda^{1/2})^{n-1} R^r(\lambda) f_{\mp}(\lambda, x) e^{\mp i\lambda^{1/2}x}], \quad n \in \mathbb{N}_0, \ x \in \mathbb{R}. \tag{2.44}$$

In particular, for $n = 0$ one obtains from (2.41) and (2.44) an alternative trace formula for $V(x)$,

$$V(x) = \mp 2 \sum_{j\in J} c_{\mp,j} \frac{d}{dx}[\psi_j(x) e^{\pm\kappa_j x}]$$

$$\mp 2\pi^{-1} \int_0^{\infty} d\lambda (2\lambda^{1/2})^{-1} \mathrm{Re}\left\{ R^r(\lambda) \frac{d}{dx}[f_{\mp}(\lambda, x) e^{\mp i\lambda^{1/2}x}]\right\}. \tag{2.45}$$

The trace formulas (2.35), (2.36), (2.44), and (2.45) take on particularly simple forms in the special case of N-soliton potentials $V_N(x)$ where H has N negative eigenvalues (i.e., $J = \{1, 2, \dots, N\}$) characterized by the vanishing of the corresponding reflection coefficients $R_N^{r(\ell)}(\lambda)$ for all $\lambda > 0$.

Thus far all of our trace formulas explicitly involved either the transmission coefficient $T(\lambda)$ or the reflection coefficients $R^{\ell(r)}(\lambda)$ from left (respectively right) incidence and hence are restricted to short-range potentials. Next we shall indicate how a variant of (2.35) extends to a surprisingly large class of potentials. In fact, as shown in some detail in the next section, a proper version of these trace formulas extends to any smooth potential bounded from below. For simplicity we only consider the case where H has no eigenvalues, i.e., where

$$\sigma_p(H) = \emptyset, \quad J = \emptyset \tag{2.46}$$

and the $\ell = 0$ case in (2.35), i.e., only the trace formula (2.36) for $V(x)$. Both restrictions will be removed in Section 3.

Under the assumption (2.46) it was noted in [7] that we may replace $[T(z)f_+(z,x)f_-(z,x) - 1]$ by

$$g(T(z)f_+(z,x)f_-(z,x)) \tag{2.47}$$

as long as g is analytic in $\mathbb{C}\backslash(-\infty, 0]$ with

$$g(\zeta) \underset{\zeta \to 1}{=} (\zeta - 1) + 0((\zeta - 1)^2). \tag{2.48}$$

The choice

$$g(\zeta) = \ln(\zeta) \tag{2.49}$$

then yields the following result upon integrating $\ln[T(z)f_+(z,x)f_-(z,x)]$ over $\Gamma_{R,\epsilon}$:

$$0 = \int_R^0 d\lambda \ln[T(\lambda - i\epsilon)f_+(\lambda - i\epsilon, x)f_-(\lambda - i\epsilon, x)]$$

$$+ \int_0^R d\lambda \ln[T(\lambda + i\epsilon)f_+(\lambda + i\epsilon, x)f_-(\lambda + i\epsilon, x)]$$

$$+ \int_0^{2\pi} iRe^{i\varphi}\, d\varphi \ln[T(Re^{i\varphi})f_+(Re^{i\varphi}, x)f_-(Re^{i\varphi}, x)] + 0(\epsilon)$$

$$= \int_0^R d\lambda \ln\{[T(\lambda + i\epsilon)f_+(\lambda + i\epsilon, x)f_-(\lambda + i\epsilon, x)]/\overline{[T(\lambda + i\epsilon)f_+(\lambda + i\epsilon, x)f_-(\lambda + i\epsilon, x)]}\}$$

$$+ \int_0^{2\pi} iRe^{i\varphi}\, d\varphi \ln\left[1 + \frac{V(x)}{2Re^{i\varphi}} + 0(R^{-3})\right] + 0(\epsilon) \tag{2.50}$$

$$= \int_0^R d\lambda \ln\{[-iG(\lambda + i\epsilon, x, x)]/\overline{[-iG(\lambda + i\epsilon, x, x)]}\}$$

$$+ \int_0^{2\pi} d\varphi\left[\frac{iV(x)}{2} + 0(R^{-1})\right] + 0(\epsilon)$$

$$\underset{\substack{R\uparrow\infty \\ \epsilon\downarrow 0}}{=} i\pi V(x) - 2i\pi \int_0^\infty d\lambda\{(1/2) - \pi^{-1}\arg[G(\lambda + i0, x, x)]\}.$$

Thus

$$V(x) = \int_0^\infty d\lambda[1 - 2\xi(\lambda, x)] \quad \text{if } \sigma_p(H) = \emptyset, \tag{2.51}$$

where we denoted

$$\xi(\lambda, x) = \pi^{-1}\arg[G(\lambda + i0, x, x)] \tag{2.52}$$

$$= 2^{-1} + \pi^{-1}\operatorname{Im}\{\ln[1 + \overset{\ell}{R^r}(\lambda)f_\mp(\lambda, x)^2|f_\mp(\lambda, x)|^{-2}]\}, \quad \lambda > 0,\ x \in \mathbb{R}. \tag{2.53}$$

Equation (2.53) follows from (2.20) and (2.10) and proves, in particular, that

$$\left[\xi(\,.\,,x) - \frac{1}{2}\right] \in L^1((0,\infty);d\lambda) \tag{2.54}$$

using (2.37). While (2.53) still invokes the reflection coefficients $R^{\ell(r)}(\lambda)$ and hence reveals its short-range nature, the original definition (2.52) just relies on the Green's function $G(z,x,x)$ of H and hence offers some hope for greatly increasing the domain of validity of trace formulas beyond the special short-range case. A successful realization of these ideas will be presented in Section 3 (see (3.32)).

Returning to the trace formulas (2.35), we note that integrating both sides over $x \in \mathbb{R}$ results after some manipulations (see [15]) in

$$\int_{\mathbb{R}} dx\omega_{2\ell+2}(x) = 2^{2\ell+3}\sum_{j\in J}\kappa_j^{2\ell+1}$$
$$+ (-1)^\ell 2^{2\ell+2}\pi^{-1}(2\ell+1)\int_0^\infty d\lambda\lambda^{\ell-1/2}\ln[\|T(\lambda)\|], \quad \ell \in \mathbb{N}_0. \tag{2.55}$$

Introducing

$$\Omega_{2\ell+2} := \int_{\mathbb{R}} dx\omega_{2\ell+2}(x), \quad I_{2\ell+1} := \int_{\mathbb{R}} dx\phi_{+,2\ell+1}, \quad \ell \in \mathbb{N}_0, \tag{2.56}$$

the identity (2.24) then yields for the KdV invariants $I_{2\ell+1}$,

$$I_{2\ell+1} = -(4\ell+2)^{-1}\Omega_{2\ell+2}, \quad \ell \in \mathbb{N}_0. \tag{2.57}$$

Now we are finally in position to briefly recall the connection between trace formulas and the Korteweg-de Vries (KdV) invariants. Let $V \in C^\infty(\mathbb{R}^2)$ be a real-valued solution of the KdV equation

$$V_t - 6VV_x + V_{xxx} = 0, \quad (x,t) \in \mathbb{R}^2 \tag{2.58}$$

which results from the usual Lax pair $(H(t), B(t))$ by means of the relation

$$\frac{d}{dt}H(t) - [B(t), H(t)] = 0 \tag{2.59}$$

$([.\,,.]$ the commutator) where $H(t)$ and $B(t)$ denote the differential expressions

$$H(t) = -\partial_x^2 + V(x,t), \quad B(t) = -4\partial_x^3 + 6V(x,t)\partial_x + 3V_x(x,t). \tag{2.60}$$

If $V(x,t)$ satisfies the KdV equation (2.55) and the short-range conditions (2.1) for each fixed $t \in \mathbb{R}$ then the substitution $V(x) \equiv V(x,0) \to V(x,t)$ is well-known to imply the corresponding substitutions

$$T(\lambda) \to T(\lambda,t) = T(\lambda), \quad \overset{r}{R^\ell}(\lambda) \to \overset{r}{R^\ell}(\lambda,t) = \overset{r}{R^\ell}(\lambda)e^{\pm i8\lambda^{3/2}t},$$
$$\kappa_j \to \kappa_j(t) = \kappa_j, \quad c_{j,\pm} \to c_{j,\pm}(t) = c_{j,\pm}e^{\pm 4\kappa_j^3 t}, \quad j \in J. \tag{2.61}$$

Replacing accordingly

$$\omega_{2\ell+2}(x) \to \omega_{2\ell+2}(x,t), \quad \ell \in \mathbb{N}_0, \tag{2.62}$$
$$\phi_{+,2\ell+1}(x) \to \phi_{+,2\ell+1}(x,t), \quad \ell \in \mathbb{N}_0 \tag{2.63}$$

we arrive at the infinite sequence of (polynomial) conservation laws for the KdV hierarchy

$$\frac{d}{dt}I_{2\ell+1} = \int_{\mathbb{R}} dx\phi_{+,2\ell+1,t}(x,t) = 0, \quad \ell \in \mathbb{N}_0 \tag{2.64}$$

since both κ_j and $T(\lambda)$ in (2.55) are t-independent by (2.61).

We conclude with a series of remarks pointing out relevant references and possible extensions for the material in this section.

Remark 2.5. (i). A detailed discussion of the results in this section, with the exception of Theorem 2.3, Corollary 2.4 and the facts (2.47)–(2.54), together with an extensive bibliography can be found in [15]. Corollary 2.2 is stated in [5] without proof in connection with the classical moment problem. The particular trace formula (2.36) for $V(x)$ was used in [7] as the basis for a solution of the inverse scattering problem. Formulas (2.11)–(2.15) and (2.21)–(2.24) can be found in [12], the integrated KdV densities (2.55) have first been derived in [32] (a different derivation can be found in [15]). Theorem 2.3 and Corollary 2.4 appear to be new, the special case (2.45) is closely related to the solution of the inverse scattering method by the Marchenko formalism ([27], Chapter 6, [28], Section 3.5). The trace formula (2.51) (for positive potentials of compact support) has first been derived in [31] by entirely different methods. However, the identification of $\pi\xi(\lambda, x)$ with the argument of $G(\lambda + i0, x, x)$ has not been attempted in [31].
(ii). A generalization of Theorem 2.1 invoking the off-diagonal Green's function $G(z, x, x')$ of H rather than its diagonal form as in (2.31) (see (2.20)) was first proved in [15]. Therein the reader can also find an extension to certain cases where H has infinitely-many negative eigenvalues accumulating at zero (including the infinite soliton limits described in [18], [19]).
(iii). Even though the trace formulas (2.32) (in contrast to the ones in (2.31)) do not contribute to the KdV conservation laws (2.64), they nevertheless give rise to interesting sum rules in scattering theory. More precisely, they lead to (higher-order) Levinson-type theorems relating the scattering phase shift for the pair (H, H_0) with the point spectrum $\sigma_p(H)$ of H (see [15]).

3   TRACE FORMULAS AND KDV INVARIANTS. II. THE GENERAL CASE

Throughout this section we shall assume that

$$V \in C^\infty(\mathbb{R}) \text{ is real-valued and bounded from below.} \qquad (3.1)$$

We then define the self-adjoint Schrödinger operator

$$Hf = \tau f, \ \tau = -\frac{d^2}{dx^2} + V(x),$$
$$\mathcal{D}(H) = \{g \in L^2(\mathbb{R}; dx) | g, g' \in AC_{\mathrm{loc}}(\mathbb{R}), \ \tau g \in L^2(\mathbb{R}; dx)\} \qquad (3.2)$$

and the corresponding self-adjoint Dirichlet Schrödinger operator

$$H_y^D f = \tau f, \quad y \in \mathbb{R}, \qquad (3.3)$$
$$\mathcal{D}(H_y^D) = \{g \in L^2(\mathbb{R}; dx) | g, g' \in AC_{\mathrm{loc}}(\mathbb{R}\backslash\{y\}), \ \lim_{\epsilon\downarrow0} g(y \pm \epsilon) = 0, \ \tau g \in L^2(\mathbb{R}; dx)\}$$

in $L^2(\mathbb{R}; dx)$. (Here $AC_{\mathrm{loc}}(\Omega)$ denotes the set of locally absolutely continuous functions on $\Omega \subseteq \mathbb{R}$.) In order to gain information about $V(x)$ we shall compare H and H_x^D which are "close" in the sense that their resolvents differ only by a rank-one perturbation. More precisely, if $G(z, x, x')$ and $G_y^D(z, x, x')$ denote the Green's functions of H and H_y^D then

$$G_y^D(z, x, x') = G(z, x, x') - G(z, y, y)^{-1} G(z, x, y) G(z, y, x'),$$
$$x, x', y \in \mathbb{R}, \ z \in \mathbb{C} \backslash \sigma(H_y^D), \tag{3.4}$$

or equivalently,

$$(H_y^D - z)^{-1} = (H - z)^{-1} - G(z, y, y)^{-1} (\overline{G(z, y, \cdot)}, \cdot) G(z, \cdot, y),$$
$$y \in \mathbb{R}, \ z \in \mathbb{C} \backslash \{\sigma(H_y^D) \cup \sigma(H)\}, \tag{3.5}$$

where $(\cdot, \cdot)$ denotes the scalar product in $L^2(\mathbb{R}; dx)$. Thus one computes for the trace of (3.5)

$$\text{Tr}[(H_x^D - z)^{-1} - (H - z)^{-1}] = -\frac{d}{dz} \ln[G(z, x, x)], \quad x \in \mathbb{R}. \tag{3.6}$$

Since we shall need a high-energy asymptotic expansion of (3.6) we recall the following result.

Lemma 3.1.

$$\text{Tr}[e^{-\tau H_x^D} - e^{-\tau H}] \underset{s \downarrow 0}{\sim} \sum_{\ell=0}^{\infty} s_\ell(x) \tau^\ell, \tag{3.7}$$

$$\text{Tr}[(H_x^D - z)^{-1} - (H - z)^{-1}] \underset{\substack{|z| \to \infty \\ z \in \mathbb{C} \backslash C_\epsilon}}{\sim} \sum_{j=0}^{\infty} r_j(x) z^{-j-1}, \tag{3.8}$$

where C_ϵ is a cone with apex at $E_0 := \inf[\sigma(H)]$ and opening angle $\epsilon > 0$ (arbitrarily small) along the semi-axis $[E_0, \infty)$ and

$$s_\ell(x) = (-1)^{\ell+1} (\ell!)^{-1} r_\ell(x), \quad \ell \in \mathbb{N}_0, \tag{3.9}$$

$$r_0(x) = \frac{1}{2}, \ r_1(x) = \frac{1}{2} V(x), \tag{3.10}$$

$$r_j(x) = (-1)^{j+1} 2^{1-2j} j \phi_{2j-1}(x) + \sum_{\ell=1}^{j-1} (-1)^{j-\ell+1} 2^{1-2(j-\ell)} \phi_{2(j-\ell)-1}(x) r_\ell(x), \quad j \geq 2,$$

$$\phi_1(x) = V(x), \ \phi_2(x) = -V'(x),$$

$$\phi_{j+1}(x) = -\phi_j'(x) - \sum_{\ell=1}^{j-1} \phi_\ell(x) \phi_{j-\ell}(x), \quad j \geq 2. \tag{3.11}$$

The asymptotic expansion (3.7) is proved in [21] by means of the Feynman-Kac formula, (3.8) is then a consequence of (3.7). We note that $\phi_j(x)$ in (3.11) formally coincide with $\phi_{+,j}(x)$ in (2.14). Moreover, one can prove that

$$r_j'(x) = (-1)^j 2^{1-2j} j \phi_{2j}(x), \quad j \in \mathbb{N}. \tag{3.12}$$

Explicitly, one computes

$$r_0(x) = \frac{1}{2}, \ r_1(x) = \frac{1}{2} V(x), \ r_2(x) = \frac{1}{2} V(x)^2 - \frac{1}{4} V''(x), \text{ etc,} \tag{3.13}$$

$$s_0(x) = -\frac{1}{2}, \ s_1(x) = \frac{1}{2} V(x), \ s_2(x) = \frac{1}{8} V''(x) - \frac{1}{4} V(x)^2, \text{ etc.} \tag{3.14}$$

A comparison of (3.10) and (2.22) shows that $\omega_{2\ell}(x)$ and $r_\ell(x)$ respectively $s_\ell(x)$, $\ell \in \mathbb{N}_0$ are multiples of each other. In particular, the analog of (2.24) also holds for $r_\ell(x)$ and $s_\ell(x)$. In view of (2.56) one also calls $r_\ell(x)$ and $s_\ell(x)$ conserved KdV densities.

Next we employ the fact that for all $x \in \mathbb{R}$, $G(z,x,x)$ is a Herglotz function with respect to z, i.e., $G(.,x,x)$ is analytic in $\mathbb{C}_+ := \{z \in \mathbb{C}\,|\,\mathrm{Im}(z) > 0\}$ and

$$\mathrm{Im}[G(z,x,x)] > 0 \text{ for all } z \in \mathbb{C}_+. \tag{3.15}$$

Hence $G(.,x,x)$ allows the representations (see, e.g., [1])

$$G(z,x,x) = a + bz + \int_{\mathbb{R}} \left[\frac{1}{\lambda - z} - \frac{\lambda}{1 + \lambda^2}\right] d\mu(\lambda, x) \tag{3.16}$$

$$= \exp\left\{c + \int_{\mathbb{R}} \left[\frac{1}{\lambda - z} - \frac{\lambda}{1 + \lambda^2}\right] \xi(\lambda, x)\, d\lambda\right\}, \tag{3.17}$$

where

$$a \in \mathbb{R},\ b \geq 0,\ \int_{\mathbb{R}} \frac{d\mu(\lambda, x)}{1 + \lambda^2} < \infty \tag{3.18}$$

and

$$c \in \mathbb{R},\ \xi(.,x) \in L^1_{\mathrm{loc}}(\mathbb{R}; d\lambda) \text{ is real-valued,}$$
$$\int_{\mathbb{R}} \frac{\xi(\lambda, x)\, d\lambda}{1 + \lambda^2} < \infty. \tag{3.19}$$

Here $d\mu(x)$ denotes the spectral measure of H and in our particular case where $G(z,x,x)$ is the Green's function of H, one can show that $b = 0$. We emphasize that $\xi(\lambda, x)\, d\lambda$ is an absolutely continuous measure with respect to Lebesgue measure $d\lambda$, whereas $d\mu(\lambda, x)$ in general consists of pure point, absolutely and singularly continuous components. Fatou's lemma then implies the existence of

$$\xi(\lambda, x) = \pi^{-1} \lim_{\epsilon \downarrow 0} \mathrm{Im}\{\ln[G(\lambda + i\epsilon, x, x)]\}$$
$$= \pi^{-1} \arg[G(\lambda + i0, x, x)] \text{ for a.e. } \lambda \in \mathbb{R} \tag{3.20}$$

for all $x \in \mathbb{R}$. Moreover,

$$\text{for all } x \in \mathbb{R}: \quad 0 \leq \xi(\lambda, x) \leq 1 \text{ for a.e. } \lambda \in \mathbb{R} \tag{3.21}$$

due to the Herglotz property of $G(z,x,x)$. It can be shown that for $\lambda \in \sigma_{ac}(H)$, the absolutely continuous spectrum of H, $\xi(\lambda, x)$ is essentially the scattering phase shift for the pair (H_x^D, H) since

$$\det[S(\lambda, H_x^D, H)] = \frac{\overline{G(\lambda + i0, x, x)}}{G(\lambda + i0, x, x)} = e^{-2\pi i \xi(\lambda, x)}, \quad \lambda \in \sigma_{ac}(H), \tag{3.22}$$

where $S(\lambda, H_x^D, H)$ denotes the unitary scattering matrix in $\mathbb{C}^2$ for the pair (H_x^D, H). In particular, $\xi(\lambda, x)$ is Krein's spectral shift function [24] for the pair (H_x^D, H). Before we continue it might be appropriate to give the following simple example.

Example 3.2. $V_0(x) \equiv 0$, $H_0 = -\frac{d^2}{dx^2}$, $G_0(z, x, x') = \frac{i}{2z^{1/2}} e^{iz^{1/2}|x - x'|}$, $\mathrm{Im}(z^{1/2}) \geq 0$. Then

$$\xi_0(\lambda, x) = \pi^{-1} \arg[G_0(\lambda + i0, x, x)] = \begin{cases} 1/2, & \lambda > 0 \\ 0, & \lambda < 0. \end{cases} \tag{3.23}$$

From now on we normalize $\xi(\lambda, x)$ to be zero below the spectrum of H, i.e.,

$$\xi(\lambda, x) = 0, \ \lambda < E_0 = \inf[\sigma(H)]. \tag{3.24}$$

This is consistent with the fact that

$$H_x^D \geq H \text{ and hence } \inf[\sigma(H_x^D)] \geq \inf[\sigma(H)] = E_0 \tag{3.25}$$

and that

$$G(\lambda + i0, x, x) > 0, \ \lambda < E_0, \quad x \in \mathbb{R}. \tag{3.26}$$

Inserting (3.17) into (3.6) yields

$$\text{Tr}[(H_x^D - z)^{-1} - (H - z)^{-1}] = -\int_{E_0}^{\infty} \frac{d\lambda \xi(\lambda, x)}{(\lambda - z)^2}, \quad z \in \mathbb{C}\backslash\sigma(H_x^D). \tag{3.27}$$

More generally, one can show that

$$\text{Tr}[F(H_x^D) - F(H)] = \int_{E_0}^{\infty} d\lambda F'(\lambda)\xi(\lambda, x) \tag{3.28}$$

whenever

$$F \in C^2(\mathbb{R}), \ (1 + \lambda^2)F^{(j)} \in L^2((0, \infty)), \quad j = 1, 2. \tag{3.29}$$

In particular,

$$\text{Tr}[e^{-\tau H_x^D} - e^{-\tau H}] = -\tau \int_{E_0}^{\infty} d\lambda e^{-\tau\lambda}\xi(\lambda, x), \quad \tau > 0. \tag{3.30}$$

Hence combining (3.7), (3.14), and (3.30) yields

$$\begin{aligned}
\text{Tr}[e^{-\tau H_x^D} - e^{-\tau H}] &\underset{\tau \downarrow 0}{=} -\frac{1}{2} + \frac{1}{2}V(x)\tau + 0(\tau^2) \\
&= -\tau \int_{E_0}^{\infty} d\lambda e^{-\tau\lambda}\xi(\lambda, x) = -\frac{1}{2} + \frac{\tau}{2}\int_{E_0}^{\infty} d\lambda e^{-\tau\lambda}[e^{\tau E_0} - 2\xi(\lambda, x)]
\end{aligned} \tag{3.31}$$

implying the general trace formula for $V(x)$

$$V(x) = \lim_{\tau \downarrow 0} \int_{E_0}^{\infty} d\lambda e^{-\tau\lambda}[e^{\tau E_0} - 2\xi(\lambda, x)]. \tag{3.32}$$

Thus the Abelian limit in (3.32) yields the proper extension of the special short-range trace formula (2.51) (in which $E_0 = 0$ since we assumed $\sigma_p(H) = \emptyset$). The computation (3.31) easily generalizes to all higher KdV invariants:

Theorem 3.3.

$$s_0(x) = -\frac{1}{2},$$

$$s_\ell(x) = \frac{(-1)^{\ell-1}}{\ell!}\left\{\frac{E_0^\ell}{2} + \ell\lim_{t \downarrow 0}\int_{E_0}^{\infty} d\lambda e^{-t\lambda}\lambda^{\ell-1}\left[\tfrac{1}{2} - \xi(\lambda, x)\right]\right\}, \quad \ell \in \mathbb{N}. \tag{3.33}$$

Explicitly, this yields

$$s_0(x) = -\frac{1}{2},$$

$$s_1(x) = \frac{1}{2}V(x) = \frac{E_0}{2} + \lim_{t\downarrow 0}\int_{E_0}^{\infty} d\lambda\, e^{-t\lambda}\left[\tfrac{1}{2} - \xi(\lambda, x)\right],$$

$$s_2(x) = \frac{1}{8}V''(x) - \frac{1}{4}V(x)^2 \tag{3.34}$$

$$= -\frac{E_0^2}{4} - \lim_{t\downarrow 0}\int_{E_0}^{\infty} d\lambda\, e^{-t\lambda}\lambda[\tfrac{1}{2} - \xi(\lambda, x)],$$

etc.

The behavior of $\xi(\lambda, x)$ for λ in spectral gaps of H is particularly simple. In fact, since V is real-valued, $G(\lambda + i0, x, x)$ is real-valued for $\lambda \in \mathbb{R}\backslash\sigma(H)$ and hence

$$\xi(\lambda, x) = \pi^{-1}\arg[G(\lambda + i0, x, x)] = 1 \text{ or } 0 \text{ for a.e. } \lambda \in \mathbb{R}\backslash\sigma(H). \tag{3.35}$$

More precisely, suppose $(\lambda_1, \lambda_2) \subset \mathbb{R}\backslash\sigma(H)$ is a spectral gap of H and $\mu(x) \in (\lambda_1, \lambda_2)$ is a (Dirichlet) eigenvalue of H_x^D, i.e., $\mu(x) \in \sigma_p(H_x^D)$. Then one easily verifies

$$\xi(\lambda, x) = \begin{cases} 1, & \lambda_1 < \lambda < \mu(x) \\ 0, & \mu(x) < \lambda < \lambda_2. \end{cases} \tag{3.36}$$

We note that since $(H_x^D - z)^{-1} - (H - z)^{-1}$ is of rank one, there is at most one Dirichlet eigenvalue (i.e., an eigenvalue of H_x^D) in each spectral gap of H. One then has in general

$$\sigma(H_x^D) = \sigma(H) \cup \{\mu_j(x)\}_{j\in J}, \tag{3.37}$$

where J is a countable index set (possibly empty).

A particularly interesting special case is the one where $V(x)$ is periodic of period $a > 0$

$$V(x + a) = V(x), \quad x \in \mathbb{R}. \tag{3.38}$$

Standard Floquet theory then yields

$$\sigma(H) = \bigcup_{n=1}^{\infty}[E_{2(n-1)}, E_{2n-1}], \quad E_0 < E_1 \leq E_2 < E_3 \leq \cdots \tag{3.39}$$

and

$$\sigma(H_x^D) = \sigma(H) \cup \{\mu_n(x)\}_{n\in\mathbb{N}}, \quad E_{2n-1} \leq \mu_n(x) \leq E_{2n}, \; x \in \mathbb{R}, \; n \in \mathbb{N}. \tag{3.40}$$

We then obtain a new proof of the famous trace formula for periodic potentials $V(x)$ originally due to Gelfand and Levitan [13] (see also [6], [8]–[11], [22], [23], [26]–[30]).

Theorem 3.4. Let $V \in C^{\infty}(\mathbb{R})$, $V(x + a) = V(x)$, $x \in \mathbb{R}$ for some $a > 0$. Then

$$2(-1)^{\ell+1}\ell! s_\ell(x) = E_0^\ell + \sum_{n=1}^{\infty}[E_{2n-1}^\ell + E_{2n}^\ell - 2\mu_n(x)^\ell], \quad \ell \in \mathbb{N}, \; x \in \mathbb{R}. \tag{3.41}$$

Since in the periodic case $G(\lambda + i0, x, x)$ is purely imaginary for $\lambda \in \sigma(H)$, one infers

$$\xi(\lambda, x) = \begin{cases} 0, & \lambda < E_0, \; \mu_n(x) < \lambda < E_{2n}, \; n \in \mathbb{N} \\ 1, & E_{2n-1} < \lambda < \mu_n(x), \; n \in \mathbb{N} \\ \dfrac{1}{2}, & E_{2(n-1)} < \lambda < E_{2n-1}, \; n \in \mathbb{N} \end{cases} \tag{3.42}$$

(and similarly for limiting cases where $\mu_n(x) \in \{E_{2n-1}, E_{2n}\}$ for some $n \in \mathbb{N}$). Insertion of the step function (3.42) into (3.33) proves (3.41). In particular, for $\ell = 1$ this yields the trace formula for periodic potentials $V(x)$,

$$V(x) = E_0 + \sum_{n=1}^{\infty} [E_{2n-1} + E_{2n} - 2\mu_n(x)]. \tag{3.43}$$

We note in passing that (3.43) extends to periodic potentials satisfying $V \in H^{1,2}([0,a])$ since then the total gap length is finite, i.e., $\sum_{n=1}^{\infty} |E_{2n} - E_{2n-1}| < \infty$. Moreover, as discussed in [6], [16], [21], [26], [27], the trace formulas (3.41) and especially (3.43) extend to certain classes of almost periodic potentials including, in particular, all quasi-periodic finite-gap potentials with purely absolutely continuous spectra obtained by algebro-geometric methods in connection with hyperelliptic curves. In addition, $\xi(\lambda, x)$ is a pure step function in the case of soliton potentials and solitons relative to the almost periodic potentials mentioned above (constructed, e.g., by the double commutation method [14]) and in the case of purely discrete spectra of H where $V(x) \xrightarrow[|x|\to\infty]{} \infty$ (in the latter case $\xi(\lambda, x) = 1$ or 0 for all $x \in \mathbb{R}$ and a.e. $\lambda \in \mathbb{R}$).

We have used the Dirichlet boundary condition at $x \in \mathbb{R}$ mainly for convenience. In the remainder of this section we shall briefly indicate how to replace it by other self-adjoint boundary conditions. Hence we replace (3.3) by

$$H_y^\beta f = \tau f, \quad y \in \mathbb{R}, \ \beta \in \mathbb{R},$$
$$\mathcal{D}(H_y^\beta) = \{g \in L^2(\mathbb{R}; dx) \mid g, g' \in AC_{\text{loc}}(\mathbb{R}\backslash\{y\}), \tag{3.44}$$
$$\lim_{\epsilon\downarrow 0}[g'(y \pm \epsilon) + \beta g(y \pm \epsilon)] = 0, \ \tau g \in L^2(\mathbb{R}; dx)\}.$$

(The particular case $\beta = 0$ now represents a Neumann boundary condition at $y \in \mathbb{R}$.) The Green's function $G_y^\beta(z, x, x')$ of H_y^β is now given by

$$G_y^\beta(z, x, x') = G(z, x, x') - [(\beta + \partial_1)(\beta + \partial_2)G(z, y, y)]^{-1} \bullet$$
$$\bullet [(\beta + \partial_2)G(z, x, y)][(\beta + \partial_1)G(z, y, x')], \tag{3.45}$$

where

$$\partial_1 G(z, y, x') := \partial_x G(z, x, x')|_{x=y}, \ \partial_2 G(z, x, y) := \partial_{x'} G(z, x, x')|_{x'=y}, \text{ etc.} \tag{3.46}$$

One verifies that

$$\partial_1 G(z, y, x) = \partial_2 G(z, x, y), \quad x \neq y \tag{3.47}$$

which proves that the rank-one integral kernel on the right-hand-side of (3.45) generates a self-adjoint operator for $z < \inf[\sigma(H_y^\beta)]$. The role of the Herglotz function $G(z, x, x)$ in the Dirichlet case is now played by $(\beta + \partial_1)(\beta + \partial_2)G(z, x, x)$. Since

$$(\beta + \partial_1)(\beta + \partial_2)G(\lambda + i0, x, x) < 0 \text{ for } -\lambda > 0 \text{ large enough} \tag{3.48}$$

and

$$\zeta_{\beta,0} := \inf[\sigma(H_x^\beta)] \leq \inf[\sigma(H)] = E_0, \tag{3.49}$$

the analogs of (3.20), (3.21), (3.24), and (3.28) then read

$$\xi_\beta(\lambda, x) = \pi^{-1} \lim_{\epsilon\downarrow 0} \text{Im}\{\ln[(\beta + \partial_1)(\beta + \partial_2)G(\lambda + i\epsilon, x, x)]\} - 1,$$

$$\beta \in \mathbb{R}, \ x \in \mathbb{R}, \ \text{a.e. } \lambda \in \mathbb{R}, \tag{3.50}$$

$$-1 \le \xi_\beta(\lambda, x) \le 0, \quad x \in \mathbb{R}, \text{ a.e. } \lambda \in \mathbb{R}, \tag{3.51}$$

$$\xi_\beta(\lambda, x) = 0, \quad \lambda < \zeta_{\beta,0} = \inf[\sigma(H_x^\beta)], \tag{3.52}$$

$$\mathrm{Tr}[F(H_x^\beta) - F(H)] = \int_{\zeta_{\beta,0}(x)}^\infty d\lambda F'(\lambda) \xi_\beta(\lambda, x). \tag{3.53}$$

Example 3.2 now turns into

Example 3.5. $V_0(x) \equiv 0$, $H_0 = -\frac{d^2}{dx^2}$,

$$(\beta + \partial_1)(\beta + \partial_2)G_0(z, x, x) = \frac{i}{2}\left[\frac{\beta^2}{z^{1/2}} + z^{1/2}\right], \quad \mathrm{Im}(z^{1/2}) \ge 0, \tag{3.54}$$

$$\zeta_{\beta,0}(x) = -\beta^2, \quad x \in \mathbb{R}, \tag{3.55}$$

$$\xi_\beta(\lambda, x) = \begin{cases} 0, & \lambda < -\beta^2 \\ -1, & -\beta^2 < \lambda < 0 \\ -\dfrac{1}{2}, & \lambda > 0 \end{cases}, \quad \beta \in \mathbb{R}\backslash\{0\}, \; x \in \mathbb{R}, \tag{3.56}$$

$$\xi_0(\lambda, x) = \begin{cases} 0, & \lambda < 0 \\ -\dfrac{1}{2}, & \lambda > 0 \end{cases}, \quad x \in \mathbb{R}, \tag{3.57}$$

$$\mathrm{Tr}[e^{-\tau H_{0,x}^\beta} - e^{-\tau H_0}] = -\frac{1}{2} + e^{\tau\beta^2}, \quad \tau > 0, \tag{3.58}$$

$$\mathrm{Tr}[(H_{0,x}^\beta - z)^{-1} - (H_0 - z)^{-1}] = \frac{\beta^2 - z}{2z(z + \beta^2)}, \quad z \in \mathbb{C}\backslash\sigma(H_{0,x}^\beta). \tag{3.59}$$

In analogy to Lemma 3.1 one obtains

$$\mathrm{Tr}[e^{-\tau H_x^\beta} - e^{-tH}] \underset{\tau\downarrow 0}{\sim} \sum_{\ell=0}^\infty s_{\beta,\ell}(x)\tau^\ell, \tag{3.60}$$

where the $s_{\beta,\ell}(x)$ can again be computed recursively. Since the actual recursion relations are somewhat involved (see [21]) we confine ourselves to the first three coefficients

$$\begin{aligned} s_{\beta,0}(x) &= \frac{1}{2}, \quad s_{\beta,1}(x) = \beta^2 - \frac{1}{2}V(x), \\ s_{\beta,2}(x) &= \frac{1}{2}\beta^4 - \beta^2 V(x) - \frac{1}{2}\beta V'(x) + \frac{1}{4}V(x)^2 - \frac{1}{8}V''(x), \text{ etc.} \end{aligned} \tag{3.61}$$

The analog of Theorem 3.3 then reads

$$s_{\beta,\ell}(x) = \frac{(-1)^{\ell-1}}{\ell!}\left\{\frac{\zeta_{\beta,0}(x)^\ell}{2} + \ell\lim_{t\downarrow 0}\int_{\zeta_{\beta,0}(x)}^\infty d\lambda e^{-t\lambda}\lambda^{\ell-1}\left[\tfrac{1}{2} - \xi_\beta(\lambda, x)\right]\right\}, \quad \ell \in \mathbb{N}. \tag{3.62}$$

Explicitly,

$$s_{\beta,0}(x) = \frac{1}{2},$$

$$s_{\beta,1}(x) = \beta^2 - \frac{1}{2}V(x) = \lim_{\tau\downarrow 0}\int_{\zeta_{\beta,0}(x)}^\infty d\lambda_1 e^{-\tau\lambda_1}\left[-\tfrac{1}{2}e^{\tau\zeta_{\beta,0}(x)} - \xi_\beta(\lambda_1, x)\right], \tag{3.63}$$

etc.

Finally, if $V(x)$ is periodic of period a as in (3.38), one infers in analogy to (3.40) that

$$\sigma(H_x^\beta) = \sigma(H) \cup \{\zeta_{\beta,n}(x)\}_{n\in\mathbb{N}_0},$$
$$\zeta_{\beta,0}(x) \le E_0, \quad E_{2n-1} \le \zeta_{\beta,n}(x) \le E_{2n}, \quad x \in \mathbb{R}, \ n \in \mathbb{N} \tag{3.64}$$

with $\sigma(H)$ given as in (3.39). Equation (3.42) then turns into

$$\xi_\beta(\lambda, x) = \begin{cases} 0, & \lambda < \zeta_{\beta,0}(x), \ E_{2n-1} < \lambda < \zeta_{\beta,n}(x), \ n \in \mathbb{N} \\ -1, & \zeta_{\beta,n}(x) < \lambda < E_{2n}, \ n \in \mathbb{N}_0 \\ -\dfrac{1}{2}, & E_{2(n-1)} < \lambda < E_{2n-1}, \ n \in \mathbb{N} \end{cases} \tag{3.65}$$

and the periodic trace formulas (3.41) become

$$2(-1)^\ell \ell! s_{\beta,\ell}(x) = 2\zeta_{\beta,0}(x)^\ell - E_0^\ell + \sum_{n=1}^\infty [2\zeta_{\beta,n}(x)^\ell - E_{2n-1}^\ell - E_{2n}^\ell], \quad \ell \in \mathbb{N}, \ x \in \mathbb{R}. \tag{3.66}$$

In particular, for $\ell = 1$ the trace formula for $V(x)$ reads

$$V(x) = 2\beta^2 + 2\zeta_{\beta,0}(x) - E_0 + \sum_{n=1}^\infty [2\zeta_{\beta,n}(x) - E_{2n-1} - E_{2n}]. \tag{3.67}$$

A detailed treatment of the results in this section will appear in [21].

4   A BRIEF OUTLOOK

The methods of Section 3 allow a variety of further ramifications. Since several of these are presently under investigation we confine ourselves to a series of brief comments.

A). Since (2.51) only describes the short-range case without eigenvalues of H we now briefly describe the general short-range case where

$$V \in H^{2,1}(\mathbb{R}) \text{ is real-valued.} \tag{4.1}$$

In this case H may have finitely or infinitely-many negative eigenvalues (accumulating at zero in the latter case). Hence we denote

$$\sigma_p(H) = \{e_j\}_{j\in J_0}, \quad e_\ell < e_{\ell+1}, \tag{4.2}$$

where

$$J_0 = \begin{cases} \emptyset \\ \{0, 1, \ldots, N_0\} \\ \mathbb{N}_0 \end{cases} \tag{4.3}$$

is an appropriate index set. Let $J = J_0 \backslash \{0\}$ if $J \ne \emptyset$ and

$$J_+ = \begin{cases} \{1, \ldots, N_0, N_0 + 1\}, \ e_{N_0+1} = 0 \\ \mathbb{N} \end{cases} \tag{4.4}$$

according to whether J_0 is finite or infinite (but non empty). Using (2.7), (2.8), (2.52), (2.53) and the Riemann-Lebesgue lemma one proves that

$$\left[\xi(\lambda, x) - \frac{1}{2}\right] \underset{\lambda\to\infty}{=} o(\lambda^{-3/2}) \tag{4.5}$$

uniformly with respect to $x \in \mathbb{R}$. This allows one to dispense with the Abelian limit in (3.32) and together with the step function behavior (3.36) then yields the proper generalization of (2.51), viz.,

$$V(x) = 2e_0 + 2 \sum_{j \in J_+} [e_j - \mu_j(x)] + \int_0^\infty d\lambda[1 - 2\xi(\lambda, x)]. \tag{4.6}$$

The trace formula (4.6) easily extends to the case where $V(x)$ tends to different asymptotes $V_\pm$ as $x \to \pm\infty$ in the sense that

$$\int_{-\infty}^0 dx|V(x) - V_-| + \int_0^\infty dx|V(x) - V_+| < \infty, \quad V_- < V_+,$$
$$V', V'' \in L^1(\mathbb{R}; dx). \tag{4.7}$$

This case is of some interest since H has uniform spectral multiplicity one in (V_-, V_+) and two in (V_+, ∞). Hence total reflection from the left occurs for energies $\lambda \in (V_-, V_+)$, whereas for energies $\lambda \in (V_+, \infty)$ one encounters reflection and transmission from left and right incidence.

These considerations extend to impurity scattering in one-dimensional crystals where the Hamiltonian is of the type $H = -\frac{d^2}{dx^2} + V + W$, with W the short-range impurity potential and $V \in H^{1,2}([0, a])$ the periodic background potential of period $a > 0$.

B). Assume that

$$V \in C(\mathbb{R}), \quad V(x) \xrightarrow[|x|\to\infty]{} \infty. \tag{4.8}$$

Then $H = -\frac{d^2}{dx^2} + V$ (defined, e.g., as a form sum) has purely discrete spectrum

$$\sigma(H) = \{e_n\}_{n \in \mathbb{N}_0}, \quad e_m < e_{m+1} \tag{4.9}$$

and

$$\sigma(H_x^D) = \{\mu_n(x)\}_{n \in \mathbb{N}}, \quad e_{m-1} \leq \mu_m(x) \leq e_m. \tag{4.10}$$

Since $|1 - 2\xi(\lambda, x)| = 1$ in this case, the Abelian limit in (3.32) represents a summability result and cannot be removed as in the cases studied under part A). Combining (3.32) and (3.36) then yields

$$V(x) = e_0 + \lim_{\tau \downarrow 0} \tau^{-1} \sum_{n=1}^\infty [2e^{-\tau\mu_n(x)} - e^{-\tau e_{n-1}} - e^{-\tau e_n}]. \tag{4.11}$$

For an explicit example consider, e.g., the harmonic oscillator $V(x) = x^2 - 1$ and place the Dirichlet boundary condition at $x = 0$. Then

$$e_n = 2n, \ n \in \mathbb{N}_0, \ \mu_n(0) = \begin{cases} 2n, & n \text{ odd} \\ 2(n-1), & n \text{ even}, n \geq 2 \end{cases} \tag{4.12}$$

and

$$\xi(\lambda, 0) = \begin{cases} 1, & \lambda \in \bigcup_{n \in \mathbb{N}_0}(4n, 4n+2) \\ 0, & \lambda \in \bigcup_{n \in \mathbb{N}_0}(4n+2, 4(n+1)). \end{cases} \tag{4.13}$$

Thus formally,

$$\int_0^\infty d\lambda[1 - 2\xi(\lambda, 0)] = -2 + 2 - 2 + \cdots \tag{4.14}$$

and (4.11) is precisely the corresponding Abelian sum which is -1 and hence coincides with $V(0)$.

C). Everything we have done so far extends to Jacobi operators h in $\ell^2(\mathbb{Z})$ of the form

$$(hu)(n) = u(n+1) + u(n-1) + v(n)u(n), \quad n \in \mathbb{Z} \tag{4.15}$$

with $v \in \ell^\infty(\mathbb{Z})$ real-valued. In particular, the analog of (3.32) then reads

$$v(n) = \frac{1}{2}(E_- + E_+) + \frac{1}{2}\int_{E_-}^{E_+} d\lambda[1 - 2\xi(\lambda, n)], \tag{4.16}$$

where

$$E_- := \inf[\sigma(H)], \ E_+ := \sup[\sigma(H)] \tag{4.17}$$

and $\xi(.,n)$ is normalized as follows

$$\xi(\lambda, n) = \begin{cases} 0, & \lambda < E_- \\ 1, & \lambda > E_+. \end{cases} \tag{4.18}$$

D). The function $\xi(\lambda, x)$ yields the following criterion for the absolutely continuous spectrum $\sigma_{ac}(H)$ of H:

$$\sigma_{ac}(H) = \overline{\{\lambda \in \mathbb{R}| 0 < \xi(\lambda, x) < 1\}}^{\text{ess}}, \tag{4.19}$$

where

$$\overline{\Omega}^{\text{ess}} := \{\lambda \in \mathbb{R} \mid m(\Omega \cap (\lambda - \epsilon, \lambda + \epsilon)) > 0 \text{ for all } \epsilon > 0\} \tag{4.20}$$

denotes the essential closure of a subset $\Omega \subset \mathbb{R}$ (and $m(.)$ abbreviates the Lebesgue measure). In particular, $\sigma_{ac}(H) = \emptyset$ if and only if $\xi(.,x_0)$, for some fixed $x_0 \in \mathbb{R}$, is equivalent to the characteristic function of a measurable subset of $\mathbb{R}$.

As a consequence of (4.19) in connection with second-order finite-difference operators one can prove a slight strengthening of a recent result of Last [25] for the almost Mathieu equation or Harper's model (see, e.g., [2]) with $v(n) = \lambda \cos(\pi\alpha n)$, $\alpha, \lambda \in \mathbb{R}$. In fact, we can prove

$$m(\sigma_{ac}(h)) \geq 4 - 2|\lambda| \tag{4.21}$$

which recovers Last's result that $\sigma_{ac}(h) \neq \emptyset$ if $|\lambda| < 2$ and α is a Liouville number.

E). Our formalism yields new insight into inverse spectral problems. For this purpose it is convenient to introduce Dirichlet data

$$(\mu_n(x_0), \sigma_n(x_0)), \ \sigma_n(x_0) \in \{+, -\}, \quad \lambda_1 < \mu_n(x_0) < \lambda_2, \ x_0 \in \mathbb{R} \text{ fixed} \tag{4.22}$$

instead of merely Dirichlet eigenvalues $\mu_n(x_0)$ in a spectral gap (λ_1, λ_2) of H. Here $\sigma_n(x_0) = +/-$ records whether $\mu_n(x_0)$ is a right/left Dirichlet eigenvalue of H (i.e., a Dirichlet eigenvalue on the right (respectively left) half-axis (x_0, ∞) (respectively $(-\infty, x_0)$) for some fixed $x_0 \in \mathbb{R}$). Then a famous uniqueness theorem of Borg [3] for periodic potentials can be rephrased as follows.

Theorem 4.1 ([3]). Let $V_1(x)$, $V_2(x)$ be periodic of period $a > 0$. Then $V_1(x) = V_2(x)$ for a.e. $x \in \mathbb{R}$ if and only if for some fixed $x_0 \in \mathbb{R}$,

$$\xi_1(\lambda, x_0) = \xi_2(\lambda, x_0), \ \lambda \in \mathbb{R} \text{ and } \sigma_{1,j}(x_0) = \sigma_{2,j}(x_0), \quad j \in I, \tag{4.23}$$

where $I \subseteq \mathbb{N}$ denotes all those indices $j \in \mathbb{N}$ such that $\mu_{\ell,j}(x_0) \notin \sigma(H_\ell)$, $\ell = 1, 2$.

Here, in obvious notation, $\xi_\ell(\lambda, x)$, $\sigma_{\ell,n}(x)$ denote the corresponding quantities for $V_\ell(x)$, $\ell = 1, 2$. Theorem 4.1 extends to algebro-geometric quasi-periodic finite-gap potentials and certain classes of almost-periodic potentials.

One of our new results in this context provides an analog of Borg's theorem for potentials with purely discrete spectra. Let

$$V(x) \xrightarrow[|x|\to\infty]{} \infty, \quad V(x) \geq c, \; x \in \mathbb{R} \tag{4.24}$$

and

$$\sigma(H) = \{e_n\}_{n\in\mathbb{N}_0}, \; \sigma(H_{x_0}^D) = \{\mu_n(x_0)\}_{n\in\mathbb{N}}, \quad e_{m-1} \leq \mu_m(x_0) \leq e_m, \; m \in \mathbb{N}. \tag{4.25}$$

In addition to the Dirichlet data

$$\{(\mu_j(x_0), \sigma_j(x_0))\}_{j\in I} \tag{4.26}$$

we also need the norming constants

$$c_{j'}^+(x_0) > 0, \; c_{j'}^-(x_0) > 0, \; j' \in I' := \mathbb{N}\backslash I \tag{4.27}$$

associated with those Dirichlet eigenvalues $\mu_{j'}(x_0)$ that coincide with $e_{j'-1}$ or $e_{j'}$. Here $c_{j'}^+(x_0)$ (respectively $c_{j'}^-(x_0)$) is the norming constant associated with the right (respectively left) half-line (x_0, ∞) (respectively $(-\infty, x_0)$) and is given as minus (respectively plus) the residue of the corresponding Weyl m-function $m^+(z, x_0)$ (respectively $m^-(z, x_0)$) at $z = \mu_{j'}(x_0)$. We have

$$G(z, x_0, x_0) = -[m^+(z, x_0) - m^-(z, x_0)]^{-1}. \tag{4.28}$$

Theorem 4.2. Let $V_\ell(x) \xrightarrow[|x|\to\infty]{} \infty$, $V_\ell(x) \geq c$, $\ell = 1, 2$ for some $c \in \mathbb{R}$. Then $V_1(x) = V_2(x)$ for a.e. $x \in \mathbb{R}$ if and only if for some fixed $x_0 \in \mathbb{R}$,

$$\begin{aligned}
&\xi_1(\lambda, x_0) = \xi_2(\lambda, x_0), \; \lambda \in \mathbb{R}, \; \sigma_{1,j}(x_0) = \sigma_{2,j}(x_0), \; j \in I, \\
&c_{1,j'}^+(x_0) = c_{2,j'}^+(x_0), \; c_{1,j'}^-(x_0) = c_{2,j'}^-(x_0), \; j' \in I'.
\end{aligned} \tag{4.29}$$

In particular, if $V_\ell(x) = V_\ell(-x)$, $\ell = 1, 2$ then $m_\ell^+(z, 0) = -m_\ell^-(z, 0)$. Hence $e_{1,n} = e_{2,n}$ for all $n \in \mathbb{N}_0$ then implies $\xi_1(\lambda, 0) = \xi_2(\lambda, 0)$ and $c_{1,n}^\pm(0) = c_{2,n}^\pm(0)$, $n \in \mathbb{N}$ and one recovers another result of Borg [4].

F). Suppose $V(x, t)$ is a solution of the KdV equation (2.58) which is short-range with respect to x in the sense of Section 2 for all fixed $t \in \mathbb{R}$. If $\xi(\lambda, x, t)$ denotes the xi-function associated with $V(x, t)$ one can show that

$$\xi_t(\lambda, x, t) = 2[V(x, t) + 2\lambda]\xi_x(\lambda, x, t), \quad \lambda > 0, \; (x, t) \in \mathbb{R}^2 \tag{4.30}$$

and

$$\xi_x(\lambda, x, t) = \lambda^{1/2}\pi^{-1}[|f_+(\lambda, x, t)|^{-2} - |f_-(\lambda, x, t)|^{-2}], \quad \lambda > 0 \tag{4.31}$$

with $f_\pm(\lambda, x, t)$ the Jost solutions of $H(t)$ as in (2.3). Equation (4.30) in the short-range case resembles in a remarkable way the differential equation

$$\mu_{n,t}(x, t) = 2[V(x, t) + 2\mu_n(x, t)]\mu_{n,x}(x, t) \tag{4.32}$$

for Dirichlet eigenvalues associated with spatially periodic, algebro-geometric finite-gap, and certain almost periodic solutions of the KdV equation (see, e.g., [27], Chapter 12). Hence we expect (4.30) to hold for $\lambda \in \sigma_{ac}(H)$ for a large class of solutions $V(x, t)$ with empty singularly continuous spectrum, $\sigma_{sc}(H(t)) = \emptyset$ and to play a vital role in the solution of the corresponding initial value problem for the KdV equation.

A detailed account of Part A) will appear in [17]. Parts C), D), and E) are discussed in [16].

Acknowledgments. I am indebted to H. Holden, B. Simon, and Z. Zhao for joint collaborations which led to most of the results presented in this contribution. I would like to thank the organizers for their kind invitation to a most stimulating conference and for the extraordinary hospitality extended to us during five wonderful days.

REFERENCES

1. N. Aronszajn and W. F. Donoghue, On exponential representations of analytic functions in the upper half-plane with positive imaginary part, *J. Anal. Math.*, *5*: 321-388 (1957).
2. J. Avron, P.H.M. v. Mouche, and B. Simon, On the measure of the spectrum for the almost Mathieu operator, *Commun. Math. Phys.*, *132*: 103-118 (1990).
3. G. Borg, Eine Umkehrung der Sturm-Liouvilleschen Eigenwertaufgabe, *Acta Math.*, *78*: 1-96 (1946).
4. G. Borg, Uniqueness theorems in the spectral theory of $y'' + (\lambda - q(x))y = 0$, *Den 11te Skandinaviske Matematikerkongress i Trondheim 1949*, Johan Grundt Tanums Forlag, Oslo, (1952), pp. 276-287.
5. D. V. Chudnovsky, One and multidimensional completely integrable systems arising from the isospectral deformation, in *Complex Analysis, Microlocal Calculus and Relativistic Quantum Theory*, D. Iagolnitzer (ed.), Lecture Notes in Physics, Vol. **126**, Springer, Berlin, 1980, pp. 352-416.
6. W. Craig, The trace formula for Schrödinger operators on the line, *Commun. Math. Phys.*, *126*: 379-407 (1989).
7. P. Deift and E. Trubowitz, Inverse scattering on the line, *Commun. Pure Appl. Math.*, *32*: 121-251 (1979).
8. L. A. Dikii, Trace formulas for Sturm-Liouville differential operators, *Amer. Math. Soc. Transl. Ser. (2)*, *18*: 81-115 (1961).
9. B. A. Dubrovin, Periodic problems for the Korteweg-de Vries equation in the class of finite band potentials, *Funct. Anal. Appl.*, *9*: 215-223 (1975).
10. H. Flaschka, On the inverse problem for Hill's operator, *Arch, Rat. Mech. Anal, 59*: 293-309 (1975).
11. I. M. Gelfand, On identities for the eigenvalues of a second-order differential operator, *Uspehi Mat. Nauk, 11:1*: 191-198 (1956). (Russian).
12. I. M. Gelfand and L. A. Dikii, Asymptotic behaviour of the resolvent of Sturm-Liouville equations and the algebra of the Korteweg-de Vries equations, *Russ. Math. Surv., 30:5*: 77-113 (1975).
13. I. M. Gelfand and B. M. Levitan, On a simple identity for the eigenvalues of a second-order differential operator, *Dokl. Akad. Nauk SSSR, 88*: 593-596 (1953). (Russian).
14. F. Gesztesy, A complete spectral characterization of the double commutation method, *J. Funct. Anal., 117*: 401-446 (1993).
15. F. Gesztesy and H. Holden, Trace formulas and conservation laws for nonlinear evolution equations, *Rev. Math. Phys.* (to appear).
16. F. Gesztesy and B. Simon, The xi function, in preparation.
17. F. Gesztesy, H. Holden, and B. Simon, Absolute summability of the trace relation for certain Schrödinger operators, in preparation.

18. F. Gesztesy, W. Karwowski, and Z. Zhao, New types of soliton solutions, *Bull. Amer. Math. Soc. (New Series), 27*: 266–272 (1992).
19. F. Gesztesy, W. Karwowski, and Z. Zhao, Limits of soliton solutions, *Duke Math. J., 68*: 101–150 (1992).
20. F. Gesztesy, H. Holden, B. Simon, and Z. Zhao, Trace formulae and inverse spectral theory for Schrödinger operators, *Bull. Amer. Math. Soc. (New Series), 29*: 250–255 (1993).
21. F. Gesztesy, H. Holden, B. Simon, and Z. Zhao, Higher order trace relations for Schrödinger operators, *Commun. Pure Appl. Math.* (to appear).
22. H. Hochstadt, On the determination of a Hill's equation from its spectrum, *Arch. Rat. Mech. Anal., 19*: 353–362 (1965).
23. S. Kotani and M. Krishna, Almost periodicity of some random potentials, *J. Funct. Anal, 78*: 390–405 (1988).
24. M. G. Krein, Perturbation determinants and a formula for the traces of unitary and self-adjoint operators, *Sov. Math. Dokl., 3*: 707–710 (1962).
25. Y. Last, A relation between a.c. spectrum of ergodic Jacobi matrices and the spectra of periodic approximants, *Commun. Math. Phys., 151*: 183–192 (1993).
26. B. M. Levitan, On the closure of the set of finite-zone potentials, *Math. USSR Sbornik, 51*: 67–89 (1985).
27. B. M. Levitan, *Inverse Sturm-Liouville Problems*, VNU Science Press, Utrecht, 1987.
28. V. A. Marchenko, *Sturm-Liouville Operators and Applications*, Birkhäuser, Basel, 1986.
29. H. P. McKean and P. van Moerbeke, The spectrum of Hill's equation, *Invent. Math., 30*: 217–274 (1975).
30. E. Trubowitz, The inverse problem for periodic potentials, *Commun. Pure Appl. Math., 30*: 321–337 (1977).
31. S. Venakides, The infinite period limit of the inverse formalism for periodic potentials, *Commun. Pure Appl. Math., 41*: 3–17 (1988).
32. V. E. Zakharov, L. D. Faddeev, Korteweg-de Vries equation: a completely integrable Hamiltonian system, *Funct. Anal. Appl., 5*: 280–287 (1971).

Picard and Finite-Gap Potentials

F. Gesztesy University of Missouri, Columbia, Missouri

R. Weikard University of Alabama-Birmingham, Birmingham, Alabama

In the last 25 years the KdV equation

$$q_t = \frac{1}{4}q_{xxx} + \frac{3}{2}qq_x, \tag{1}$$

originally studied by Korteweg and de Vries [18] in 1895 in order to describe shallow water waves, has received a lot of attention in connection with its complete integrability as an infinitely-dimensional Hamiltonian system.

The Cauchy problem of the KdV equation for spatially rapidly decaying initial conditions was solved in 1967 by Gardner, Greene, Kruskal and Miura [10], [11] using the inverse scattering technique. The corresponding Cauchy problem for spatially periodic initial conditions turned out to be of an entirely different nature and its solution was achieved by algebro-geometric techniques in 1975 by Its and Matveev [17]. While we will briefly indicate the main points in the solution of the latter problem we refer the reader to the book by Novikov et al. [24] for further details and a more complete bibliography.

The starting point is the definition of the so called KdV hierarchy. Let L be the differential expression

$$L = \frac{\partial^2}{\partial x^2} + q, \tag{2}$$

where the potential q depends on $x \in \mathbb{R}$ and, in addition, on a (deformation) parameter $t \in \mathbb{R}$. It is well known (see, e.g., Wilson [30]) that the coefficients p_j in

$$P_{2g+1} = \frac{\partial^{2g+1}}{\partial x^{2g+1}} + p_{2g-1}\frac{\partial^{2g-1}}{\partial x^{2g-1}} + ... + p_0 \tag{3}$$

can be chosen in such a way that P_{2g+1} and L are almost commuting, i.e., that their commutator $[P_{2g+1}, L]$ is a multiplication operator. More specifically, the p_j have to be certain polynomials in q and its x-derivatives. (P_{2g+1}, L) is then called a Lax pair and the equation

$$q_t = [P_{2g+1}, L] \tag{4}$$

is a nonlinear evolution equation for q. The collection of all these equations for all possible choices of P_{2g+1} and all nonnegative integers g is called the KdV hierarchy. In particular, the choice

$$P_3 = \frac{\partial^3}{\partial x^3} + \frac{3}{2}q\frac{\partial}{\partial x} + \frac{3}{4}q_x, \tag{5}$$

yields the KdV equation (1).

Novikov [25] showed that for all periodic stationary (i.e., t-independent) solutions of any equation in the KdV hierarchy the spectrum of the self-adjoint operator associated with L (assuming that q is real-valued and smooth) consists of finitely many intervals $[E_{2g-1}, E_{2g-2}]$, ..., $[E_1, E_0]$, the so called spectral bands, and a semi-axis $(-\infty, E_{2g}]$. These are separated by open intervals (E_{2g}, E_{2g-1}), ..., (E_2, E_1) which are called spectral gaps. Hence any periodic stationary solution of any equation in the KdV hierarchy may be called a finite-gap or finite-band potential. (Note that generically the spectrum of a periodic potential consists of infinitely-many spectral bands separated by infinitely-many spectral gaps.) Conversely, it was shown by Dubrovin [8] that any finite-gap potential satisfies some stationary equation of the KdV hierarchy.

Since general periodic smooth potentials may be approximated uniformly by finite-gap potentials (see, e.g., Levitan [19], [20], Marchenko and Ostrovsky [21]) many efforts concentrated on solving the Cauchy problem of the KdV equation with finite-gap initial conditions.

The first to succeed for an explicit example were Dubrovin and Novikov [9]. Under the initial condition $q(x, 0) = -6\mathcal{P}(x)$ (where $\mathcal{P}$ denotes Weierstrass' elliptic function) they obtained

$$q(x, t) = -2\sum_{j=1}^{3} \mathcal{P}(x - x_j(t)) \tag{6}$$

which solves the KdV equation (1) for appropriate functions $x_j(t)$, $j = 1, 2, 3$.

Subsequently, Its and Matveev [17] developed their celebrated formula which gives the solution of the KdV equation with any finite-gap potential as initial condition. They proved that every g-gap potential can be expressed by

$$q(x) = C + 2\frac{d^2}{dx^2}\log\Theta(\mathbf{a}x + \mathbf{b}) \tag{7}$$

for suitable $\mathbf{a}, \mathbf{b} \in \mathbb{C}^g$ and $C \in \mathbb{R}$, where Θ denotes Riemann's Θ-function associated with the underlying compact hyperelliptic Riemann surface of genus g. The solution of the KdV equation is then given by

$$q(x, t) = C + 2\frac{d^2}{dx^2}\log\Theta(\mathbf{a}x + \mathbf{b} + \mathbf{c}t) \tag{8}$$

where $\mathbf{c} \in \mathbb{C}^g$ is another suitable vector which may be determined from characteristics of the underlying hyperelliptic curve and thus from the initial data $q(x, 0)$.

The representation of the KdV equation using a Lax pair uncovered another important feature, namely that the spectrum of the operator L in (2) is independent of the (isospectral) deformation parameter t if q evolves according to the KdV equation. In particular, if q satisfies the KdV equation then $q(\cdot, t)$ stays in the isospectral manifold of $q(\cdot, 0)$. Here the isospectral manifold of a potential q_0 is defined by

$$I(q_0) = \{q \in C^\infty(\mathbb{R}) : E_i(q) = E_i(q_0), i = 0, 1, 2, ...\} \tag{9}$$

where the $E_i(q)$ denote the edges of the spectral bands of the operator $\partial^2/\partial x^2 + q$.

Finally it should be emphasized that the stationary KdV hierarchy, since it can be written as $[P_{2g+1}, L] = 0$, is connected to the question of commutativity of ordinary differential expressions. In this context a result by Burchnall and Chaundy [5], [6] implies that P_{2g+1} and L satisfy an algebraic relation of the form

$$P_{2g+1}^2 = R_{2g+1}(L) = \prod_{n=0}^{2g}(L - E_n). \tag{10}$$

The locations of the finite branch points of the associated compact hyperelliptic Riemann surface $y^2 = R_{2g+1}(z)$ are precisely the band edges of the operator L whenever the potential q is real-valued and smooth. Because of these facts it is common to define q as a finite-gap potential if it satisfies one (and hence infinitely-many) equation(s) of the stationary KdV hierarchy, or equivalently, if there is a hyperelliptic curve with finitely many branch points associated to it in the manner described above. (This characterization, in particular, extends to complex-valued, quasi-periodic, and meromorphic potentials q, where the branch points are not necessarily located on the real axis.) One calls q a g-gap potential if and only if the underlying Riemann surface has (arithmetic) genus g.

For these reasons finite-gap potentials and isospectral sets of such potentials have been studied thoroughly. Without even attempting to list a complete bibliography we mention, e.g., the following references: McKean and van Moerbeke [23] (in the case of finitely-many gaps), McKean and Trubowitz [22] (in connection with infinitely-many gaps), Airault, McKean and Moser [1] (in the case of rational and elliptic potentials), Birnir [3], [4] (in connection with complex and singular potentials).

The number of explicit examples of finite-gap potentials, however, is surprisingly small. The first such example was the so called Lamé potential

$$q(x) = -\alpha \mathcal{P}(x + \omega_3) \tag{11}$$

where $\mathcal{P}$ is Weierstrass' elliptic function with a real period $2\omega_1$ and a purely imaginary period $2\omega_3$ and where α is real. Ince [15] showed that this potential, which is real continuous and periodic with period $2\omega_1$ for real x, is an s-gap potential if and only if $\alpha = s(s+1)$ for some integer s.

The next example was that given by Dubrovin and Novikov mentioned in (6) who used Lamé's potential with $\alpha = 2(2+1) = 6$, a 2-gap potential, as an initial condition for the KdV equation. Since the the KdV flow leaves the spectrum invariant, all the potentials in (6) are 2-gap potentials.

But until 1987 no other explicit examples of finite-gap potentials other than Lamé's were known. (The Its-Matveev formula (8) of course contains all such examples but is not very explicit in this respect.) A variety of new examples were recently found by Verdier

[29] and Treibich and Verdier [26], [27], [28]. They are of the type

$$q(x) = -\sum_{j=1}^{4} s_j(s_j + 1)\mathcal{P}(x - \omega_j) \tag{12}$$

($\omega_2 = \omega_1 + \omega_3$, $\omega_4 = 0$), where the s_j are nonnegative integers.

The theory of elliptic finite-gap potentials and isospectral sets thereof is still incomplete to date and the following pertinent questions immediately arise:

- When are elliptic potentials finite gap?

- What are the band edges or – at least – what is their number?

- How can isospectral manifolds of elliptic finite-gap potentials be classified?

The last point warrants additional comment. Airault, McKean and Moser [1] have shown that every element in the isospectral manifold of an elliptic finite-gap potential can be expressed as

$$q(x) = -2\sum_{j=1}^{M} \mathcal{P}(x - x_j) \tag{13}$$

for suitable (not necessarily distinct) numbers x_j. The number M is constant throughout the isospectral set and is greater or equal than the genus g of the underlying hyperelliptic curve.

Now consider the potentials (the periods $2\omega_1$, $2\omega_3$ are fixed in the following)

$$\begin{aligned}
q_1 &= -20\mathcal{P}(x + \omega_3) \\
q_2 &= -20\mathcal{P}(x + \omega_3) - 12\mathcal{P}(x + \omega_1 + \omega_3) \\
q_3 &= -30\mathcal{P}(x + \omega_3) - 2\mathcal{P}(x + \omega_1 + \omega_3).
\end{aligned}$$

It turns out that the potential q_1 has $g = 4$ and $M = 10$, the potential q_2 has $g = 4$ and $M = 16$, and the potential q_3 has $g = 5$ and $M = 16$. Hence it is certainly not sufficient to specify either g or M alone in order to determine the isospectral set of an elliptic finite-gap potential. Although the specification of g and M suffices to distinguish between the above examples, it follows from our investigations in [13] that the specification of g and M is insufficient in general, i.e., there exist nonisospectral elliptic finite-gap potentials with identical g and M. A complete classification of elliptic finite-gap potentials (with fixed periods $2\omega_1$ and $2\omega_3$) will therefore necessarily involve a further invariant of isospectral manifolds in addition to g and M.

For these reasons we started to develop our own approach toward a solution of this problem. Our techniques rely on entirely different ideas; in particular, our main new strategy is based on a systematic use of a powerful theorem of Picard (see Theorem 1) concerning ordinary differential equations with elliptic coefficients. This approach not only recovers the results of Ince and of Treibich and Verdier [12], [13] but also extends to all even Picard potentials (see Definition 2) [14].

An appropriate version of the above mentioned theorem of Picard for second-order equations is the following (see, e.g., Ince [16], p.375-376).

Theorem 1 *Consider the differential equation*

$$y''(z) + Q(z)y(z) = 0, \qquad (14)$$

where Q is an elliptic function with fundamental periods $2\omega_1$ and $2\omega_3$. Suppose the general solution of this equation is meromorphic. Then there exists at least one solution ψ_1 which is elliptic of the second kind, i.e., which is meromorphic and satisfies

$$\psi_1(z + 2\omega_j) = \rho_j \psi_1(z), \quad j = 1, 3 \qquad (15)$$

for some constants $\rho_1, \rho_3 \in \mathbb{C}$.

Picard's theorem then motivates the definition of Picard potentials.

Definition 2 *Let q be an elliptic function. Then q is called a **Picard** potential if and only if the equation*

$$y''(E, z) + q(z)y(E, z) = Ey(E, z) \qquad (16)$$

has a meromorphic fundamental system of solutions for each $E \in \mathbb{C}$.

Details of our investigations for the general case of any Picard potential will appear elsewhere [12], [13], [14]. For brevity we can outline our approach here only in the simplest case of Lamé's potential, i.e.,

$$q(x) = -s(s+1)\mathcal{P}(x + \omega_3), \quad s \in \mathbb{N} \cup \{0\}, \qquad (17)$$

where the periods $2\omega_1$ and $2\omega_3$ are real and purely imaginary, respectively. In this case, as mentioned above, q is real-valued, continuous, and periodic with period $2\omega_1$ as a function of x on the real line. Investigating the differential equation (16) and the associated self-adjoint operator L then recovers Ince's celebrated result [15] as follows.

Theorem 3 *The Lamé potential (17) is a Picard potential and the self-adjoint operator L associated with the differential equation (16) has the finite-gap spectrum*

$$\sigma(L) = (-\infty, E_{2s}] \cup [E_{2s-1}, E_{2s-2}] \cup ... \cup [E_1, E_0]. \qquad (18)$$

The band edges E_j, $j = 0, ..., 2s$, are given as eigenvalues of certain Jacobi matrices.

Proof: Any solution of (16) is analytic near any ordinary point of the equation. Consider any of the (regular) singular points. The exponents of singularity at this point are $s + 1$ and $-s$ respectively, i.e., they are integers. Hence we immediately have the existence of one solution which is analytic (since $s + 1 > 0$) near this point. The Frobenius method (see, e.g., Coddington and Levinson [7], Section 4.8) also shows, however, that the solution associated with the exponent $-s$ is meromorphic. Hence Lamé's potential is a Picard potential.

By Picard's theorem equation (16) has at least one solution $\psi_a(E, z)$ which is elliptic of the second kind. By the theory of elliptic functions any function which is elliptic of the second kind is of the form

$$Ce^{\lambda z} \frac{\prod_{j=1}^{n} \sigma(z - a_j)}{\prod_{j=1}^{n} \sigma(z - b_j)}. \qquad (19)$$

(Here $\sigma(z)$ is the Weierstrass σ-function.) Inserting this into the differential equation shows that $n = s$ and that $b_1 = ... = b_s = 0$. In addition, one finds conditions on the a_j and on λ. Altogether we infer that (16) has a solution of the form

$$\psi_a(E, z) = e^{\lambda_a(E)z} \frac{\prod_{j=1}^s \sigma(z - a_j(E))}{\sigma(z)^s} \tag{20}$$

and that a function of the form (20) satisfies (16) if and only if

$$E = (2s - 1) \sum_{j=1}^s \mathcal{P}(a_j(E)), \tag{21}$$

$$\lambda_a(E) = \sum_{j=1}^s \zeta(a_j(E)), \tag{22}$$

$$0 = \sum_{\substack{\ell=1 \\ \ell \neq j}}^s (\zeta(a_\ell(E) - a_j(E)) - \zeta(a_\ell(E)) + \zeta(a_j(E)))$$

$$= \frac{1}{2} \sum_{\substack{\ell=1 \\ \ell \neq j}}^s \frac{\mathcal{P}'(a_\ell(E)) + \mathcal{P}'(a_j(E))}{\mathcal{P}(a_\ell(E)) - \mathcal{P}(a_j(E))}, \quad 1 \leq j \leq s, \tag{23}$$

where $\zeta(z)$ denotes the Weierstrass ζ-function. The equations (23) prove in particular that $a_\ell(E) \neq a_j(E)$ for $\ell \neq j$.

Equations (21), (22), and (23) show that together with ψ_a also ψ_{-a} is a solution of (16) for the same value of E, where $-a(E) = (-a_1(E), ..., -a_s(E))$. This fact is due to the reflection symmetry of the potential, i.e., due to $q(-z) = q(z)$.

There is always at least one solution of the form ψ_a and by the preceding paragraph we now have established the existence of even two such solutions except when these two are linearly dependent. In order to determine when the latter case occurs we now compute the Wronskian $W(E) = W(\psi_a(E), \psi_{-a}(E))$. One finds

$$W(E) = -\sum_{j=1}^s \mathcal{P}'(a_j) \prod_{\substack{\ell=1 \\ \ell \neq j}}^s (\mathcal{P}(z) - \mathcal{P}(a_\ell)). \tag{24}$$

Since $W(E)$ is independent of z we may evaluate it at $z = a_j$ to obtain

$$W(E) = -\mathcal{P}'(a_j) \prod_{\substack{\ell=1 \\ \ell \neq j}}^s (\mathcal{P}(a_j) - \mathcal{P}(a_\ell)), \quad 1 \leq j \leq s. \tag{25}$$

We now apply the fact that the band edges of L are located precisely at those E-values where there is only one Floquet solution (apart from constant multiples), i.e., only one solution satisfying $y(z + 2\omega_1) = \rho(E)y(z)$ for some constant $\rho(E)$. Such a $\rho(E)$ is called a Floquet multiplier. The Floquet multipliers are obtained as eigenvalues of $\Phi(2\omega_1)$ where $\Phi(z)$ is the fundamental matrix (monodromy matrix) with $\Phi(0)$ equal to the identity. Together with $\rho(E)$ also $1/\rho(E)$ is a Floquet multiplier since the determinant of $\Phi(z)$ does not depend on z and hence is equal to 1. Therefore, if $\rho(E)$ is different from ± 1 then there exist two linearly independent Floquet solutions. If $\rho(E)$ is equal to 1 or -1, however, then it might happen that there is only one Floquet solution. The values of E where this occurs are precisely the band edges we are interested in.

Note that both $\psi_a(E, z)$ and $\psi_{-a}(E, z)$ are Floquet solutions; therefore we have two Floquet solutions for all values of E where $W(E)$ does not vanish. On the other hand, if $W(E)$ vanishes then ψ_a and ψ_{-a} are the same solution and one can show in addition that no other Floquet solution exists. Hence the band edges of L are exactly those points on the E-axis where $W(E)$ vanishes.

Now $W(E)$ vanishes if any of the factors on the right hand side of (25) vanishes. Note that $\mathcal{P}'(a_j) = 0$ if and only if a_j is a half-period (i.e., equal to ω_1, ω_2 or ω_3 modulo the fundamental period parallelogram (f.p.p.)). Since $a_j \neq a_\ell$ for $j \neq \ell$ we infer that $\mathcal{P}(a_j) - \mathcal{P}(a_\ell) = 0$ if and only if $a_j = -a_\ell$ (modulo the f.p.p.). Since (25) has to be satisfied for each $j = 1, ..., s$ we obtain that for even s either

$$\{a_1, \ldots, a_s\} = \{a_{j_1}, -a_{j_1}, a_{j_2}, -a_{j_2}, \ldots, a_{j_{s/2}}, -a_{j_{s/2}}\} \tag{26}$$

or

$$\{a_1, \ldots, a_s\} = \{\omega_{\ell_1}, \omega_{\ell_2}, a_{j_1}, -a_{j_1}, \ldots, a_{j_{(s-1)/2}}, -a_{j_{(s-1)/2}}\}, \tag{27}$$

where $\omega_{\ell_1}, \omega_{\ell_2} \in \{\omega_1, \omega_2, \omega_3\}$ and $\omega_{\ell_1} \neq \omega_{\ell_2}$. Similarly one has for odd s either

$$\{a_1, \ldots, a_s\} = \{\omega_{\ell_1}, a_{j_1}, -a_{j_1}, \ldots, a_{j_{(s-1)/2}}, -a_{j_{(s-1)/2}}\} \tag{28}$$

with $\omega_{\ell_1} \in \{\omega_1, \omega_2, \omega_3\}$ or

$$\{a_1, \ldots, a_s\} = \{\omega_1, \omega_2, \omega_3, a_{j_1}, -a_{j_1}, \ldots, a_{j_{(s-3)/2}}, -a_{j_{(s-3)/2}}\}. \tag{29}$$

Next we consider the case (26) in more detail. In this case $\psi_a(E, z)$ becomes

$$\psi_a(E, z) = \frac{\prod_{j=1}^{s/2} \sigma(z - a_j)\sigma(z + a_j)}{\sigma(z)^s}$$

$$= \prod_{j=1}^{s/2}(-\sigma(a_j)^2(\mathcal{P}(z) - \mathcal{P}(a_j))) = \sum_{j=0}^{s/2} \hat{c}_j \mathcal{P}(z)^j \tag{30}$$

for appropriate constants $\hat{c}_j$ using the addition theorem for σ-functions.

The coefficients $\hat{c}_j$ are in a one-to-one correspondence with the pairs $(a_{j_k}, -a_{j_k})$, $k = 1, ..., s/2$. Hence we may determine the coefficients $\hat{c}_j$ instead of the a_j. This is done by inserting the right hand side of (30) or, in order to get a simpler result, the slightly refined ansatz

$$\psi_a(E, z) = \sum_{j=0}^{s/2} c_j(\mathcal{P}(z) - \mathcal{P}(\omega_2))^j \tag{31}$$

into (16). Defining $c_{-1} = c_{(s+2)/2} = 0$ and using

$$\mathcal{P}'^2 = 4(\mathcal{P} - e_2)^3 + 12e_2(\mathcal{P} - e_2)^2 + 4(e_2 - e_1)(e_2 - e_3)(\mathcal{P} - e_2) \tag{32}$$

$(e_j = \mathcal{P}(\omega_j), j = 1, 2, 3)$, then yields

$$\sum_{j=0}^{s/2}\{(1 - 2j - s)(2 - 2j + s)c_{j-1} + (e_2(12j^2 - s(s + 1)) - E)c_j$$

$$+ 2(e_2 - e_1)(e_2 - e_3)(1 + j)(1 + 2j)c_{j+1}\}(\mathcal{P}(z) - e_2)^j = 0. \tag{33}$$

This is equivalent to the eigenvalue problem

$$J\underline{c} = E\underline{c}, \quad \underline{c} = (c_{s/2}, ..., c_0)^T, \tag{34}$$

where J is the $(1 + s/2) \times (1 + s/2)$ Jacobi matrix of the form

$$J = \begin{pmatrix} \beta_d & \alpha_d & 0 & \cdots & \cdots & 0 \\ \gamma_{d-1} & \beta_{d-1} & \alpha_{d-1} & \ddots & & \vdots \\ 0 & \gamma_{d-2} & \beta_{d-2} & \ddots & \ddots & \vdots \\ \vdots & \ddots & \ddots & \ddots & \ddots & 0 \\ \vdots & & & \ddots & \gamma_1 & \beta_1 & \alpha_1 \\ 0 & \cdots & \cdots & 0 & \gamma_0 & \beta_0 \end{pmatrix} \tag{35}$$

(abbreviating $s/2$ by d) with

$$\alpha_j = (1 - 2j - s)(2 - 2j + s), \tag{36}$$
$$\beta_j = e_2[12j^2 - s(s + 1)], \tag{37}$$
$$\gamma_j = 2(e_2 - e_1)(e_2 - e_3)(1 + j)(1 + 2j), \tag{38}$$

$j = 1, ..., s/2$. We note that $\gamma_j \alpha_{j+1} > 0$ since $e_3 < e_2 < e_1$. This implies that all eigenvalues of J are distinct (see, e.g., Arscott [2], p. 21).

The case (27) actually represents three cases since we have three choices to pick ω_{ℓ_1} and ω_{ℓ_2} from $\{\omega_1, \omega_2, \omega_3\}$. In each of these cases the ansatz (31) has to be slightly modified to obtain

$$\psi_a(E, z) = \frac{\sigma(z - \omega_{\ell_1})\sigma(z - \omega_{\ell_2})}{\sigma(z)^2} \sum_{j=0}^{(s-2)/2} c_j(\mathcal{P}(z) - e_2)^j. \tag{39}$$

Again we get eigenvalue problems for Jacobi matrices but this time their size is only $s/2 \times s/2$. As before, the eigenvalues of each of these matrices are distinct. Due to the different forms of the functions in (31) and (39) we would have the existence of two linearly independent Floquet solutions whenver an eigenvalue occurs simultaneously in any two of the four associated matrices. But this is impossible since, by construction, all the eigenvalues of these matrices refer to points where only one Floquet solution exists. This proves that all eigenvalues of the four matrices associated with (31) and (39) are distinct.

Hence if s is even there is one eigenvalue problem for a matrix of size $((s+2)/2) \times ((s+2)/2)$ and three eigenvalue problems for matrices of size $s/2 \times s/2$. Altogether this yields $(s + 2)/2 + 3(s/2) = 2s + 1$ solutions, i.e., $2s + 1$ distinct eigenvalues and each of these is a band edge of the spectrum of the operator L.

If s is odd, similar considerations give the existence of $3(s + 1)/2 + (s - 1)/2 = 2s + 1$ band edges. Thus the theorem is proved.   ■

A similar but more intricate analysis is possible for the case of all even Picard potentials where even the assumption that q is real-valued and smooth can be dropped if one uses the generalized definition of the term "finite-gap". One then obtains the following result [14].

Theorem 4 *Every even Picard potential is finite-gap in the sense that it satisfies a stationary KdV equation $[P_{2g+1}, L] = 0$ for some differential expression P_{2g+1} of the form (3). The $2g + 1$ branch points of the associated hyperelliptic curve (cf. (10)) are obtained from certain linear algebraic eigenvalue problems subject to additional constraints.*

In the case of a Treibich-Verdier potential

$$q(x) = -\sum_{i=1}^{4} s_i(s_i + 1)\mathcal{P}(x - \omega_i), \quad s_i \in \mathbb{N} \cup \{0\}, \tag{40}$$

this analysis provides the number of finite branch points of the Riemann surface associated with q (or the number of band edges of L in the real-valued and smooth case). The following table, where $s = s_1 + s_2 + s_3 + s_4$ (and, without loss of generality, $s_1 \geq s_2 \geq s_3 \geq s_4 \geq 0$), summarizes our results for (40):

s		# of finite branch points
even	$s_2 + s_3 \leq s_1 + s_4$	$2s_1 + 1$
even	$s_2 + s_3 \geq s_1 + s_4$	$s_1 + s_2 + s_3 - s_4 + 1$
odd	$s_2 + s_3 + s_4 < s_1$	$2s_1 + 1$
odd	$s_2 + s_3 + s_4 > s_1$	$s_1 + s_2 + s_3 + s_4 + 2$

ACKNOWLEDGMENTS

We want to thank the organizers for their kind invitation to a very exciting conference. R.W. also thanks the Alabama EPSCoR program for financial support.

REFERENCES

[1] H. Airault, H. P. McKean, and J. Moser. Rational and elliptic solutions of the Korteweg-de Vries equation and a related many-body problem. *Commun. Pure Appl. Math.*, 30:95–148, 1977.

[2] F. M. Arscott. *Periodic Differential Equations*. MacMillan, New York, 1964.

[3] B. Birnir. Complex Hill's equation and the complex periodic Korteweg-de Vries equations. *Commun. Pure Appl. Math.*, 39:1–49, 1986.

[4] B. Birnir. Singularities of the complex Korteweg-de Vries flow. *Commun. Pure Appl. Math.*, 39:283–305, 1986.

[5] J. L. Burchnall and T. W. Chaundy. Commutative ordinary differential operators. *Proc. London Math. Soc. Ser. 2*, 21:420–440, 1923.

[6] J. L. Burchnall and T. W. Chaundy. Commutative ordinary differential operators. *Proc. Roy. Soc. London*, A118:557–583, 1928.

[7] E. A. Coddington and N. Levinson. *Theory of Ordinary Differential Equations*. McGraw-Hill, New York, 1955.

[8] B. A. Dubrovin. Periodic problems for the Korteweg-de Vries equation in the class of finite band potentials. *Funct. Anal. Appl.*, 9:215–223, 1975.

[9] B. A. Dubrovin and S. P. Novikov. Periodic and conditionally periodic analogs of the many-soliton solutions of the Korteweg-de Vries equation. *Sov. Phys.-JETP*, 40:1058–1063, 1975.

[10] C. S. Gardner, J. M. Greene, M. D. Kruskal, and R. M. Miura. Method for solving the Korteweg-deVries equation. *Phys. Rev. Lett.*, 19:1095– 1097, 1967.

[11] C. S. Gardner, J. M. Greene, M. D. Kruskal, and R. M. Miura. Korteweg-deVries equation and generalizations. VI. Methods for exact solution. *Commun. Pure Appl. Math.*, 27:97–133, 1974.

[12] F. Gesztesy and R. Weikard. Lamé potentials and the stationary (m)KdV hierarchy. Preprint, 1993.

[13] F. Gesztesy and R. Weikard. Treibich-Verdier potentials and the stationary (m)KdV hierarchy. Preprint, 1993.

[14] F. Gesztesy and R. Weikard. In preparation.

[15] E. L. Ince. Further investigations into the periodic Lamé functions. *Proc. Roy. Soc. Edinburgh*, 60:83–99, 1940.

[16] E. L. Ince. *Ordinary Differential Equations*. Dover, New York, 1956.

[17] A. R. Its and V. B. Matveev. Schrödinger operators with finite-gap spectrum and N-soliton solutions of the Korteweg-de Vries equation. *Theoret. Math. Phys.*, 23:343–355, 1975.

[18] D. J. Korteweg and G. de Vries. On the change of form of long waves advancing in a rectangular canal, and on a new type of long stationary waves. *Phil. Mag.*, 39:422 – 443, 1895.

[19] B. M. Levitan. Approximation of infinite zone potentials by finite-zone potentials. *Math. USSR Izv.*, 20:55–87, 1983.

[20] B. M. Levitan. On the closure of the set of finite-zone potentials. *Math. USSR Sbornik*, 51:67–89, 1985.

[21] V. A. Marchenko and I. V. Ostrovsky. Approximation of periodic by finite-zone potentials. *Sel. Math. Sov.*, 6:101-136, 1987.

[22] H. P. McKean and E. Trubowitz. Hill's operator and hyperelliptic function theory in the presence of infinitely many branch points. *Commun. Pure Appl. Math.*, 29:143–226, 1976.

[23] H. P. McKean and P. van Moerbeke. The spectrum of Hill's equation. *Invent. Math.*, 30:217–274, 1975.

[24] S. Novikov, S.V. Manakov, L.P. Pitaevskii, and V.E. Zakharov. *Theory of Solitons*. Consultants Bureau, New York and London, 1984.

[25] S. P. Novikov. The periodic problem for the Korteweg-de Vries equation. *Funct. Anal. Appl.*, 8:236–246, 1974.

[26] A. Treibich and J.-L. Verdier. Revêtements tangentiels et sommes de 4 nombres triangulaires. *C. R. Acad. Sci. Paris, Série I*, 311:51–54, 1990.

[27] A. Treibich and J. L. Verdier. Solitons elliptiques. In P. Cartier, L. Illusie, N. M. Katz, G. Laumon, Y. Manin, and K. A. Ribet, editors, *The Grothendieck Festschrift, Volume III*, pages 437–480. Birkhäuser, Boston, Basel, Berlin, 1990.

[28] A. Treibich and J.-L. Verdier. Revêtements exceptionelles et sommes de 4 nombres triangulaires. *Duke Math. J.*, 68:217–236, 1992.

[29] J.-L. Verdier. New elliptic solitons. In M. Kashiwara and T. Kawai, editors, *Algebraic Analysis, Papers Dedicated to Professor Mikio Sato on the Occasion of His Sixtieth Birthday, Vol. II*, pages 901–910. Academic Press, 1988.

[30] G. Wilson. Commuting flows and conservation laws for Lax equations. *Math. Proc. Camb. Phil. Soc.*, 86:131–143, 1979.

Commutator Bounds for Eigenvalues
of Some Differential Operators

Evans M. Harrell II[*] and Patricia L. Michel[**] School of Mathematics, Georgia Institute
of Technology, Atlanta, Georgia

1 INTRODUCTION

Consider the evolution due to the linear diffusion equation

$$u_t = \kappa \Delta u$$

or to similar equations with popular additional complications, such as

$$u_t(x,t) = \operatorname{div}\left(P(x)\operatorname{grad} u(x,t)\right) + V(x)u(x,t),$$

the heat equation with inhomogeneities and endothermic reactions, or

$$i\hbar u_t(x,t) = (\mathbf{p} - \mathbf{A}(x))^2 u(x,t) + V(x)u(x,t),$$

where

$$\mathbf{p} := -i\hbar\nabla,$$

the Schrödinger equation with a magnetic field. Here *div* and *grad* may be tensorially
defined.

[*]Partially supported by the US NSF through grant DMS-9211624.

[**]Part of this research was conducted during the US-Sweden Workshop on Spectral Methods sponsored
by the NSF under grant INT-9217529.

Work related to Doctoral thesis in preparation at Georgia Institute of Technology.

Typically the generator of the semigroup $\exp(-tH)$, where H is the linear operator on the right side, has two important positivity properties, i.e., H is semibounded and $\exp(-tH)$ is positivity improving, which is essentially equivalent to having a positive integral kernel. In addition, often at least the lower portion of the spectrum is discrete, especially if the equation is defined on a bounded domain or manifold Ω.

The positivity properties of the evolution imply in a standard way that the lowest eigenvalue is nondegenerate and has a positive eigenfunction. Well-known arguments establish that estimates of gaps between eigenvalues correspond to exponential rates of convergence of the system to equilibrium: If the initial condition is $u(x,0)$, then, for example for the heat equation with Neumann boundary conditions:

$$u(x,t) - \langle u, u_0 \rangle u_0 = \sum_{n=1}^{\infty} \exp(-\mu_n t)\langle u, u_n \rangle u_n = \mathrm{O}(\exp(-\mu_1 t)),$$

where $\langle \cdot, \cdot \rangle$ denotes the inner product on $L^2(\Omega)$ and the eigenfunctions are denoted u_n, and in this case $\lambda_0 = 0$, with $u_0 = 1/\sqrt{|\Omega|}$. Analogous things occur for other problems. Estimates of this kind have been familiar at least since [14].

The work discussed here concerns upper bounds on eigenvalue gaps (and, implicitly, on rates of convergence of evolution equations), using a theorem we recently proved [9]. This in turn is closely related to earlier work of Hook [13], Harrell [7], and Davies and Harrell (appendix to [8]).

The essential reason for the association between eigenvalue gaps and commutators is elementary. Let H be a self-adjoint operator such that $Hu_k = \lambda_k u_k$, and suppose G is an auxiliary operator. Then a formal calculation (ignoring domain questions) shows that

$$\langle u_k, [H,G]u_j \rangle = (\lambda_k - \lambda_j)\langle u_k, Gu_j \rangle.$$

A great many of the good estimates known for eigenvalue gaps can be derived from this formula. As with all variational techniques, the test function – in this case G – must be chosen cleverly in order to get a good estimate. For further discussion of these points, see [8]. A novel feature of the bounds we discuss in this article is that they involve the interplay between first and second commutators. As was shown in [7], when particularized to $H =$ the Laplace-Beltrami operator on a Riemannian manifold, similar bounds show how the spectral gaps are controlled by the curvature. One application we make below is to sharpen some of the estimates of that paper. Other applications are made to bounds for Neumann boundary conditions and for bi-Laplacians.

The operator under study is a self-adjoint operator H, and there are two families of symmetric "test operators," which we call G_j and Π_j. (The Π_j's are often analogues of the momentum operator of quantum mechanics, accounting for our notation. Hook [13] earlier had a similar theorem, and a rough correlation with his notation is that our G's correspond to his B's and our Π's correspond to his T's times i.)

THEOREM 1.1 *Let H be self-adjoint on a Hilbert space $\mathcal{H}$, and suppose that the lower portion of its spectrum consists of discrete eigenvalues $\lambda_1 \le \lambda_2 \le \ldots \le \lambda_n < \lambda_{n+1} \le \ldots$ Let P be the spectral projection for $\lambda_1, \lambda_2, \ldots, \lambda_n$, and let $\{G_j\}$ and $\{\Pi_j\}$ be two families* *of symmetric operators such that all products of the form $\Pi_j G_j$, $G_j \Pi_j$, $G_j^2 H$, HG_j^2, and $G_j H G_j$ are well defined. Then*

$$\sum_{j=1}^{m} Tr\left((\lambda_{n+1} I - H)^{-1} P\Pi_j^2\right) \geq \frac{\left|\sum_{j=1}^{m} Tr\left(P[\Pi_j, G_j]\right)\right|^2}{2\sum_{j=1}^{m} Tr\left(P[G_j, [H, G_j]]\right)}, \tag{1.1}$$

assuming that these three traces are finite and nonzero.

REMARKS: Note that because of the hypothesis $\lambda_{n+1} > \lambda_n$, the operator $\lambda_{n+1} I - H$ is uniquely invertible from the range of P to itself.

The natural setting for this theorem is that of C^*-algebras, in which the assumption on products of operators is unnecessary. Here, however, we are interested in unbounded operators, so domain questions must be considered carefully. While there is a certain amount of freedom in the choice of the auxiliary operators Π_j and G_j, it is important that the G'_js are chosen in such a way that HG_j is defined on the given domain, i.e. for $u \in \mathcal{D}(H)$ we must have $G_j u \in \mathcal{D}(H)$. Similarly for $G_j^2 u$.

An upper bound for $\lambda_{n+1} - \lambda_n$ is obtained if the left hand side is increased to $(\lambda_{n+1} - \lambda_n)^{-1} \sum_{j=1}^{m} Tr(P\Pi_j^2)$ and then one solves for the gap $\lambda_{n+1} - \lambda_n$. The resulting bounds are analogous to the classic Payne-Pólya-Weinberger bounds [16] (see [1] for a discussion of this bound) and those derived in [7]. In most applications, m will be taken to be ν, the dimension of the underlying space, but other choices are possible and will in some cases give better results.

2 EXAMPLES

In this section we list a few applications of our technique, some of which are to a certain extent already known in some form, but which we feel we obtain more efficiently. Theorem 1.1 is an abstract version of the Hile–Protter inequality [10]. Hook's earlier abstract Hile–Protter bound in [13] contains a free parameter, which he needs to optimize in order to get some of the applications which we obtain directly below.

As in [9] we define two universal constants C_{PPW} and C_{HP} as follows

$$C_{PPW}(\mathcal{M}) := \sup_{n, \Omega \subset \mathcal{M}} \frac{\lambda_{n+1} - \lambda_n}{\langle \lambda_\ell \rangle_{\ell \leq n}}$$

and

$$C_{HP}(\mathcal{M}) := \sup_{n, \Omega \subset \mathcal{M}} \left\langle \frac{\lambda_\ell}{\lambda_{n+1} - \lambda_\ell} \right\rangle_{\ell \leq n}^{-1}$$

where $\langle f_\ell \rangle_{\ell \leq n}$ denotes the average of an expression involving eigenvalues over all $\ell \leq n$. Thus, for example $\langle \lambda_\ell \rangle_{\ell \leq n} = \frac{1}{n} \sum_{\ell \leq n} \lambda_\ell$. Or, as will occur later, if the sum starts at $\ell = 0$ then we divide by $n + 1$. According to these definitions we always have $C_{HP} \geq C_{PPW}$.

Example 1: Operators with continuous spectra

We begin with a remark about operators with continuous spectra, which immediately follows from the original estimates of Payne, Pólya, and Weinberger [16] and of Hile and Protter [10], but which does not appear to have been appreciated. Under wide circumstances it is known that continuous spectra of Schrödinger operators, Sturm–Liouville operators, Jacobi matrices, etc. can be quantified by using a measure known as the density of states, defined by cut-off procedures (for example, see [4], [15], and [6]).

Let us call the density of states measure $dn(\lambda)$, and assume that it is well–defined (as a weak-$*$ limit and has bounded first moment) for the operator H, which is a local operator on $L^2(\mathbf{R}^\nu)$ or $L^2(\mathbf{Z}^\nu)$. Suppose that $\Gamma = (a, b)$ is a gap in the spectrum of H (i.e., an open interval of length $|\Gamma|$ in the resolvent set, the ends of which are in the spectrum), and, moreover, that when H is restricted to a bounded set with Dirichlet boundary conditions (vanishing conditions, in the discrete case), there is a finite C_{HP}. Then, when we restrict H to rectangles of side R and let $R \to \infty$, we get:

$$\frac{|\Gamma| \int_{-\infty}^{a} dn(\lambda)}{\int_{-\infty}^{a} \lambda \, dn(\lambda)} \leq C_{PPW} \leq C_{HP}.$$

For example, if H is a Schrödinger operator in ν dimensions with a smooth, periodic, non-negative potential, then

$$|\Gamma| \leq \frac{4 \int_{-\infty}^{a} \lambda \, dn(\lambda)}{\nu \int_{-\infty}^{a} dn(\lambda)} \leq \frac{4a}{\nu}.$$

In particular cases, where estimates can be made of the density of states, much sharper bounds than $4a/\nu$ are possible.

Example 2: Algebraic formulation of quantum mechanics

In quantum mechanics it is not strictly necessary to represent the algebra of observables with differential operators; for many purposes an abstract algebra is quite sufficient (cf. [17]).

In this setting, the fundamental phenomena of quantum mechanics have their origin in the quantum canonical commutation relations, which state that if H is a Hamiltonian operator, then canonically conjugate classical variables satisfy the Heisenberg uncertainty principle. In the Heisenberg picture, the operators X, Π are a canonically conjugate pair when

$$\dot{X} = \frac{\Pi}{m} = \frac{i}{\hbar}[H, X],$$

where m is the mass of the particle and $\hbar$ is Planck's constant divided by 2π. In this case, the commutation relations state that $[X, \Pi] = i\hbar$.

If we identify X with G, and assume that $\Pi^2 \leq \beta H$, then the bound on C_{HP} in this case becomes:

$$C_{HP} \leq \frac{2\beta n}{m Tr(P)} = \frac{2\beta}{m}.$$

If there are M canonical momenta, such that

$$\sum_{j=1}^{M} \Pi_j^2 \leq \beta H \,,$$

then, similarly,

$$C_{HP} \leq \frac{2\beta n}{mMTr(P)} = \frac{2\beta}{mM} \,. \tag{2.1}$$

With the usual Euclidean momenta, $\beta = 2m$, and a Hile–Protter type bound results. In the language of quantum mechanics, inequality (2.1) states that the ratio of the gap excitation energy to the average of the energy below an energy gap is bounded by $2\beta/mM$ divided by the trace of the density matrix for the unexcited states.

Example 3: Schrödinger operators with magnetic fields and Dirichlet boundary conditions

Compare with [13] p. 628. Consider the case when $H = (\mathbf{p} - \mathbf{A}(\mathbf{x}))^2 + V(\mathbf{x})$ on $\mathbf{L}^2(\Omega)$ for Ω a bounded domain in $\mathbf{R}^\nu$. And where $\mathbf{p} = -i\nabla$ and $\mathbf{A}(\mathbf{x}) = (A_1(\mathbf{x}), A_2(\mathbf{x}), \ldots, A_\nu(\mathbf{x}))$ is a magnetic vector potential for the magnetic field $\mathbf{B}(\mathbf{x})$. Suppose $Hu_i = \lambda_i u_i$. The operator H has some discrete eigenvalues under very wide circumstances (see [4]).

Assume that $V(\mathbf{x}) \geq -M$. The special choices to be made for the auxiliary operators whose indices will run from $1, \ldots, \nu + 1$ are

$$\Pi_j = -i\frac{\partial}{\partial x_j} - A_j(\mathbf{x}) \,, \qquad j = 1, \ldots, \nu$$

$$\Pi_{\nu+1} = (V(\mathbf{x}) + M)^{\frac{1}{2}}$$

$$G_j = x_j \,, \qquad j = 1, \ldots, \nu$$

$$G_{\nu+1} = 1.$$

It is easy to see that $\sum_{j=1}^{\nu+1} \Pi_j^2 = H + M$. So we see that the left hand side of inequality (1.1) is,

$$\sum_{i=1}^{n} (\lambda_{n+1} - \lambda_i)^{-1} \langle u_i, (H + M)u_i \rangle = \sum_{i=1}^{n} (\lambda_{n+1} - \lambda_i)^{-1}(\lambda_i + M).$$

For the right hand side we must compute the commutators

$$[\Pi_j, G_j] = [-i\frac{\partial}{\partial x_j}, x_j] = -i \,, \qquad j = 1, \ldots, \nu$$

$$[\Pi_{\nu+1}, G_{\nu+1}] = 0 \,.$$

And the double commutators,

$$\left[G_j, [H, G_j]\right] = \left[x_j, [-\Delta + i\nabla \cdot \mathbf{A} + i\mathbf{A} \cdot \nabla, x_j]\right]$$

$$= \left[x_j, (-2\frac{\partial}{\partial x_j} + 2iA_j)\right]$$

$$= 2, \quad j = 1, \ldots, \nu$$

$$\left[G_{\nu+1}, [H, G_{\nu+1}]\right] = 0.$$

Thus, the right hand side of inequality (1.1) is

$$\frac{\left|\sum_{j=1}^{\nu} Tr(P)\right|^2}{2 \sum_{j=1}^{\nu} Tr\left(P \cdot 2\right)} = \frac{\nu n}{4}$$

and by the theorem,

$$\sum_{i=1}^{n} \frac{\lambda_i + M}{\lambda_{n+1} - \lambda_i} \geq \frac{\nu n}{4}. \quad \square$$

Example 4: The bi-Laplacian

The operator $H = \Delta^2$ on $L^2(\Omega)$ where $\Omega \subset \mathbf{R}^\nu$ is a bounded domain with smooth boundary has applications in the theory of elasticity. Compare with [3] and [13] p. 633. In particular, the problem

$$\Delta^2 u_i = \mu_i u_i \quad \text{in} \ \Omega$$

$$u_i = \frac{\partial u_i}{\partial n} = 0 \quad \text{on} \ \partial\Omega$$

is related to the modes of vibration of a clamped plate. This problem was studied by Payne, Pólya, and Weinberger in [16] where they obtained the bound

$$\mu_{n+1} - \mu_n \leq \frac{8(\nu + 2)}{\nu^2 n} \sum_{i=1}^{n} \mu_i \tag{2.2}$$

and later by Hile and Yeh in [11] where the result was improved to

$$\sum_{i=1}^{n} \frac{\sqrt{\mu_i}}{\mu_{n+1} - \mu_i} \geq \frac{\nu^2 n^{\frac{3}{2}}}{8(\nu + 2)} \left(\sum_{i=1}^{n} \mu_i\right)^{-\frac{1}{2}}. \tag{2.3}$$

We now recover the result of [11] by applying Theorem 1.1. Assume that the eigenvalues are enumerated in the usual way. The choice for the auxiliary operators for this case is

$$\Pi_j \;=\; -i\frac{\partial}{\partial x_j}$$

$$G_j \;=\; x_j$$

where $j = 1, \ldots, \nu$.

The left hand side of equation (1.1) is then

$$\sum_{i=1}^{n}(\mu_{n+1} - \mu_i)^{-1}\langle u_i, -\Delta u_i\rangle \leq \sum_{i=1}^{n}(\mu_{n+1} - \mu_i)^{-1}\langle u_i, u_i\rangle^{\frac{1}{2}}\langle u_i, \Delta^2 u_i\rangle^{\frac{1}{2}}$$

$$= \sum_{i=1}^{n}(\mu_{n+1} - \mu_i)^{-1}\mu_i^{\frac{1}{2}}$$

where the Schwarz inequality and the boundary conditions were used to obtain the inequality.

For the right hand side we again have $[\Pi_j, G_j] = -i$. And now the double commutators are

$$\big[G_j, [H, G_j]\big] = \big[x_j, [\Delta^2, x_j]\big] = \big[x_j, 4\partial_j\Delta\big] = -4(\Delta + 2\partial_j^2)$$

so that

$$\sum_{j=1}^{\nu} \big[G_j, [H, G_j]\big] = 4(\nu + 2)(-\Delta).$$

Then, again applying the Schwarz inequality and the boundary conditions we have

$$\sum_{i=1}^{n}\langle u_i, -\Delta u_i\rangle \leq \left(\sum_{i=1}^{n}\langle u_i, u_i\rangle\right)^{\frac{1}{2}}\left(\sum_{i=1}^{n}\langle u_i, \Delta^2 u_i\rangle\right)^{\frac{1}{2}}.$$

So for the right hand side of equation (1.1) we have

$$\frac{(\nu n)^2}{8(\nu + 2)n^{\frac{1}{2}}}\left(\sum_{i=1}^{n}\mu_i\right)^{-\frac{1}{2}}$$

and so by the theorem we have

$$\sum_{i=1}^{n}\frac{\mu_i^{\frac{1}{2}}}{\mu_{n+1} - \mu_i} \geq \frac{\nu^2 n^{\frac{3}{2}}}{8(\nu + 2)}\left(\sum_{i=1}^{n}\mu_i\right)^{-\frac{1}{2}}$$

as in [11]. $\square$

3   NEUMANN BOUNDS

As a final use of our techniques, we obtain an entirely new bound for the spectral geometry
of the Laplacian with Neumann boundary conditions. We shall show that gaps in the
spectrum of the Laplacian are controlled, inversely, by the inradius, which by definition is
the radius of the largest inscribed ball. The inradius is known to control the ground state
of the Dirichlet Laplacian, and thus to play an important role in the spectral geometry for
that operator (cf. [2] and [5]). On the other hand, we believe that the constant we obtain
is far from optimal.

Since the Neumann Laplacian has a zero eigenvalue, and in our technique we need it to
dominate another expression, we let $H_M := -\Delta + M$, where M is a positive constant to be
specified below. The addition of the constant M obviously does not affect any commutators
with H. Let $\tilde{u}$ be the first nonconstant radial Neumann L^2 – normalized eigenfunction for
the unit ball and let $\tilde{\lambda}$ be the corresponding eigenvalue.

PROPOSITION 3.1 *Let Ω be a domain in R^ν with piecewise smooth boundary and
inradius r, and let λ_k be the eigenvalues of the Neumann Laplacian for Ω. Then*

$$\left\langle \frac{\lambda_\ell + M}{\lambda_{n+1} - \lambda_\ell} \right\rangle_{\ell \leq n} \geq \frac{K}{n+1}\,,$$

where:

$$M := \frac{\tilde{\lambda}^2 \|\tilde{u}\|_\infty^2}{4r^2 \|\nabla \tilde{u}\|_\infty^2}$$

and

$$K := \frac{r^\nu \tilde{\lambda}}{8\,|\Omega|\,\|\nabla \tilde{u}\|_\infty^2}\,.$$

REMARK: This complicated expression implies a somewhat simpler bound of the form

$$\lambda_{n+1} - \lambda_n \leq (n+1)\left(A \langle \lambda_\ell \rangle_{\ell \leq n} + B \right),$$

where A and B depend only on the inradius r and Ω.

PROOF: The special choices are $G := \tilde{u}$, centered at the center of the inscribed ball and
with variable $|x|/r$ in the ball and extended as a constant outside the ball, and $\Pi :=
i[H, G] = i(-\Delta G - 2\nabla G \cdot \nabla)$.

According to our usual calculation,

$$\left\langle \frac{\lambda_\ell + M}{\lambda_{n+1} - \lambda_\ell} \right\rangle_{\ell \leq n} \geq \frac{|Tr\left(P[G, [H, G]]\right)|^2}{2(n+1)\beta Tr\left(P[G, [H, G]]\right)} = \frac{Tr\left(P\,|\nabla G|^2\right)}{(n+1)\beta}\,, \tag{3.1}$$

where β is any constant such that

$$\|(\Delta G)\zeta + 2\nabla G \cdot \nabla \zeta\|^2 \leq \beta\left(\|\nabla \zeta\|^2 + M\|\zeta\|^2\right)$$

for all ζ in the quadratic form domain of the Neumann Laplacian.

Since the left side of this expression is

$$\left\| \frac{\tilde{\lambda}}{r^2}\tilde{u}\left(\frac{x}{r}\right)\zeta + \frac{2}{r}\nabla\tilde{u}\left(\frac{x}{r}\right)\cdot\nabla\zeta \right\|^2 \leq \frac{2\tilde{\lambda}^2\|\tilde{u}\|_\infty^2}{r^4}\|\zeta\|^2 + \frac{8}{r^2}\|\nabla\tilde{u}\|_\infty^2\|\nabla\zeta\|^2$$

(the cross term has been estimated by $2ab \leq a^2 + b^2$), we can take

$$\beta := \frac{8}{r^2}\|\nabla\tilde{u}\|_\infty^2 \quad \text{and} \quad M := \frac{\tilde{\lambda}^2\|\tilde{u}\|_\infty^2}{4r^2\|\nabla\tilde{u}\|_\infty^2}.$$

Since the ground state is $1/\sqrt{|\Omega|}$, we estimate the numerator of equation (3.1) by

$$Tr(P\,|\nabla G|^2) \geq \int |\nabla G|^2 \frac{1}{|\Omega|},$$

which

$$= \frac{r^\nu\tilde{\lambda}}{r^2\,|\Omega|},$$

and the claim follows. $\square$

ACKNOWLEDGEMENTS

The first author is grateful to the Georgia Tech Foundation for support and to the Centre de Physique Théorique, Luminy, France, and the Université de Toulon, France, for hospitality while part of this work was performed.

The second author would like to thank the Mittag-Leffler Institute, Djursholm, Sweden, for their hospitality while some of this work was carried out.

REFERENCES

[1] M.S. Ashbaugh and R.D. Benguria, A sharp bound for the ratio of the first two eigenvalues of Dirichlet Laplacians and extensions, *Annals Math.* **135** (1992) 601-628.

[2] I. Chavel, *Eigenvalues in Riemannian Geometry*, Academic Press, New York, 1984.

[3] Z.C. Chen, Inequalities for eigenvalues of a class of polyharmonic operators, *Applicable Analysis* **27** (1988) 289-314.

[4] H.L. Cycon, R.G. Froese, W. Kirsch, and B. Simon, *Schrödinger Operators*, Texts and Monographs in Physics, Springer, New York, 1987.

[5] E.B. Davies, *Heat Kernels and Spectral Theory*, Cambridge Tracts in Mathematics 92, Cambridge University Press, Cambridge, 1989.

[6] J. Geronimo, E.M. Harrell II and W. Van Assche, On the density of states for some banded matrices, *Constructive Approximation,* **4** (1988) 403-417.

[7] E.M. Harrell II, Some geometric bounds on eigenvalue gaps, *Comm. Part. Diff. Eqns.,* **18** (1993) 179-198.

[8] E.M. Harrell II, General bounds for the eigenvalues of Schrödinger operators, in *Maximum Principles and Eigenvalue Problems in Partial Differential Equations*, P.W. Schaefer, ed., Pitman Research Notes in math. Series **175**, Longman Scientific and Technical, Harlow, Essex, U.K., 1988, p. 146-166.

[9] E.M. Harrell II and P.L. Michel, Commutator bounds for eigenvalues, with applications to spectral geometry, *preprint*, 1993.

[10] G.N. Hile and M.H. Protter, Inequalities for eigenvalues of the Laplacian, *Indiana U. Math. J.* **29** (1980) 523-538.

[11] G.N. Hile and R.Z. Yeh, Inequalities for eigenvalues of the biharmonic operator, *Pacific J. Math.* **112** (1984) 115-133.

[12] S.M. Hook, Inequalities for eigenvalues of self-adjoint operators, *Trans. Amer. Math. Soc.* **318** (1990) 237-259.

[13] S.M. Hook, Domain-independent upper bounds for eigenvalues of elliptic operators, *Trans. Amer. Math. Soc.* **318** (1990) 615-642.

[14] M. Kac, Mathematical mechanisms of phase transitions, *Brandeis Univ. Summer Institute in Theoretical Physics 1966* I, M. Chrétien, E.P. Gross, and S. Desai, eds., Gordon & Breach, New York, 1968.

[15] W. Kirsch, Random Schrödinger operators and the density of states, in *Stochastic Aspects of Classical and Quantum Systems*, Lecture Notes in Mathematics, vol. 1109, S. Albeverio, Ph. Combe, and M. Sirugue-Collin, eds., Springer, Berlin and New York, 1985.

[16] L.E. Payne, G. Pólya, and H.F. Weinberger, On the ratio of consecutive eigenvalues, *J. Math. and Phys.* **35** (1956) 289-298.

[17] W. Thirring, *Course in Mathematical Physics, vol. III: Quantum Mechanics of Atoms and Molecules*, Springer-Verlag, New York and Vienna, 1981.

A Trace Regularity Result for Solutions to Plate Equations with Simply Supported Boundary Conditions

Mary Ann Horn   University of Minnesota, Minneapolis, Minnesota

1   INTRODUCTION

Stabilization of plate equations is a problem on which much attention has been focused in recent years. Both the linear Kirchhoff plate and the nonlinear von Kármán plate have been of particular interest to control theorists. One of the techniques used to prove such plates are stabilizable is that of multipliers, which can be described as choosing appropriate multipliers, e.g. position, velocity, etc., integrating by parts, and using the resulting identities to derive inequalities which are then used to prove stabilization does indeed hold. However, difficulties are often encountered in going from the result arising from the integration by parts to the desired inequality. To facilitate the proof of the final inequality, trace regularity results for the solution of the plate equation are often needed.

This material is based upon work partially supported under a National Science Foundation Mathematical Sciences Postdoctoral Research Fellowship.

In this paper, one such result is discussed.

1.1   Statement of the Problem

Let Ω be an open bounded domain in R^2 with a sufficiently smooth (e.g., C^∞), boundary, Γ. In Ω, we consider the following plate model with homogeneous Dirichlet boundary conditions and feedback control acting through a second order boundary condition (as a moment):

$$w_{tt} - \gamma^2 \Delta w_{tt} + \Delta^2 w = f \quad \text{in } Q_T = (0,T) \times \Omega \tag{1.1.a}$$

$$\left.\begin{array}{c} w(0,\cdot) = w_0 \\[2mm] w_t(0,\cdot) = w_1 \end{array}\right\} \quad \text{in } \Omega \tag{1.1.b}$$

$$w = 0 \quad \text{on } \Sigma_T = (0,T) \times \Gamma \tag{1.1.c}$$

$$\Delta w + (1-\mu)\mathcal{B}w = -\frac{\partial}{\partial\nu}w_t \quad \text{on } \Sigma_T = (0,T) \times \Gamma, \tag{1.1.d}$$

where $0 < \mu < \frac{1}{2}$ is Poisson's ratio. In (1.1.d), the operator $\mathcal{B}$ is given by

$$\mathcal{B}w \equiv -\frac{\partial^2}{\partial\tau^2}w - k\frac{\partial}{\partial\nu}w = -k\frac{\partial}{\partial\nu}w, \tag{1.2}$$

where the second equality follows because $w \equiv 0$ on the boundary. The parameter, γ, is proportional to the thickness of the plate and is therefore assumed to be small.

Remark 1: Although f is essentially arbitrary, in order to obtain the existence of a solution with some degree of regularity, this function must, in turn, possess a certain amount of regularity. Since our desire is to work with solutions in $H^2(\Omega) \times H_0^1(\Omega)$, a natural energy space associated with this type of plate equation, our choice of f must have sufficient regularity to guarantee solutions lie in this space. Indeed, this is satisfied for many different plate models which are obtained by an appropriate choice of f. Of particular interest are the following two cases:

   i.) Kirchhoff plate equation: $f \equiv 0$;

   ii.) von Kármán plate equation with light internal damping: $f \equiv -b(x)w_t + [w,\chi]$,

   where $b(x) \in L^\infty(\Omega)$ satisfies $b(x) > 0$ *a.e.* in Ω and $\chi(w)$ satisfies the system of equations:

$$\left.\begin{array}{rl} \Delta^2\chi & = -[w,w] \quad in \ \Omega \\[2mm] \chi & = \frac{\partial}{\partial\nu}\chi = 0 \quad on \ \Gamma \end{array}\right\}$$

where

$$[\phi, \psi] = \frac{\partial^2 \phi}{\partial x^2}\frac{\partial^2 \psi}{\partial y^2} + \frac{\partial^2 \phi}{\partial y^2}\frac{\partial^2 \psi}{\partial x^2} - 2\frac{\partial^2 \phi}{\partial x \partial y}\frac{\partial^2 \psi}{\partial x \partial y}.$$

For both of these cases, the following wellposedness results have been established.

THEOREM 1.1 *(i) (See [8].) Assume $f \equiv 0$. Let $(w_0, w_1) \in H^2(\Omega) \cap H_0^1(\Omega) \times H_0^1(\Omega)$. Then there exists a unique solution (in the sense of distributions), $(w(t), w_t(t)) \in C([0, T]; H^2(\Omega) \cap H_0^1(\Omega) \times H_0^1(\Omega))$, satisfying system (1.1).*

(ii) (See [3].) Assume $f \equiv -b(x)w_t + [w, \chi]$. Let $(w_0, w_1) \in H^2(\Omega) \cap H_0^1(\Omega) \times H_0^1(\Omega)$. Then there exists a unique solution (in the sense of distributions), $(w(t), w_t(t)) \in C([0, T]; H^2(\Omega) \cap H_0^1(\Omega) \times H_0^1(\Omega))$, satisfying system (1.1).

In the proof of boundary stabilization for both of these models, a critical problem is the need for estimates for traces of the solution. Semigroup theory (see [10]) can be used to show that proving these plate models are uniformly stable is equivalent to proving the following estimate holds: $\exists T_0 > 0$ such that $\forall T > T_0$,

$$E(T) \leq C_T(1 + \gamma^2)\|\frac{\partial}{\partial \nu}w_t\|^2_{L_2(\Sigma_T)}, \tag{1.3}$$

where $E(T)$ is the energy of the system at time T. As expected, the definition of $E(T)$ will depend on the system under consideration. In the above two cases, the energy is defined by:

Kirchhoff plate:

$$E(t) \equiv \frac{1}{2}\int_\Omega \{|w_t|^2 + \gamma^2|\nabla w_t|^2 + |\Delta w|^2\}dt, \tag{1.4}$$

von Kármán plate:

$$E(t) = \frac{1}{2}\int_\Omega \{|w_t|^2 + \gamma^2|\nabla w_t|^2 + |\Delta w|^2 + |\Delta\chi(w)|^2\}d\Omega. \tag{1.5}$$

When multiplier methods are used to obtain the above estimate, a boundary integral term arises which must be bounded in terms of the norm of the control, the norm of the right-hand side of (1.1), and lower order terms. To do so, the following trace regularity result is required.

THEOREM 1.2 *Let w be the solution to (1.1) and $0 < \alpha < T/2$. Then w satisfies the following inequality.*

$$\|\frac{\partial}{\partial \nu}\Delta w\|^2_{H^{-1}(\alpha, T-\alpha; L_2(\Gamma))} \leq C_T(1 + \gamma^2)\|\frac{\partial}{\partial \nu}w_t\|^2_{L_2(\Sigma_T)} + C(T+1)(1 + \gamma^2)\|w\|^2_{L_2(0, T; H^{2-\epsilon}(\Omega))}$$

$$+ C\|f\|^2_{L_2(0, T; H^{-1/2+\epsilon}(\Omega))}, \tag{1.6}$$

where $0 < \epsilon < 1/2$ and the constants C_T and C are independent of γ.

Remark 2: Notice that to obtain an estimate for the norm of the trace over the time interval $(\alpha, T - \alpha)$, we must "observe" the solution for a slightly longer time. This is the trade-off that must be made in order to obtain a result which is considerably better than that found directly by using trace theory. If trace theory were applied, we would find

$$\|\frac{\partial}{\partial \nu}\Delta w\|_{L_2(\Gamma)} \leq C\|w\|_{H^{7/2+\epsilon}(\Omega)},$$

a space far higher than that of the energy of the system.

Trace regularity results have been used in proving boundary stabilization of many plate models. Results which have been proven by using microlocal analysis techniques can be found in [6] (Lemma 2.2) and [9] (Theorem 2.1). Two different sets of boundary conditions are considered in these papers. In [6], the boundary conditions considered are the same as (1.1.c) and (1.1.d). Higher order boundary conditions, with control acting through both bending moments and torques, are considered in [9]. These results have been applied to both the Kirchhoff and von Kármán plate models (see [1], [2], [6], [7]).

Theorem 1.2 has been used to prove uniform stabilization for a Kirchhoff plate model in [6] and for the von Kármán model in [1].

2 PROOF OF THEOREM 1.2

Step 1: Let $\psi(t) \in C_0^\infty(R)$ be a cutoff function defined such that $0 \leq \psi(t) \leq 1 \; \forall t$ and

$$\psi(t) = \begin{cases} 1 & \text{in } (\alpha, T - \alpha) \\ 0 & \text{outside } (\frac{\alpha}{2}, T - \frac{\alpha}{2}). \end{cases}$$

Define $w_c(t, \cdot) \equiv \psi(t)w(t, \cdot)$, where $w(t, \cdot)$ is our original solution to (1.1). Then $w_c(t, \cdot)$ satisfies

$$w_{c,tt} - \gamma^2 \Delta w_{c,tt} + \Delta^2 w_c = [\tilde{P}, \psi]w + \psi f \quad \text{in } Q_\infty$$

$$\left.\begin{array}{r} w_c(0) = 0 \\ w_{c,t}(0) = 0 \end{array}\right\} \quad \text{in } \Omega \qquad (2.1)$$

$$w_c = 0 \quad \text{on } \Sigma_\infty$$

$$\Delta w_c - (1-\mu)k\frac{\partial}{\partial\nu}w_c = -\psi\frac{\partial}{\partial\nu}w_t \quad \text{on } \Sigma_\infty$$

where $[A, B]$ denotes the commutator of two operators, A and B and, in our case, $\tilde{P}w = w_{tt} - \gamma^2 \Delta w_{tt} + \Delta^2 w$. Rewriting this equation for w_c as

$$w_{c,tt} + (I - \gamma^2 \Delta)^{-1} \Delta^2 w_c = (I - \gamma^2 \Delta)^{-1} [\tilde{P}, \psi]w + (I - \gamma^2 \Delta)^{-1} \psi f \quad \text{in } Q_\infty$$

$$\left. \begin{aligned} w_c(0) &= 0 \\ w_{c,t}(0) &= 0 \end{aligned} \right\} \quad \text{in } \Omega \qquad (2.2)$$

$$w_c = 0 \quad \text{on } \Sigma_\infty$$

$$\Delta w_c - (1 - \mu)k\frac{\partial}{\partial\nu}w_c = -\psi\frac{\partial}{\partial\nu}w_t \quad \text{on } \Sigma_\infty,$$

we see that $w_c(t, \cdot)$ can be written implicitly using the variation of parameters formula as follows:

$$w_c(t) = \frac{1}{2i}\sqrt{A_\gamma}^{-1} \int_0^t (e^{i\sqrt{A_\gamma}(t-\tau)} - e^{-i\sqrt{A_\gamma}(t-\tau)})F(\tau)d\tau,$$

where

$$Ah \equiv \Delta^2 h, \quad \mathcal{D}(A) = \{h \in H^4(\Omega) \cap H_0^1(\Omega) : \Delta h|_\Gamma = 0\},$$

$$A_\gamma \equiv (I + \gamma^2 A^{1/2})^{-1} A,$$

$$F(t) \equiv (I + \gamma^2 A^{1/2})^{-1}\{A^{1/2}D((1 - \mu)k\frac{\partial}{\partial\nu}w_c - \psi\frac{\partial}{\partial\nu}w_t(t)) + [\tilde{P}, \psi]w(t) + \psi f\},$$

and Dg is defined to be the harmonic extension of the function g from the boundary into the interior, i.e.,

$$Dg = v \quad \Longleftrightarrow \quad \left\{ \begin{aligned} \Delta v &= 0 \quad \text{in } \Omega \\ v &= g \quad \text{on } \Gamma. \end{aligned} \right.$$

Define

$$\begin{aligned} w_1(t) &\equiv \tfrac{1}{2i}\sqrt{A_\gamma}^{-1} \int_0^t (e^{i\sqrt{A_\gamma}(t-\tau)} - e^{-i\sqrt{A_\gamma}(t-\tau)})(I + \gamma^2 A^{1/2})^{-1}\psi f(\tau)d\tau, \\ w_2(t) &\equiv \tfrac{1}{2i}\sqrt{A_\gamma}^{-1} \int_0^t (e^{i\sqrt{A_\gamma}(t-\tau)} - e^{-i\sqrt{A_\gamma}(t-\tau)})(I + \gamma^2 A^{1/2})^{-1}A^{1/2}D((1 - \mu)k\frac{\partial}{\partial\nu}w_c)d\tau \\ &\quad - \tfrac{1}{2i}\sqrt{A_\gamma}^{-1} \int_0^t (e^{i\sqrt{A_\gamma}(t-\tau)} - e^{-i\sqrt{A_\gamma}(t-\tau)})(I + \gamma^2 A^{1/2})^{-1}A^{1/2}D(\psi\frac{\partial}{\partial\nu}w_t(\tau))d\tau, \\ w_3(t) &\equiv \tfrac{1}{2i}\sqrt{A_\gamma}^{-1} \int_0^t (e^{i\sqrt{A_\gamma}(t-\tau)} - e^{-i\sqrt{A_\gamma}(t-\tau)})(I + \gamma^2 A^{1/2})^{-1}[\tilde{P}, \psi]w(\tau)d\tau. \end{aligned}$$

Step 2: Expressions for $w_1(t)$, $w_2(t)$ and $w_3(t)$. Since w_c is compactly supported on $(0, T)$,

$$\|w_c\|_{H^{-1}(0, T)} = \|\int_0^t w_c(\tau)d\tau\|_{L_2(0, T)}.$$

Hence,

$$\|\frac{\partial}{\partial\nu}\Delta w_c\|_{H^{-1}(0, \infty; L_2(\Gamma))} = \|\frac{\partial}{\partial\nu}A^{1/2}\int_0^t (w_1 + w_2 + w_3)(\tau)d\tau\|_{L_2(\Sigma_\infty)}.$$

Integrating $w_1(t)$ from 0 to t gives

$$\frac{\partial}{\partial \nu} A^{1/2} \int_0^t w_1(\tau) d\tau$$
$$= -\frac{1}{2}\frac{\partial}{\partial \nu} A^{-1/2} \int_0^t (e^{i\sqrt{A_\gamma}(t-\tau)} + e^{-i\sqrt{A_\gamma}(t-\tau)} - 2)\psi f(\tau) d\tau. \tag{2.3}$$

Differentiating $w_1(t)$ with respect to t, we find

$$w_1'(t) = \frac{1}{2}(I + \gamma^2 A^{1/2})^{-1} \int_0^t (e^{i\sqrt{A_\gamma}(t-\tau)} + e^{-i\sqrt{A_\gamma}(t-\tau)})\psi f(\tau) d\tau, \tag{2.4}$$

where, for convenience, we have denoted differentiation with respect to time by $'$. By noting that

$$A^{1/2} A_\gamma^{-1}(I + \gamma^2 A^{1/2})^{-1} A^{1/2} = I, \tag{2.5}$$

and

$$\gamma^2 A^{1/2}(I + \gamma^2 A^{1/2})^{-1} - I = (I + \gamma^2 A^{1/2})^{-1}, \tag{2.6}$$

we can compare these two expressions and find that

$$\frac{\partial}{\partial \nu} A^{1/2} \int_0^t w_1(\tau) d\tau$$
$$= -\gamma^2 \frac{\partial}{\partial \nu} w_1'(t) + \frac{\partial}{\partial \nu} A^{-1/2} \int_0^t \psi f(\tau) d\tau$$
$$+ \frac{1}{2}\frac{\partial}{\partial \nu}(I + \gamma^2 A^{1/2})^{-1} A^{-1/2} \int_0^t (e^{i\sqrt{A_\gamma}(t-\tau)} + e^{-i\sqrt{A_\gamma}(t-\tau)})\psi f(\tau) d\tau. \tag{2.7}$$

Proceeding in the same manner as for $w_1(t)$, we can find corresponding identities for $w_2(t)$ and $w_3(t)$:

$$\frac{\partial}{\partial \nu} A^{1/2} \int_0^t w_2(\tau) d\tau$$
$$= -\gamma^2 \frac{\partial}{\partial \nu} w_2'(t) + \frac{\partial}{\partial \nu} \int_0^t \psi(\tau) D(\frac{\partial}{\partial \nu} w_t(\tau)) d\tau$$
$$- \frac{1}{2}\frac{\partial}{\partial \nu} \int_0^t (e^{i\sqrt{A_\gamma}(t-\tau)} + e^{-i\sqrt{A_\gamma}(t-\tau)})(I + \gamma^2 A^{1/2})^{-1}\psi(\tau) D(\frac{\partial}{\partial \nu} w_t(\tau)) d\tau, \tag{2.8}$$

$$\frac{\partial}{\partial \nu} A^{1/2} \int_0^t w_3(\tau) d\tau$$
$$= -\gamma^2 \frac{\partial}{\partial \nu} w_3'(t) + \frac{\partial}{\partial \nu} \int_0^t A^{-1/2}[\tilde{P}, \psi] w(\tau) d\tau$$
$$- \frac{1}{2}\frac{\partial}{\partial \nu} \int_0^t (e^{i\sqrt{A_\gamma}(t-\tau)} + e^{-i\sqrt{A_\gamma}(t-\tau)})(I + \gamma^2 A^{1/2})^{-1} A^{-1/2}[\tilde{P}, \psi] w(\tau) d\tau. \tag{2.9}$$

Combining the above two expressions with (2.7), we find

$$\left\|\frac{\partial}{\partial\nu}A^{1/2}w_c\right\|_{H^{-1}(0,\infty;L_2(\Gamma))}$$
$$\leq \gamma^2\left\|\frac{\partial}{\partial\nu}(w_c)_t\right\|_{L_2(\Sigma_\infty)} + \left\|\frac{\partial}{\partial\nu}\int_0^t A^{-1/2}D(k\frac{\partial}{\partial\nu}w_c(\tau))d\tau\right\|_{L_2(\Sigma_\infty)}$$
$$+\frac{1}{2}\left\|\frac{\partial}{\partial\nu}\int_0^t(e^{i\sqrt{A_\gamma}(t-\tau)} + e^{-i\sqrt{A_\gamma}(t-\tau)})(I+\gamma^2A^{1/2})^{-1}D(k\frac{\partial}{\partial\nu}w_c(\tau))d\tau\right\|_{L_2(\Sigma_\infty)}$$
$$+\left\|\frac{\partial}{\partial\nu}\int_0^t\psi(\tau)D(\frac{\partial}{\partial\nu}w_t(\tau))d\tau\right\|_{L_2(\Sigma_\infty)}$$
$$+\frac{1}{2}\left\|\frac{\partial}{\partial\nu}\int_0^t(e^{i\sqrt{A_\gamma}(t-\tau)} + e^{-i\sqrt{A_\gamma}(t-\tau)})(I+\gamma^2A^{1/2})^{-1}\psi(\tau)D(\frac{\partial}{\partial\nu}w_t(\tau))d\tau\right\|_{L_2(\Sigma_\infty)}$$
$$+\left\|\frac{\partial}{\partial\nu}\int_0^t A^{-1/2}[\tilde{P},\psi]w(\tau)d\tau\right\|_{L_2(\Sigma_\infty)}$$
$$+\frac{1}{2}\left\|\frac{\partial}{\partial\nu}\int_0^t(e^{i\sqrt{A_\gamma}(t-\tau)} + e^{-i\sqrt{A_\gamma}(t-\tau)})(I+\gamma^2A^{1/2})^{-1}A^{-1/2}[\tilde{P},\psi]w(\tau)d\tau\right\|_{L_2(\Sigma_\infty)}$$
$$+\left\|\frac{\partial}{\partial\nu}\int_0^t A^{-1/2}\psi f(\tau)d\tau\right\|_{L_2(\Sigma_\infty)}$$
$$+\frac{1}{2}\left\|\frac{\partial}{\partial\nu}\int_0^t(e^{i\sqrt{A_\gamma}(t-\tau)} + e^{-i\sqrt{A_\gamma}(t-\tau)})(I+\gamma^2A^{1/2})^{-1}A^{-1/2}\psi f(\tau)d\tau\right\|_{L_2(\Sigma_\infty)}. \tag{2.10}$$

Step 3: Bounding Terms By Trace Theory. The second term on the right-hand side of (2.10) can be bounded directly by using trace theory and the properties of Sobolev spaces:

$$\left\|\frac{\partial}{\partial\nu}\int_0^t A^{-1/2}D(k\frac{\partial}{\partial\nu}w_c(\tau))d\tau\right\|_{L_2(\Sigma_\infty)} \leq C\left\|\frac{\partial}{\partial\nu}A^{-1/2}D(k\frac{\partial}{\partial\nu}w_c)\right\|_{L_2(\Sigma_\infty)}$$
$$\leq C\left\|D(k\frac{\partial}{\partial\nu}w_c)\right\|_{L_2(0,\infty;L_2(\Omega))} \leq C\left\|\frac{\partial}{\partial\nu}w_c\right\|_{L_2(0,\infty;L_2(\Gamma))}$$
$$\leq C\left\|\frac{\partial}{\partial\nu}w\right\|_{L_2(0,T;L_2(\Gamma))} \leq C\|w\|_{L_2(0,T;H^{3/2+\epsilon}(\Omega))}. \tag{2.11}$$

By writing $[\tilde{P},\psi]w$ explicitly as

$$[\tilde{P},\psi]w \equiv -2\gamma^2 A^{1/2}(\psi'w)' - \psi''w + 2(\psi'w)' + \gamma^2\psi''A^{1/2}w, \tag{2.12}$$

the sixth term on the right-hand side of (2.10) can be bounded using trace theory as follows:

$$\left\|\frac{\partial}{\partial\nu}\int_0^t A^{-1/2}[\tilde{P},\psi]w(\tau)d\tau\right\|_{L_2(\Sigma_\infty)}$$
$$= \left\|\frac{\partial}{\partial\nu}A^{-1/2}[\tilde{P},\psi]w\right\|_{H^{-1}(0,\infty;L_2(\Gamma))}$$
$$\leq C\{\left\|\frac{\partial}{\partial\nu}A^{-1/2}\gamma^2 A^{1/2}(\psi'w)'\right\|_{H^{-1}(0,\infty;L_2(\Gamma))} + \left\|A^{-1/2}\psi''w\right\|_{H^{-1}(0,\infty;H^{3/2+\epsilon}(\Omega))}$$
$$+\left\|A^{-1/2}\psi'w\right\|_{H^{-1}(0,\infty;H^{3/2+\epsilon}(\Omega))} + \gamma^2\left\|\frac{\partial}{\partial\nu}w\right\|_{L_2(\Sigma_T)}\} \tag{2.13}$$
$$\leq C\{\gamma^2\left\|\frac{\partial}{\partial\nu}(\psi'w)'\right\|_{H^{-1}(0,\infty;L_2(\Gamma))} + \|w\|_{L_2(0,T;L_2(\Omega))} + \gamma^2\left\|\frac{\partial}{\partial\nu}w\right\|_{L_2(\Sigma_T)}\}$$
$$\leq C\{\gamma^2\left\|\frac{\partial}{\partial\nu}w\right\|_{L_2(\Sigma_T)} + \|w\|_{L_2(0,T;L_2(\Omega))}\}.$$

Finally, the second to last term on the right-hand side of (2.10) is bounded by:

$$\left\|\frac{\partial}{\partial\nu}\int_0^t A^{-1/2}\psi f(\tau)d\tau\right\|_{L_2(\Sigma_\infty)} \leq C\left\|A^{-1/2}\psi f\right\|_{H^{-1}(0,T;H^{3/2+\epsilon}(\Omega))} \leq C\|f\|_{L_2(0,T;H^{-1/2+\epsilon}(\Omega))}. \tag{2.14}$$

Step 4: To bound the fourth term on the right-hand side of (2.10), we note that

$$\int_0^t \psi(\tau)D(\frac{\partial}{\partial \nu}w_t(\tau))d\tau = \psi(t)D(\frac{\partial}{\partial \nu}w(t)) - \int_0^t \psi'(\tau)D(\frac{\partial}{\partial \nu}w(\tau))d\tau.$$

Thus,

$$\|\frac{\partial}{\partial \nu}\int_0^t \psi(\tau)D(\frac{\partial}{\partial \nu}w_t(\tau))d\tau\|_{L_2(\Sigma_T)} \leq C_\psi\|\frac{\partial}{\partial \nu}D(\frac{\partial}{\partial \nu}w)\|_{L_2(\Sigma_T)}.$$

But, we also know, by elliptic regularity, that

$$\|\tfrac{\partial}{\partial\nu}D(\tfrac{\partial}{\partial\nu}w)\|_{L_2(\Sigma_T)} \leq C\|D(\tfrac{\partial}{\partial\nu}w)\|_{L_2(0,\,T;\,H^{3/2}(\Omega))}$$
$$\leq C\|\tfrac{\partial}{\partial\nu}w\|_{L_2(0,\,T;\,H^1(\Gamma))} \leq C\|\tfrac{\partial}{\partial\tau}\tfrac{\partial}{\partial\nu}w\|_{L_2(\Sigma_T)}. \tag{2.15}$$

Step 5: Bounding terms using trace regularity results for the Schrödinger equation. Let $\mathcal{T}_i$, $i = 3, 5, 7, 9$ respectively denote the third, fifth, seventh, and last terms on the right-hand side of (2.10). In section 3, we prove that these terms can be bounded as follows:

$$\mathcal{T}_3 \leq C_T\|\frac{\partial}{\partial \nu}w_t\|_{L_2(\Sigma_T)}, \tag{2.16}$$

$$\mathcal{T}_5 \leq C_T\|\frac{\partial}{\partial \nu}w_t\|_{L_2(\Sigma_T)}, \tag{2.17}$$

$$\mathcal{T}_7 \leq C(T+1)(1+\gamma^2)\|w\|_{L_2(0,\,T;\,H_0^1(\Omega))}, \tag{2.18}$$

$$\mathcal{T}_9 \leq C(T+1)\|f\|_{L_2(0,\,T;\,H^{-1}(\Omega))}. \tag{2.19}$$

Step 6: Final Estimates. Substituting the estimates we have obtained, (2.11), and (2.13)-(2.19), into (2.10), we obtain

$$\|\tfrac{\partial}{\partial\nu}A^{1/2}w\|_{H^{-1}(\alpha,\,T-\alpha;\,L_2(\Gamma))} \leq C_T(1+\gamma^2)\|\tfrac{\partial}{\partial\nu}w_t\|_{L_2(\Sigma_T)} + C(T+1)\|f\|_{L_2(0,\,T;\,H^{-1/2+\epsilon}(\Omega))}$$
$$+C(T+1)(1+\gamma^2)\|w\|_{L_2(0,\,T;\,H^{2-\epsilon}(\Omega))} + C\|\tfrac{\partial}{\partial\tau}\tfrac{\partial}{\partial\nu}w\|_{L_2(\Sigma_T)}. \tag{2.20}$$

To achieve our desired result, we must remove the last term on the right-hand side of the above inequality. To do so, we require the following trace regularity result which is an adaptation of Proposition 3.1 in [6]:

LEMMA 2.1 *Let w be the solution to (1.1). Then w satisfies the following estimate:*

$$\|\frac{\partial}{\partial \tau}\frac{\partial}{\partial \nu}w\|^2_{L_2(\alpha,\,T-\alpha;\,\Gamma)} \leq C\{\|\frac{\partial}{\partial \nu}w_t\|^2_{L_2(\Sigma_T)} + \|f\|^2_{L_2(0,\,T;\,H^{-3/2+\epsilon}(\Omega))} + (1+\gamma^2)\|w\|^2_{L_2(0,\,T;\,H^{2-\epsilon}(\Omega))}\} \tag{2.21}$$

Considering the estimate, (2.20), over the time interval $(\alpha, T - \alpha)$ instead of $(0, T)$ and applying the above lemma, we arrive at our desired result, (1.6). $\square$

3   TRACE REGULARITY FOR A SCHRÖDINGER EQUATION

In the proof of trace regularity for the plate equation, we use the following result which was proven in [6].

PROPOSITION 3.1 *Consider the following two "abstract" Schrödinger problems:*

$$z_t = i\sqrt{A_\gamma}z + (I + \gamma^2 A^{1/2})^{-1}g$$
$$z(0) = 0, \tag{3.1}$$

$$z_t = i\sqrt{A_\gamma}z + (I + \gamma^2 A^{1/2})^{-1}Dg$$
$$z(0) = 0. \tag{3.2}$$

Then, assuming g has sufficient regularity,

i.) If $z(t)$ satisfies (3.1), then

$$\|\frac{\partial}{\partial\nu}z\|_{L_2(\Sigma_T)} \leq C(T+1)\|A^{1/4}(I + \gamma^2 A^{1/2})^{-1}g\|_{L_1(0,\,T;\,L_2(\Omega))}, \tag{3.3}$$

which, in turn, implies both

$$\|\frac{\partial}{\partial\nu}z\|_{L_2(\Sigma_T)} \leq C(T+1)\|g\|_{L_1(0,\,T;\,H^1(\Omega))} \tag{3.4}$$

$$and \quad \|\frac{\partial}{\partial\nu}z\|_{L_2(\Sigma_T)} \leq \frac{C(T+1)}{\gamma^2}\|g\|_{L_1(0,\,T;\,H^{-1}(\Omega))}, \tag{3.5}$$

where C is independent of γ.

ii.) If $z(t)$ satisfies (3.2), then

$$\|\frac{\partial}{\partial\nu}z\|_{L_2(\Sigma_T)} \leq C_T\|f\|_{L_2(\Sigma_T)}, \tag{3.6}$$

where C_T is independent of γ.

Remark 3: Results in the same spirit as those in the above proposition were proven in [5], however, the main differences are that in the present context:

i.) An "abstract" Schrödinger equation ($A_\gamma^{1/2}$ is a pseudodifferential operator) rather than a standard Schrödinger operator, $i\Delta$, as in [5], appears in our equations;

ii.) the dependence on γ is explicitly stated, so that the final constants, C and C_T, are independent of γ.

Taking advantage of the form of the integrals in the third, fifth, seventh, and last terms on the right-hand side of (2.10), we can use Proposition 3.1 to bound these terms.

Bounding the seventh term: To bound the seventh term on the right-hand side of (2.10), it is enough to estimate

$$\left\| \frac{\partial}{\partial \nu} \int_0^t e^{i\sqrt{A_\gamma}(t-\tau)}(I + \gamma^2 A^{1/2})^{-1} A^{-1/2}[\tilde{P}, \psi]w(\tau)d\tau \right\|_{L_2(\Sigma_\infty)},$$

since the argument for the $e^{-i\sqrt{A_\gamma}}$ term is the same. Recalling (2.12) and applying part $i.)$ of Proposition 2.1 to the second and fourth terms in (2.12) with $g \equiv A^{-1/2}w\psi''$ and $g \equiv w\psi''$, respectively, we find

$$\begin{aligned}
\Big\| \frac{\partial}{\partial \nu} &\int_0^t e^{i\sqrt{A_\gamma}(t-\tau)}(I + \gamma^2 A^{1/2})^{-1} A^{-1/2}[\tilde{P}, \psi]w(\tau)d\tau \Big\|_{L_2(\Sigma_\infty)} \\
&\leq 2\Big\| \frac{\partial}{\partial \nu} \int_0^t e^{i\sqrt{A_\gamma}(t-\tau)}\gamma^2 A^{1/2}(I + \gamma^2 A^{1/2})^{-1} A^{-1/2}(\psi' w)' d\tau \Big\|_{L_2(\Sigma_\infty)} \\
&\quad + 2\Big\| \frac{\partial}{\partial \nu} \int_0^t e^{i\sqrt{A_\gamma}(t-\tau)}(I + \gamma^2 A^{1/2})^{-1} A^{-1/2}(\psi' w)' d\tau \Big\|_{L_2(\Sigma_\infty)} \\
&\quad + C(T+1)\gamma^2 \|w\|_{L_2(0,T; H^1(\Omega))} \equiv \mathcal{P}_1 + \mathcal{P}_2 + C(T+1)\gamma^2\|w\|_{L_2(0,T; H_0^1(\Omega))}.
\end{aligned} \tag{3.7}$$

Remark 4: In the remainder of this section, we will frequently be using the estimate,

$$\|\gamma^2 A^{1/2}(I + \gamma^2 A^{1/2})^{-1} x\|_{\mathcal{H}} \leq C\|x\|_{\mathcal{H}} \quad uniformly \ in \ \gamma, \tag{3.8}$$

where $\mathcal{H}$ is any Sobolev space, and also the result by Grisvard (see [4]),

$$\mathcal{D}(A^\theta) \sim H_0^{4\theta}(\Omega) \ \forall \theta < \frac{1}{2}. \tag{3.9}$$

Bound for $\mathcal{P}_1$: Integrating the integral in $\mathcal{P}_1$ by parts, we find

$$\begin{aligned}
\int_0^t e^{i\sqrt{A_\gamma}(t-\tau)} &\gamma^2 A^{1/2}(I + \gamma^2 A^{1/2})^{-1} A^{-1/2}(\psi' w)' d\tau \\
&= (I + \gamma^2 A^{1/2})^{-1}\gamma^2(\psi' w)(t) \\
&\quad - \int_0^t e^{i\sqrt{A_\gamma}(t-\tau)}(I + \gamma^2 A^{1/2})^{-1/2} A^{1/2}\gamma^2(I + \gamma^2 A^{1/2})^{-1}(\psi' w)d\tau \\
&\equiv \mathcal{P}_{11} + \mathcal{P}_{12}.
\end{aligned} \tag{3.10}$$

$\mathcal{P}_{11}$ can be bounded using trace theory, (3.8), and (3.9):

$$\begin{aligned}
\Big\| \frac{\partial}{\partial \nu} \mathcal{P}_{11} \Big\|_{L_2(\Sigma_\infty)} &\leq C\|A^{3/8+\epsilon}\mathcal{P}_{11}\|_{L_2(Q_\infty)} \\
&\leq C\|A^{3/8+\epsilon}\gamma^2(I + \gamma^2 A^{1/2})^{-1}(\psi' w)\|_{L_2(Q_\infty)} \\
&\leq C\|\gamma^2 A^{1/2}(I + \gamma^2 A^{1/2})^{-1}(\psi' w)\|_{L_2(Q_\infty)} \leq C\|w\|_{L_2(Q_T)}.
\end{aligned} \tag{3.11}$$

To bound $\mathcal{P}_{12}$, we use part $i.)$ of Proposition 2.1 with $g \equiv \sqrt{A_\gamma}\gamma^2\psi' w$ to show

$$\begin{aligned}
\Big\| \frac{\partial}{\partial \nu} \mathcal{P}_{12} \Big\|_{L_2(\Sigma_\infty)} &\leq C(T+1)\|A^{1/4}(I + \gamma^2 A^{1/2})^{-1}g\|_{L_2(0,\infty; L_2(\Omega))} \\
&\leq C(T+1)\|A^{1/4}(I + \gamma^2 A^{1/2})^{-1}\gamma^2 A^{1/2}(I + \gamma^2 A^{1/2})^{-1/2}w\|_{L_2(0,T; L_2(\Omega))} \\
&\leq C(T+1)\|A^{1/4}w\|_{L_2(0,T; L_2(\Omega))} \leq C(T+1)\|w\|_{L_2(0,T; H_0^1(\Omega))}.
\end{aligned} \tag{3.12}$$

Collecting (3.10)-(3.12), we obtain

$$\mathcal{P}_1 \le C(T+1)\|w\|_{L_2(0,\infty;H_0^1(\Omega))}. \tag{3.13}$$

Bound for $\mathcal{P}_2$: As for $\mathcal{P}_1$, we integrate the integral inside the norm by parts to find

$$\begin{aligned}
\int_0^t e^{i\sqrt{A_\gamma}(t-\tau)}&(I+\gamma^2 A^{1/2})^{-1}A^{-1/2}(\psi'w)'d\tau \\
&= (I+\gamma^2 A^{1/2})^{-1}A^{-1/2}(\psi'w)(t) \\
&\quad - \int_0^t e^{i\sqrt{A_\gamma}(t-\tau)}(I+\gamma^2 A^{1/2})^{-1}(I+\gamma^2 A^{1/2})^{-1/2}(\psi'w)d\tau \\
&\equiv \mathcal{P}_{21}+\mathcal{P}_{22}.
\end{aligned} \tag{3.14}$$

To bound $\mathcal{P}_{21}$, we apply trace theory and (3.9) in a straightforward manner to find

$$\left\|\frac{\partial}{\partial\nu}\mathcal{P}_{21}\right\|_{L_2(\Sigma_\infty)} \le C\|w\|_{L_2(Q_T)}. \tag{3.15}$$

For $\mathcal{P}_{22}$, we apply part $i.)$ of Proposition 2.1 with $g \equiv (I+\gamma^2 A^{1/2})^{-1/2}\psi'w$. Since

$$\left\|(I+\gamma^2 A^{1/2})^{-1/2}\psi'w\right\|_{L_2(0,\infty;H^1(\Omega))} \le C\|w\|_{L_2(0,T;H_0^1(\Omega))},$$

we find

$$\left\|\frac{\partial}{\partial\nu}\mathcal{P}_{22}\right\|_{L_2(\Sigma_\infty)} \le C(T+1)\|w\|_{L_2(0,T;H_0^1(\Omega))}. \tag{3.16}$$

Combining the above inequalities, we obtain

$$\mathcal{P}_2 \le C(T+1)\|w\|_{L_2(0,T;H_0^1(\Omega))}. \tag{3.17}$$

Therefore, from (3.7), (3.13), and (3.17), we obtain

$$\begin{aligned}
\left\|\frac{\partial}{\partial\nu}\int_0^t (e^{i\sqrt{A_\gamma}(t-\tau)}+e^{-i\sqrt{A_\gamma}(t-\tau)})(I+\gamma^2 A^{1/2})^{-1}A^{-1/2}[\tilde{P},\psi]w(\tau)d\tau\right\|_{L_2(\Sigma_\infty)} \\
\le C(T+1)(1+\gamma^2)\|w\|_{L_2(0,T;H_0^1(\Omega))}.
\end{aligned} \tag{3.18}$$

Bounding the last term: To bound the last term on the right-hand side of (2.10), we apply part $i.)$ of Proposition 2.1 to find

$$\begin{aligned}
\left\|\frac{\partial}{\partial\nu}\int_0^t (e^{i\sqrt{A_\gamma}(t-\tau)}\right. &\left.+e^{-i\sqrt{A_\gamma}(t-\tau)})(I+\gamma^2 A^{1/2})^{-1}A^{-1/2}\psi f(\tau)d\tau\right\|_{L_2(\Sigma_\infty)} \\
&\le C(T+1)\|A^{-1/2}\psi f\|_{L_1(0,\infty;H_0^1(\Omega))} \\
&\le C(T+1)\|f\|_{L_1(0,T;H^{-1}(\Omega))} \le C(T+1)\|f\|_{L_2(0,T;H^{-1}(\Omega))}.
\end{aligned} \tag{3.19}$$

Bounding the fifth term: Applying part $ii.)$ of Proposition 2.1 with $g \equiv \psi\frac{\partial}{\partial\nu}w_t$ to the fifth term on the right-hand side of (2.10), we find

$$\begin{aligned}
\left\|\frac{\partial}{\partial\nu}\int_0^t (e^{i\sqrt{A_\gamma}(t-\tau)}\right. &\left.+e^{-i\sqrt{A_\gamma}(t-\tau)})(I+\gamma^2 A^{1/2})^{-1}\psi(\tau)D(\frac{\partial}{\partial\nu}w_t(\tau))d\tau\right\|_{L_2(\Sigma_\infty)} \\
&\le C_T\left\|\frac{\partial}{\partial\nu}w_t\right\|_{L_2(\Sigma_T)}.
\end{aligned} \tag{3.20}$$

Bounding the third term: Finally, applying part *ii.)* of Proposition 2.1 with $g \equiv k\frac{\partial}{\partial\nu}w_c$ to the third term on the right-hand side of (2.10) gives us our final estimate,

$$\|\frac{\partial}{\partial\nu}\int_0^t (e^{i\sqrt{A_\gamma}(t-\tau)} + e^{-i\sqrt{A_\gamma}(t-\tau)})(I + \gamma^2 A^{1/2})^{-1}D(k\frac{\partial}{\partial\nu}w_c(\tau))d\tau\|_{L_2(\Sigma_\infty)}$$
$$\leq C_T\|\frac{\partial}{\partial\nu}w_t\|_{L_2(\Sigma_T)}. \tag{3.21}$$

References

[1] M. E. Bradley and M. A. Horn. Global stabilization of the von Kármán plate with boundary feedback acting via bending moments only. Submitted.

[2] M. E. Bradley and I. Lasiecka. Global decay rates for the solutions to a von Kármán plate without geometric conditions. *Journal of Mathematical Analysis and Applications.* To appear.

[3] A. Favini and I. Lasiecka. Well-posedness and regularity of second-order abstract equations arising in hyperbolic-like problems with nonlinear boundary conditions. *Osaka Journal of Mathematics.* To appear.

[4] P. Grisvard. Caracterization de quelques espaces d'interpolation. *Archive for Rational Mechanics and Analysis*, 25:40–63, 1967.

[5] M. A. Horn. Uniform decay rates for the solutions to the Euler-Bernoulli plate equation with boundary feedback acting via bending moments. *Differential and Integral Equations*, 5(5):1121–1150, September 1992.

[6] M. A. Horn and I. Lasiecka. Asymptotic behavior with respect to thickness of boundary stabilizing feedback for the Kirchoff plate. *Journal of Differential Equations.* To appear.

[7] M. A. Horn and I. Lasiecka. Nonlinear boundary stabilization of a von Kármán plate equation. In *Differential Equations and Geometric Dynamics: Control Science and Dynamical Systems*, Lecture Notes in Pure and Applied Mathematics. Mercel Dekker, 1992. To appear.

[8] I. Lasiecka and R. Triggiani. Exact controllability of the Euler-Bernoulli equation with boundary controls for displacement and moment. *Journal of Mathematical Analysis and Applications*, 146:1–33, 1990.

[9] I. Lasiecka and R. Triggiani. Sharp trace estimates of solutions to Kirchoff and Euler-Bernoulli equations. *Applied Mathematics and Optimization*, 28:277–306, 1993.

[10] A. Pazy. *Semigroups of Linear Operators and Applications to Partial Differential Equations.* Springer-Verlag, New York, 1983.

Subharmonic Bifurcations of Periodic Perturbations

L. Ke   University of Southwestern Louisiana, Lafayette, Louisiana

1. INTRODUCTION

Consider

$$x' = \omega A x + a x_1 x_2 + f(t, x, \lambda) \tag{1.1}$$

where $\omega > 0$, $a = (a_1, a_2)^{\mathrm{T}}$ is a 2-dimensional vector, $\lambda \in \mathbb{R}^m$ is an m-dimensional vector parameter, $A = \{a_{ij}\}$ is a 2×2 matrix with $a_{11} = a_{22} = 0$, $a_{21} = -a_{12} = 1$, and $f \in C^3(\mathbb{R} \times \mathbb{R}^2 \times \mathbb{R}^m, \mathbb{R}^2)$, $f(t+1, x, \lambda) \equiv f(t, x, \lambda)$ for $(t, x, \lambda) \in \mathbb{R} \times \mathbb{R}^2 \times \mathbb{R}^m$, and $f(t, x, 0) \equiv 0$ for $(t, x) \in \mathbb{R} \times \mathbb{R}^2$.

Sufficient conditions for existence and nonexistence of periodic solutions of the system (1.1) and the stability of the periodic bifurcations are obtained. These criteria can be applied to Lotka-Volterra systems and epidemic disease models. The methods used are motivated by Smith [8].

2. SUBHARMONIC BIFURCATIONS

First, we consider the case $\omega \cong 2\pi/n$ for some integer $n > 1$ and investigate n-subharmonic bifurcations from the equilibrium $x = 0$ of the unperturbed system

$$x' = \omega Ax + ax_1 x_2. \tag{2.1}$$

Expand $f(t, x, \lambda)$ at $(x, \lambda) = (0, 0)$:

$$f(t, x, \lambda) = f_\lambda^0(t)\lambda + f_{x\lambda}^0(t)\lambda x + \cdots$$

where superscript ^{0} presents the evaluations at $(x, \lambda) = (0, 0)$, namely $f_\lambda^0(t) = f_\lambda(t, 0, 0)$ and $f_{x\lambda}^0(t) = f_{x\lambda}(t, 0, 0)$. If $\omega \neq 2\pi$, then for sufficiently small λ, the system (1.1) has a unique 1-periodic solution $\tilde{x}(\lambda)$. Then $\tilde{x}(\lambda)$ can be written as follow:

$$\tilde{x}(\lambda) = X\lambda + O(|\lambda|^2 + |\eta|^2)$$

where $\eta = \omega - 2\pi/n$,

$$X(t) = \Phi(t)\left\{[I - \Phi(1)]^{-1}\int_0^1 \Phi^{-1}(\tau)f_\lambda^0(\tau)d\tau + \int_0^t \Phi^{-1}(\tau)f_\lambda^0(\tau)d\tau]\right\},$$

and $\Phi(t)$ is the standard fundamental matrix of $x' = \frac{2\pi}{n}Ax$. Then $|\eta| << 1$. Let $x - \tilde{x}(\lambda) = \Phi(t)y$. Denote the first row of Φ by Φ_1 and the second row by Φ_2. The system (1.1) becomes

$$\begin{aligned}
y' =& \eta Ay + \Phi^{-1}(t)a\left[\frac{1}{2}(y_1^2 - y_2^2)\sin\frac{4\pi t}{n} + y_1 y_2\cos\frac{4\pi t}{n}\right] + \Phi^{-1}(t)ay^{\mathrm{T}}\begin{pmatrix}\Phi_2(t)\\ \Phi_1(t)\end{pmatrix}X\lambda \\
& + \Phi^{-1}(t)f_{x\lambda}^0(t)\lambda\Phi(t)y + O(|\lambda|^2 + |\lambda||y|^2) \\
\equiv& F(t, y, \lambda, \eta).
\end{aligned} \tag{2.2}$$

Denote $\mathcal{P}_n$ the Banach space of 2-dimensional vectors of continuous n-periodic functions with norm

$$\|\varphi\| = \left[\left(\max_{1\leq t\leq n}|\varphi_1(t)|\right)^2 + \left(\max_{1\leq t\leq n}|\varphi_2(t)|\right)^2\right]^{\frac{1}{2}} \text{ for any } \varphi = (\varphi_1, \varphi_2)^{\mathrm{T}} \in \mathcal{P}_n.$$

Define operators P and L on $\mathcal{P}_n$ as follows: for any $\psi \in \mathcal{P}_n$

$$P\psi = \frac{1}{n}\int_0^n \psi(t)dt, \quad L\psi = (I - P)\int_0^n \psi(t)dt,$$

where I is the identity operator on $\mathcal{P}_n$. Then both P and L are linear and P is a projection from $\mathcal{P}_n$ into its subspace $\mathbb{R}^2$. The equation (2.2) has small n-periodic solutions if and only if the following equations can be solved: (Hale, [5])

$$y = Py + L(I - P)F(t, y, \lambda, \eta), \tag{2.3}$$

$$PF(t, y, \lambda, \eta) = 0. \tag{2.4}$$

Let $b = Py$. By the implicit function theorem, for small λ and η there exists a solution $y^*(\lambda, \eta)$ of (2.3) satisfying $y^*(0, 0) = 0$. Since there is no nontrivial n-periodic solution of (2.2) for $\lambda = 0$, let

$$y^*(\lambda, \eta) = b + V_1(t)\lambda b + V_2(t)\eta b + Q(b, b) + \cdots. \tag{2.5}$$

By (2.3), we obtain

$$V_1(t)\lambda b = \boldsymbol{L}(\boldsymbol{I} - \boldsymbol{P})\left\{\Phi^{-1}(t)ab^{\mathrm{T}}\begin{pmatrix}\Phi_1\\\Phi_2\end{pmatrix}X\lambda + \Phi^{-1}(t)f_{x\lambda}^0(t)\lambda\Phi(t)b\right\},$$

$$V_2(t) = 0,$$

$$Q(b,b) = (Q_1, Q_2)^{\mathrm{T}}$$

where

$$Q_1 = \frac{1}{3}\left\{(b_1^2 - b_2^2)\left[-a_1(\cos\frac{2\pi t}{n})^3 + a_2(\sin\frac{2\pi t}{n})^3\right]\right.$$
$$\left. + b_1 b_2\left(a_1\sin\frac{2\pi t}{n} - a_2\cos\frac{2\pi t}{n}\right)\cos\frac{4\pi t}{n}\right\},$$

$$Q_2 = \frac{1}{3}\left\{-(b_1^2 - b_2^2)\left[a_1(\sin\frac{2\pi t}{n})^3 + a_2(\cos\frac{2\pi t}{n})^3\right]\right.$$
$$\left. + b_1 b_2\left(a_1\cos\frac{2\pi t}{n} + a_2\sin\frac{2\pi t}{n}\right)\cos\frac{4\pi t}{n}\right\}.$$

Let

$$C(\lambda) = \frac{1}{n}\int_0^n\left[-\Phi^{-1}(t)a\lambda^{\mathrm{T}}X^{\mathrm{T}}\Phi^{\mathrm{T}}(t) + \Phi^{-1}(t)f_{x\lambda}^0(t)\lambda\Phi(t)\right]dt.$$

By substituting $y^*(\lambda, \eta)$, (2.4) becomes

$$(\eta A + C(\lambda))b - Ab\bar{a}(b_1^2 + b_2^2) + O(|\lambda|^2|b| + |b|^4 + |\eta|^2|b|) = 0, \tag{2.6}$$

where $\bar{a} = (a_1^2 + a_2^2)/24$.

THEOREM 1. (1). *If the system* (1.1) *has nontrivial n-subharmonic bifurcations, then,*

$$(C_{12}(\lambda) - C_{21}(\lambda))^2 - 4\det C(\lambda) \geq 0. \tag{2.7}$$

(2). *If* $\det(\eta A + C(\lambda)) < 0$, *then the system* (1.1) *has two nontrivial n-subharmonic bifurcations.*

(3). *If* $\det(\eta A + C(\lambda)) > 0$, $\eta > [C_{12}(\lambda) - C_{21}(\lambda)]/2$, *and* (2.7) *holds, then the system* (1.1) *has four nontrivial n-subharmonic bifurcations.*

Except for the cases of either two or four nontrivial bifurcations, the system (1.1) *has no nontrivial n-subharmonic bifurcations.*

PROOF: Let $b_1 = r\cos\theta$ and $b_2 = r\sin\theta$. Since we are only interested in nontrivial bifurcations near the equilibrium $x = 0$, we assume $r > 0$ and omit the higher order terms in (2.6). We have

$$\begin{cases} C_{11}(\lambda)\cos\theta + (-\eta + C_{12}(\lambda) + Ar^2)\sin\theta = 0, \\ (\eta + C_{21}(\lambda) - Ar^2)\cos\theta + C_{22}(\lambda)\sin\theta = 0. \end{cases}$$

Solve for r^2, we have

$$r_\pm^2 = \frac{1}{2A}\left[[2\eta + C_{21}(\lambda) - C_{12}(\lambda)] \pm \sqrt{(2\eta + C_{21}(\lambda) - C_{12}(\lambda))^2 - 4\det(\eta R + C(\lambda))}\right].$$

Thus, the equation have real number solution only if

$$(2\eta + C_{21}(\lambda) - C_{12}(\lambda))^2 - 4\det(\eta R + C(\lambda)) = (C_{12}(\lambda) - C_{21}(\lambda))^2 - 4\det C(\lambda) \geq 0.$$

If $\det(\eta R + C(\lambda)) < 0$, then

$$\begin{aligned}
&[2\eta + C_{21}(\lambda) - C_{12}(\lambda)] + \sqrt{(2\eta + C_{21}(\lambda) - C_{12}(\lambda))^2 - 4\det(\eta R + C(\lambda))} \\
> \ &[2\eta + C_{21}(\lambda) - C_{12}(\lambda)] + |2\eta + C_{21}(\lambda) - C_{12}(\lambda)| \\
\geq \ &0,
\end{aligned}$$

and

$$\begin{aligned}
&[2\eta + C_{21}(\lambda) - C_{12}(\lambda)] - \sqrt{(2\eta + C_{21}(\lambda) - C_{12}(\lambda))^2 - 4\det(\eta R + C(\lambda))} \\
< \ &[2\eta + C_{21}(\lambda) - C_{12}(\lambda)] - |2\eta + C_{21}(\lambda) - C_{12}(\lambda)| \\
\leq \ &0.
\end{aligned}$$

Hence r_+^2 is the only positive solution. If $\det(\eta R + C(\lambda)) \geq 0$, $\eta > [C_{12}(\lambda) - C_{21}(\lambda)]/2$, and (2.7) holds. then

$$\begin{aligned}
&[2\eta + C_{21}(\lambda) - C_{12}(\lambda)] \pm \sqrt{(2\eta + C_{21}(\lambda) - C_{12}(\lambda))^2 - 4\det(\eta R + C(\lambda))} \\
\geq \ &[2\eta + C_{21}(\lambda) - C_{12}(\lambda)] - \sqrt{(2\eta + C_{21}(\lambda) - C_{12}(\lambda))^2 - 4\det(\eta R + C(\lambda))} \\
> \ &0.
\end{aligned}$$

Hence, there exist two different positive solutions r_+ and r_-. Since the nontrivial n-periodic solutions of (2.2) are equivalent to the n-subharmonic bifurcations of (1.1), the theorem is proved.

THEOREM 2. *Nontrivial n-subharmonic bifurcations of the system* (1.1) *appear in pairs, and the two of each pair have the same stability.*

PROOF: From the proof of Theorem 2.1, the first part of this theorem is obvious.

The stability of a nontrivial bifurcation is equivalent to the stability of its corresponding solution of (2.3) and (2.4), namely the stability of the equilibrium of the following equation (see [7]),

$$\frac{db}{d\tau} = \boldsymbol{P}F(t, y, \lambda, \eta) \equiv G(b, \lambda, \eta),$$

Denote $J(b)$ the variation matrix with respect to an equilibrium b of the above equation. Then

$$J(b) = A\begin{pmatrix} -2b_1 b_2, & -b_1^2 - 3b_2^2 \\ 3b_1^2 + b_2^2, & 2b_1 b_2 \end{pmatrix} + C(\lambda) + \eta R.$$

Thus,

$$\text{trace}[J(b)] = \text{trace}[C(\lambda)] = C_{11}(\lambda) + C_{22}(\lambda),$$

$$\det[J(b)] = \frac{Ar_\pm^2}{C_{11}(\lambda) + [Ar_\pm^2 - C_{12}(\lambda)]^2}$$
$$\Big\{ -[C_{12}(\lambda) + C_{21}(\lambda)][(C_{12}(\lambda) + C_{21}(\lambda))^2 - 2C_{11}(\lambda)C_{22}(\lambda)]$$
$$\pm [(C_{12}(\lambda) + C_{21}(\lambda))^2 + 2C_{11}(\lambda)(C_{22}(\lambda) - C_{11}(\lambda))]$$
$$\sqrt{[C_{21}(\lambda) - C_{12}(\lambda)]^2 - 4\det C(\lambda)} \Big\}$$

where $r_\pm$ is the value corresponding to b. We know that the equilibrium b is asymptotically stable when $\text{trace}[J(b)] < 0$ and $\det[J(b)] > 0$, and b unstable when $\text{trace}[J(b)] > 0$ or $\det[J(b)] < 0$. Since $\text{trace}[J(b)]$ and $\det[J(b)]$ involve $r_\pm$ but not b, it follows that the stability is determined by $r_\pm$.

In the next part, we consider changes and bifurcations of the n-periodic orbit of the unperturbed system. Let us first give an important result, which is motivated by Hsu [6], about the system

$$x' = \omega Ax + ax_1x_2. \tag{2.8}$$

LEMMA 3. *For any $T \in (2\pi/\omega, +\infty)$, the equation (2.8) has exactly one periodic solution with minimal positive period T. Furthermore, if $x(c)|_{t=0} = (c,0)^{\mathrm{T}}$ where $c > 0$, and $T(c)$ is the minimal positive period of $x(c)$, then $T(c)$ is strictly increasing with respect to c in $(0, +\infty)$ and $\lim_{c\to+\infty} T(c) = +\infty$.*

PROOF: Let us prove the last two parts first. Let $\tau = \omega t$, (2.8) can be rewritten as

$$\begin{cases} x_1' = -x_2 + \alpha_1 x_1 x_2 \\ x_2' = x_1 + \alpha_2 x_1 x_2 \end{cases} \tag{2.9}$$

where $\alpha_i = a_i/\omega$, $i = 1,2$. For convenience, we denote α_i by a_i. We need only to consider the following two different cases since others can be easily changed into one of these two: case (1) $a_1 < 0$, $a_2 > 0$; case (2) $a_1 = 0$, $a_2 > 0$.

Case 1: The equation (2.9) is changed into

$$x_1'' + \frac{a_1 x_1'^2}{(1 - a_1 x_1)} - a_2 x_1 x_1' + x_1(1 - a_1 x_1) = 0.$$

Let $e^z = 1 - a_1 x_1$, then it is changed into

$$z'' - a_2 z'(e^z - 1) + (e^z - 1) = 0.$$

By the results of [8], the last two parts of the lemma is proved.

Case 2: Suppose the trajectory starts at the x-axis with $x_{01} > 0$ and $x_{02} = 0$. (2.9) has solutions of the form

$$\frac{1}{2}x_1^2 + \frac{x_2}{a_2} - \left[\frac{1}{a_2}\right]^2 \ln(1 + a_2 x_2) = \frac{1}{2}x_{01}^2. \tag{2.10}$$

Let

$$F(x_2) = \frac{1}{a_2}x_2 - \left[\frac{1}{a_2}\right]^2 \ln(1 + a_2 x_2).$$

Then, (2.10) can be written as

$$F(x_2) = \frac{1}{2}(x_{01}^2 - x_1^2). \tag{2.11}$$

Denote the restriction of $F(x_2)$ on $[0, +\infty)$ by $F_1(x_2)$, and the restriction on $(-1/a_2, 0)$ by $F_2(x_2)$. It is obvious that F_1 and F_2 are one-one maps on $[0, +\infty)$ and on $(-1/a_2, 0)$ respectively. Let T_1 be the time elapse of the system spending on trajectory section Γ_{cd} ,and T_2 be the time elapse on trajectory section Γ_{bc}. Then

$$T_1 = \int_0^{x_{01}} \frac{dx_1}{F_1^{-1}[\frac{1}{2}(x_{01}^2 - x_1^2)]}, \quad T_2 = \int_{x_{01}}^0 \frac{dx_1}{F_2^{-1}[\frac{1}{2}(x_{01}^2 - x_1^2)]}.$$

The minimal positive period of the closed trajectory Γ_{abcda} is

$$T(x_{01}) = T_{ab} + T_{bc} + T_{cd} + T_{da} = 2[T_1(x_{01}) + T_2(x_{01})].$$

Since Γ is symmetric with respect to y-axis, $T_{cd} = T_{da} = T_1(x_{01})$, and $T_{ab} = T_{bc} = T_2(x_{01})$. Therefore, we need only to show that $T_1(x_{01})$ and $T_2(x_{01})$ are strictly increasing on $[0, +\infty)$. Let us consider $T_1(x_{01})$ first. Let

$$F_1^{-1}[\frac{1}{2}(x_{01}^2 - x_1^2)] = F_1^{-1}[\frac{1}{2}x_{01}^2]\sin\xi, \ 0 < \xi < \frac{\pi}{2}$$

We have

$$\frac{1}{2}(x_{01}^2 - x_1^2) = F_1[F_1^{-1}[\frac{1}{2}x_{01}^2]\sin\xi]$$

Hence,

$$x_1 = \left\{x_{01}^2 - F_1[F_1^{-1}(\frac{1}{2}x_{01}^2)\sin\xi]\right\}^{\frac{1}{2}}$$

$$= \left\{x_{01}^2 - \frac{1}{a_2}F_1^{-1}(\frac{1}{2}x_{01}^2)\sin\xi + 2\left[\frac{1}{a_2}\right]^2 \ln[1 + a_2 F_1^{-1}(\frac{1}{2}x_{01}^2)\sin\xi]\right\}^{\frac{1}{2}}$$

$$\equiv g_1(x_{01}, \xi).$$

From (2.11), we have

$$-x_1 dx_1 = \frac{dF_1}{dx_2}dx_2 = \frac{x_2}{1 + a_2 x_2}dx_2 = \frac{F_1^{-1}(\frac{1}{2}x_{01}^2)\sin\xi}{1 + a_2 F_1^{-1}(\frac{1}{2}x_{01}^2)\sin\xi}F_1^{-1}(\frac{1}{2}x_{01}^2)\cos\xi d\xi$$

Thus,

$$T_1(x_{01}) = \int_0^{\frac{\pi}{2}} \frac{F_1^{-1}(\frac{1}{2}x_{01}^2)\cos\xi}{1 + a_2 F_1^{-1}(\frac{1}{2}x_{01}^2)\sin\xi} \frac{d\xi}{g_1(x_{01}, \xi)} \equiv \int_0^{\frac{\pi}{2}} k_1(x_{01}, \xi)\cos\xi d\xi.$$

We have the following inequalities

$$\frac{1}{x_1} > 0, \quad \frac{\partial}{\partial x_{01}}\left(\frac{1}{x_1}\right) = -x_1^{-2}\frac{\partial x_1}{\partial x_{01}} > 0, \quad \frac{F_1^{-1}(\frac{1}{2}x_{01}^2)}{1 + a_2 F_1^{-1}(\frac{1}{2}x_{01}^2)\sin\xi} > 0.$$

Since $F_1(x_2)$ is strictly increasing on $[0, +\infty)$, we have

$$\frac{\partial}{\partial x_{01}}\left(\frac{F_1^{-1}(\frac{1}{2}x_{01}^2)}{1 + a_2 F_1^{-1}(\frac{1}{2}x_{01}^2)\sin\xi}\right) = \frac{1}{\left[1 + a_2 F_1^{-1}(\frac{1}{2}x_{01}^2)\sin\xi\right]^2}\frac{\partial}{\partial x_{01}}\left(F_1^{-1}(\frac{1}{2}x_{01}^2)\right) > 0.$$

Therefore,

$$\frac{\partial k_1(x_{01},\xi)}{\partial x_{01}} = \frac{1}{x_1}\frac{\partial}{\partial x_{01}}\left(\frac{F_1^{-1}(\frac{1}{2}x_{01}^2)}{1 + a_2 F_1^{-1}(\frac{1}{2}x_{01}^2)\sin\xi}\right) + \frac{F_1^{-1}(\frac{1}{2}x_{01}^2)}{1 + a_2 F_1^{-1}(\frac{1}{2}x_{01}^2)\sin\xi}\frac{\partial}{\partial x_{01}}\left(\frac{1}{x_1}\right) > 0.$$

It follows that

$$\frac{\partial T_1(x_{01})}{\partial x_{01}} = \int_0^{\frac{\pi}{2}}\frac{\partial k_1(x_{01},\xi)}{\partial x_{01}}\cos\xi\, d\xi > 0,$$

and hence, $T_1(x_{01})$ is strictly increasing on $(0, +\infty)$.

Similarly, we can prove $T_2(x_{01})$ is strictly increasing on $(0, +\infty)$. Let

$$F_2^{-1}[\frac{1}{2}(x_{01}^2 - x_1^2)] = F_2^{-1}[\frac{1}{2}x_{01}^2]\sin\xi, \quad 0 < \xi < \frac{\pi}{2},$$

$$g_2(x_{01},\xi) = \left\{x_{01}^2 - F_2[F_2^{-1}(\frac{1}{2}x_{01}^2)\sin\xi]\right\}^{\frac{1}{2}}.$$

Then

$$T_2(x_{01}) = -\int_0^{\frac{\pi}{2}}\frac{F_2^{-1}(\frac{1}{2}x_{01}^2)\cos\xi}{[1 + a_2 F_2^{-1}(\frac{1}{2}x_{01}^2)\sin\xi]g_2(x_{01},\xi)}d\xi \equiv \int_0^{\frac{\pi}{2}}k_2(x_{01},\xi)\cos\xi\, d\xi.$$

We have the following inequalities

$$\frac{1}{x_1} > 0, \quad \frac{\partial}{\partial x_{01}}\left(\frac{1}{x_1}\right) = -x_1^{-2}\frac{\partial x_1}{\partial x_{01}} > 0, \quad \frac{F_2^{-1}(\frac{1}{2}x_{01}^2)}{1 + a_2 F_2^{-1}(\frac{1}{2}x_{01}^2)\sin\xi} < 0,$$

Since $F_2(x_2)$ is strictly decreasing on $(-\frac{1}{a_2}, +\infty)$, we have

$$\frac{\partial}{\partial x_{01}}\left(\frac{F_2^{-1}(\frac{1}{2}x_{01}^2)}{1 + a_2 F_2^{-1}(\frac{1}{2}x_{01}^2)\sin\xi}\right) = \frac{1}{[1 + a_2 F_2^{-1}(\frac{1}{2}x_{01}^2)\sin\xi]^2}\frac{\partial}{\partial x_{01}}\left(F_2^{-1}(\frac{1}{2}x_{01}^2)\right) < 0.$$

Therefore,

$$\frac{\partial k_2(x_{01},\xi)}{\partial x_{01}} = \frac{1}{x_1}\frac{\partial}{\partial x_{01}}\left(\frac{F_2^{-1}(\frac{1}{2}x_{01}^2)}{1 + a_2 F_2^{-1}(\frac{1}{2}x_{01}^2)\sin\xi}\right) + \frac{F_2^{-1}(\frac{1}{2}x_{01}^2)}{1 + a_2 F_2^{-1}(\frac{1}{2}x_{01}^2)\sin\xi}\frac{\partial}{\partial x_{01}}\left(\frac{1}{x_1}\right) < 0.$$

It follows that

$$\frac{\partial T_2(x_{01})}{\partial x_{01}} = -\int_0^{\frac{\pi}{2}}\frac{\partial k_2(x_{01},\xi)}{\partial x_{01}}\cos\xi\, d\xi > 0,$$

and hence, $T_2(x_{01})$ is strictly increasing on $(0, +\infty)$. Since $T(x_{01}) = 2[T_1(x_{01}) + T_2(x_{01})]$, $T(x_{01})$ is strictly increasing on $(0, +\infty)$.

To prove $\lim_{x\to+\infty} T(x_{01}) = +\infty$, it is sufficient to show that $\lim_{x\to+\infty} T_1(x_{01}) = +\infty$. Since $1/F_1^{-1}[\frac{1}{2}(x_{01}^2 - x_1^2)]$ is a strictly increasing function with respect to x_1 on $[0, x_{01}]$,

$$T_1(x_{01}) = \int_0^{x_{01}}\frac{dx_1}{F_1^{-1}[\frac{1}{2}(x_{01}^2 - x_1^2)]} > \int_0^{x_{01}}\frac{dx_1}{F_1^{-1}[0]} = \frac{x_{01}}{F_1^{-1}[0]}$$

Thus, $\lim_{x\to+\infty} T_1(x_{01}) = +\infty$.

To prove the first part of this lemma, let $x(t, c)$ be a solution of (2.2.1) satisfying $x(0, c) = (c, 0)^{\mathrm{T}}$, where $c > 0$. The Poincare-Lindstedt series for $x(t, c)$ is

$$x_1(t, c) = c \cos \beta(c)t + \frac{c^2 \{2a_1[\cos \beta(c)t - \cos 2\beta(c)t] + a_2[-2\sin \beta(c)t + \sin 2\beta(c)t]\}}{6\beta(c)} + \cdots$$

$$x_2(t, c) = c \sin \beta(c)t + \frac{c^2 \{a_1[2\sin \beta(c)t - \sin 2\beta(c)t] + 2a_2[\cos \beta(c)t + \cos 2\beta(c)t]\}}{6\beta(c)} + \cdots$$

$$\beta(c) = \omega \left[1 - \frac{1}{24\omega^2}(a_1^2 + a_2^2)c^2\right] + o(c^4).$$

The conclusion of the lemma follows from this series.

For $n > 2\pi/\omega$, let $x_n(t)$ be the n-periodic solution with the minimal positive period n of the equation (2.8) with $x_{n1}(0) > 0$ and $x_{n2}(0) = 0$. Then the variation matrix of the equation (2.8) with respect to x_n is written as

$$J(x) = \begin{pmatrix} a_1 x_{n2}, & -\omega + a_1 x_{n1} \\ \omega + a_2 x_{n2}, & a_2 x_{n1} \end{pmatrix}.$$

The adjoint equation $u' = -uJ(x_n)$ of the variation equation of (2.8) with respect to x_n has one nontrivial n-periodic solution:

$$\tilde{u} = \left(\frac{x_{n1}}{\omega + a_1 x_{n1}}, \frac{x_{n2}}{\omega + a_2 x_{n2}}\right).$$

Define an operator P on $\mathcal{P}_n$ as

$$Pu = \frac{\tilde{u}^{\mathrm{T}}(t) \int_0^n \tilde{u}(t)\varphi(t)dt}{\int_0^n \tilde{u}(t)\tilde{u}^{\mathrm{T}}(t)dt} \quad \text{for any } \varphi \in \mathcal{P}_n.$$

It can be seen that P is a continuous projection on $\mathcal{P}_n$, and the equation

$$v' = J(x_n)v + \psi(t) \tag{2.12}$$

has an n-periodic solution $\varphi(t, \psi)$ if and only if $P\psi \equiv 0$ where $\psi \in \mathcal{P}_n$; furthermore, any n-periodic solution of (2.12) can be represented as

$$v = \beta\tilde{v} + \varphi(t, \psi), \ \beta \in \mathbb{R}$$

where $\tilde{v} = (-\omega x_{n2} + a_1 x_{n1} x_{n2}, \omega x_{n1} + a_2 x_{n1} x_{n2})^{\mathrm{T}}$. To make $v(0)$ and $\tilde{v}(0)$ orthogonal, let

$$\beta(\psi) = \frac{\tilde{v}(0)\varphi(0, \psi)}{|\tilde{v}(0)|^2}.$$

we define $L : (I - P)\mathcal{P}_n \to \mathcal{P}_n$ by

$$(L\psi)(t) = \beta(\psi)\tilde{v}(t) + \varphi(t, \psi), \forall \psi \in (I - P)\mathcal{P}_n.$$

Consider the n-periodic solution $x(t)$ of the equation (2.1) near the trajectory $\Gamma(x_n)$. If $x(-\alpha) = x_n(0) + v(0)$, $v_1(0) = b > 0$ and $v_2(0) = 0$, then $x(t - \alpha) = x_n(t) + v(t)$, where $v(t)$ satisfies

$$v' = J(x_n)v + av_1v_2 + f(t - \alpha, x_n + v, \lambda). \tag{2.13}$$

It is sufficient to consider the n-periodic solutions of the equation (2.13) instead of those of (2.1). Based on Chow and Hale [1], it is known that (2.13) has an n-periodic solution $v(t)$ such that $v(0)$ and $\tilde{v}(0)$ are orthogonal if and only if the equations,

$$v = \boldsymbol{L}(\boldsymbol{I} - \boldsymbol{P})[av_1v_2 + F(t, v, \lambda, \alpha)]$$

$$\boldsymbol{P}[av_1v_2 + F(t, v, \lambda, \alpha)] = 0 \tag{2.14}$$

have nontrivial solutions, where $F(t, v, \lambda, \alpha) = f(t - \alpha, x_n + v, \lambda)$. Let

$$h_n(\alpha) = \int_0^n \bar{u}(t) \frac{\partial f(t - \alpha, x_n(t), 0)}{\partial \lambda} dt.$$

Then (2.14) can be written as

$$G(\alpha, \lambda) = h_n(\alpha)\lambda + \tilde{G}(\alpha, \lambda) = 0, \tag{2.15}$$

where $\tilde{G}(\alpha, \lambda) = O(|\lambda|^2)$.

THEOREM 4. *If the dimension of λ is greater than one, and $h_n''(\alpha)\theta \neq 0$ for the values satisfying $h_n(\alpha)h_n'(\alpha)\theta = 0$ and $|\theta| = 1$, then for sufficiently small λ, there is a nontrivial solution of (2.15) satisfying $\alpha \in [0, 1)$ and $\alpha^2 + |\theta| \neq 0$, and (2.13) has at least $2n$ n-periodic solutions. If λ is a scalar, and there is an $\alpha \in [0, 1]$ such that*

$$h_n(\alpha) = 0, \ h_n'(\alpha) \neq 0,$$

then for sufficiently small λ, (2.15) has the solution $\alpha(\lambda) \in [0, 1)$, and (2.13) has at least n n-periodic solutions.

3. EXAMPLES

In this section, we consider two applications: Lotka-Volterra systems and epidemic disease models. Both of them can be transformed into the form of (1.1) and be studied in a uniform way.

3.1. Lokta-Volterra Systems

Let us consider the following Lokta-Volterra system:

$$\begin{aligned} x_1' &= x_1(\alpha - \beta x_2) + \epsilon f_1(t, x_1, x_2), \\ x_2' &= x_2(-\gamma + \delta x_1) + \epsilon f_2(t, x_1, x_2), \end{aligned} \tag{3.1}$$

where $\epsilon \in \mathbb{R}$ and $f_i(t+1, x_1, x_2) \equiv f_i(t, x_1, x_2)$ for $i = 1, 2$. Let

$$y_1 = \sqrt{\alpha\gamma}(x_1 - \frac{\gamma}{\delta}),$$

$$y_2 = \frac{\gamma\beta}{\delta}(x_2 - \frac{\alpha}{\beta}),$$

$$\omega = \sqrt{\alpha\gamma},$$

$$a = (-\frac{\delta}{\gamma}, \frac{\delta}{\omega})^{\mathrm{T}},$$

$$g_1(t, y) = \omega f_1(t, \frac{\gamma}{\delta} + \frac{\delta}{\omega}y_1, \frac{\alpha}{\beta} + \frac{\delta}{\gamma\beta}y_2),$$

$$g_2(t, y) = \frac{\gamma\beta}{\delta} f_2(t, \frac{\gamma}{\delta} + \frac{\delta}{\omega}y_1, \frac{\alpha}{\beta} + \frac{\delta}{\gamma\beta}y_2).$$

Then, the equation (3.1) becomes

$$y' = \omega Ay + ay_1y_2 + \epsilon g(t, y). \tag{3.2}$$

We can then apply our results to problem (3.2). Near the equilibrium $y = 0$ with $\omega \cong 2\pi/n(n > 1)$, the bifurcation matrix $C(\lambda)$ is in the following form

$$C(\lambda) = \epsilon C = \frac{\epsilon}{n} \int_0^n \Phi^{-1}(t) \left[g_y^0 \Phi(t) + \int_0^t (g^0(\tau))^t \Phi(\tau) d\tau \right] dt.$$

The bifurcation equation is

$$(\eta A + \epsilon C)b - A\bar{b}\bar{a}(b_1^2 + b_2^2) + O(\epsilon^2|b| + |b|^4 + |\eta|^2|b|) = 0,$$

where $\bar{a} = (a_1^2 + a_2^2)/24$.

Near the n-periodic solution y_n of the unperturbed system under perturbation, the bifurcation equation is

$$H(\xi, \epsilon) = h_n(\xi) + \tilde{H}(\xi, \epsilon) = h_n(\xi) + o(\epsilon).$$

3.2. Epidemic Models

Dietz,K. [2] established the following differential equation model for infectious diseases. Let S, I and R represent the proportion of sensitive people, infectious people and recovered people in the population respectively. Suppose that the population is constant, both birth rate and death rate are $\mu > 0$, recovered rate is $\gamma > 0$, and a sensitive person has the capability of infecting as he is infected. Then, the model is given as follows:

$$S' = \mu(1 - S) + \beta(t)SI,$$

$$I' = \beta(t)SI - (\mu + \gamma)I,$$

$$R' = \gamma I - \mu R, \tag{3.3}$$

$$S + I + R = 1,$$

$$S, \ I, \ R \geq 0$$

where $\beta(t) = \beta_0(1 + \delta e(t))$, $e(t+1) \equiv e(t)$ and $|e(t)| << 1$.

It is sufficient to consider the first two equations of the model (3.3) in the region $S+I \le 1$, S, $I \ge 0$. Smith [8] studied the subharmonic bifurcations for a special case $e(t) = \sin 2\pi t$ of this problem.

If $\delta = 0$, then (3.3) has an equilibrium

$$S^* = \frac{1}{Q}, \quad I^* = \frac{\mu(Q-1)}{\beta_0}, \quad Q^* = \frac{\beta_0}{(\mu+\gamma)}.$$

Let

$$\epsilon^2 = \frac{\mu}{(\mu+\gamma)},$$

$$K^2 = Q - 1,$$

$$\tilde{\delta} = \epsilon\delta,$$

$$\omega = [\mu(\beta_0 - \mu - \gamma)],$$

$$S = Q^{-1}(1 + \epsilon K x_1),$$

$$I = Q^{-1}\epsilon^2 K(1 + x_2).$$

Then, we have

$$
\begin{aligned}
x_1' &= -\omega[x_2 + \frac{\epsilon}{K}x_1(1 + K^2 + K^2 x_2) + \epsilon\delta(1 + x_2)(1 + \epsilon x_1)e(t)] \\
&\equiv -\omega x_2 + f_1(x_1, x_2, \epsilon, \delta), \\
x_2' &= \omega(1 + x_2)[x_1 + \delta(\frac{1}{K} + \epsilon x_1)e(t)] \\
&\equiv \omega x_1 + \omega x_1 x_2 + f_2(x_1, x_2, \epsilon, \delta).
\end{aligned}
\tag{3.4}
$$

Then the results in section 2 can be applied. Suppose ϵ and δ are independent small parameters. Based on the data from Fine and Clarkson [3] and London and Yorke [7], consider the case of $\omega \cong \pi$, and we have the bifurcation matrix

$$D = (d_{ij})_{2\times 2} = \frac{1}{2}\int_0^2 V(t)dt.$$

where

$$V(t) = \frac{1}{K}\begin{bmatrix} H_1 \sin \pi t, & H_2 \sin \pi t \\ -H_1 \cos \pi t, & -H_2 \cos \pi t \end{bmatrix},$$

$$\begin{bmatrix} H_1 \\ H_2 \end{bmatrix} = \frac{\pi}{4}\begin{bmatrix} \sin 2\pi t, & \cos 2\pi t \\ -\cos 2\pi t, & \sin 2\pi t \end{bmatrix}\begin{bmatrix} 2E_1(t) - E_1(1) \\ 2E_2(t) - E_2(2) \end{bmatrix} + \begin{bmatrix} \sin \pi t \\ \cos \pi t \end{bmatrix}e(t),$$

$$E_1(t) = \int_0^t e(\tau)\sin \pi\tau d\tau, \quad E_2(t) = \int_0^t e(\tau)\cos \pi\tau d\tau,$$

and, the bifurcation equation of (3.4) is

$$(\eta A - l\epsilon I + \delta D)b + \frac{1}{24}(b_1^2 + b_2^2)Ab + O((\epsilon^2 + \delta^2)|b| + |b|^4 + |\eta|^2|b|) = 0,$$

where $l = (K^2 + 1)/(2K)$.

References

1. Chow, S. N. and J. K. Hale, *Methods of bifurcation theory*, Springer-Verlag, New York, Berlin Heidelberg (1982).

2. Dietz, K., The incidence of infectious diseases under the influence of seasonal fluctuations, In: *Lecture Notes in Biomath.* **11**, 1-15, Springer-Verlag, New York, Berlin Heidelberg (1976).

3. Fine, P. and J. Clarkson, Measles in England and Wales I. An analysis of factors underlying seasonal patterns, *Inter. J. Epidemiol.* **11**, 5-14 (1982).

4. Hale, J. K., Stability from the bifurcation function, In: *Differential Equations.* 23-30, New York, Academic Press (1980).

5. Hale, J. K., *Ordinary differential equations*, Robert E. Krieger Publishing Company, Huntington, New York (1980).

6. Hsu, S. B., A remark on the period of the periodic solution in the Lotka-Volterra systems, *J. Math Anal.* **95**, 428-436 (1983).

7. London, W. P. and J. A. Yorke, Recurrent outbreaks of measles, chickenpox and mumps I. Seasonal variation in contact rates, *Amer. J. Epidemiol.* **98**, 453-468 (1973).

8. Smith, H. L., Subharmonic bifurcation in an SIR epidemic model, *J. Math. Biol.* **17**, 163-178 (1983).

On Comparison Results for Classical and Viscosity Solutions in Elliptic Systems

Suzanne Lenhart and Philip Schaefer   University of Tennessee, Knoxville, Tennessee

1. INTRODUCTION.

The classical maximum principle for weakly coupled systems of elliptic partial differential equations is contained in the following

THEOREM [15]. *Let the function* $\vec{u} \in \left(C^2(\Omega) \cap C^0(\bar{\Omega})\right)^k$ *satisfy the weakly coupled system of elliptic inequalities*

$$\sum_{i,j=1}^{n} a_{ij}^{(\mu)} \frac{\partial^2 u_\mu}{\partial x_i \partial x_j} + \sum_{i=1}^{n} b_i^{(\mu)} \frac{\partial u_\mu}{\partial x_i} + \sum_{\gamma=1}^{k} h_{\mu\gamma} u_\gamma \geq 0, \quad \mu = 1, 2, \ldots, k$$

in a domain $\Omega \subset R^n$, *where the coefficients are bounded in* Ω *and*

$$h_{\mu\gamma} \geq 0 \text{ for } \mu \neq \gamma, \quad \mu, \gamma = 1, 2, \ldots, k, \tag{1.1}$$

$$\sum_{\gamma=1}^{k} h_{\mu\gamma} \leq 0, \qquad \mu = 1, 2, \ldots, k. \tag{1.2}$$

If $\vec{u} \leq \vec{M}$ *componentwise on the boundary* $\partial\Omega$ *with* $\vec{M} \geq 0$, *then* $\vec{u} \leq \vec{M}$ *in* Ω.

We note that the system contains k inequalities with different elliptic operators and that the off–diagonal condition (1.1) and the row sum condition (1.2) are sufficient to deduce a maximum principle.

Recently, there has been much interest in maximum principles in general and for elliptic systems ([2–5], [7–9], [11], [13], [16–17], [19–20]) and for parabolic systems ([10–11], [18], [21]) in particular. The use of maximum principles for uniqueness, comparison, stability, and existence results as well as for a priori bounds in elliptic boundary value problems is, of course, well known.

The study of sufficient conditions on the coupling matrix to obtain comparison results has occurred for classical solutions of elliptic and parabolic systems [4,18,21] as well as for viscosity solutions of Hamilton–Jacobi systems [12,14]. The different viewpoints and different methods of proof for the comparison results lead to some new results in each area.

In this note we feature a "delicate balance" between the matrix of differential operators in the system and the coupling matrix when attempting to decouple the system or otherwise transform the coupling matrix by a nonsingular transformation when the coupling matrix does not satisfy both the off–diagonal and the row sum conditions of the classical theorem. Briefly, we find that when the off–diagonal condition and an M-matrix condition [1] hold, the system may involve k different operators, but when the off–diagonal condition fails, the system is restricted to involve the same differential operators in each equation of the system.

In section 2 we discuss coupling matrices and comparison results from the classical viewpoint while in section 3 we discuss the viscosity solution viewpoint.

2. COMPARISON RESULTS FOR SYSTEMS OF CLASSICAL SOLUTIONS

In their study of a semilinear parabolic system with application to biology, de Figueiredo and Mitidieri [7] consider the Dirichlet problem

$$\left.\begin{array}{l} \Delta u + au + bv + f(x) = 0 \\ \Delta v + cu + dv + g(x) = 0 \end{array}\right\} \text{ in } \Omega, \qquad u = v = 0 \text{ on } \partial\Omega, \qquad (2.1)$$

where Δ is the n-dimensional Laplacian and a, b, c, d are real constants. They cited conditions on the coupling matrix and the inhomogeneous terms of the system such that the solution functions u and v are positive in Ω. The result is obtained by decoupling the system, i.e., diagonalizing the coupling matrix, which requires that the elliptic operators in (2.1) be the same but permits the off–diagonal elements of the coupling matrix to be of opposite sign.

In [3], Cosner and Schaefer examined the idea of decoupling a system of the form (2.1) more completely and were able to deduce sufficient conditions under which one could determine the algebraic sign of the solution functions throughout the domain (see Table 2 in [3]). In the process, they improved somewhat a condition in [7].

The idea of decoupling a system of two elliptic equations when the coupling matrix has variable elements was undertaken by Cosner and Schaefer in [4]. In particular, they considered the system

$$\begin{array}{l} Lu + a(x)u + b(x)v + f(x) = 0 \\ Lv + c(x)u + d(x)v + g(x) = 0, \end{array} \qquad (2.2)$$

in Ω, where

$$L := \sum_{i,j=1}^{n} a_{ij}(x)\frac{\partial^2}{\partial x_i \partial x_j} + \sum_{i=1}^{n} b_i(x)\frac{\partial}{\partial x_i}. \tag{2.3}$$

It was determined that a nonsingular constant matrix S could be found which would diagonalize the coupling matrix C only if

$$C = \begin{pmatrix} a(x) & b^*\,[a(x) - d(x)] \\ c^*\,[a(x) - d(x)] & d(x) \end{pmatrix}, \qquad a(x) - d(x) \not\equiv 0, \quad b^*, c^* \text{ constants}$$

or

$$C = \begin{pmatrix} a(x) & b^\star r(x) \\ c^\star r(x) & a(x) \end{pmatrix}, \qquad b^\star, c^\star > 0, \text{ constant.}$$

The results in [3] were able to be extended to the system (2.2), albeit only for a system of two elliptic equations.

Although the coupling matrix in the Dirichlet problem

$$\left.\begin{array}{c} \Delta u - u - v + f(x) = 0 \\ \Delta u + u - 3v + g(x) = 0 \end{array}\right\} \text{ in } \Omega, \qquad u = v = 0 \text{ on } \partial\Omega,$$

does not satisfy the conditions for the classical maximum principle nor allow decoupling, it can be shown that the solutions are positive in Ω when $f(x) \geq g(x) \geq 0$ in Ω. This is a consequence of the results in [5] where a constant matrix S is sought which will transform the variable coupling matrix in (2.2) into one for which the classical theorem applies. This argument permits the off–diagonal elements to change sign in Ω, i.e.,

$$|b(x)| \leq b_0, \qquad |c(x)| \leq c_0,$$

in which case it is asked that

$$a(x) - d(x) \geq 2\sqrt{b_0 c_0}, \qquad a(x) \leq \sqrt{b_0 c_0}.$$

The method of transforming the coupling matrix into the domain of the classical systems result can be extended to a system of k equations

$$\mathcal{L}\vec{w} + C\vec{w} = \vec{f},$$

where $\mathcal{L}$ is a scalar matrix of the operator L given by (2.3). Although this approach allows extension to k equations in the system and a variable transforming matrix $S(x)$, the determination of S when $k > 2$ is rather overwhelming and the elliptic operator in each equation must be the same.

The methods of decoupling and transforming the coupling matrix into the classical conditions permit one to consider systems of elliptic equations where conditions (1.1) and (1.2) do not both hold. However, one is restricted to a system where each equation involves the same elliptic operator and usually only two equations are involved. We now show that if condition (1.1) and a "weighted row sum condition" hold for the elements in the coupling matrix, then one can permit k different elliptic operators in the matrix of operators.

We consider the system of inequalities

$$-L_i u_i + \sum_{j=1}^{N} c_{ij}(x) u_j \geq 0 \text{ in } \Omega, \qquad u_i \geq 0 \text{ on } \partial\Omega, \tag{2.4}$$

for different elliptic operators L_i, $i = 1, 2, \ldots, N$ of the form given in (2.3) and elements of the coupling matrix which satisfy

$$m_{ij} \leq c_{ij}(x) \leq 0, \qquad i \neq j, \qquad 0 \leq m_{ii} \leq c_{ii}(x). \tag{2.5}$$

We note that the matrix $-C = (-c_{ij})$ satisfies the off–diagonal condition (1.1) but need not satisfy the row sum condition (1.2). Further, we assume the bounding matrix $\mathcal{M} = (m_{ij})$ is an M-matrix, i.e., that $\mathcal{M}^{-1}$ has all nonnegative entries, denoted $\mathcal{M}^{-1} \geq 0$. This is one of many equivalent conditions for a M-matrix cited in [1]. The use of the M-matrix condition for comparison results in partial differential equations was first introduced in the case of first order Hamilton–Jacobi equations by Engler and Lenhart[12].

If we let D be the diagonal matrix

$$D = \text{diag}\,(\vec{d}), \qquad \vec{d} = M^{-1}(1, 1, \ldots, 1)^T > 0$$

and introduce the transformation $\vec{u} = D\vec{v}$, then we obtain the system

$$-L_i d_i v_i + \sum_{j=1}^{N} c_{ij}(x) d_j v_j \geq 0 \text{ in } \Omega, \qquad d_i v_i \geq 0 \text{ on } \partial\Omega,$$

for $i = 1, 2, \ldots, N$, where for each $i = 1, 2, \ldots, N$

$$\sum_{j=1}^{N} c_{ij} d_j = c_{ii}(x) d_i + \sum_{j=1, j \neq i}^{N} c_{ij}(x) d_j \geq m_{ii} d_i + \sum_{j=1, j \neq i}^{N} m_{ij} d_j = 1 > 0. \tag{2.6}$$

Condition (2.6) is a weighted row sum condition on C. Consequently, the transformed system

$$L_i d_i v_i + \sum_{j=1}^{N} [-c_{ij}(x)]\, d_j v_j \leq 0 \text{ in } \Omega, \qquad d_i v_i \geq 0 \text{ on } \partial\Omega,$$

for $i = 1, 2, \ldots, N$ "fits" a variation of the classical result and implies that $v_i \geq 0$ in $\bar{\Omega}$ or, ultimately, the minimum principle, $u_i \geq 0$ in $\bar{\Omega}$. A reversal of relevant inequality signs results in a maximum principle. Thus we have the following result.

THEOREM. If $\vec{u} = (u_1, \ldots, u_N)$ satisfies system (2.4), where the coupling matrix satisfies (2.5) and $\mathcal{M}$ is an M-matrix, then $u_i \geq 0$, $i = 1, 2, \ldots, N$ in $\overline{\Omega}$. If the inequality signs in (2.4) are reversed, then $u_i \leq 0$, $i = 1, 2, \ldots, N$ in $\overline{\Omega}$.

As a consequence of the above, we see that if it is possible to transform a coupling matrix into an M-matrix which satisfies a weighted row sum condition, then it is also possible to transform the coupling matrix into one which satisfies the off–diagonal and row sum conditions (1.1), (1.2). We also note that the M-matrix condition can be extended to bounded nonlinear coupling of lower order terms as will be illustrated in the next section. Finally, we note that the theorem above can also be obtained as a corollary of results in [9] or [15] by choosing supersolutions in an appropriate way.

3. COMPARISON RESULTS FOR SYSTEMS OF VISCOSITY SOLUTIONS.

In the development of viscosity solutions for first order Hamilton- - Jacobi systems [12], the M-matrix condition arose as a natural assumption for comparison results. In the simplest Hamilton–Jacobi systems, where derivative terms are not present, a simple comparison result is

$$C\vec{u} = \vec{f} \geq 0 \text{ implies } \vec{u} \geq \vec{0} \text{ (componentwise)}.$$

However, if C^{-1} has some negative elements, there exist $\vec{f} \geq 0$ such that $C\vec{u} = \vec{f} \geq 0$ and $u = C^{-1}\vec{f} \not\geq 0$. Thus the M-matrix assumption of "C^{-1} has all nonnegative entries" cannot be dropped.

Let us now illustrate how the M-matrix condition appears in the nonlinear case. Consider the system

$$H^i\left(x, \vec{u}(x), \nabla u_i(x)\right) = 0 \text{ in } \Omega, \qquad i = 1,\ldots,k, \tag{3.1}$$

where $\vec{u} = (u_1,\ldots,u_k)$ and there exists constants m_{ij} such that

$$m_{ij} \leq \frac{\partial H^i}{\partial u_j} \leq 0, \qquad i \neq j, \qquad m_{ii} \leq \frac{\partial H^i}{\partial u_i}. \tag{3.2}$$

Condition (3.2) for the matrix $\mathcal{M} = (m_{ij})$ together with $\mathcal{M}^{-1}$ having all nonnegative entries constitutes the M-matrix condition. A standard comparison result would then be: if u is a viscosity subsolution of (3.1) on Ω and v is a viscosity supersolution of (3.1) on Ω with $u \leq v$ on $\partial\Omega$, then $u \leq v$ in Ω.

Viscosity solutions need not be differentiable; the differential inequalities for subsolution and supersolution are satisfied with the gradient of a test function φ substituted for ∇u at points where $u_i - \varphi$ attain a maximum or minimum (see [6] for background on this approach). We note, further, that the viscosity solution approach allows uniqueness results from comparison results for first order Hamilton–Jacobi systems, a consequence which was not possible for other previous notions of weak solution.

Recently, comparison results for degenerate elliptic systems of the form

$$H^k\left(x, \vec{u}(x), \nabla u_k(x), D^2 u_k(x)\right) = 0 \text{ in } \Omega$$

were obtained by Ishii and Koike [14]. Under the assumptions the Hamiltonians H^k are continuous in all of their arguments and Lipschitz continuous in the "u" position, the M-matrix condition (together with standard regularity assumptions for the derivative positions) implies comparison results for degenerate elliptic systems. Comparison results are extremely important for viscosity solutions since many existence results depend on Perron's method which requires comparison results.

The decoupling and transforming techniques of [4,5] yield new results in the case of a two equation, first order Hamilton–Jacobi system or degenerate elliptic system. If the coupling matrix is not an M-matrix but does satisfy the structure assumptions for decoupling or transforming in [4,5], then one can obtain comparison results for the transformed system and hence for the original system. Since the off–diagonal elements in the coupling matrix are allowed to change sign, conditions which were not allowed in the quasi– monotone [15] or M-matrix conditions for viscosity solutions, one obtains truly new comparison results. For example, comparison results hold for systems of viscosity solutions with coupling matrices

$$C_1 = \begin{pmatrix} -1 & -1 \\ 1 & -3 \end{pmatrix}, \qquad C_2 = \begin{pmatrix} -1 & -1 \\ 1 & -4 \end{pmatrix}.$$

The matrices C_1 and C_2 are not M-matrices, but C_2 is decoupable and C_1 is transformable to the conditions of the classical principle. However, the fact that the differential operators in the system must be the same is an off–setting restriction.

4. CONCLUDING REMARKS.

The "delicate balance" between the matrix of differential operators and the coupling matrix is a consequence of the technique of transforming the coupling matrix. This technique of decoupling or transforming permits one to go beyond the classical component–wise maximum or minimum principle, i.e., to determine the (possible differing) algebraic sign of the solution components when a component–wise maximum principle does not exist. Clearly, other techniques will lead to other variations of a maximum principle (such as for the sup norm of the vector solution of the system, see [11] and [19]) where the form of the matrix of differential operators is not as important as here.

If we let $\mathcal{D}$ denote the set of coupling matrices which are diagonalizable (decouplable), $\mathcal{T}$ the set which are transformable to the classical system conditions, and $\mathcal{M}$ the set which satisfy (1.1) and have an M-matrix bound, it is not difficult to formulate 2×2 coupling matrices which are in $\mathcal{D}-(\mathcal{T}\cup\mathcal{M})$, $\mathcal{D}\cap\mathcal{T}$, $\mathcal{T}-(\mathcal{D}\cup\mathcal{M})$, $\mathcal{D}\cap\mathcal{M}$, and $\mathcal{M}-(\mathcal{T}\cup\mathcal{D})$. Thus the set $\mathcal{D}$ overlaps $\mathcal{T}$ and $\mathcal{M}$ but the intersection $\mathcal{T}\cap\mathcal{M}$ is empty.

ACKNOWLEDGEMENT: *Lenhart's research was partially supported by National Science Foundation grant DMS – 8906226 and the University of Tennessee Science Alliance Research Award.*

REFERENCES

1. A. Berman and R.J. Plemmons, *Nonnegative Matrices in the Mathematical Sciences*, Academic Press, New York, 1979.
2. G. Caristi and E. Mitidieri, *Maximum principles for a class of non cooperative elliptic systems*, Delft Prog. Rep. **14** (1990), 33–56.
3. C. Cosner and P. W. Schaefer, *Sign–definite solutions in elliptic systems with constant coefficients*, Differential Equations and Applications, vol. II, Editor R. Aftabizadeh, Ohio U. Press, 1988, pp. 379–384.
4. _______, *Sign–definite solutions in some linear elliptic systems*, Proc. Royal Soc. Edinburgh **111A** (1989), 347–358.
5. _______, *Some comparison principles for weakly coupled systems of elliptic equations*, Diff. and Integral Eqs. **4** (1991), 555–562.
6. M.G. Crandall, L.C. Evans, and P.L. Lions, *Some properties of viscosity solutions of Hamilton–Jacobi equations*, Trans AMS **282** (1984), 487–502.
7. D.G. deFigueiredo and E. Mitidieri, *A maximum principle for an elliptic system and applications to semilinear problems*, SIAM J. Math. Anal **17** (1986), 836–849.

8. ______, *Maximum principles for cooperative elliptic systems*, C.R. Acad. Sci. Paris **310** Serie I (1990), 49–52.

9. ______, *Maximum principles for linear elliptic systems*, Rendiconti dell'Istituto di Matematica dell'Università di Trieste **XXII** (1990), 36–66.

10. M. A. Dow, *Strong maximum principles for weakly coupled systems of quasilinear parabolic inequalities*, J. Austral. Math. Soc. **19** (1975), 103–120.

11. D.R. Dunninger, *On Liouville–type theorems for elliptic and parabolic systems*, J. Math. Anal. Appl. **72** (1979), 413–417.

12. H. Engler and S. Lenhart, *Viscosity solutions for weakly coupled systems of Hamilton–Jacobi equations*, Proc. London Math. Soc. **63** (1991), 212–240.

13. J. Hernández, *Maximum principles and decoupling for positive solutions of reaction–diffusion systems*, preprint.

14. H. Ishii and S. Koike, *Viscosity solutions for monotone systems of second order elliptic PDE's*, Comm. Partial Diff. Eq. (to appear).

15. M. H. Protter and H.F. Weinberger, *Maximum Principles in Differential Equations*, Prentice–Hall, Englewood Cliffs, N. J., 1967.

16. G. Sweers, *A strong maximum principle for a noncooperative elliptic system*, SIAM J. Math. Anal. **20** (1989), 367–371.

17. ______, *Strong positivity in $C(\Omega)$ for elliptic systems*, Math Z. **209** (1992), 251–271.

18. H.F. Weinberger, *Invariant sets for weakly coupled parabolic and elliptic systems*, Rendiconti di Matematica **8** (1975), 295–310.

19. M.J. Winter and Pui-Kei Wong, *Comparison and maximum theorems for systems of quasilinear elliptic differential equations*, J. Math. Anal. Appl. **40** (1972), 634–642.

20. C. Zhou, *Maximum principles for elliptic systems*, J. Math. Anal. Appl. (to appear).

21. ______, *Maximum principles for weakly coupled parabolic systems*, preprint.

On a Coagulation–Fragmentation Equation

A. C. McBride, D. J. McLaughlin, and W. Lamb   University of Strathclyde, Glasgow, Scotland

1. INTRODUCTION

In this paper we shall discuss an initial-value problem of the form

$$\frac{\partial}{\partial t}u(x,t) = (Au(x,t)); u(x,0) = u_0(x)$$

for a function $u(x,t)$ which measures the number of particles of size x at time t during a reaction. The operator A is a non-linear integral operator involving kernels which govern the rate at which particles coalesce or break up during the reaction.

Several authors have developed existence and uniqueness theorems for this problem in the case of various special kernels. We shall review this work before going on to consider the reformulation of the initial-value problem as an abstract Cauchy problem and to obtain some existence and uniqueness theorems via semigroup theory.

More precisely, we shall consider the non-linear equation

$$\frac{\partial u}{\partial t}(x,t) = \frac{1}{2}\int_0^x K(x-y,y)u(x-y,t)u(y,t)dy - \int_0^\infty K(x,y)u(x,t)u(y,t)dy$$

$$- \frac{1}{2}\int_0^x F(x-y,y)u(x,t)dy + \int_0^\infty F(x,y)u(x+y,t)dy \qquad (1.1)$$

subject to the initial condition

$$u(x,0) = u_0(x). \qquad (1.2)$$

We seek $u(x,t)$ ($x \geq 0$, $t \geq 0$) satisfying (1.1) for $x \geq 0$, $t > 0$ and (1.2) for $x \geq 0$ (almost everywhere). The function u_0 in (1.2) is non-negative (a.e.).

The equation has been used to model a system of particles which can coalesce to form larger particles or fragment to form smaller particles. As already indicated, for $x \geq 0$ and $t \geq 0$, $u(x,t)$ measures the number of particles of size x at time t and the evolution of u is governed by the coagulation kernel K and the fragmentation kernel F.

Each of the four terms on the right-hand side of (1.1) has a physical interpretation. For example, the coagulation kernel $K(x,y)$ measures the rate at which particles of size x coalesce with particles of size y and the second of the four terms represents the loss of particles of size x owing to coalescence with particles of size y. Similarly the fragmentation kernel $F(x,y)$ is the rate at which particles of size $x+y$ fragment to form particles of sizes x and y so that the last term on the right-hand side of (1.1) gives the increase in particles of size x owing to such fragmentation. For further discussion see McLaughlin, Lamb and McBride (1993a).

We shall assume that K and F are symmetric, i.e.

$$K(x,y) = K(y,x) \quad \text{and} \quad F(x,y) = F(y,x) \quad \text{for all } x,y \geq 0 \qquad (1.3)$$

and that K and F are both continuous and non-negative. Initially, as suggested by the notation, we think of K and F as being independent of time t.

The problem represented by (1.1) and (1.2) has been studied by a variety of authors, notably Aizenman and Bak (1979) and Stewart (1989, 1990a, b, 1991). Their papers contain many references which need not be repeated here. Suffice to say that areas of application include reacting polymers, aerosols, meteorology, astrophysics, aggregation of blood-cells and birth-death processes.

Aizenman and Bak (1979) considered the case when both K and F are constant. Melzak (1957) had previously considered certain bounded kernels. In both cases theorems concerning global existence and uniqueness of solutions of (1.1) and (1.2) were obtained. Stewart (1989) went on to consider unbounded kernels and used the setting of Hausdorff gauge spaces, such spaces having a topology generated by a separating family of pseudometrics (semimetrics). Stewart's kernels satisfied growth conditions of the form

$$K(x,y) \leq C_1[(1+x)^\alpha + (1+y)^\alpha] \quad (x,y \geq 0)$$
$$F(x,y) \leq C_2(1+x+y)^\beta \quad (x,y \geq 0) \qquad (1.4)$$

for constants α and β in the range $0 \le \alpha, \beta < 1$. The problem (1.1) and (1.2) was then studied with K and F replaced by the truncated kernels K_n and $F_n (n = 1, 2, \cdots)$, where

$$K_n(x,y) = \begin{cases} K(x,y) & x + y \le n \\ 0 & x + y > n \end{cases} \tag{1.5}$$

and F_n is defined similarly. The "truncated problem" for K_n and F_n had a unique global non-negative solution u_n. By using a weak compactness argument, Stewart extracted from the sequence $\{u_n\}_{n=1}^{\infty}$ a subsequence $\{u_{n_k}\}_{k=1}^{\infty}$ which converged weakly in L^1 to a function u. It was then shown that u was a solution of (1.1) and (1.2) with the original K and F. We have detailed Stewart's approach as we shall also make use of truncation (but in a different way) in our approach.

For physical reasons, it is often required to find solutions such that

$$\int_0^{\infty} x u(x,t) dx = \int_0^{\infty} x u_0(x) dx \tag{1.6}$$

for all $t \ge 0$. This corresponds to conservation of mass or density (according to the context). See Stewart (1991) for further details of this and the related concept of gelation. We shall restrict attention in the sequel to such mass-conserving solutions.

2. A SEMIGROUP APPROACH: PURE FRAGMENTATION

To use the theory of semigroups we must recast the problem (1.1) and (1.2) as an abstract Cauchy problem (ACP). We begin with the case when

$$K(x,y) \equiv 0 \quad \text{(pure fragmentation)}.$$

Then (1.1) becomes a <u>linear</u> equation in u, giving rise to a linear ACP. Having dealt with the linear part, we shall proceed in the next section to study the full non-linear problem by a standard perturbation technique.

As usual we shall use $u(x,t)$ to generate a function u from $[0, \infty)$ into some Banach space X of functions of x. For mass conservation the appropriate choice of X, suggested by (1.6), is the weighted L^1-space

$$X = \{ f : \int_0^{\infty} x \mid f(x) \mid dx < \infty \} \tag{2.1}$$

with

$$\| f \| = \int_0^{\infty} x \mid f(x) \mid dx \quad (f \in X). \tag{2.2}$$

Given $u(x,t)$ $(x \ge 0,\ t \ge 0)$, define $u : [0, \infty) \to X$ by

$$[u(t)](x) = u(x,t) \quad (x \ge 0,\ t \ge 0). \tag{2.3}$$

No confusion should arise from using u (rather than, say, $\bar{u}$) for this X-valued function. This leads to the following linear ACP.

PROBLEM 2.1   Find $u : [0, \infty) \to X$ such that

$$\frac{du}{dt} = Au \quad (t > 0); \quad u(0) = u_0 \tag{2.4}$$

where, for $t \geq 0$, $(Au)(t)$ is defined by

$$[(Au)(t)](x) = -\frac{1}{2} \int_0^\infty F(x - y, y)[u(t)](x)dy$$
$$+ \int_0^\infty F(x, y)[u(t)](x + y)dy. \tag{2.5}$$

We study Problem 2.1 via truncation. Whereas Stewart truncated his kernels via (1.5), we shall truncate our functions by letting

$$X_n = \{f \in X : f(x) = 0 \text{ (a.e.) for } x \geq n\} \tag{2.6}$$

for $n = 1, 2, \cdots$. Since F is continuous on the compact set $[0, n] \times [0, n]$, the following result is easily obtained (McLaughlin, Lamb and McBride, 1993a).

LEMMA 2.2   The restriction of A to X_n generates a C_0-semigroup $\{T_n(t)\}_{t \geq 0}$ on X_n.

It now follows that for each $u_0 \in X_n$, Problem 2.1 has a unique solution in X_n. However, to ensure mass conservation, rather more has to be proved.

THEOREM 2.3

(i) For any $n \geq 1$ and $u_0 \in X_n$, the ACP (2.4) has a unique solution in X_n given by $u(t) = T_n(t)u_0$.

(ii) The semigroups $\{T_n(t)\}_{t \geq 0}$ are consistent in the sense that, if $n > m$ (so that $X_m \subset X_n$) then
$$T_n(t)f = T_m(t)f \quad (f \in X_m, t \geq 0). \tag{2.7}$$

(iii) For each $t \geq 0$, the operator $T_n(t)$ is a positivity-preserving contraction on X_n, i.e. if $f \in X_n$ then

$$f(x) \geq 0 \text{ (a.e.)} \Rightarrow [T_n(t)f](x) \geq 0 \text{ (a.e.)}$$
$$\| T_n(t)f \| \leq \| f \| .$$

(iv) For $f \in X_n$ and $t \geq 0$,

$$\int_0^\infty x \mid [T_n(t)f](x) \mid dx = \int_0^\infty x \mid f(x) \mid dx. \tag{2.8}$$

Proof: See Lemma 1.2 in McLaughlin, Lamb and McBride (1993a).

We would merely comment that the hardest work goes into (iii). The gist is to show that each of the two integral operators on the right-hand side of (2.2) generates a positivity-preserving semigroup on X_n. The required result then follows from the Trotter product formula (Goldstein, 1985, p.53).

To obtain results for X from those for X_n we shall let

$$P_n = \text{ the natural projection from } X \text{ onto } X_n \ (n \geq 1). \tag{2.9}$$

We can obtain a limiting semigroup on X from those on X_n as follows.

THEOREM 2.4   For $t \geq 0$ define the operator $T(t)$ on X by

$$T(t)f = \lim_{n \to \infty} T_n(t)P_n f \quad (f \in X). \tag{2.10}$$

Then $\{T(t)\}_{t \geq 0}$ is a strongly continuous, positivity-preserving, contraction semigroup of bounded linear operators on X. Furthermore for $f \in X$ and $t \geq 0$

$$\int_0^\infty x \mid [T(t)f](x) \mid dx = \int_0^\infty x \mid f(x) \mid dx. \tag{2.11}$$

Proof: See Theorem 1.3 in McLaughlin, Lamb and McBride (1993a).

As might be imagined, the proof merely consists of justifying passage to the limit as $n \to \infty$. In particular (2.11) follows from (2.8) by application of the Lebesgue Dominated Convergence Theorem.

We return to Problem 2.1. The condition that F be continuous on $[0, \infty) \times [0, \infty)$ is not sufficient to guarantee that A is a bounded linear operator on all of X. To relate the semigroup $\{T(t)\}_{t \geq 0}$ obtained in Theorem 2.4 to A, it is necessary to use the concept of a core (Goldstein, 1985, p.44). As an illustration, assume that F is such that

$$F(x,y) \leq C(x + y)^\beta \tag{2.12}$$

where C is a positive constant and $0 \leq \beta < 1$. (Recall (1.4) above.)

LEMMA 2.5   Let $\mathcal{A}$ be the generator of the semigroup $\{T(t)\}_{t \geq 0}$ in (2.10). Then a core for $\mathcal{A}$ is given by the set

$$D_\beta(\mathcal{A}) = \{f \in X : \int_0^\infty x^{2+\beta} \mid f(x) \mid dx < \infty\}. \tag{2.13}$$

Proof: See Theorem 1.4(i) in McLaughlin, Lamb and McBride (1993a).

COROLLARY 2.6   Let F satisfy (2.12) and let $u_0 \in D_\beta(\mathcal{A})$, as defined in (2.13). Then Problem 2.1 has at most one solution u such that $u(t) \in D_\beta(\mathcal{A})$ for all $t \geq 0$.

NOTE 2.7 If $D(\mathcal{A})$ denotes the domain of the generator $\mathcal{A}$, then it can be shown that $D_\beta(\mathcal{A}) \subset D(\mathcal{A})$. If in Corollary 2.6 we replace $D_\beta(\mathcal{A})$ by $D(\mathcal{A})$ throughout, we are, of course, guaranteed existence and uniqueness.

3. A SEMIGROUP APPROACH: THE FULL EQUATION

As previously indicated, we shall now treat the full equation (1.1) by means of a corresponding ACP which we regard as a non-linear perturbation of the linear Problem 2.1. With the notation of the previous section, we shall consider

PROBLEM 3.1 Find $u : [0, \infty) \to X$ such that

$$\frac{du}{dt} = Au + Nu \quad (t > 0); \quad u(0) = u_0 \tag{3.1}$$

where, for $t \geq 0$, $(Au)(t)$ is defined by (2.5) and $(Nu)(t)$ by

$$[(Nu)(t)](x) = \frac{1}{2} \int_0^x K(x - y, y)\{[u(t)](x - y)\}\{[u(t)](y)\}dy$$

$$- \int_0^\infty K(x, y)\{[u(t)](x)\}\{[u(t)](y)\}dy. \tag{3.2}$$

To make progress we shall consider functions $f \in X$ which also belong to (unweighted) L^1, i.e. we use the Banach space $(Y, \| \ \|_Y)$ where

$$Y = X \cap L^1(0, \infty), \quad \| f \|_Y = \int_0^\infty (1 + x) \mid f(x) \mid dx. \tag{3.3}$$

LEMMA 3.2 Assume that the coagulation kernel K is bounded i.e.

$$\sup\{\mid K(x, y) \mid : x \geq 0, y \geq 0\} = M < \infty. \tag{3.4}$$

Then N defines a continuous mapping from Y into Y and satisfies

$$\| Nf_1 - Nf_2 \|_Y \leq 2M(\| f_1 \|_Y + \| f_2 \|_Y) \| f_1 - f_2 \|_Y \tag{3.5}$$

for all $f_1, f_2 \in Y$ (i.e. N is locally Lipschitz).

Proof: See McLaughlin, Lamb and McBride (1993b).

The usual variation of parameters formula when applied to (3.1) suggests that, under appropriate conditions,

$$u(t) = T(t)u_0 + \int_0^t T(t - s) [(Nu)(s)]ds \tag{3.6}$$

where $\{T(t)\}_{t\geq 0}$ is as in the preceding section. Any function u satisfying (3.6) can be regarded as a mild solution of Problem 3.1. For existence and uniqueness of solutions to (3.6) we apply the Contraction Mapping Theorem to obtain

THEOREM 3.3   Given $u_0 \in Y$, $\exists\, \tau(0 < \tau \leq \infty)$ for which (3.6) has a unique solution u satisfying

$$u(0) = u_0; \quad u(t) \in Y \text{ for } 0 \leq t \leq \tau.$$

Further τ^*, the maximal choice of τ, is finite if and only if

$$\lim_{t \to \tau^*} \| u(t) \|_Y = \infty.$$

Proof: The main ingredients are Lemma 3.2 above (which is used to generate the required contraction) and Theorem 2.6 in Goldstein (1985, p.90). For full details see McLaughlin, Lamb and McBride (1993b).

We have therefore shown that under relatively weak conditions on K and F the full ACP (3.1) has a unique mild solution. Further work is needed to check positivity preservation and mass conservation.

4. TIME-DEPENDENT KERNELS

We can consider the possibility of the kernels K and F depending on time t. Here we shall restrict attention to the simple case in which $K \equiv 0$ and $F \equiv F(x,y,t)$ ($x \geq 0$, $y \geq 0$, $t \geq 0$) and indicate briefly what happens. Full details can be found in McLaughlin, Lamb and McBride (1993c).

We have to solve an analogue of Problem 2.1 in which the operator A now depends on time, having the form (2.5) but with a time-dependent kernel. We assume that, for each fixed $t \geq 0$, F is continuous in x and y, symmetric in x and y (recall (1.3)) and non-negative. We start again in the space X_n defined via (2.6).

Fix $s \geq 0$. By our previous theory, the operator

$$A_n(s) = A(s)\,|_{X_n} = \text{ the restriction of } A(s) \text{ from } X \text{ to } X_n \tag{4.1}$$

generates a positivity-preserving contraction semigroup $\{T_{n,s}(t)\}_{t\geq 0}$ on X_n. By Theorem 2.3, the semigroups $\{T_{n,s}(t)\}$ for $n \geq 1$ are consistent and

$$\int_0^\infty x \mid [T_{n,s}(t)f](x) \mid dx = \int_0^\infty x \mid f(x) \mid dx$$

for all $f \in X_n$ and all $n \geq 1, s \geq 0$. To solve the time-dependent version of Problem 2.1, we assume that, for fixed n,

$$A_n(s) \text{ is continuous with respect to } s\ (\geq 0) \tag{4.2}$$

in the uniform operator topology. We obtain an evolution system $U_n(t,s)$ on X_n (Pazy, 1983, p.129) which is unique in the sense stated in the next theorem.

THEOREM 4.1   For each fixed $\tau \geq 0$, there is a unique evolution system $U_n^\tau(t,s)$ on X_n $(0 \leq s \leq t \leq \tau)$ such that

$\quad$ (i) $\quad \| U_n^\tau(t,s) \| \leq 1$

$\quad$ (ii) $\quad \dfrac{\partial^+}{\partial t} U_n^\tau(t,s)f \mid_{t=s} = A_n(s)f$ for all $f \in X_n$

$\quad$ (iii) $\quad \dfrac{\partial}{\partial s} U_n^\tau(t,s)f = -U_n^\tau(t,s)A_n(s)f \quad$ for all $f \in X_n$.

Proof: The main tool is Theorem 3.1 in Pazy (1983). See McLaughlin, Lamb and McBride (1993c).

COROLLARY 4.2   The "truncated" ACP

$$\frac{du}{dt} = A_n(t)u(t) \ (t > 0); \quad u(0) = u_0 \in X_n$$

has a unique solution u_n in X_n such that

$$u_n(t) = U_n^\tau(t,0)u_0 \quad \text{for } 0 \leq t \leq \tau.$$

Proof: This is immediate from properties of $U_n^\tau(t,s)$.

It can be proved that for each fixed s and t $(0 \leq s \leq t \leq \tau)$ the evolutionary systems $\{U_n^\tau(t,s)\}$ for $n \geq 1$ are consistent and for each fixed n, the operator $U_n^\tau(t,s)$ is positivity-preserving on X_n. Further there is mass conservation over the interval $[0,\tau]$. The operation is now completed via an analogue of (2.10).

THEOREM 4.3.   For $0 \leq s \leq t \leq \tau$ define the operator $U^\tau(t,s)$ on X by

$$U^\tau(t,s)f = \lim_{n \to \infty} U_n^\tau(t,s)P_n f \quad (f \in X) \tag{4.3}$$

with P_n as in (2.9). Then

$\quad$ (i) $\{U^\tau(t,s) : 0 \leq s \leq t \leq \tau\}$ is a positivity-preserving, contraction evolution system on X, and

$\quad$ (ii) for $0 \leq s \leq t \leq \tau$ and $f \in X$

$$\int_0^\infty x \mid [U^\tau(t,s)f](x) \mid dx = \int_0^\infty x \mid f(x) \mid dx. \tag{4.4}$$

Proof: This is similar to that of Theorem 2.4. See McLaughlin, Lamb and McBride (1993c).

Finally we can consider whether or not analogues of Lemma 2.5 and Corollary 2.6 hold provided that

$$F(x,y,t) \leq C(t)(x+y)^{\beta}$$

where $C(t)$ is a constant depending on t and $0 \leq \beta < 1$. We have to investigate if the set (2.13) then turns out to be a core for each $\mathcal{A}(s)$ where, for appropriate functions $f \in X$,

$$\mathcal{A}(s)f = \lim_{t \to 0^+} \frac{U^{\tau}(t+s,s)f - f}{t} \quad (0 \leq s \leq \tau).$$

Work is in progress to use these results for the linear time-dependent fragmentation problem to derive results for the full non-linear time-dependent equation corresponding to $K(x,y,t) \not\equiv 0$.

ACKNOWLEDGEMENT

The work of the second author was supported by a Research Studentship from the Science and Engineering Research Council.

REFERENCES

Aizenman, M. and Bak, T. A. (1979). Convergence to equilibrium in a system of reacting polymers, Commun. Math. Phys., 65: 203.

Goldstein, J. A. (1985). Semigroups of linear operators and applications, Oxford University Press, Oxford.

McLaughlin, D. J., Lamb, W. and McBride, A.C. (1993a). A coagulation-fragmentation equation: pure fragmentation, Department of Mathematics Departmental Report , University of Strathclyde, Glasgow.

McLaughlin, D. J., Lamb, W. and McBride, A.C. (1993b). A coagulation-fragmentation equation: full equation, Department of Mathematics Departmental Report , University of Strathclyde, Glasgow.

McLaughlin, D. J., Lamb, W. and McBride, A.C. (1993c). A time-dependent fragmentation equation, Department of Mathematics Departmental Report , University of Strathclyde, Glasgow.

Melzak, Z. A. (1957). A scalar transport equation, Trans. Amer. Math. Soc., 85: 547.

Pazy, A. (1983). Semigroups of linear operators and applications to partial differential equations, Springer-Verlag, Berlin.

Stewart, I. W. (1989). A global existence theorem for the general coagulation -fragmentation equation with unbounded kernels. Math. Meth. in the Appl. Sci., 11: 627.

Stewart, I. W. (1990a). A uniqueness theorem for the coagulation-fragmentation equation, Math. Proc. Camb. Phil. Soc., 107: 573.

Stewart, I. W. (1990b). On the coagulation-fragmentation equation, ZAMP, 41: 917.

Stewart, I. W. (1991). Density conservation for a coagulation equation, ZAMP, 42: 746.

A Distributional Version of Spectral Theory
for Normal Operators

D. F. McGhee University of Strathclyde, Glasgow, Scotland

F. Baptiste University of Wisconsin-Milwaukee, Milwaukee, Wisconsin

W. Lamb University of Strathclyde, Glasgow, Scotland

1. INTRODUCTION

In a recent paper [3], Lamb and McGhee extended the work of Zemanian [6] and Judge [2] to develop a spectral theory for the extension, to an appropriate space of distributions, of an arbitrary self-adjoint operator in a separable Hilbert space. A functional calculus was shown to be valid in this setting and it was applied, via semi-group theory, to the solution of distributional initial-boundary value problems.

Here, we consider a similar approach for a normal operator in a separable Hilbert space. It could be argued that a normal operator is a complex function of an appropriate self-adjoint operator and so could be dealt with by the self-adjoint theory. However, there are two reasons why we wish to consider the normal operator case in its own right. Firstly, considering the normal operator as a function of a self-adjoint operator would not give rise to a natural functional calculus for the normal operator and, for the applications that we have in mind, we consider this to be one of the essential features of our development. Secondly, the definition we use here for integrals with respect to vector valued measures which is due to Weidmann [5] is more general than the essentially Riemann integral approach that was adopted in [3].

2. PRELIMINARIES

Let H be a separable complex Hilbert space with inner product $(\cdot, \cdot)$, linear in its first argument, and norm $\| \cdot \| := \sqrt{(\cdot, \cdot)}$. The following development of a strong version of the spectral theorem for a normal operator is taken from [5].

Let $L(H)$ denote the space of all continuous linear operators on H. A function $E : \mathbf{C} \to L(H)$ is called a complex spectral family if there are real spectral families E_1 and E_2 such that

$$E(s + it) = E_1(s)E_2(t) \quad \forall s, t \in \mathbf{R}.$$

(It is assumed that the reader is familiar with the concept of a real spectral family.)

Let $J := \{z = s + it \in \mathbf{C} : s \in J_1,\ t \in J_2\} \subset \mathbf{C}$, where J_1 and J_2 are arbitrary intervals in $\mathbf{R}$, be called an interval in $\mathbf{C}$. Then, using E_j, $j = 1, 2$, to denote the associated spectral measures on $\mathbf{R}$ as well as the real spectral families,

$$E(J) := E_1(J_1)E_2(J_2)$$

is an operator valued function defined on arbitrary intervals $J \in \mathbf{C}$. This can be extended to a projection valued measure on the Borel subsets of $\mathbf{C}$.

For every $\varphi \in H$,

$$E_\varphi(J) := \|E(J)\varphi\|^2$$

defines a measure on the Borel subsets of $\mathbf{C}$. A function is said to be *E-measurable* if it is E_φ-measurable for all $\varphi \in H$.

Throughout this paper, integrals are over all of $\mathbf{C}$.

THEOREM 2.1 ([5], Theorem 7.32) *Let T be a normal operator on the complex Hilbert space H. Then, there exists a unique complex spectral family, $E : \mathbf{C} \to L(H)$, such that*

$$T\varphi = \int z\, dE(z)\varphi$$

and

$$\mathcal{D}(T) = \{\varphi \in H : \int |z|^2 dE_\varphi(z) < \infty\}.$$

Moreover, for any E-measurable function $F : \mathbf{C} \to \mathbf{C}$, there exists a unique normal operator $F(T)$ defined by

$$\mathcal{D}(F(T)) := \{\varphi \in H : \int |F(z)|^2 dE_\varphi(z) < \infty\}$$

$$F(T)\varphi := \int F(z)\, dE(z)\varphi.$$

Here, the vector-valued integrals are defined as follows:

For a step function $s : \mathbf{C} \to \mathbf{C}$, $s(z) = \sum_{j=1}^n c_j \mathcal{X}_{J_j}(z)$, where $J_j, j = 1,\ldots,n$, are intervals in $\mathbf{C}$,

$$\int s(z)\, dE(z) := \sum_{j=1}^n c_j E(J_j).$$

For any step function $s : \mathbf{C} \to \mathbf{C}$ and for any $\varphi \in H$, a simple calculation gives

$$\| \int s(z)\, dE(z)\varphi \|^2 = \int |s(z)|^2 \, dE_\varphi(z).$$

Now, if $L^2(\mathbf{C}, E_\varphi)$ denotes the complex L^2-space of functions that are square integrable on $\mathbf{C}$ with respect to the measure E_φ, then

$$\varphi \in \mathcal{D}(F(T)) \iff F \in L^2(\mathbf{C}, E_\varphi),$$

so that, for $\varphi \in \mathcal{D}(F(T))$, there exists a sequence of step functions (s_n) such that $s_n \to F$ in $L^2(\mathbf{C}, E_\varphi)$. Then,

$$\| \int s_m(z)\, dE(z)\varphi - \int s_n(z)\, dE(z)\varphi \|^2 = \int |s_m(z) - s_n(z)|^2 \, dE_\varphi(z) \to 0 \text{ as } m, n \to \infty,$$

which shows that the sequence $(\int s_n\, dE(z)\varphi)$ is Cauchy in H. Therefore, we define

$$\int F(z)\, dE(z)\varphi := \lim_{n\to\infty} \int s_n(z)\, dE(z)\varphi.$$

This definition is independent of the choice of sequence (s_n), and we have

$$\| \int F(z)\, dE(z)\varphi \|^2 = \int |F(z)|^2 \, dE_\varphi(z).$$

3. SPACES OF TEST FUNCTIONS AND DISTRIBUTIONS

Given a normal operator $T : \mathcal{D} \subset H \to H$, we consider the set

$$\mathcal{D}(T^\infty) := \bigcap_{k \in \mathbf{N}_0} \mathcal{D}(T^k) \subset H \qquad (\mathbf{N}_0 := \{0, 1, 2, 3, \ldots\}),$$

and introduce a countable family of inner products $\{(\cdot, \cdot)_k \mid k \in \mathbf{N}_0\}$ and associated norms on $\mathcal{D}(T^\infty)$ by

$$(\varphi, \psi)_k := \sum_{j=0}^{k} (T^j \varphi, T^j \psi), \qquad \|\varphi\|_k = \sqrt{(\varphi, \varphi)_k}, \qquad k \in \mathbf{N}_0.$$

DEFINITION 3.1 We define $\mathcal{D}_\infty$ as the countably pre-Hilbert space (see[1], p57) consisting of the set $\mathcal{D}(T^\infty)$ with topology generated by the countable family of inner products $\{(\cdot, \cdot)_k \mid k \in \mathbf{N}_0\}$.

Sequential convergence in $\mathcal{D}_\infty$ is described as follows: A sequence (φ_n) in $\mathcal{D}_\infty$ converges to φ in $\mathcal{D}_\infty$ if and only if $\|\varphi_n - \varphi\|_k \longrightarrow 0$ as $n \to \infty$ for every $k \in \mathbf{N}_0$.

THEOREM 3.2 $\mathcal{D}_\infty$ *is a countably Hilbert space.*

Proof: We need only show completeness, and this follows easily from the fact that the operators T^j, $j \in \mathbf{N}_0$, are all closed.

The next result is well known but is included for reference.

LEMMA 3.3 *Let* $F : \mathbf{C} \longrightarrow \mathbf{C}$, $G : \mathbf{C} \longrightarrow \mathbf{C}$, *be two E-measurable functions.*
If $\varphi \in \mathcal{D}(F(T))$, *then*

$$F(T)\varphi \in \mathcal{D}(G(T)) \iff \varphi \in \mathcal{D}(FG(T)),$$

where $FG(z) = F(z)G(z)$.

Proof: See [4], Theorem 13.24(b).

THEOREM 3.4 $\mathcal{D}_\infty$ *is dense in* H.

Proof: By [4, p364] we know that $T = UP = PU$ where U is unitary, P is positive and self-adjoint and $\mathcal{D}(T) = \mathcal{D}(P)$. Thus,

$$\mathcal{D}(T^k) = \mathcal{D}((UP)^k) = \mathcal{D}(U^k P^k) = \mathcal{D}(P^k) \ \ \forall k \in \mathbf{N}_0.$$

Thus, $\mathcal{D}(T^\infty) = \mathcal{D}(P^\infty)$. But, P is self-adjoint, hence $\mathcal{D}(P^\infty)$ is dense in H (see[3], Theorem 2.4), and the result follows.

DEFINITION 3.5 The space of all continuous linear functionals on $\mathcal{D}_\infty$ is denoted by $\mathcal{D}'_\infty$. We impose the weak* topology on $\mathcal{D}'_\infty$. Thus, sequential convergence is given by

$$f_n \longrightarrow f \text{ as } n \to \infty \text{ in } \mathcal{D}'_\infty \iff \langle f_n, \varphi \rangle \longrightarrow \langle f, \varphi \rangle \text{ as } n \to \infty \; \forall \varphi \in \mathcal{D}_\infty,$$

where $\langle f, \varphi \rangle$ denotes the action of $f \in \mathcal{D}'_\infty$ on $\varphi \in \mathcal{D}_\infty$.

THEOREM 3.6 $T^r : \mathcal{D}_\infty \longrightarrow \mathcal{D}_\infty$ *is continuous for all* $r \in \mathbf{N}_0$.

Proof: This is trivial when $r = 0$.

Let $k \in \mathbf{N}_0$, $r \in \mathbf{N}$ be arbitrary, and let $\varphi_n \longrightarrow 0$ as $n \to \infty$ in $\mathcal{D}_\infty$. Then,

$$\|T^r \varphi_n\|_k^2 = \sum_{j=0}^{k} \|T^j T^r \varphi_n\|^2 = \sum_{j=0}^{k} \|T^{j+r} \varphi_n\|^2$$

$$= \sum_{j=r}^{k+r} \|T^j \varphi_n\|^2 = \sum_{j=0}^{k+r} \|T^j \varphi_n\|^2 - \sum_{j=0}^{r-1} \|T^j \varphi_n\|^2$$

$$= \|\varphi_n\|_{k+r}^2 - \|\varphi_n\|_{r-1}^2 \longrightarrow 0 \text{ as } n \to \infty$$

since $\varphi_n \longrightarrow 0$ as $n \to \infty$ in $\mathcal{D}_\infty$.

The next result follows immediately from [1, p49].

THEOREM 3.7 $\mathcal{D}'_\infty$ *is sequentially weak*-complete.*

The proof of the next result is straightforward.

THEOREM 3.8 *Each* $\eta \in H$ *generates, uniquely, a continuous linear functional* $\tilde{\eta} \in \mathcal{D}'_\infty$ *via the formula*

$$\langle \tilde{\eta}, \varphi \rangle := (\varphi, \eta) \; , \; \varphi \in \mathcal{D}_\infty.$$

As a consequence of Theorem 3.8 we can regard H as a subspace of $\mathcal{D}'_\infty$ and thus we have the triplet

$$\mathcal{D}_\infty \hookrightarrow H \hookrightarrow \mathcal{D}'_\infty$$

where $\hookrightarrow$ denotes a continuous embedding: the continuity of the embeddings is easily verified.

We now turn our attention to the problem of obtaining a continuous extension $\widetilde{T}$ of T to $\mathcal{D}'_\infty$. In particular, we require $\widetilde{T}\widetilde{\eta} = \widetilde{T\eta}$ $\forall \eta \in \mathcal{D}(T)$, that is

$$\langle \widetilde{T}\widetilde{\eta}, \varphi \rangle = \langle \widetilde{T\eta}, \varphi \rangle = (\varphi, T\eta)$$
$$= (T^*\varphi, \eta) = \langle \widetilde{\eta}, T^*\varphi \rangle \quad \forall \varphi \in \mathcal{D}_\infty, \quad \forall \eta \in \mathcal{D}(T).$$

Note that $\varphi \in \mathcal{D}_\infty \Longrightarrow \varphi \in \mathcal{D}(T) = \mathcal{D}(T^*)$ since T is a normal operator.

Thus, we are led to the following definition of $\widetilde{T}$:

DEFINITION 3.9 The extension of T from $\mathcal{D}(T)$ to $\mathcal{D}'_\infty$, denoted by $\widetilde{T}$, is defined by

$$\langle \widetilde{T}f, \varphi \rangle := \langle f, T^*\varphi \rangle \quad \forall f \in \mathcal{D}'_\infty, \ \varphi \in \mathcal{D}_\infty.$$

That is, $\widetilde{T} := (T^*)'$, the $\mathcal{D}_\infty$-adjoint of the Hilbert-adjoint T^* of T.

This should be compared with the self-adjoint case [3] where we have $\widetilde{T} := T'$, that is, the extension of T to $\mathcal{D}'_\infty$ is the $\mathcal{D}_\infty$-adjoint of T.

LEMMA 3.10 $T^* \in L(\mathcal{D}_\infty)$.

Proof: Since T is normal, it follows that T^k is normal for all $k \in \mathbf{N}$ so that $\mathcal{D}(T^k) = \mathcal{D}((T^k)^*) = \mathcal{D}((T^*)^k)$ for all $k \in \mathbf{N}$, and it is easily shown that T^* maps $\mathcal{D}_\infty$ into $\mathcal{D}_\infty$.

Let (φ_n) be a sequence in $\mathcal{D}_\infty$ such that $\varphi_n \longrightarrow 0$ as $n \to \infty$. Then

$$\|\varphi_n\|_k^2 \longrightarrow 0 \text{ as } n \to \infty \ \forall k \in \mathbf{N}_0$$
$$\Longrightarrow T\varphi_n \longrightarrow 0 \text{ as } n \to \infty \text{ in } \mathcal{D}_\infty \text{ since } T \in L(\mathcal{D}_\infty),$$

and therefore

$$\|T^*\varphi_n\|_k^2 = \sum_{j=0}^{k} \|T^j T^* \varphi_n\|^2 = \sum_{j=0}^{k} \|T^* T^j \varphi_n\|^2$$
$$= \sum_{j=0}^{k} \|T T^j \varphi_n\|^2 = \sum_{j=0}^{k} \|T^j T \varphi_n\|^2$$
$$= \|T\varphi_n\|_k^2 \longrightarrow 0 \text{ as } n \to \infty \ \forall k \in \mathbf{N}_0,$$

so that $T^*\varphi_n \longrightarrow 0$ in $\mathcal{D}_\infty$ as $n \longrightarrow \infty$.

THEOREM 3.11 $\widetilde{T} \in L(\mathcal{D}'_\infty)$.

Proof: This follows immediately from Definition 3.9 and Lemma 3.10 since the $\mathcal{D}_\infty$-adjoint of a continuous mapping is continuous (see [1, p65]).

Defining $\widetilde{T^k}$ the extension of T^k via

$$\langle \widetilde{T^k}f, \varphi \rangle := \langle f, (T^k)^*\varphi \rangle \quad \forall \varphi \in \mathcal{D}_\infty, f \in \mathcal{D}_\infty,$$

we obtain the following result:

COROLLARY 3.12 $\widetilde{T^k} = \tilde{T}^k \ \ \forall k \in \mathbf{N}$.

Proof:
$$\langle \widetilde{T^k} f, \varphi \rangle := \langle f, (T^k)^* \varphi \rangle = \langle f, (T^*)^k \varphi \rangle$$
$$= \langle \tilde{T}^k f, \varphi \rangle \ \ \forall f \in \mathcal{D}'_\infty, \ \varphi \in \mathcal{D}_\infty, \ k \in \mathbf{N}.$$

An exact characterisation of the elements of $\mathcal{D}'_\infty$ is given by the following theorem:

THEOREM 3.13 $f \in \mathcal{D}'_\infty$ *if and only if there exist a non-negative integer* n *and a sequence* $(\eta_r)^n_{r=0} \subset H$ *(both dependent upon* f*) such that*

$$f = \sum_{r=0}^n \tilde{T}^r \widetilde{\eta_r}.$$

Proof: Sufficiency follows easily from Theorems 3.8 and 3.11 and Corollary 3.12.

For necessity, let let $f \in \mathcal{D}'_\infty$. Then, by [1, p34], $\exists c > 0$ and an integer $n \in N_0$ such that

$$|\langle f, \varphi \rangle| \leq c \|\varphi\|_n = c \left[\sum_{k=0}^n \|T^k \varphi\|^2 \right]^{\frac{1}{2}} \ \ \forall \varphi \in \mathcal{D}_\infty . \tag{3.1}$$

Let $H^{n+1} := \overbrace{H \times \ldots \times H}^{n+1 \ \text{times}}$ with an inner product defined by

$$(\Phi, \Psi) := \sum_{i=0}^n (\varphi_i, \psi_i),$$

where

$$\Phi = (\varphi_0, \varphi_1, \ldots, \varphi_n) \in H^{n+1}, \quad \Psi = (\psi_0, \psi_1, \ldots, \psi_n) \in H^{n+1}.$$

Let $P : \mathcal{D}_\infty \longrightarrow H^{n+1}$ be defined by

$$P\varphi := (\varphi, T^* \varphi, (T^*)^2 \varphi, \ldots, (T^*)^n \varphi), \ \varphi \in \mathcal{D}_\infty.$$

Clearly, $P\varphi = P\psi \Longleftrightarrow \varphi = \psi$, so that P is injective.

Define Q on $P(\mathcal{D}_\infty)$ by

$$Q : P(\mathcal{D}_\infty) \longrightarrow \mathbf{C} : \ \ Q(\varphi, T^* \varphi, (T^*)^2 \varphi, \ldots, (T^*)^n \varphi) := \langle f, \varphi \rangle.$$

Then, by (3.1),

$$\left|Q(\varphi, T^*\varphi, (T^*)^2\varphi, \ldots, (T^*)^n\varphi)\right| = |\langle f, \varphi\rangle|$$

$$\leq c\left[\sum_{k=0}^{n}\|T^k\varphi\|^2\right]^{\frac{1}{2}} = c\left[\sum_{k=0}^{n}\|(T^*)^k\varphi\|^2\right]^{\frac{1}{2}} = c\|P\varphi\|,$$

so that Q is a continuous linear functional on $P(\mathcal{D}_\infty)$. By the Hahn-Banach theorem, we can extend Q to a continuous linear functional on H^{n+1}; we continue to denote this extension by Q.

Now, by the Riesz representation theorem, there exists an $\boldsymbol{\eta} = (\eta_0, \eta_1, \ldots, \eta_n) \in H^{n+1}$ such that

$$Q(\Phi) = (\Phi, \boldsymbol{\eta}) = \sum_{r=0}^{n}(\varphi_r, \eta_r) \ \ \forall \Phi \in H^{n+1}.$$

Hence,

$$\langle f, \varphi\rangle = Q(\varphi, T^*\varphi, (T^*)^2\varphi, \ldots, (T^*)^n\varphi) = \sum_{r=0}^{n}(T^{*^r}\varphi, \eta_r)$$

$$= \sum_{r=0}^{n}\langle \widetilde{\eta}_r, T^{*^r}\varphi\rangle = \sum_{r=0}^{n}\langle \widetilde{T^r\widetilde{\eta}_r}, \varphi\rangle \ \ \forall \varphi \in \mathcal{D}_\infty,$$

so that $f = \sum_{r=0}^{n}\widetilde{T^r\widetilde{\eta}_r}$, as required.

4. SPECTRAL THEORY IN $\mathcal{D}_\infty$ AND $\mathcal{D}_\infty'$

We have seen in Section 2 that each $\varphi \in H$ can be represented in the form

$$\varphi = \int dE(z)\varphi, \tag{4.1}$$

where E is the complex spectral family associated with the normal operator T, and the integral exists strongly in the sense defined in Section 2. Similarly,

$$T^r\varphi = \int z^r \, dE(z)\varphi \ \ \forall \varphi \in \mathcal{D}(T^r), \ r \in \mathbf{N}, \tag{4.2}$$

and, for a given E-measurable function F, an operator $F(T)$ may be defined by:

$$\mathcal{D}(F(T)) := \{\varphi \in H \mid \int |F(z)|^2 \, dE_\varphi(z) < \infty\}$$

$$F(T)\varphi := \int F(z) \, dE(z)\varphi. \tag{4.3}$$

Our aim in this section is to obtain generalisations of (4.1)–(4.3) to the spaces $\mathcal{D}_\infty$ and $\mathcal{D}'_\infty$. In particular, we must show that the integrals with respect to the vector valued measures can be well defined in $\mathcal{D}_\infty$ and $\mathcal{D}'_\infty$.

We begin with the space $\mathcal{D}_\infty$ and examine the complex spectral measure E.

LEMMA 4.1 *For any Borel set $J \subset \mathbf{C}$, $E(J) \in L(\mathcal{D}_\infty)$.*

Proof: We know that, for any Borel set J, $E(J) \in L(H)$ and is in fact a self-adjoint projection operator. Therefore, for any $\varphi \in \mathcal{D}_\infty$,

$$\|E(J)\varphi\|_k^2 = \sum_{j=0}^k \|T^j E(J)\varphi\|^2 = \sum_{j=0}^k \|E(J)T^j\varphi\|^2$$
$$\leq \sum_{j=0}^k \|T^j\varphi\|^2 = \|\varphi\|_k^2,$$

from which the continuity of $E(J)$ on $\mathcal{D}_\infty$ follows.

We now define integration in $\mathcal{D}_\infty$. For a step function $s(z) = \sum_{j=1}^n c_j \mathcal{X}_{J_j}(z)$ and $\varphi \in \mathcal{D}_\infty$, we define

$$\mathcal{D}_\infty - \int s(z)dE(z)\varphi := \sum_{j=1}^n c_j E(J_j)\varphi.$$

Then, for a bounded E–measurable function F and any $\varphi \in \mathcal{D}_\infty$ there exists a bounded sequence of step functions $(s_n)_{n=1}^\infty$,

$$s_n(z) = \sum_{j=1}^{N_n} c_{nj} \mathcal{X}_{J_{nj}}(z)$$

such that

$$s_n \to F \text{ in } L^2(\mathbf{C}, E_\varphi)$$
$$\implies s_n \to F \text{ a.e. with respect to } E_\varphi \text{ (since } s_n, F \text{ are bounded)}$$
$$\implies s_n \to F \text{ a.e. with respect to } E_{T^r\varphi} \quad \forall r \in \mathbf{N};$$

the last implication follows since, for any Borel set $J \subset \mathbf{C}$,

$$E_\varphi(J) = \|E(J)\varphi\|^2 = 0 \implies E_{T^r\varphi}(J) = \|E(J)T^r\varphi\|^2 = \|T^r E(J)\varphi\|^2 = 0.$$

From the Lebesgue dominated convergence theorem we deduce that

$$s_n \to F \text{ in } L^2(\mathbf{C}, E_{T^r\varphi}) \quad \forall r \in \mathbf{N}.$$

Since F is a bounded function, $\varphi \in \mathcal{D}_\infty$ implies $F(T)\varphi \in \mathcal{D}_\infty$ and we have

$$\|F(T)\varphi - \int s_n(z)dE(z)\varphi\|_k^2 = \sum_{r=0}^{k} \|T^r(F(T)\varphi - \sum_{j=0}^{N_n} c_{nj}E(J_{nj})\varphi)\|^2$$

$$= \sum_{r=0}^{k} \|F(T)T^r\varphi - \sum_{j=0}^{N_n} c_{nj}E(J_{nj})T^r\varphi\|^2 = \sum_{r=0}^{k} \|F(T)T^r\varphi - \int s_n(z)dE(z)T^r\varphi\|^2$$

$$\to 0 \text{ as } n \to \infty.$$

Therefore, for a bounded E-measurable function F, we define

$$\mathcal{D}_\infty - \int F(z)dE(z)\varphi := \mathcal{D}_\infty - \lim_{n\to\infty} \int s_n(z)dE(z)\varphi = F(T)\varphi.$$

Finally, we consider spectral integrals of unbounded functions. The next theorem gives necessary and sufficient conditions on F for the operator $F(T)$ defined in H by (4.3) to be a continuous operator on $\mathcal{D}_\infty$.

THEOREM 4.2 *Let F be a E-measurable function defined on* **C**. *Then, $F(T)$ defined by (4.3) is a continuous operator on $\mathcal{D}_\infty$ if and only if there exist $m \in \mathbf{N}_0$ and $M < \infty$ such that*

$$\sup_{z\in\sigma(T)} |F(z)|(1 + |z|^2)^{-m} = M, \tag{4.4}$$

that is, if and only if $F(T)(I + T^*T)^{-m} \in L(H)$, where I is the identity operator on H.

Proof: *Sufficiency:* Suppose F satisfies (4.4). Then, for any $\varphi \in \mathcal{D}_\infty$ and any $j \in \mathbf{N}_0$,

$$\int |z^j F(z)|^2 \, dE_\varphi(z) = \int |z|^{2j}|F(z)|^2 \, dE_\varphi(z)$$

$$\leq M^2 \int \left|(1 + |z|^2)^m z^j\right|^2 \, dE_\varphi(z) = M^2\|T^j(I + T^*T)^m\varphi\|^2 < \infty.$$

Hence, defining $G_j(z) := z^j$, it follows that

$$\varphi \in \mathcal{D}((FG_j)(T)) \ \ \forall \varphi \in \mathcal{D}_\infty, \ j \in \mathbf{N}_0.$$

By Lemma 3.3 we deduce that

$$\varphi \in \mathcal{D}(F(T)T^j) \cap \mathcal{D}(T^j F(T)) \ \ \forall j \in \mathbf{N}_0,$$

and, for each $j \in \mathbf{N}_0$, we have

$$\|T^j F(T)\varphi\|^2 \leq M^2\|T^j(I + T^*T)^m\varphi\|^2.$$

Thus, for all $\varphi \in \mathcal{D}_\infty$,

$$\|F(T)\varphi\|_k^2 = \sum_{j=0}^{k} \|T^j F(T)\varphi\|^2 \leq \sum_{j=0}^{k} M^2 \|T^j(I+T^*T)^m\varphi\|^2 = M^2 \sum_{j=0}^{k} \|T^j(I+T^*T)^m\varphi\|^2.$$

Since $(I + T^*T)^m \in L(\mathcal{D}_\infty)$ for each $m \in \mathbf{N}_0$, it follows that $F(T) \in L(\mathcal{D}_\infty)$.

Necessity: Suppose that $F(T) \in L(\mathcal{D}_\infty)$. Then, by [6], Lemma 1.10-1, there exist $c > 0$ and $m \in \mathbf{N}_0$ such that

$$\|F(T)\varphi\|^2 = \|F(T)\varphi\|_0^2 \leq c^2 \max_{0 \leq j \leq m} \{\|\varphi\|_j^2\} \quad \forall \varphi \in \mathcal{D}_\infty.$$

Hence,

$$\begin{aligned}
\|F(T)(I + T^*T)^{-m}\varphi\|^2 &\leq c^2 \max_{0 \leq j \leq m} \{\|T^j(I + T^*T)^{-m}\varphi\|^2\} \\
&= c^2 \max_{0 \leq j \leq m} \left\{ \int \left| z^j(1 + |z|^2)^{-m} \right|^2 dE_\varphi(z) \right\} \\
&\leq c^2 \int dE_\varphi = c^2\|\varphi\|^2.
\end{aligned}$$

Thus, $F(T)(I + T^*T)^{-m} \in L(H)$ and so (4.4) holds.

Now, for any F satisfying (4.4), we define cut-off functions F_n by

$$F_n(z) := \begin{cases} F(z) & \text{if } |F(z)| < n, \\ 0 & \text{otherwise,} \end{cases} \tag{4.5}$$

and we define

$$\begin{aligned}
\mathcal{D}_\infty - \int F(z)dE(z)\varphi &:= \mathcal{D}_\infty - \lim_{n \to \infty} \int F_n(z)dE(z)\varphi \\
&= \mathcal{D}_\infty - \lim_{n \to \infty} F_n(T)\varphi.
\end{aligned} \tag{4.6}$$

THEOREM 4.3 *Given a function* $F : \mathbf{C} \to \mathbf{C}$ *satisfying (4.4) and cut-off functions* F_n *defined by (4.5), then (4.6) makes sense, and*

$$\mathcal{D}_\infty - \int F(z)dE(z)\varphi = F(T)\varphi.$$

Proof: For any $\varphi \in \mathcal{D}_\infty$,

$$\begin{aligned}
\|F(T)\varphi - F_n(T)\varphi\|_k^2 &= \sum_{r=0}^{k} \|T^r(F(T)\varphi - F_n(T)\varphi)\|^2 \\
&= \sum_{r=0}^{k} \|F(T)T^r\varphi - F_n(T)T^r\varphi\|^2 \to 0 \text{ as } n \to \infty
\end{aligned}$$

since, for all $\varphi \in \mathcal{D}_\infty$ and $r \in \mathbf{N}_0$,

$$F(T)T^r\varphi = \int F(z)dE(z)T^r\varphi = H - \lim_{n\to\infty} \int F_n(z)dE(z)T^r\varphi = H - \lim_{n\to\infty} F_n(T)T^r\varphi.$$

Thus,

$$F_n(T)\varphi \to F(T)\varphi \text{ in } \mathcal{D}_\infty.$$

We now turn our attention to the space $\mathcal{D}'_\infty$. Firstly, we extend the spectral measure E to $\mathcal{D}'_\infty$.

DEFINITION 4.4 For any Borel set $J \subset \mathbf{C}$ the extension of $E(J)$ to $\mathcal{D}'_\infty$, denoted by $\widetilde{E}(J) : \mathcal{D}'_\infty \to \mathcal{D}'_\infty$ is defined by

$$\langle \widetilde{E}(J)f, \varphi \rangle := \langle f, E(J)\varphi \rangle \quad f \in \mathcal{D}'_\infty, \; \varphi \in \mathcal{D}_\infty.$$

It is clear that $\widetilde{E}$ is an operator valued measure defined on the Borel subsets of $\mathbf{C}$.

Following our development of spectral integrals in $\mathcal{D}_\infty$, we now consider integrals in $\mathcal{D}'_\infty$. For a step function $s(z) = \sum_{j=1}^n c_j \mathcal{X}_{J_j}(z)$ we define $\int s(z)d\widetilde{E}(z) : \mathcal{D}'_\infty \to \mathcal{D}'_\infty$ by

$$\langle \int s(z)d\widetilde{E}(z)f, \varphi \rangle := \langle \sum_{j=1}^n c_j \widetilde{E}(J_j)f, \varphi \rangle = \langle f, \sum_{j=1}^n c_j E(J_j)\varphi \rangle$$
$$= \langle f, \int s(z)dE(z)\varphi \rangle = \langle f, s(T)\varphi \rangle \quad f \in \mathcal{D}'_\infty, \; \varphi \in \mathcal{D}_\infty.$$

For a bounded E-measurable function F and any $\varphi \in \mathcal{D}_\infty$, we have, as before, a bounded sequence of step functions $(s_n)_{n=1}^\infty$ converging to F in $L^2(\mathbf{C}, E_{T^r\varphi})$ and a.e. with respect to $E_{T^r\varphi}$ for all $r \in \mathbf{N}_0$. We define $\int F(z)d\widetilde{E}(z) : \mathcal{D}'_\infty \to \mathcal{D}'_\infty$ by

$$\langle \int F(z)d\widetilde{E}(z)f, \varphi \rangle := \lim_{n\to\infty} \langle \int s_n(z)d\widetilde{E}(z)f, \varphi \rangle$$
$$= \lim_{n\to\infty} \langle f, \int s_n(z)dE(z)\varphi \rangle$$
$$= \langle f, \lim_{n\to\infty} \int s_n(z)dE(z)\varphi \rangle \quad \text{by the continuity of } f$$
$$= \langle f, \int F(z)dE(z)\varphi \rangle = \langle f, F(T)\varphi \rangle.$$

Finally, for an unbounded E-measurable function F satisfying (4.4), we use the cut-off functions F_n given in (4.5) and define $\int F(z)d\widetilde{E}(z) : \mathcal{D}'_\infty \to \mathcal{D}'_\infty$ by

$$\left\langle \int F(z)d\widetilde{E}(z)f, \varphi \right\rangle := \lim_{n \to \infty} \left\langle \int F_n(z)d\widetilde{E}(z)f, \varphi \right\rangle$$

$$= \lim_{n \to \infty} \langle f, F_n(T)\varphi \rangle$$

$$= \langle f, \lim_{n \to \infty} F_n(T)\varphi \rangle \quad \text{by the continuity of } f$$

$$= \langle f, F(T)\varphi \rangle = \langle f, \int F(z)dE(z)\varphi \rangle$$

since f is continuous and $F_n(T)\varphi \to F(T)\varphi$ in $\mathcal{D}_\infty$.

With this definition of integration in $\mathcal{D}'_\infty$ we can now define a functional calculus in $L(\mathcal{D}'_\infty)$; this, of course, must be consistent with Definition 3.9.

DEFINITION 4.5 For any E-measurable function $F : \mathbf{C} \to \mathbf{C}$ which satisfies (4.4), we define

$$F(\widetilde{T}) := \int F(\overline{z})d\widetilde{E}(z).$$

Thus,

$$\langle F(\widetilde{T})f, \varphi \rangle = \left\langle \int F(\overline{z})d\widetilde{E}(z)f, \varphi \right\rangle = \langle f, \int F(\overline{z})dE(z)\varphi \rangle$$

$$= \langle f, F(T^*)\varphi \rangle \quad f \in \mathcal{D}'_\infty, \ \varphi \in \mathcal{D}_\infty,$$

that is

$$F(\widetilde{T}) = (F(T^*))', \tag{4.7}$$

the $\mathcal{D}_\infty$-adjoint of $F(T^*)$. As such, $F(\widetilde{T})$ is a continuous operator on $\mathcal{D}_\infty$. In particular,

$$\widetilde{T} = \int \overline{z}d\widetilde{E}(z) = (T^*)'$$

in agreement with Definition 3.10.

That we have a good functional calculus in $L(\mathcal{D}'_\infty)$ is shown by the following:

THEOREM 4.6 *Let F and G be E-measurable functions on $\mathbf{C}$ which satisfy (4.4) for suitable values of m and M. Then,*
(a) $(\alpha F + \beta G)(\widetilde{T}) = \alpha F(\widetilde{T}) + \beta G(\widetilde{T}) \in L(\mathcal{D}'_\infty) \ \forall \alpha, \ \beta \in \mathbf{C}$;
(b) $(FG)(\widetilde{T}) = F(\widetilde{T})G(\widetilde{T}) \in L(\mathcal{D}'_\infty)$.

Proof: These results are straightforward calculations from (4.7).

5. CONCLUSION

For examples that illustrate the application of distributional spectral theory to initial-boundary value problems we refer to [3] where evolution equations involving self-adjoint operators and distributional initial data are considered. The theory developed here enables a similar treatment to be given to evolution equations involving normal operators.

Work currently in progress is considering a normal realisation of the differential operator $x\dfrac{d}{dx}$. Preliminary results indicate that the spaces $\mathcal{D}_\infty$ and $\mathcal{D}'_\infty$ generated by this operator correspond to certain classical spaces of test functions and distributions.

REFERENCES

[1] I.M. Gel'fand and G.E. Shilov, "Generalized Functions", Vol2, Academic Press, New York (1968).

[2] D. Judge, On Zemanian's distributional eigenfunction transforms, *J. Math. Anal. Appl.*, 34: 187–201 (1971).

[3] W. Lamb and D.F. McGhee, Spectral theory and functional calculus on spaces of generalized functions, *J. Math. Anal. Appl.*, 163: 238–260 (1992).

[4] W. Rudin, "Functional Analysis", McGraw-Hill, New Delhi (1973).

[5] J. Weidmann, "Linear Operators in Hilbert Space", Springer, New York (1980).

[6] A. Zemanian, "Generalized Integral Transformations", Interscience, New York (1968).

Semigroup Methods for Nonautonomous Cauchy Problems

Rainer Nagel   Mathematisches Institut, Universität Tübingen, Tübingen, Germany

ABSTRACT: It is well known that – by adding a new variable – a non-autonomous Cauchy problem

$$\dot{u}(t) = A(t)u(t), \quad u(s) = x_s \qquad\qquad (nCP)$$

can be made into an autonomous problem, which then can be solved by semigroup methods. We survey some recent results on wellposedness and hyperbolicity using this approach.

1 SEMIGROUPS SOLVE AUTONOMOUS CAUCHY PROBLEMS (AND VICE VERSA)

It was G. Peano who in 1888 had the courage to write the solutions of the Cauchy problem for a system of linear ordinary differential equations with constant coefficients as the exponential function

$$t \mapsto e^{tA} := \sum_{n=0}^{\infty} \frac{t^n A^n}{n!}$$

of the coefficient matrix A. The case of linear partial differential equations required an extension of that definition to infinite dimensional spaces and unbounded linear operators. The cornerstones for such an extension were laid in the 40's by E. Hille, K. Yosida and

---

This paper is part of a research project supported by the Deutsche Forschungsgemeinschaft

many others. By now there exists a fairly complete theory being elegant and applicable at the same time. We refer to the standard monographs (e.g., [Go], [Na], [Pa]) and quote two main results dealing with the existence and the qualitative behavior of the solutions of the autonomous Cauchy problem

$$\begin{aligned} \dot{u}(t) &= Au(t), \quad t \geq 0, \\ u(0) &= x_0. \end{aligned} \qquad (aCP)$$

Here and in the following X is a complex Banach space, $u(\cdot)$ a function from $\mathbb{R}_+$ into X and A a linear operator on X with domain $D(A)$. We call the Cauchy problem (aCP) *well posed* if there exist unique solutions for sufficiently many initial values x_0 and if these solutions depend continuously on the initial values (see [Neu] or [Go] for a more precise statement). In that case the solution of (aCP) is given by

$$u(t) := T(t)x_0,$$

where $(T(t))_{t \geq 0}$ is a *strongly continuous semigroup* of bounded, linear operators on X. Conversely, every such semigroup has a *generator*

$$Ax := \lim_{t \downarrow 0} \frac{1}{t}(T(t)x - x)$$

with domain $D(A) := \{x \in X : \lim_{t \downarrow 0} \frac{1}{t}(T(t)x - x) \text{ exists } \}$ and $(T(t))_{t \geq 0}$ *solves* the (aCP) (or: is the *solution semigroup*) corresponding to this operator A. These facts might be expressed briefly as follows.

WELL POSEDNESS THEOREM 1.1 *Every well posed Cauchy problem (aCP) is solved by a strongly continuous semigroup and every such semigroup is the solution semigroup of a well posed Cauchy problem.*

All this is explained in detail in the standard books on operator semigroups and the famous Hille-Yosida theorem then characterizes those operators A yielding well posed Cauchy problems (or equivalently: strongly continuous semigroups).

Having established well posedness of (aCP) it is most interesting to discuss the qualitative behavior of the solutions, i.e., of the solution semigroup. Clearly there is a broad range of possible phenomena and we concentrate here on a notion whose fundamental importance for linear and nonlinear dynamical systems is already evident from the finite dimensional situation (see, e.g., [Am]).

DEFINITION 1.2 *A strongly continuous semigroup $(T(t))_{t \geq 0}$ on the Banach space X is called* hyperbolic *if X can be written as a direct sum $X = X_0 \oplus X_1$ of $(T(t))$-invariant, closed subspaces X_0, X_1 such that the restrictions $T_0(t) := T(t)|_{X_0}$ and $T_1(t) := T(t)|_{X_1}$ satisfy the following: The operators $T_1(t)$ are invertible on X_1 and the semigroups*

$$(T_0(t))_{t \geq 0} \quad and \quad (T_1(t)^{-1})_{t \geq 0}$$

are exponentially stable, i.e. there exists $M \geq 1$, $\epsilon > 0$ such that for all $t \geq 0$

$$\|T_0(t)\| \leq Me^{-\epsilon t} \quad and \quad \|T_1(t)^{-1}\| \leq Me^{-\epsilon t}.$$

As in the finite dimensional case one is now looking for characterizations of hyperbolicity in terms of the spectrum $\sigma(A)$ of the generator A. One might hope that – as in finite dimensions, see [Am] – the condition

$$\text{``}\sigma(A) \cap i\mathbb{R} = \emptyset\text{''}$$

will do the job. While this is not true in general (see [Na], A-III for counterexamples), one understands the situation quite well and has the following result.

HYPERBOLICITY THEOREM 1.3   *For a strongly continuous semigroup $(T(t))_{t\geq 0}$ with generator A consider the following statements.*

(a) $(T(t))_{t\geq 0}$ *is hyperbolic.*

(b) $\sigma(A) \cap i\mathbb{R} = \emptyset$.

Then (a) always implies (b), while (b) implies (a) if the spectral mapping theorem

$$\sigma(T(t)) \setminus \{0\} = e^{t \cdot \sigma(A)}, \quad t \geq 0, \tag{SMT}$$

holds.

Proof.   By standard functional calculus one shows that hyperbolicity of $(T(t))_{t\geq 0}$ is equivalent to

(a*) $\sigma(T(t_0)) \cap \Gamma = \emptyset$

for one (hence, for all) $t_0 > 0$ (here Γ denotes the unit circle). We then use that for strongly continuous semigroups the spectral radius of the operators $T(t)$ satisfies

$$r(T(t)) = e^{\omega t}, \quad t \geq 0,$$

where ω is the *growth bound* of $(T(t))_{t\geq 0}$. See [Na], A-III for details.

Now (a) implies (b) by the Spectral Inclusion Theorem ([Na], A-III.6.2) and the converse holds under the assumption (SMT). $\square$

From this and A-III, Theorem 6.6 in [Na] we obtain a large class of semigroups fulfilling our expectations.

COROLLARY 1.4   *If the semigroup $(T(t))_{t\geq 0}$ is eventually norm continuous then it is hyperbolic if and only if*

$$\sigma(A) \cap i\mathbb{R} = \emptyset.$$

These few results should be sufficient to show that for (aCP) the theory concerning well posedness and asymptotic behavior is clean and clear. This changes dramatically if we pass to non-autonomous Cauchy problems.

2 EVOLUTION FAMILIES SOLVING NON-AUTONOMOUS CAUCHY PROBLEMS

We now assume that the linear operator A varies with t and hence consider *non-autonomous Cauchy problems*

$$\begin{aligned} \dot{u}(t) &= A(t)u(t) \text{ for } t \geq s \in \mathbb{R}, \\ u(s) &= x_s. \end{aligned} \qquad (nCP)$$

It is clear that the solution of (nCP) at "time" t will depend on the "initial time" s and the "initial value" x_s. The solutions – if they exist uniquely and depend continuously on these initial data – are therefore obtained by a family of operators satisfying the following.

DEFINITION 2.1   (i) *A family* $(U(t,s))_{s \leq t}$ *of bounded, linear operators on* X *is called a* strongly continuous evolution family *(or:* propagator*) if*

(i) $U(t,r)U(r,s) = U(t,s), U(s,s) = Id$ *for* $s \leq r \leq t$,

(ii) $(t,s) \mapsto U(t,s)$ *is strongly continuous and*

(iii) $\|U(t,s)\| \leq Me^{\omega(t-s)}$ *for some* $M \geq 1$, $\omega \in \mathbb{R}$ *and all* $s \leq t$.

(ii) *The evolution family* $(U(t,s))_{s \leq t}$ *is then said to* solve *(nCP) (or: to be the* solution family*) if* $u(\cdot) := U(\cdot,s)x_s$ *is the unique solution of (nCP) for every* $x_s \in D(A(s))$, *i.e.,* $u(\cdot)$ *is differentiable,* $u(t) \in D(A(t))$ *for* $t \geq s$ *and (nCP) holds.*

Observe that every strongly continuous semigroup $(T(t))_{t \geq 0}$ defines an evolution family by

$$U(t,s) := T(t-s),$$

but that the exponential boundedness (iii), which is automatic for semigroups, does not follow from (i) and (ii).

Moreover the reader should not be disturbed by the fact that most authors (e.g., [Fa], [Da-Ko]) consider evolution families for $0 \leq s \leq t$ only. Defining

$$V(t,s) := U(\max(t,0), \max(s,0))$$

one easily obtains a canonical extension of the evolution family to all $t \geq s \in \mathbb{R}$.

If all operators $A(t)$ are bounded it is quite well understood under which additional assumptions on the function

$$t \mapsto A(t)$$

one can obtain an evolution family solving (nCP). This evolution family will then be norm continuous and is usually obtained by a (Banach) fixed point argument. We refer to [Da-Kr] or [Fa] for the details.

For unbounded operators $A(t)$ the situation is quite complex. Following the fundamental papers by Kato, Sobolevskii and Tanabe in the 50's and 60's there are available by now numerous different conditions on $A(t)$, $t \geq 0$, implying wellposedness of (nCP). We subsume all this conditions under the name "K-S-T type" and refer to [Da-Ko], [Fa] or [Ac-Te] for precise statements.

WELL POSEDNESS PRINCIPLE 2.2   *If the operators $A(t)$ satisfy conditions of the "K-S-T type" then the nonautonomous Cauchy problem*

$$\begin{aligned} \dot{u}(t) &= A(t)u(t) \ \ for \ t \geq s \in \mathbb{R}, \\ u(s) &= x_s, \end{aligned} \qquad (nCP)$$

is well posed and the solutions are given by a strongly continuous evolution family $(U(t,s))_{s \leq t}$.

The reason for stating this "Principle" is to point out the difference to the Well Posedness Theorem 1.1 in the semigroup case. There the "Hille-Yosida Conditions" are necessary and sufficient for well posedness and to every strongly continuous semigroup corresponds a well posed Cauchy problem. Here the strong continuity of the evolution family does not imply any differentiability. This can already occur in the one-dimensional Banach space $X = \mathbb{C}$.

COUNTEREXAMPLE 2.3   Take a continuous function $p : \mathbb{R} \to [\frac{1}{2}, 1]$ and define

$$U(t,s) := \frac{p(t)}{p(s)}$$

for $s \leq t$. Then $(U(t,s))_{s \leq t}$ is a strongly continuous evolution family on $\mathbb{C}$ which is not differentiable if we choose p appropriately.

On the other hand, the above evolution family is differentiable if we can write it as

$$U(t,s) = \exp\left(\int_s^t a(\sigma)\, d\sigma \right)$$

for some locally integrable function $a(\cdot)$. Clearly this evolution family $(U(t,s))_{s \leq t}$ solves the corresponding non-autonomous Cauchy problem (nCP) almost everywhere.

Observe also that this representation formula for the evolution family does not extend to higher dimensions due to the noncommutativity of the operators $A(t)$.

Forced to leave open the characterization of well posedness of (nCP) we now have a short look at the qualitative behavior of strongly continuous evolution families $(U(t,s))_{s \leq t}$. Assuming differentiability, i.e., the existence of operators $A(t)$ such that the evolution family solves (nCP) and remembering the semigroup results (see Section 1) one might hope that the spectra $\sigma(A(t))$ have some influence on the asymptotic behavior of $\|U(t,s)\|$ as $t \to \infty$. Again this is far from being true.

COUNTEREXAMPLE 2.4   Take $X = \mathbb{C}^2$, $A := \begin{pmatrix} -1 & 2 \\ 0 & -1 \end{pmatrix}$, $A^* := \begin{pmatrix} -1 & 0 \\ 2 & -1 \end{pmatrix}$ and define

$$A(t) := \begin{cases} A & \text{for } 2n \leq t < 2n + 1 \\ A^* & \text{for } 2n + 1 \leq t < 2n + 2. \end{cases}$$

Writing $T(t) := \begin{pmatrix} e^{-t} & 2te^{-t} \\ 0 & e^{-t} \end{pmatrix}$ and $T(t)^* := \begin{pmatrix} e^{-t} & 0 \\ 2te^{-t} & e^{-t} \end{pmatrix}$ we obtain (some of) the evolution operators corresponding to $A(\cdot)$ as

$$U(t,0) = \begin{cases} T(t - 2n)[T(1)^*T(1)]^n & \text{for } 2n \leq t < 2n + 1 \\ T(t - 2n - 1)^*T(1)[T(1)^*T(1)]^n & \text{for } 2n + 1 \leq t < 2n + 2. \end{cases}$$

Since the spectral radius of $[T(1)^*T(1)]$ is greater than 1 it follows that $\lim_{n\to\infty} \|U(2n,0)\| = \lim_{n\to\infty} \|[T(1)^*T(1)]^n\| = \infty$, while $\sigma(A(t)) = \{-1\}$ for all t.

Consequently a simple spectral theory (in analogy to the situation in Section 1) seems not to be available for evolution families. But it is quite natural how to define "hyperbolicity".

DEFINITION 2.5   *A strongly continuous evolution family $(U(t,s))_{s\leq t}$ on X is called hyperbolic if for each $s \in \mathbb{R}$ the Banach space X decomposes into*

$$X = X_0(s) \oplus X_1(s)$$

such that the corresponding projections $P(s)$ are uniformly bounded, strongly measurable in s and the operators

$$U_0(t,s) := U(t,s)P(s) \quad and \quad U_1(t,s) := U(t,s)(Id - P(s))$$

satisfy the following: The operators $U_1(t,s)$ are invertible and there exist constants $M \geq 1$, $\epsilon > 0$ such that

$$\|U_0(t,s)\| \leq Me^{-\epsilon(t-s)} \quad and \quad \|U_1(t,s)^{-1}\| \leq Me^{-\epsilon(t-s)}$$

for $t \geq s$.

For periodic operator families $A(\cdot)$ "Floquet theory" is a spectral theory allowing an analysis of the qualitative behavior of evolution families (see, e.g., [Ko-Da]). We show how semigroup theory can be used to treat the general case.

3 EVOLUTION SEMIGOUPS ON FUNCTION SPACES

For a given Banach space X we consider the space

$$E := \mathcal{C}_0(\mathbb{R}, X)$$

of all continuous, X-valued functions on $\mathbb{R}$ vanishing at infinity. This is a Banach space for the supremum norm, and the (right) translations

$$T_0(t)f(s) := f(s - t) \quad \text{for} \quad s, t \in \mathbb{R},$$

define a strongly continuous (semi)group on E. This group is well understood and its generator G_0 is

$$G_0 f := -f' \quad \text{for} \quad D(G_0) := \{f \in E \cap \mathcal{C}^1(\mathbb{R}, X) : f' \in E\}.$$

We now ask in which way the translation semigroup $(T_0(t))_{t \geq 0}$ can be perturbed by "multiplication operators". More precisely, for which operator valued functions

$$(t, s) \mapsto U(t, s) \in \mathcal{L}(X)$$

does the definition

$$T(t)f(s) := U(s, s - t)f(s - t) \quad \text{for} \quad s \in \mathbb{R}, \quad t \geq 0, \tag{$*$}$$

yield a strongly continuous semigroup on E? It is easy to show that the semigroup property "$T(t + s) = T(t)T(s)$" is equivalent to "$U(t, r)U(r, s) = U(t, s)$". Looking then carefully at the continuity properties of $(T(t))_{t \geq 0}$ and $(U(t, s))_{s \leq t}$ one obtains the following characterization.

PROPOSITION 3.1 *With the above definitions $(T(t))_{t \geq 0}$ is a strongly continuous semigroup on E if and only if $(U(t, s))_{s \leq t}$ is a strongly continuous evolution family on X.*

The definition $(*)$, or similar ones, is based on the classical method of adding a new variable in order to convert a nonautonomous into an autonomous Cauchy problem. In infinite dimensional Banach spaces it has been proposed by, e.g., [Ev], [Ho], [Ne], [Pag]). Since the evolution family $(U(t, s))_{s \leq t}$ can be retrieved from the semigroup by applying $T(t)$ to functions being constant on large intervals the semigroup is in unique correspondence with the evolution family. We therefore use the following terminology.

DEFINITION 3.2 *The semigroup $(T(t))_{t \geq 0}$ on E defined by*

$$T(t)f(s) := U(s, s - t)f(s - t) \tag{$*$}$$

for $f \in E$, $s \in \mathbb{R}$, $0 \leq t$, is called the evolution semigroup *corresponding to the evolution family $(U(t, s))_{s \leq t}$ on X. If it is strongly continuous, we denote its generator by G with domain $D(G)$ in E.*

VARIATIONS 3.3   Given an evolution family $(U(t, s))_{s \le t}$ on X there are other canonical construcions of associated "evolution semigroups". We mention a few.

(i)   If $(U(t, s))_{s \le t}$ is norm continuous we can define

$$\tilde{T}(t)F(s) := U(s, s - t)F(s - t)$$

for operator valued functions $F \in \tilde{E} := C_0(\mathbb{R}_+, \mathcal{L}(X))$.

(ii)   Instead of functions on $\mathbb{R}$ one might consider functions on $\mathbb{R}_+$ and then take the left translations

$$T(t)f(s) := U(s, s + t)f(s + t)$$

if the evolution family $(U(t, s))$ is also defined for $t \le s$.

(iii)   Under a minor additional assumptions on the evolution family the definition $(*)$ can be extended to the space $C_{v_0}(\mathbb{R}, X)$ of all continuous functions f from $\mathbb{R}$ into X such that vf vanishes at infinity, where $v(s) = e^{-|s|}$, $s \in \mathbb{R}$, endowed with the following Norm (see [Na-Rh2])

$$\|f\|_v := \sup_{s \in \mathbb{R}} \|f(s)v(s)\| .$$

(iv)   Another extension is possible to the spaces $L^p(\mathbb{R}, X)$, $1 \le p < \infty$, when $(*)$ is taken almost everywhere. See [Ra1,2].

In this last case the strong continuity of $(U(t, s))_{s \le t}$ is not necessary in order to obtain strong continuity of the evolution semigroup. Therefore it might be reasonable to weaken condition (ii) in Definition 2.1 and adapt it to L^p-evolution semigroups. This is also suggested by the following counterexample which once again demonstrates the different behavior of evolution families compared to semigroups.

COUNTEREXAMPLE 3.4   Take X to be a reflexive Banach space and a function Q from $\mathbb{R}$ into the bounded, invertible operators on X. If Q and Q^{-1} are bounded and strongly continuous, but the adjoint function Q^* with values in $\mathcal{L}(X^*)$ is not strongly continuous then

$$U(t, s) := Q(t)Q(s)^{-1}$$

defines a strongly continuous evolution family on X such that the adjoint evolution family

$$V(t, s) := U(t, s)^*$$

is not strongly continuous on X^*.

As a concrete example one can take $X := L^2(\mathbb{R}_+)$ and

$$Q(s) := \begin{cases} Id - \frac{1}{2}V(\frac{1}{|s|}) & \text{for } s \ne 0 \\ Id & \text{for } s = 0, \end{cases}$$

where $V(t)f(r) := f(r + t)$ for $f \in X$, $t \in \mathbb{R}$, $r \geq 0$.

Having defined – in one way or another – an evolution semigroup associated to an evolution family and keeping in mind the powerful theory available for semigroups (see Section 1) it is now tempting to study the evolution semigroup in order to deduce information on the evolution family. In particular this might be a way to overcome the pitfalls indicated in Counterexample 2.3 and 2.4. So it seems to be worthwhile to study the following problems.

PROBLEM 3.5 Let $(T(t))_{t \geq 0}$ be a strongly continuous evolution semigroup on some appropriate function space. Characterize the generator G of $(T(t))_{t \geq 0}$. In particular, find conditions implying the existence of operators $A(t)$, such that

$$Gf(s) = -f'(s) + A(s)f(s)$$

in a properly choosen domain. Finally show in what sense the corresponding evolution family solves a non-autonomous Cauchy problem (nCP).

Besides this question our interest focusses again on the relation of spectral theory and asymptotic behavior. Among the many interesting questions we state the following.

PROBLEM 3.6 Let $(T(t))_{t \geq 0}$ be a strongly continuous evolution semigroup and let G be its generator. Find conditions on the spectrum $\sigma(G)$ characterizing hyperbolicity of the evolution semigroup (in the sense of Definition 1.2) and of the evolution family (in the sense of Definition 2.5).

In the next two sections we indicate some recent answers to the above questions.

4 GENERATORS OF EVOLUTION SEMIGROUPS

An evolution semigroup $(T(t))_{t \geq 0}$ is a kind of multiplicative perturbation of the translation semigroup $(T_0(t))_{t \geq 0}$ by an evolution family $(U(t, s))_{s \leq t}$. Intuitively it might be expected that its generator is an additive perturbation of the generator of the translation semigroup by an operator valued function. Counterexample 2.3 showed that we have to be careful with such an assertion and we therefore proceed differently.

Let $G_0 := -\frac{d}{ds}$ be the generator of the translation semigroup in the X-valued funtion space $E := \mathcal{C}_{v_0}(\mathbb{R}, X)$. Every operator valued function $A(\cdot) \in \mathcal{C}_b(\mathbb{R}, \mathcal{L}_s(X))$, i.e., $A(\cdot)x \in \mathcal{C}_b(\mathbb{R}, X)$ for every $x \in X$, defines a bounded linear operator in E by

$$(A(\cdot)f)(s) := A(s)f(s) \quad \text{for} \quad s \in \mathbb{R}.$$

Therefore the sum

$$G := G_0 + A(\cdot)$$

is the generator of a strongly continuous semigroup which turns out to be an evolution semigroup.

PROPOSITION 4.1 *Let $A(\cdot) \in C_b(\mathbb{R}, \mathcal{L}_s(X))$. Then there exists a strongly continuous evolution family $(U(t,s))_{s \leq t}$ such that*

$$G := G_0 + A(\cdot) \quad \text{with domain} \quad D(G) = D(G_0)$$

is the generator of the corresponding evolution semigroup.

Proof. That G is a generator follows from the bounded perturbation theorem (see [Go], Theorem I.6.4). By the Dyson-Phillips expansion ([Go], Theorem I.6.5) the new semigroup $(T(t))_{t \geq 0}$ is given by

$$T(t) = \sum_{n=0}^{\infty} T_n(t),$$

Where $T_n(t)f := \int_0^t T_0(t - \sigma)A(\cdot)T_{n-1}(\sigma)f \, d\sigma$ for every $f \in E$ and $n \in \mathbb{N}$. If $T_n(t)$ is of the form

$$T_n(t)f(s) = Q_n(s, s - t)f(s - t) \quad \text{for} \quad f \in E, \quad s \in \mathbb{R}$$

and for some function $Q_n(\cdot, \cdot - t) \in C_b(\mathbb{R}, \mathcal{L}_s(E))$ we obtain

$$
\begin{aligned}
T_{n+1}(t)f(s) &= \int_0^t T_0(t - \sigma)A(s)Q_n(s, s - \sigma)f(s - \sigma) \, d\sigma \\
&= \int_0^t A(s - t - \sigma)Q_n(s - t - \sigma, s - t) \, d\sigma \cdot f(s - t) \\
&=: Q_{n+1}(s, s - t)f(s - t).
\end{aligned}
$$

From the convergence of $\sum_{n=0}^{\infty} T_n(t)$ in operator norm it follows that

$$U(t, s) := \sum_{n=0}^{\infty} Q_n(t, s)$$

is well defined and hence $T(t)f(s) = U(s, s - t)f(s - t)$ for every $f \in E$. Since $(T(t))_{t \geq 0}$ is a strongly continuous semigroup it follows from Proposition 3.1 that $(U(t,s))_{s \leq t}$ is a strongly continuous evolution family. $\square$

Differentiating $t \mapsto T(t)(\mathbf{1} \otimes x)$ for the constant functions $\mathbf{1} \otimes x$, which all belong to the domain $D(G)$, we immediately see that the evolution family $(U(t,s))_{s \leq t}$ solves the (nCP) corresponding to $A(t)$. So well posedness has been obtained from the bounded perturbation theorem of semigroup theory. Applying the Chernoff product formula (see

[Go], Theorem I.8.4) to operators $V(t) \in \mathcal{L}(E)$ defined as

$$(V(t)f)(s) := \exp\left(\int_{s-t}^{s} A(\tau)\, d\tau\right) f(s-t)$$

we even arrive at a formula for the evolution family.

COROLLARY 4.2 *Under the above assumptions the evolution family* $(U(t,s))_{s \leq t}$ *is given by*

$$U(s, s-t)x = \lim_{n \to \infty} \prod_{k=0}^{n-1} \exp\left(\int_{\frac{kt}{n}}^{\frac{(k+1)t}{n}} A(s-\tau)\, d\tau\right)x$$

for all $x \in X$.

The evolution families we obtained in this way are all norm continuous and differentiable and we refer to [Na-Rh2] for more details. Here we pursue the idea of applying perturbation theorems in order to obtain evolution semigroups and hence evolution families.

THEOREM 4.3 *Let the Banach space* X *have the Radon-Nikodym property and take* $A(\cdot) : \mathbb{R} \to \mathcal{L}(X)$ *such that* $A(\cdot)x \in L^{\infty}(\mathbb{R}, X)$ *for all* $x \in X$. *Then the operator*

$$\begin{aligned} Gf &:= -f' + A(\cdot)f \ \ \text{with domain} \\ D(G) &:= \{f \in E : f \ \text{a.e. differentiable,} \ f' \in L^{\infty}(\mathbb{R}, X) \ \text{and} \ f' + A(\cdot)f \in E\} \end{aligned}$$

is the generator of a strongly continuous evolution semigroup on E.

Proof. Since X has the Radon-Nikodym property one shows as in [Na-Si] that $L^{\infty}(\mathbb{R}, X)$ is the extrapolated Favard class of the translation semigroup on E. The operator $A(\cdot)$ is now bounded from E into $L^{\infty}(\mathbb{R}, X)$, hence it satisfies the assumptions of the Desch-Schappacher perturbation theorem (see [De-Sch]). This proves that G is a generator. The assertion that the semigroup is an evolution semigroup can be shown as in Proposition 4.1. $\square$

Again, a more detailed discussion of the properties of the evolution family corresponding to the above evolution semigroup can be found in [Na-Rh2]. In a subsequent paper we will treat functions $A(\cdot)$ of unbounded operators $A(t)$ and show how they yield semigroup generators of the form

$$Gf := -f' + A(\cdot)f$$

on X-valued function spaces.

5 HYPERBOLIC EVOLUTION SEMIGROUPS

In this section we look for a spectral characterization of hyperbolicity (see Definition 2.5) of evolution families and report on some recent work by Y. Latushkin-A. Stepin and R. Rau in which they generalized earlier results (see [Da-Kr], [BA-Go-Ka]) to infinite dimensional Banach spaces. It turns out that again the corresponding evolution semigroup and its generator yield the appropriate tools.

So let $(U(t,s))_{s \leq t}$ be a strongly continuous evolution family on X and denote by $(T(t))_{t \geq 0}$ its evolution semigroup on any of the X-valued function spaces E mentioned in Section 3. Then the spectra of the operators $T(t)$ as well as the spectrum of the generator G have remarkable symmetry properties.

PROPOSITION 5.1   *For the evolution semigroup $(T(t))_{t \geq 0}$ and its generator G the following holds.*

(i) $\sigma(T(t)) = \Gamma \cdot \sigma(T(t))$ *for all $t > 0$.*

(ii) $\sigma(G) = \sigma(G) + i\mathbb{R}$.

Proof.   Each function ϵ_μ for $\mu \in \mathbb{R}$ with $\epsilon_\mu(s) := e^{i\mu s}$, $s \in \mathbb{R}$, defines an invertible multiplication operator M_μ on the function space E. It then follows that

$$M_{-\mu}T(t)M_\mu = e^{i\mu t} \cdot T(t) \quad \text{and} \quad M_{-\mu}GM_\mu = G + i\mu,$$

which proves the assertions. $\square$

It is even more remarkable that the evolution semigroups always satisfy the Spectral Mapping Theorem (SMT) (see Theorem 1.4). In the infinite dimensional case this has been proved by R. Rau [Ra1] for Hilbert spaces. Recently, Y. Latushkin-S. Montgomery-Smith announced a proof for Banach spaces in [La-Mo].

SPECTRAL MAPPING THEOREM 5.2   (Latushkin-Montgomery) *Let $(T(t))_{t \geq 0}$ be an evolution semigroup with generator G on $L^p(\mathbb{R}, X)$. Then*

$$\sigma(T(t)) \setminus \{0\} = e^{t \cdot \sigma(G)} \qquad\qquad (SMT)$$

for all $t \geq 0$.

As a consequence of the two results above we obtain that the spectrum $\sigma(G)$, and consequently $\sigma(T(t))$, is completely determined by the set

$$\Sigma := \sigma(G) \cap \mathbb{R}.$$

In particular if some real number μ is not contained in Σ then the circle

$$e^{i\mu} \cdot \Gamma$$

belongs to the resolvent set of the operators $T(t)$ and separates $\sigma(T(t))$ into two disjoint subsets. Therefore there exists a spectral projection splitting E into the direct sum of two $(T(t))$-invariant subspaces. If the above holds for $\mu = 0$ we obtain that $(T(t))_{t\geq 0}$ is a hyperbolic semigroup on E (according to Definition 1.3). Still it is not evident that this spectral projection is of the form as required in Definition 2.5 for hyperbolic evolution families. At least for invertible evolution families R. Rau was able to obtain the desired spectral characterization (see [Ra1-3]).

HYPERBOLICITY THEOREM 5.3 (Rau) *For a strongly continuous, invertible evolution family $(U(t,s))_{t,s\in\mathbb{R}}$ on X and its evolution group $(T(t))_{t\in\mathbb{R}}$ on $E = C_0(\mathbb{R}, X)$ the following assertions are equivalent*

(i) The evolution family is hyperbolic.

(ii) 0 does not belong to the spectrum of the generator of the evolution group.

As an immediate consequence we observe that the evolution family $(U(t,s))_{t,s\in\mathbb{R}}$ is *exponentially stable*, i.e.,

$$\|U(t,s)\| \leq Me^{-\epsilon(t-s)} \text{ for } t \geq s \text{ and constants } M \geq 1, \quad \epsilon > 0$$

if and only if

$$\sigma(G) \cap [0, \infty) = \emptyset$$

for the generator G of the evolution group.

If we assume in addition X to be a separable Hilbert space (and write H instead) we can simplify this characterization again.

To this purpose we use notions and results from the theory of von Neumann algebras (see [Sa]). Consider the von Neumann algebra $M := \mathcal{L}(H)$ and its predual M_* consisting of all trace class operators on H. Then the space $L^1(\mathbb{R}, M_*)$ of all Bochner-integrable functions with values in M_* is isomporphic to the (completed) tensor product $L^1(\mathbb{R})\bar{\otimes}M_*$. Its dual is the tensor product $L^\infty(\mathbb{R})\bar{\otimes}M$ which will be denoted by $L^\infty(\mathbb{R}, M)$ and which can be identified with the weak*-measurable, bounded functions from $\mathbb{R}$ into M. On this space we define a new evolution semigroup

$$\tilde{T}(t)F(s) := U(s, s - t)F(s - t)U(s, s - t)^*$$

for $F \in L^\infty(\mathbb{R}, M)$ and $s, t \in \mathbb{R}$. The operators $\tilde{T}(t)$ are positive for the canonical order in the von Neumann algebra $L^\infty(\mathbb{R}, M)$ and the function $t \mapsto \tilde{T}(t)$ is *weak**-continuous. Therefore the resolvent integral

$$\int_0^\infty e^{-\lambda t}\tilde{T}(t)\, dt$$

exists for large $\lambda \in \mathbb{R}$ and corresponds to a "resolvent positive" operator $\tilde{G}$ (see [Ar] or [Na-Rh1]). To this operator we now apply the theorem from [Na-Rh1] and obtain our final result.

STABILITY THEOREM 5.4   *Let $(U(s,t))_{s,t\in\mathbb{R}}$ be a strongly continuous evolution family on the separable Hilbert space H and consider the evolution (semi)groups $(T(t))_{t\in\mathbb{R}}$ on $L^2(\mathbb{R}, H)$ and $(\tilde{T}(t))_{t\in\mathbb{R}}$ on $L^\infty(\mathbb{R}, \mathcal{L}(H))$ with generators G and $\tilde{G}$, resp.. Then the following assertions are equivalent.*

(i)   *The evolution family $(U(s,t))_{s,t\in\mathbb{R}}$ is exponentially stable.*

(ii) $\sigma(G) \cap [0,\infty) = \emptyset.$

(iii) $0 \notin \sigma(\tilde{G})$ *and* $\tilde{G}^{-1} \leq 0.$

(iv)   *There exists* $0 \leq \tilde{F} \in D(\tilde{G})$ *such that*

$$\tilde{G}\tilde{F}(s) = -Id_H \quad \text{for a.e. } s \in \mathbb{R}.$$

FINAL COMMENT If the generator G is of the form $-\frac{d}{ds} + A(\cdot)$ for some operator family $A(\cdot)$ (see Section 4) then condition (iv) is equivalent to an infinite dimensional, linear Riccati equation

$$-\frac{d}{ds}\tilde{F}(s) + A(s)F(s) + F(s)A(s)^* = -Id_H.$$

See [Na-Rh1] for details and also the articles by Chicone-Latushkin [Ch-La1], [Ch-La2].

REFERENCES

[**Ac-Te**] Acquistapace, P. and Terreni, B.: A unified approch to abstract linear nonautonomous parabolic equations. Rend. Sem. Mat. Univ. Padova **78** (1987), 47-107.

[**Am**] Amann, H.: Gewöhnliche Differentialgleichungen. de Gruyter, Berlin 1983.

[**Ar**] Arendt, W.: Resolvent positive operators. Proc. London Math. Soc. **54** (1984), 321-349.

[**Be-Go-Ka**] Ben-Artzi, A., Gohberg, I. and Kaashoek, M.A.: Invertibility and dichotomy of differential operators on a half line. Preprint.

[**Ch-La1**] Chicone, C. and Latushkin, Y.D.: Quadratic Lyapunov functions for linear skew-product flows and weighted composition operators. Diff. Int. Eq. (to appear).

[**Ch-La2**] Chicone, C. and Latushkin, Y.D.: Hyperbolicity and dissipativity. Preprint 1993.

[**Da-Kr**] Daleckii, J.L. and Krein, M.G.: Stability of Solutions of Differential Equations in Banach Space. Amer. Math. Soc. 1974.

[**Da-Ko**] Daners, D. and Koch Medina, P.: Abstract Evolution Equations, Periodic Problems and Applications. Pitman Research Notes Math. Ser. **279**. Longman 1993.

[**De-Sch**] Desch, W. and Schappacher, W.: On relatively bounded perturbations of linear C_0-semigroups. Ann. Scuola Norm. Sup. Pisa **11** (1984), 327-341.

[**Ev**] Evans, D.E.: Time dependent perturbations and scattering of strongly continuous groups on Banach spaces. Math. Ann. **221** (1976), 275-290.

[**Fa**] Fattorini, H.O.: The Cauchy Problem. Addison-Wesley 1983.

[**Go**] Goldstein, J.A.: Semigroup of Linear Operators and Applications. Oxford 1985.

[**Ho**] Howland, J.S.: Stationary scattering theory for time-dependent Hamiltonians. Math. Ann. **207** (1974), 315-335.

[**La-Mo**] Latushkin, Y.D. and Montgomery-Smith, S.: Lyapunov theorems for Banach spaces, Preprint 1993.

[**La-St**] Latushkin, Y.D. and Stepin, A.M.: Weighted translation operators and linear extensions of dynamical systems. Russian Math. Surveys **46** (1991), 95-165.

[**Na**] Nagel, R. (ed.): One-parameter Semigroups of Positive Operators. Lecture Notes Math. **1184**, Springer-Verlag 1986.

[**Na-Rh1**] Nagel, R. and Rhandi, A.: Positivity and Lyapunov stability conditions for linear systems. Adv. Math. Sci. Appl. (to appear).

[**Na-Rh2**] Nagel, R. and Rhandi, A.: A characterization of Lipschitz continuous evolution families on Banach spaces. J. Int. Eq. Op. Theory. (to appear).

[**Na-Si**] Nagel, R. and Sinestrari, E.: Inhomogeneous Volterra integrodifferential equations for Hille-Yosida operators. Proc. Conf. Funct. Analysis, Essen 1991.

[**Ne**] Neidhardt, H.: On abstract linear evolution equations, I. Math. Nachr. **103** (1981), 283-298.

[**Neu**] Neubrander, F.: Well-posedness of abstact Cauchy problems. Semigroup Forum **29** (1984), 75-85.

[**Paq**] Paquet, L.: Semigroupes géneralisés et équations d'evolution.

[**Pa**] Pazy, A.: Semigroups of Linear Operators and Applications to Partial Differential Equations. Springer-Verlag 1983.

[**Ra1**] Rau, R.T.: Hyperbolic Evolution Semigroups. Dissertation, Tübingen, 1992.

[**Ra2**] Rau, R.T.: Hyperbolic evolution groups and dichotomic evolution families. J. Funct. Anal. (to appear).

[Ra3] Rau, R.T.: Hyperbolic evolution semigroups on vector valued function spaces. Semigroup Forum (to appear).

[Sa] Sakai, S.: C^*-Algebras and W^*-Algebras. Springer-Verlag 1971.

On KLKH Systems of Scalar-type Spectral Operators

B. Nagy   Faculty of Chemical Engineering, Technical University of Budapest, Budapest, Hungary

1. INTRODUCTION

Let G be a (closed) spectral operator of scalar type in a complex Banach space X with resolution of the identity E and with spectrum $\sigma(G)$ such that $\{\mathrm{Re}\lambda : \lambda \in \sigma(G)\}$ is not nesessarily bounded from above. (For the fundamentals on spectral operators we shall refer to [3].) Let

$$T(t) = \int_{\sigma(G)} e^{t\lambda} E(d\lambda) \quad (t \geq 0). \qquad (*)$$

The functional calculus for scalar-type operators, in particular [3; XVIII.2.11 and XVIII. 2.18], shows that
(1) every $T(t)$ is a closed linear operator (in notation: $T(t) \in C(X)$) that is scalar-type spectral, and whose resolution of the identity E_t satisfies

$$\sup\{|E_t(\cdot)|; t \geq 0\} = M < \infty;$$

Supported by an OTKA grant in Hungary

(2) $T(t + s) = T(t)T(s)$ for any $t, s \geq 0$, and $T(0) = I$;

(3) for any $s > 0$, $x \in D(T(s))$ the function $t \mapsto T(t)x$ is continuous in the norm topology of X on the interval $[0, s]$ (in notation: $T(\cdot)x \in C([0, s], X)$).

The purpose of this paper is to show that if, in the converse direction, we assume that $\{T(t);\ \ t \geq 0\}$ is a family of operators satisfying (1), (2) and (3), then there is a closed scalar-type operator G with resolution of the identity E, for which $(*)$ holds.

Semigroups of (possibly unbounded) scalar-type operators with all spectra on the real line $\mathbf{R}$ were studied by Sourour [9; Theorem 9.1] under similar conditions, and a similar representation was obtained. Semigroups of unbounded normal operators in Hilbert space were considered by Devinatz and Nussbaum [2]. However, their methods do not seem to be extendable to the Banach space case. Therefore we shall make use of some ideas in [9] and the main representation result of Lange and Nagy [7] which, in turn, extends the main result of Berkson [1] to the case of an arbitrary (rather than a weakly complete) Banach space.

The abbreviation KLKH in the title refers to the papers by Klein and Landau [6] and by Kantorovitz and Hughes [4]. In the latter (and later) paper [4] the following definition of a KLKH system (in our terminology) was given.

Let $\Delta = [0, a)$, where $0 < a \leq \infty$, and for each $t \in \Delta$ let $P(t)$ be a linear transformation in X with domain D_t satisfying

(a) $\{D_t;\ \ t \in \Delta\}$ is a nonincreasing family of sets; $P(0) = I$;

(b) $P(t + s) \subset P(t)P(s)$ if $t, s, t + s \in \Delta$;

(c) $P(\cdot)x \in C([0, t], X)$ for each $x \in D_t$.

Then $\{P(t), D_t, \Delta\}$ will be called a KLKH system.

Kantorovitz and Hughes in [4; Theorem 2.3] obtained a necessary and sufficient condition for a KLKH system of closed operators with $\Delta = [0, \infty)$ and with a certain boundedness condition on the trajectories in a reflexive Banach space to have the property that the associated local semigroup is a spectral local semigroup. The latter assertion means that the local semigroup consists of restrictions of the operators e^{tG} for some closed scalar-type spectral operator G with real spectrum. Ricker [8] extended this result to the case of an arbitrary underlying Banach space. We want to emphasize the similarity of their problems with ours as well as the essential differences: their local treatment of the problem, the different assumptions owing to the special situation $\sigma(G) \subset \mathbf{R}$ and the completely different methods.

It is clear from the considerations above that a KLKH system of closed scalar-type operators with uniformly bounded resolutions of the identity for which $\Delta = [0, \infty)$ and $P(t + s) = P(t)P(s)$ $(t, s \geq 0)$ satisfies the conditions (1), (2) and (3) above, and the converse is also valid. The representation $(*)$ in Theorem 2 of such a KLKH system is the main result of this paper. However, the main technical problems lie in showing that the nonnegative and circled parts are also such KLKH systems (see Theorem 1 below). The relevant polar decomposition of an unbounded scalar-type operator and its uniqueness will be obtained in Lemma 1.

Most notations are standard (see, e.g., [3]), or explained in the text below. In particular, $B(X)$ will denote the algebra of all bounded linear operators in the complex Banach space X, and $|\cdot|$ the norm on it. A linear operator in $C(X)$ will be called nonnegative or circled if its spectrum is a subset of $[0, \infty)$ or the unit circle, respectively. As usual, $A \subset B$ for operators A, B will mean that B is an extension of A.

2. THE RESULTS.

We shall need the following

LEMMA 1. *Let T be a closed scalar-type operator in a Banach space X with resolution of the identity E. Then there are operators R, U such that*
(a) $T = RU = UR$,
(b) $U \in B(X)$ is circled and scalar-type,
(c) $R \in C(X)$ is nonnegative and scalar-type,
(d) $UE(0) = E(0)U = E(0)$.
Further, if there are operators R_1, U_1 such that
(1) $T = U_1 R_1 \subset R_1 U_1$,
(2) $U_1 \in B(X)$ is circled,
(3) $R_1 \in C(X)$ is nonnegative and scalar-type,
hold, then $R_1 = R$. Finally, if (1), (2), (3) and
(4) $U_1 E(0) = E(0)U_1 = E(0)$
hold, then $U_1 = U$. (Here $E(0)$ stands for $E(\{0\})$).
PROOF. Let $u(\lambda) = \lambda/|\lambda|$ for $\lambda \in \mathbf{C} \setminus \{0\}$, $u(0) = 1$, and let

$$R = \int |\lambda|\, E(d\lambda), \qquad U = \int u(\lambda)\, E(d\lambda).$$

Properties (a), (b) and (c) then follow from the properties of the operational calculus for closed scalar-type operators ([3; XVIII.2.11]) and from [3; XVIII.2.17, 25], whereas (d) is evident.

Assume now (1)-(3). Then $U_1 T \subset U_1 R_1 U_1 = T U_1$, hence U_1 commutes with $E(\cdot)$, and so does U_1^{-1}. This implies

$$E(\cdot)R_1 = E(\cdot)U_1^{-1}T = U_1^{-1}E(\cdot)T \subset U_1^{-1}TE(\cdot) = R_1 E(\cdot).$$

Hence the resolution of the identity E_1 of R_1 also commutes with E. From (1) we have $D(R_1) = D(T)$, which implies $E(b)X \subset D(R_1)$ for any bounded Borel set b in $\mathbf{C}$. For any $x \in E(b)X$

$$(R_1|E(b)X)x = \int \lambda E_1(d\lambda)x,$$

where the domain of integration may be the bounded set $\sigma(R_1|E(b)X)$, which is clearly contained in $\sigma(R_1) \subset [0,\infty)$. Since $\sigma(U_1|E(b)X) \subset \sigma(U_1)$ are contained in the unit circle C_1, the circled operator $U_1|E(b)X$, the bounded nonnegative scalar-type operator $R_1|E(b)X$ and $T|E(b)X \in B(E(b)X)$ satisfy (i), (ii)' and (iii) of [9; Lemma (4.1)]. By that result, $R_1|E(b)X$ is then the uniquely determined nonnegative part of $T|E(b)X$, i.e.

$$R_1|E(b)X = R|E(b)X.$$

By definition, $x \in D(R)$ if and only if, with the notation $b_n = \{\lambda \in \mathbf{C}; |\lambda| \leq n\}$, the limit $\lim_{n\to\infty} RE(b_n)x = Rx$ exists in the norm topology of X. Since R_1 is closed, then $R_1 x = Rx$. By (1) and (a), we have $D(R_1) = D(T) = D(R)$, therefore $R_1 = R$.

Assume now (1)-(4). For any bounded Borel set b then the restrictions to the subspace $E(b)X$ satisfy (i), (ii)', (iii) and (iv) of [9; Lemma (4.1)]. Hence U_1 and U agree on each $E(b)X$ and, since they are continuous, on all of X.

REMARK. The formulae for R and U show clearly that the projections in the resolutions of the identity for R and U are contained in the family

$$\{E(b); \, b \, \text{Borel set}\},$$

i.e. in the range of the resolution of the identity for T.

THEOREM 1. *Let $\{T(t); \, t \geq 0\}$ be a KLKH system of closed scalar-type operators in a Banach space X with uniformly bounded resolutions of the identity E_t, for which*

$$T(t+s) = T(t)T(s), \quad |E_t(\cdot)| \leq M \quad (t,s \geq 0).$$

Let $R(t)$ and $U(t)$ denote the nonnegative and circled parts of $T(t)$, respectively. Then $\{U(t)\}$ is a semigroup of class (C_o), and $\{R(t)\}$ is a KLKH system of closed scalar-type operators, for which

$$R(t+s) = R(t)R(s) \quad (t,s \geq 0),$$

$$R(t) = \int_{[0,\infty)} \lambda^t F(d\lambda) = \int_{\mathbf{R}} e^{t\lambda} H(d\lambda) \quad (t \geq 0).$$

Here F and H are strongly countably additive spectral measures on the Borel sets of $[0, \infty)$ and $\mathbf{R}$, respectively, satisfying $F(\{0\}) = 0$. Further, $\{T(t), R(t), U(t); \, t \geq 0\}$ is a *commutative family in the sense that the corresponding resolutions of the identity form a commutative family.*

PROOF. At first we prove that $\{E_t(\cdot);\ t \geq 0\}$ is a commutative family. Let $t \geq 0$, $n \in \mathbf{N}$. By assumption, then

$$T(t) = T(t/n)^n = \int_C \lambda^n\, E_{t/n}(d\lambda).$$

The last equality here follows from [3; XVIII.2.9 and 16]. For every Borel set b in $\mathbf{C}$, then $E_t(b) = E_{t/n}(\{\lambda \in \mathbf{C} : \lambda^n \in b\})$, by [3; XVIII.2.17]. Hence E_t commutes with $E_{t/n}$, and

$$\int_C \lambda E_{t/n}(d\lambda) E_t(\cdot) x = E_t(\cdot) \int_C \lambda E_{t/n}(d\lambda) x$$

for each x in X for which the right-hand "integral" exists. This means $T(t/n)E_t(\cdot) \supset E_t(\cdot)T(t/n)$, which implies for every $k \in \mathbf{N}$

$$T(tk/n)E_t(\cdot) \supset E_t(\cdot)T(tk/n).$$

For every $0 < s < t$ choose a sequence $\{r_n\}$ of positive rational numbers such that $\lim r_n t = s$ from the left, and let $x_s \in D(T(s))$. Then $x_s \in D(T(r_n t))$, and

$$\lim T(r_n t)E_t(\cdot)x_s = \lim E_t(\cdot)T(r_n t)x_s = E_t(\cdot)T(s)x_s.$$

Let b be any bounded Borel set in $\mathbf{C}$. Then $E_t(b)X \subset D(T(t)) \subset D(T(s))$, hence $T(s)E_t(b)x_s = E_t(b)T(s)x_s$. Let h be an arbitrary Borel set in $\mathbf{C}$, and $\{b_n\}$ an increasing sequence of bounded Borel sets with union h. Then

$$\lim E_t(b_n)x_s = E_t(h)x_s, \quad \lim T(s)E_t(b_n)x_s = E_t(h)T(s)x_s.$$

Since $T(s)$ is closed,

$$T(s)E_t(h)x_s = E_t(h)T(s)x_s.$$

This shows that E_t commutes with $T(s)$, hence also with its resolution of the identity E_s.

We have seen that $E_t(\{0\}) = E_{t/n}(\{0\})$ for any $t \geq 0$. Let $x = E_t(\{0\})x$. Then $T(t/n)x = 0$ for every $n \in \mathbf{N}$. Since $\{T(t)\}$ is a KLKH system, this implies $x = 0$. Hence $E_t(\{0\}) = 0$ for every $t \geq 0$.

Let F_t denote the resolution of the identity for $R(t)$, and let $F = F_1$. The preceding paragraph then implies $F_t(\{0\}) = 0$ for every $t \geq 0$. Let

$$S(t) = \int_{[0,\infty)} \lambda^t F(d\lambda) = \int_{(0,\infty)} \lambda^t F(d\lambda).$$

We have seen that $T(nt) = (T(t))^n$ for any $t \geq 0$ and any positive integer n. This implies as on p. 214 of [9] that $S(t) = R(t)$ for every rational $t \geq 0$.

Let $e_k = [1/k, k]$, $X_k = F(e_k)X$, $X_\circ = \bigcup_{k=1}^\infty X_k$. Then $X_\circ$ is dense in X, $X_\circ \subset D(S(t))$ and $X_\circ$ is a core for each $S(t)$, i.e. $S(t) = \overline{S(t)|X_\circ}$. Since E_1 is the resolution of the identity for $T(1)$,

$$X_k = E_1(\{\lambda \in \mathbf{C} : 1/k \leq |\lambda| \leq k\})X \subset D((T(1))^n) = D(T(n)) \quad (k \in \mathbf{N})$$

for every positive integer n. By assumption, $D(T(t))$ is a decreasing family of sets (with t increasing), hence $X_\circ \subset D(T(t))$ for every $t \geq 0$. Further, $T(t)$ commutes with $\{F(\cdot)\}$. Introduce the notations

$$S_k(t) = S(t)|X_k, \quad T_k(t) = T(t)|X_k \quad (t \geq 0, \ k = 0, 1, \ldots).$$

For $k = 1, 2, \ldots$ they are all bounded operators; moreover, the $S_k(t)$ all have inverses in $B(X_k)$. Hence $S_\circ(t) : X_\circ \to X_\circ$ is a bijection, and the linear operators

$$V_k(t) = T_k(t)S_k(t)^{-1} : X_k \to X_k \quad (t \geq 0; \ k = 0, 1, \ldots)$$

are well-defined. For rational r we have

$$V_\circ(r) = T_\circ(r)(R(r)|X_\circ)^{-1} = U(r)|X_\circ \subset U(r)$$

where $U(r) \in B(X)$ is the circled part of $T(r)$.

Let $t \in (0, \infty)$ be irrational, and let $\{r_n\}$ be a sequence of rational numbers converging to t. For $x_k \in X_k$

$$S_\circ(r_n)^{-1} x_k = \int_{1/k}^{k} \lambda^{-r_n} F(d\lambda)x_k \to \int_{1/k}^{k} \lambda^{-t} F(d\lambda)x_k = S_\circ(t)^{-1} x_k.$$

By assumption, the function $s \mapsto T_k(s)x_k$ is continuous on $[0, \infty)$, hence for every $\alpha > 0$

$$\sup\{|T_k(t)|; \ 0 \leq t \leq \alpha\} < \infty.$$

Therefore

$$U(r_n)x_k = T_k(r_n)S_\circ(r_n)^{-1} x_k \to T_k(t)S_\circ(t)^{-1} x_k = V_\circ(t)x_k.$$

By assumption, for any x in X and $s \geq 0$

$$|U(s)x| = \left| \int u(\lambda)\, E_s(d\lambda)x \right| \leq 4M|x|.$$

Since X_o is dense in X, [3; II.1.18] shows that $U(r_n)$ converges in the strong operator topology to some $V(t) \in B(X)$ for which $|V(t)| \le 4M$ and $V(t) = \overline{V_o(t)}$.

Since the resolutions of the identity commute, and $U(r_n)x \to V(t)x$ for $x \in X$, each $V(t)$ commutes with $T(s)$ and $S(s)$. For $x_k \in X_k$ then

$$V(t+s)x_k = T_k(t)T_k(s)S_k(s)^{-1}S_k(t)^{-1}x_k = V_k(s)V_k(t)x_k = V(t)V(s)x_k.$$

The operators $V(t)$ are bounded, hence $V(t+s) = V(t)V(s)$. Now it can be proved that each $V(t)$ is circled exactly as on p. 215 in [9].

For every $t \ge 0$ we have $T(t) = \overline{T_o(t)}$. Indeed, every projection $F(\cdot)$ commutes with $T(t)$, hence for each $z \in D(T(t))$

$$z = \lim_{k \to \infty} F(e_k)z, \quad T(t)z = \lim_{k \to \infty} F(e_k)T(t)z = \lim_{k \to \infty} T_o(t)F(e_k)z,$$

which proves our claim.

Since $V(t) \supset T_o(t)S_o(t)^{-1}$, we obtain $V(t)S_o(t) = T_o(t)$. Hence $T(t) = \overline{V(t)S_o(t)}$. Since $V(t)$ is circled, $V(t)^{-1} \in B(X)$, therefore $V(t)\overline{S_o(t)} = V(t)S(t) \in C(X)$ (cf. [5; III.5.7]). It follows that

$$T(t) \subset V(t)S(t).$$

On the other hand, if $x \in D(S(t))$, then there is a sequence $\{x_n\}$ in $D(S_o(t))$ such that

$$\lim_{n \to \infty} x_n = x, \quad \lim_{n \to \infty} S_o(t)x_n = S(t)x.$$

This implies $\lim_{n \to \infty} T_o(t)x_n = \lim V(t)S_o(t)x_n = V(t)S(t)x$. Since $T(t)$ is the closure of $T_o(t)$, we obtain $x \in D(T(t))$. Hence

$$T(t) = V(t)S(t).$$

Since $V(t)$ commutes with $S(t)$, we obtain

$$T(t) = V(t)S(t) \subset S(t)V(t) \quad (t \ge 0).$$

The uniqueness assertion of Lemma 1 shows that for $t \ge 0$ (recall that $E_t(\{0\}) = 0$) we have

$$U(t) = V(t), \quad R(t) = S(t) = \int \lambda^t F(d\lambda).$$

The latter equalities and the operational calculus for scalar operators show at once that $R(t+s) \supset R(t)R(s)$ and $D(R(t)R(s)) = D(R(t+s)) \cap D(R(s))$. Since $\lambda^{t+s} > \lambda^s$ for $\lambda > 1$, $D(R(t+s)) \subset D(R(s))$. Hence

$$R(t+s) = R(t)R(s) \quad (t, s \ge 0).$$

The above representation shows also that $R(t)$ satisfies the remaining requirements for a KLKH system.Since $F(\{0\}) = 0$, defining $H(b)$ to be $F(\exp b)$ for any Borel set b in $\mathbf{R}$, H is a countably additive spectral measure satisfying

$$R(t) = \int e^{t\lambda}\, H(d\lambda).$$

Finally, since $V(t) = U(t)$, a part of the above proof shows that $U(r_n)x \to U(t)x$ for any sequence $\{r_n\} \subset [0, \infty)$ that converges to any $t \in [0, \infty)$, for any x in X. Hence $\{U(t)\}$ is of class (C_o).

The commutativity of $\{T(t),\, S(t),\, V(t);\, t \geq 0\}$, which is identical with $\{T(t),\, R(t),\, U(t);\, t \geq 0\}$, was established earlier in the proof.

THEOREM 2. *Let $\{T(t);\, t \geq 0\}$ be a KLKH system of closed scalar-type operators in a Banach space X with resolutions of the identity E_t satisfying*

$$T(t + s) = T(t)T(s) \quad (t, s \geq 0),$$

$$\sup\{|E_t(\cdot)|;\, t \geq 0\} = M < \infty.$$

Then there is a closed scalar-type operator G with resolution of the identity E, for which

$$T(t) = \int_{\sigma(G)} e^{t\lambda}\, E(d\lambda) \quad (t \geq 0). \qquad (*)$$

PROOF. By Theorem 1 and the Remark after Lemma 1, the circled parts $\{U(t);\, t \geq 0\}$ form a semigroup of class (C_o) of bounded scalar-type operators with resolutions of the identity K_t satisfying

$$\sup\{|K_t(\cdot)|;\, t \geq 0\} \leq M.$$

By [7; Theorem 1], there is a strongly countably additive spectral measure K on the Borel sets of $\mathbf{R}$ such that

$$U(t) = \int_{\mathbf{R}} e^{it\lambda}\, K(d\lambda) \quad (t \geq 0). \qquad (**)$$

Theorem 1 shows that the nonnegative parts $\{R(t);\, t \geq 0\}$ satisfy

$$R(t) = \int_{\mathbf{R}} e^{t\lambda}\, H(d\lambda) \quad (t \geq 0),$$

where H is a strongly countably additive spectral measure on the Borel sets of $\mathbf{R}$. The method of the proof of [9; Theorem (5.3)] yields that there is a strongly countably additive spectral measure E on the Borel sets of $\mathbf{C}$ satisfying $E(a \times b) = H(a)K(b)$ for any Borel sets $a, b \subset \mathbf{R}$ and $(*)$. Note that the assumption of the weak completeness of X in [9] was needed only for establishing $(**)$, which was obtained in [7; Theorem 1] for an arbitrary Banach space X.

Theorem 2 and its direct counterpart from the beginning of the paper give together a complete description of the object that we can call for short a semigroup of unbounded scalar-type operators. The question naturally arises how we can recognize that we are dealing exactly with this object without verifying that each operator in the semigroup is of scalar type, which can indeed be an arduous task. In this direction we have

THEOREM 3. *Let $\{T(t); t \geq 0\}$ be a KLKH system of closed linear operators in a Banach space X satisfying*

$$T(t + s) = T(t)T(s) \quad (t, s \geq 0),$$

and define

$$Ax = \lim_{t \to 0+} \frac{T(t) - I}{t} x$$

for exactly those $x \in X$, for which $x \in D(T(s))$ for some $s > 0$ and the limit above exists in the norm topology of X. The system $\{T(t)\}$ is a semigroup of unbounded scalar-type operators if and only if the closure $\bar{A}$ is a spectral operator of scalar type with resolution of the identity E, the linear manifold $Q = \cup E(b)X$, where b runs through all the bounded Borel subsets of $\mathbf{C}$, is contained in $D(A)$, and Q is a core for all the operators $T(t)$.

PROOF. Assume first that $\{T(t)\}$ is a semigroup of unbounded scalar-type operators. By Theorem 2, there is a spectral operator of scalar type G with resolution of the identity E such that

$$T(t) = \int_{\sigma(G)} e^{t\lambda} E(d\lambda) \quad (t \geq 0).$$

The functional calculus for scalar-type operators shows that for any bounded Borel set $b \subset \mathbf{C}$ the restrictions $T(t)|E(b)X$ form a semigroup of bounded scalar-type operators and, since $\lim_{t \to 0+} |e^{t\lambda} - 1| = 0$ uniformly for $\lambda \in b$, this semigroup is continuous in the uniform operator topology. Hence its generator operator $A|E(b)X$ is everywhere defined, and from the integral representation of the restriction semigroup it is easily seen that $A|E(b)X = G|E(b)X$.

Assume now that there exists $Ax = y$, and $\{b_n\}$ is an increasing sequence of bounded Borel subsets of $\mathbf{C}$ with union $\mathbf{C}$. Since each $T(t)$ commutes with each $E(b_n)$, we obtain $AE(b_n)x = E(b_n)Ax$. Hence $GE(b_n)x = AE(b_n)x \to y$ as $n \to \infty$. Since $E(b_n)x \to x$ and the scalar-type operator G is closed, we obtain that $Gx = y$, hence $\bar{A} \subset G$. On the other hand, if $z \in D(G)$, then $E(b_n)z \to z$ and $AE(b_n)z = E(b_n)Gz \to Gz$. This shows that $\bar{A} = G$. The fact that Q is a core for each $T(t)$ follows from the basic properties of the standard functional calculus for scalar-type operators.

To prove the converse, assume that $G = \bar{A}$ is a spectral operator of scalar type with resolution of the identity E and that $E(b)X \subset D(A)$ for every bounded Borel set $b \subset \mathbf{C}$. Then $AE(b)x = GE(b)x$ for every $x \in X$. The operator G generates, by means of the functional calculus, a semigroup $\{S(t) = \int_{\sigma(G)} e^{t\lambda} E(d\lambda)\}$ of unbounded scalar-type operators, for which let the corresponding operator $\lim_{t \to 0+} \frac{S(t)-I}{t}$ be denoted by $A(S)$. By the first part of the present proof, each $E(b)X$ is contained in $D(A(S))$, and on each such subspace (for bounded b) $A(S)|E(b)X = G|E(b)X = A|E(b)X$. Hence the generated (semi)groups agree on each such subspace: $T(t)|E(b)X = S(t)|E(b)X$. By assumption, Q is a core for each $T(t)$. As mentioned in the preceding paragraph, Q is a core also for each $S(t)$, hence the closed operators $T(t)$ and $S(t)$ are identical for every $t \geq 0$. The proof is complete.

REFERENCES

1.E.Berkson, Semi-groups of scalar type operators and a theorem of Stone, Illinois J. Math. 10 (1966), 345-352.

2. A.Devinatz and A.E.Nussbaum, On the permutability of normal operators, Annals Math. 65 (1957), 144-152.

3. N.Dunford and J.T.Schwartz, Linear operators: Part I (1958), Part II (1963), Part III (1971), Wiley-Interscience, New York.

4. S.Kantorovitz and R.J.Hughes, Spectral representation of local semigroups, Math. Ann. 259 (1982), 455-470.

5. T.Kato, Perturbation theory for linear operators, Springer, New York, 1966.

6. A.Klein and L.Landau, Construction of a unique self-adjoint generator for a symmetric local semigroup, J. Funct. Anal. 44 (1981), 121-137.

7. R.Lange and B.Nagy, Semigroups and scalar-type operators in Banach spaces, Manuscript.

8. W.Ricker, Spectral representation of local semigroups associated with Klein-Landau systems, Math. Proc. Camb. Phil. Soc. 95 (1984), 93-100.

9. A.R.Sourour, Semigroups of scalar-type operators on Banach spaces, Trans. Amer. Math. Soc. 200 (1974), 207-232.

Weak Measurability of the Orbits
of an Adjoint Semigroup

J. M. A. M. van Neerven   Alfred P. Sloan Laboratory of Mathematics, California
Institute of Technology, Pasadena, California

B. de Pagter   Faculty of Technical Mathematics and Informatics, Delft University of
Technology, Delft, The Netherlands

A. R. Schep   University of South Carolina, Columbia, South Carolina

0. INTRODUCTION

Let $\mathbf{T} = \{T(t)\}_{t \geq 0}$ be a C_0-semigroup on a Banach space X. As usual, we denote by $\mathbf{T}^* = \{T^*(t)\}_{t \geq 0}$ the adjoint semigroup on the dual space X^*, defined by $T^*(t) := (T(t))^*$, and by $X^\odot$ the maximal subspace of X^* on which $\mathbf{T}^*$ acts in a strongly continuous way.

Let Y be a Banach space. We say that a map $f : \mathbb{R}_+ \to Y$ is weakly measurable if for each $y^* \in Y^*$ the map $f_{y^*}(t) := \langle y^*, f(t) \rangle$ is Lebesgue measurable. The purpose of this paper is to prove the following result, assuming Martin's Axiom (MA).

Theorem 0.1 (MA).   *Let $\mathbf{T} = \{T(t)\}_{t \geq 0}$ be a positive C_0-semigroup on a Banach*

---

[1] I would like to thank the Netherlands Organization for Scientific Research (NWO) for financial
support.

lattice E. *If, for some* $x^* \in E^*$, *the map* $t \mapsto T^*(t)x^*$ *is weakly measurable, then* $T^*(t)x \in (E^{\odot})^{dd}$ *for all* $t > 0$.

Here $(E^{\odot})^{dd}$ denotes the band generated by $E^{\odot}$. This theorem implies for example that translating a finite Borel measure μ on $\mathbb{R}$ is weakly measurable if and only if μ is absolutely continuous with respect to the Lebesgue measure; cf. Corollary 3.7 below.

In order to motivate our result, we will briefly sketch its history. Let us agree to call $\mathbf{T}^*$ weakly measurable if each of its orbits $t \mapsto T^*(t)x^*$ is. In Feller ([2]) it is proved that the adjoint of the translation group $\mathbf{T}$ on $C_0(\mathbb{R})$ fails to be weakly measurable. Explicitly, he shows that the map $t \mapsto T^*(t)\delta_0 = \delta_t$, the Dirac measure at t, is not weakly measurable. The proof of this is very short and is reproduced at the beginning of Section 3, for it is essentially this argument that we will generalize. In Van Neerven ([6]) the following is shown: if $\mathbf{T}$ is a positive C_0-semigroup on a space $E = C(K)$, K compact Hausdorff, then $\mathbf{T}^*$ is weakly Borel measurable if and only if $\mathbf{T}^*$ is strongly continuous for $t > 0$. Moreover, in this setting it is known that $E^{\odot} = (E^{\odot})^{dd}$. Thus, apart from the subtlety concerning Borel measurability, this result is implied by our Theorem 0.1. Both the proofs in Feller and Van Neerven avoid the use of set-theoretical axioms. Finally, Talagrand ([11]) shows, assuming Martin's Axiom, that translation of a function $f \in L^{\infty}(G)$, where G is a compact abelian group, is weakly measurable if and only if f is equal a.e. to a Riemann measurable function. In fact, it was this result which motivated us to use Martin's Axiom. Of course, his conclusion for the concrete case of translation in $L^{\infty}(G)$ is much stronger than what is implied by our general result.

1. PRELIMINARIES

In this section we recall some of the basic facts about adjoint semigroups which will be used in the sequel. Proofs of these facts can be found in Van Neerven ([6]).

Let $\mathbf{T}$ be a C_0-semigroup (i.e., a strongly continuous semigroup) on a Banach space X. Its generator will be denoted by A with domain $D(A)$. Considering the adjoint semigroup $\mathbf{T}^*$ on the dual space X^*, we define

$$X^{\odot} = \{x^* \in X^* : \lim_{t\downarrow 0} \|T^*(t)x^* - x^*\| = 0\},$$

the domain of strong continuity of $\mathbf{T}^*$. Then $X^{\odot}$ is a $\mathbf{T}^*$-invariant, norm closed, weak*-dense subspace of X^* (hence $X^{\odot} = X^*$ if X is reflexive). The space $X^{\odot}$ is precisely the norm closure of $D(A^*)$, the domain of the adjoint of A. In particular, for $\lambda \in \varrho(A) = \varrho(A^*)$ we have $R(\lambda, A^*)x^* \in X^{\odot}$ for all $x^* \in X^*$, where $R(\lambda, A^*) = R(\lambda, A)^* = (\lambda - A^*)^{-1}$ is the resolvent.

If $\mathbf{T}$ extends to a C_0-group, then the space $X^{\odot}$ with respect to the semigroup $\{T(t)\}_{t\geq 0}$ is equal to the domain of strong continuity of the group $\{T^*(t)\}_{t\in\mathbb{R}}$. In particular we mention that if $\mathbf{T}$ is the translation group on $X = C_0(\mathbb{R})$, then $X^{\odot}$ is precisely the subspace of all finite regular Borel measures on $\mathbb{R}$ which are absolutely continuous with respect to the Lebesgue measure. In this way $X^{\odot}$ can be identified canonically with $L^1(\mathbb{R})$. This classical result is due to Plessner ([9]).

Now let E be a Banach lattice and $\mathbf{T}$ a positive C_0-semigroup on E. Choose M, ω such that $\|T(t)\| \leq Me^{\omega t}$ for all $t \geq 0$. If $\lambda \in \mathbb{R}$ is such that $\lambda > \omega$, then $\lambda \in \varrho(A)$ and $R(\lambda, A) \geq 0$ (for the general theory of positive semigroups we refer to Nagel ([5])). We denote by $(E^{\odot})^d$ the disjoint complement of $E^{\odot}$ in E^*, i.e.,

$$(E^{\odot})^d = \{x^* \in E^* : x^* \perp y^{\odot} \text{ for all } y^{\odot} \in E^{\odot}\}.$$

Here $x^* \perp y^{\odot}$ means that $|x^*| \wedge |y^{\odot}| = 0$. Then $(E^{\odot})^{dd}$, the disjoint complement of $(E^{\odot})^d$, is equal to the band generated by $E^{\odot}$. We will need the following results concerning the adjoint of a positive C_0-semigroups (Van Neerven and De Pagter, [7]). Let $\mathbf{T}$ be a positive C_0-semigroup on a Banach lattice E, let ω be as above and fix $\lambda > \omega$.

Lemma 1.1 (Van Neerven and De Pagter, Section 1). *The band $(E^{\odot})^{dd}$ generated by $E^{\odot}$ is $\mathbf{T}^*$-invariant. If, moreover, $\mathbf{T}^*$ is a lattice semigroup, then also $(E^{\odot})^d$ is $\mathbf{T}^*$-invariant.*

Lemma 1.2 (Van Neerven and De Pagter, Lemma 3.1). *Let $0 \leq x \in E$ and $0 \leq x^*, y^* \in E^*$. If $\langle R(\lambda, A^*)x^* \wedge y^*, x \rangle = 0$, then $\langle T^*(t)x^* \wedge y^*, x \rangle = 0$ for almost all $t \geq 0$.*

Lemma 1.3 (Van Neerven and De Pagter, Theorem 3.5). *For any $0 \leq x^* \in E^*$ we have $x^* \perp E^{\odot}$ if and only if $x^* \perp R(\lambda, A^*)x^*$.*

We assume the reader to be familiar with the elementary theory of Banach lattices, and refer to Aliprantis and Burkinshaw ([1]) or Meyer-Nieberg ([4]) for more details. Our results are valid both for real and complex Banach lattices.

2. SOME CONSEQUENCES OF MARTIN'S AXIOM

In this section we collect some useful consequences of Martin's Axiom (MA). We let m and m^* denote the Lebesgue measure and outer Lebesgue measure, respectively. The following result can be found in Fremlin ([3]), Exercise 32P(d).

Lemma 2.1 (MA). *Let $S \subset [0,1] \times [0,1]$ be Lebesgue measurable, $m(S) = 1$. Then there is a subset $H \subset [0,1]$, $m^*(H) = 1$, such that for all $t, s \in H$, $t \neq s$, we have $(t, s) \in S$.*

In other words, $(H \times H) \backslash \Delta \subset S$, where Δ is the diagonal. The set H need not be Lebesgue measurable in general.

The next result is implicit in Talagrand ([12]), p. 99-102. (actually, this reference contains an extension of a combination of Lemmas 2.1 and 2.2).

Lemma 2.2 (MA). *Let $H \subset [0,1]$, $m^*(H) = 1$. Then $H = H_0 \cup H_1$ with $H_0 \cap H_1 = \emptyset$ and $m^*(H_0) = m^*(H_1) = 1$.*

Under assumption of the Continuum Hypothesis, Lemma 2.2 had been proved earlier by Sierpinski ([10]). If H is Lebesgue measurable, then the lemma can be proved without set-theoretical axioms. This is not very relevant for our purposes, however, since we will be interested in the case where H is non-measurable.

Lemma 2.3 (MA). *Let $H \subset [0,1]$, $m^*(H) = 1$. Then there exists a function $f : H \to [0,1]$ with the following property: whenever $g : [0,1] \to \mathbb{C}$ is a function such that $g|_H = f$, then g is not Lebesgue measurable.*

Proof: Let $H = H_0 \cup H_1$ as in Lemma 2.2 and let f be the characteristic function of H_0. Let $g : [0,1] \to \mathbb{C}$ be any function such that $g|_H = f$, and consider the set $G = \{|g| \leq \frac{1}{2}\}$. Then $H_1 \subset G$, so $m^*(G) = 1$. Also, $H_0 \subset [0,1] \backslash G$, so $m^*([0,1] \backslash G) = 1$. Therefore, G cannot be Lebesgue measurable. $\blacksquare$

3. PROOF OF THE MAIN THEOREM

We start with Feller's proof that translation of a Dirac measure is not weakly measurable in $M(\mathbb{R})$, the space of finite Borel measures on $\mathbb{R}$. Denote by $\mathbf{T} = \{T(t)\}_{t \in \mathbb{R}}$ the translation group defined by $T(t)f(\omega) = f(\omega + t)$ in the space $C_0(\mathbb{R})$ of all continuous functions on $\mathbb{R}$ vanishing at infinity. Let $g : \mathbb{R} \to [0,1]$ be any non Lebesgue measurable function. Let $Z = \text{span}\{\delta_t : t \in \mathbb{R}\}$ and define a linear form Φ on Z by $\Phi(\delta_t) := g(t)$. Because

$$\left| \Phi\left(\sum_{j=1}^{n} \alpha_j \delta_{t_j}\right) \right| = \left| \sum_{j=1}^{n} \alpha_j g(t_j) \right| \leq \sum_{j=1}^{n} |\alpha_j| = \left\| \sum_{j=1}^{n} \alpha_j \delta_{t_j} \right\|,$$

it follows that Φ extends to a bounded linear functional on the closure of Z. Letting $x^{**} \in (M(\mathbb{R}))^*$ be any Hahn-Banach extension of Φ, we have

$$\langle x^{**}, T^*(t)\delta_0 \rangle = \Phi(T^*(t)\delta_0) = \Phi(\delta_t) = g(t).$$

This shows that $t \mapsto T^*(t)\delta_0$ is not weakly Lebesgue measurable.

Inspection of the proof shows that it depends on three facts:

(i) For any two $t \neq s$, we have $\delta_t \perp \delta_s$;
(ii) For any two t, s we have $\|\delta_t\| = \|\delta_s\|$;
(iii) $M(\mathbb{R})$ is an AL-space, and therefore $\left\| \sum_{j=1}^{n} \alpha_j \delta_{t_j} \right\| = \sum_{j=1}^{n} |\alpha_j|$.

By (i), it is possible to define the form Φ and by (ii) and (iii), it is bounded. Our proof of Theorem 0.1 will be a generalization of the above argument. However, we run into a number of obstructions. The first problem is to find an analogue of (i). This is done in Lemma 3.3 below. There we obtain a certain subset of $\mathbb{R}_+ \times \mathbb{R}_+$. In order to get from this something similar to (i), this subset should contain a sufficiently large set of the form $(H \times H)\backslash\Delta$, where $H \subset \mathbb{R}_+$ and Δ is the diagonal of $\mathbb{R}_+ \times \mathbb{R}_+$. At this point we use Lemma 2.1 and Martins Axiom comes in. Secondly, in general $\mathbf{T}$ need not be isometric. In Lemma 3.4 we establish a substitute for (ii), which at the same time takes care of the fact that in general $(E^\odot)^d$ need not be $\mathbf{T}^*$-invariant. Finally, not every dual Banach lattice is an AL-space. This leads to the difficulty of proving that the linear form Φ is bounded. We overcome this problem by identifying an isomorphic copy of the AL-space $l^1(H)$ in the closed span of any orbit which does not lie in $(E^\odot)^{dd}$ for all $t > 0$. For this we use a variant of the lattice identity $|\phi + \psi| = |\phi| + |\psi|$, $\phi \perp \psi$; see Lemma 3.5.

We will work out these heuristics in a series of lemmas, which eventually culminates in a proof of Theorem 0.1. Throughout the rest of this section, $\mathbf{T}$ is a positive C_0-semigroup with generator A on a Banach lattice E. We fix constants M and ω such that $\|T(t)\| \leq Me^{\omega t}$ and fix some $\lambda > \omega$. By P we will denote the band projection onto the band $(E^\odot)^d$, the disjoint complement of $E^\odot$.

Lemma 3.1. *Let* $0 \leq x^* \in E^*$ *and let* Q *denote the band projection onto the band generated by* $R(\lambda, A^*)x^*$. *Then for all* $t \geq 0$ *we have* $(I - P)T^*(t)x^* = QT^*(t)x^*$.

Proof: Clearly $\{R(\lambda, A^*)x^*\}^{dd} \subset (E^\odot)^{dd}$, so $Q \leq I - P$ and $QT^*(t)x^* \leq (I - P)T^*(t)x^*$.

To prove the reverse inequality, fix $t \geq 0$ and put $y^* := (I - Q)(I - P)T^*(t)x^*$. Then $y^* \perp R(\lambda, A^*)x^*$. Also, from $0 \leq y^* \leq T^*(t)x^*$ we infer that

$$0 \leq R(\lambda, A^*)y^* \leq R(\lambda, A^*)T^*(t)x^* = e^{\lambda t}\int_t^\infty e^{-\lambda s}T^*(s)x^*ds \leq e^{\lambda t}R(\lambda, A^*)x^*,$$

the integral being in the weak* sense. It follows that $y^* \perp R(\lambda, A^*)y^*$. By Lemma 1.3, $y^* \perp E^\odot$. But by its definition, we also have $y^* \in (E^\odot)^{dd}$, so $y^* = 0$. Hence, $(I - P)T^*(t)x^* = Q(I - P)T^*(t)x^* \leq QT^*(t)x^*$. ∎

Lemma 3.2. *Let* $0 \leq x^* \in E^*$, $0 \leq y^* \in E^*$ *and* $0 \leq x \in E$. *Then the map* $(t, s) \mapsto \langle T^*(t)x^* \wedge PT^*(s)t^*, x \rangle$ *is Borel measurable on* $\mathbb{R}_+ \times \mathbb{R}_+$.

Proof: First,

$$\langle T^*(t)x^* \wedge T^*(s)y^*, x \rangle = \inf\{\langle T^*(t)x^*, u \rangle + \langle T^*(s)y^*, v \rangle : u, v \in [0, x]; u + v = x\},$$

so the map $(t, s) \mapsto \langle T^*(t)x^* \wedge T^*(s)y^*, x \rangle$ is Borel measurable, being the pointwise infimum of continuous functions. Therefore it remains to show that the map

$$(t, s) \mapsto \langle T^*(t)x^* \wedge (I - P)T^*(s)y^*, x \rangle$$

is Borel. Let Q be the band projection onto the band generated by $R(\lambda, A^*)y^*$. By Lemma 3.1 and general vector lattice theory,

$$(I - P)T^*(s)y^* = QT^*(s)y^* = \sup_n \{T^*(s)y^* \wedge nR(\lambda, A^*)y^*\}.$$

Hence,

$$\langle T^*(t)x^* \wedge (I - P)T^*(s)t^*, x \rangle = \sup_n \{\langle T^*(t)x^* \wedge T^*(s)y^* \wedge nR(\lambda, A^*)y^*, x \rangle\}$$

and

$$\langle T^*(t)x^* \wedge T^*(s)y^* \wedge nR(\lambda, A^*)y^*, x \rangle =$$
$$\inf\{\langle T^*(t)x^*, u \rangle + \langle T^*(s)y^*, v \rangle + \langle nR(\lambda, A^*)y^*, w \rangle : \ u, v, w \in [0, x]; u + v + w = x\}.$$

Since the latter is clearly a Borel function, the lemma follows.   ∎

Lemma 3.3. *Let $x^* \in E^*$ and $0 \leq x \in E$. The set*

$$N := \{(t, s) \in \mathbb{R}_+ \times \mathbb{R}_+ : \ \langle |T^*(t)x^*| \wedge |PT^*(s)x^*|, x \rangle > 0\}$$

is a Lebesgue measurable set of measure zero.

Proof: Put $M := \{(t, s) \in \mathbb{R}_+ \times \mathbb{R}_+ : \ \langle T^*(t)|x^*| \wedge PT^*(s)|x^*|, x \rangle > 0\}$. Then $N \subset M$, so it suffices to show that M is a Borel set of measure zero. By Lemma 3.2, M is a Borel set. Fix $s \geq 0$. We have $PT^*(s)|x^*| \in (E^{\odot})^d$, hence certainly $PT^*(s)|x^*| \wedge R(\lambda, A^*)|x^*| = 0$. Therefore, by Lemma 1.2 we see that $\langle T^*(t)|x^*| \wedge PT^*(s)|x^*|, x \rangle = 0$ for almost all $t \geq 0$. Hence, for all $s \geq 0$, the set $\{t \geq 0 : \ (t, s) \in M\}$ is a zero set. Therefore M is a zero set by the Fubini theorem.   ∎

The next lemma describes a certain weak*-continuity property of orbits which do not lie in $(E^{\odot})^{dd}$ for all $t > 0$.

Lemma 3.4. *Let $x^* \in E^*$, $0 \leq x \in E$ and $s_0 > 0$ be such that $\langle |PT^*(s_0)x^*|, x \rangle > 0$. Then there exists $0 < \epsilon < s_0$ such that*

$$\langle |PT^*(s)x^*|, x \rangle \geq \frac{1}{2}\langle |PT^*(s_0)x^*|, x \rangle, \quad \forall s \in [s_0 - \epsilon, s_0].$$

Proof: Let $C > 0$ be such that $\|PT^*(s)x^*\| \leq C$ for all $s \in [0, s_0]$. For $s \in [0, s_0]$ we write
$$T^*(s_0)x^* = T^*(s_0 - s)T^*(s)x^*$$
$$= T^*(s_0 - s)PT^*(s_0)x^* + T^*(s_0 - s)(I - P)T^*(s)x^*,$$

so
$$PT^*(s_0)x^* = PT^*(s_0 - s)PT^*(s)x^* + PT^*(s_0 - s)(I - P)T^*(s)x^*$$
$$= PT^*(s_0 - s)PT^*(s)x^*.$$

In the last identity we used the fact that $(E^\odot)^{dd}$ is $\mathbf{T}^*$-invariant (Lemma 1.1). It follows that $|PT^*(s_0)x^*| \leq T^*(s_0 - s)|PT^*(s)x^*|$. Hence,

$$\langle |PT^*(s_0)x^*|, x \rangle \leq \langle |PT^*(s)x^*|, T(s_0 - s)x \rangle$$
$$= \langle |PT^*(s)x^*|, x \rangle + \langle |PT^*(s)x^*|, T(s_0 - s)x - x \rangle$$
$$\leq \langle |PT^*(s)x^*|, x \rangle + C\|T(s_0 - s)x - x\|.$$

By taking $\epsilon > 0$ so small that $\|T(s_0 - s)x - x\| \leq (2C)^{-1} \langle |PT^*(s_0)x^*|, x \rangle$ for all $0 \leq s_0 - s \leq \epsilon$, the lemma now follows. ∎

Finally we need the following substitute for the norm additivity of disjoint vectors.

Lemma 3.5. *Suppose* $\phi_1, ... \phi_n \in E^*$ *and* $0 \leq x \in E$ *are such that* $\langle |\phi_j| \wedge |\phi_k|, x \rangle = 0$ *for all* $j \neq k$. *Then*

$$\left\langle \left|\sum_{j=1}^{n} \phi_j \right|, x \right\rangle = \sum_{j=1}^{n} \langle |\phi_j|, x \rangle.$$

Proof: Let $\phi, \psi \in E^*$ such that $\langle |\phi| \wedge |\psi|, x \rangle = 0$. Because of the lattice identity $2(|\phi| \wedge |\psi|) = |\phi| + |\psi| - ||\phi| - |\psi||$ we have $\langle ||\phi| - |\psi||, x \rangle = \langle |\phi|, x \rangle + \langle |\psi|, x \rangle$. Since $||\phi| - |\psi|| \leq |\phi + \psi| \leq |\phi| + |\psi|$ it follows that $\langle |\phi + \psi|, x \rangle = \langle |\phi|, x \rangle + \langle |\psi|, x \rangle$. Now proceed by induction on n. ∎

Proof of Theorem 0.1: Suppose $x^* \in E^*$ and $s_0 > 0$ are such that $T^*(s_0)x^* \notin (E^\odot)^{dd}$, or equivalently $PT^*(s_0)x^* \neq 0$. We will prove that the orbit $t \mapsto T^*(t)x^*$ fails to be weakly measurable.

Choose $0 \leq x \in E$, $\|x\| = 1$, such that $\langle |PT^*(s_0)x^*|, x \rangle > 0$. Let $\epsilon > 0$ be as in Lemma 3.4 and put

$$S := \{(t, s) \in [s_0 - \epsilon, s_0] \times [s_0 - \epsilon, s_0] : \langle |T^*(t)x^*| \wedge |PT^*(s)x^*|, x \rangle = 0\}.$$

By Lemma 3.3 we know that $m(S) = \epsilon^2$. Hence by Lemma 2.1, there is a subset $H \subset [s_0 - \epsilon, s_0]$ with $m^*(H) = \epsilon$ such that for all $t, s \in H$, $t \neq s$, we have $\langle |T^*(t)x^*| \wedge |PT^*(s)x^*|, x \rangle = 0$.

We claim that there exists a constant $K > 0$ such that for all $n \in \mathbb{N}$, scalars $\alpha_1, ..., \alpha_n$ and $s_1, ..., s_n \in H$ we have

$$\left\| \sum_{j=1}^{n} \alpha_j T^*(s_j)x^* \right\| \geq K \sum_{j=1}^{n} |\alpha_j|.$$

To see this, first observe that for all $t, s \geq 0$ we have $|T^*(t)x^*| \wedge |PT^*(s)x^*| \in (E^\odot)^d$. Therefore, for any two $s_j, s_k \in H$, $s_j \neq s_k$, we have

$$\langle |PT^*(s_j)x^*| \wedge |PT^*(s_k)x^*|, x \rangle = \langle P(|T^*(s_j)x^*| \wedge |PT^*(s_k)x^*|), x \rangle$$
$$= \langle |T^*(s_j)x^*| \wedge |PT^*(s_k)x^*|, x \rangle = 0.$$

Hence Lemma 3.5 applies, and by using it in tandem with the choice of ϵ we have

$$\left\|\sum_{j=1}^{n}\alpha_j T^*(s_j)x^*\right\| \geq \left\|\sum_{j=1}^{n}\alpha_j PT^*(s_j)x^*\right\| \geq \left\langle\left|\sum_{j=1}^{n}\alpha_j PT^*(s_j)x^*\right|, x\right\rangle$$

$$= \sum_{j=1}^{n}|\alpha_j|\,\langle|PT^*(s_j)x^*|, x\rangle \geq \frac{1}{2}\langle|PT^*(s_0)x^*|, x\rangle \cdot \sum_{j=1}^{n}|\alpha_j|.$$

This proves the claim.

Let $f : H \to [0,1]$ be a function as Lemma 2.3. Define

$$Z := \mathrm{span}\{T^*(s)x^* : \ s \in H\}.$$

It follows in particular from the above that the set $\{T^*(s)x^* : \ s \in H\}$ is linearly independent, so it makes sense to define a linear form Φ on Z by

$$\Phi\big(T^*(s)x^*\big) := f(s).$$

We show that Φ is bounded. Indeed, by the claim we have

$$\left|\Phi\left(\sum_{j=1}^{n}\alpha_j T^*(s_j)x^*\right)\right| = \left|\sum_{j=1}^{n}\alpha_j f(s_j)\right| \leq \sum_{j=1}^{n}|\alpha_j| \leq \frac{1}{K}\left\|\sum_{j=1}^{n}\alpha_j T^*(s_j)x^*\right\|.$$

Therefore, Φ extends to a bounded linear functional on the closure of Z in E^*. Let $x^{**} \in E^{**}$ be any Hahn-Banach extension of Φ and let the function g be defined by $g(t) = \langle x^{**}, T^*(t)x^*\rangle$. Clearly $g|_H = f$, so $g|_{[s_0-\epsilon,s_0]}$ is not Lebesgue measurable by Lemma 2.3. $\blacksquare$

Theorem 0.1 can be improved if the adjoint semigroup $\mathbf{T}^*$ is a lattice semigroup. For example, this is the case if $\mathbf{T}$ extends to a positive group.

Corollary 3.6 (MA). *Suppose $\mathbf{T}$ is a positive C_0-semigroup on E such that $\mathbf{T}^*$ is a lattice semigroup. If for some $x^* \in E^*$ the map $t \mapsto T^*(t)x^*$ is weakly measurable, then $x^* \in (E^{\odot})^{dd}$.*

Proof: Write $x^* = x_0^* + x_1^*$ with $x_0^* \in (E^{\odot})^d$, $x_1^* \in (E^{\odot})^{dd}$. By Lemma 1.1, we have for any $t \geq 0$ that $T^*(t)x_0^* \in (E^{\odot})^d$ and $T^*(t)x_1^* \in (E^{\odot})^{dd}$. Also, by Theorem 0.1, $T^*(t)x^* \in (E^{\odot})^{dd}$ for $t > 0$. Therefore we must have $T^*(t)x_0^* = 0$ for all $t > 0$. From the weak*-continuity of $t \mapsto T^*(t)x_0^*$, we conclude that $x_0^* = 0$. $\blacksquare$

If the dual space E^* has order continuous norm, then $E^{\odot} = (E^{\odot})^{dd}$ holds for any positive C_0-semigroup on E (see Van Neerven and De Pagter ([7]), Theorem 2.1 or De Pagter ([8]), Theorem 2.1). This holds in particular if $E = C_0(\Omega)$, Ω locally compact Hausdorff. In combination with the above results, this yields the following corollary.

Corollary 3.7 (MA). *Suppose that* **T** *is a positive C_0-semigroup on a Banach lattice E whose dual E^* has order continuous norm. If, for some $x^* \in E^*$, the map $t \mapsto T^*(t)x^*$ is weakly measurable, then $t \mapsto T^*(t)x^*$ is strongly continuous for $t > 0$. If moreover,* **T*** *is a lattice semigroup, then $t \mapsto T^*(t)x^*$ is strongly continuous for $t \geq 0$, i.e. $x^* \in E^\odot$.*

For the translation group in $E = C_0(\mathbb{R})$ we know that $E^\odot = (E^\odot)^{dd} = L^1(\mathbb{R})$ (cf. Section 1), which gives the following result.

Corollary 3.8 (MA). *Let μ be a finite Borel measure on $\mathbb{R}$. Then translation of μ is weakly measurable in $M(\mathbb{R})$ if and only if μ is absolutely continuous with respect to the Lebesgue measure.*

Finally, we mention one more consequence of Corollary 3.6.

Corollary 3.9 (MA). *Let* **T** *be a positive C_0-group with unbounded generator on a space $C_0(\Omega)$, Ω locally compact Hausdorff. Then* **T*** *is not weakly measurable.*

Proof: From Van Neerven and De Pagter ([7]), Corollary 3.18, we know that $E^\odot = (E^\odot)^{dd} \neq E^*$, so the result follows from Corollary 3.6. ∎

Acknowledgement - We would like to thank Ralph Howard and Jim Roberts for suggesting to us the proof of Lemma 2.3. The second named author likes to thank the members of the Department of Mathematics, University of South Carolina, Columbia, where part of the work on this paper was done, for their hospitality during his visit.

4. REFERENCES

[1] C.D. Aliprantis, O. Burkinshaw, *Positive Operators*, Pure and Applied Math., vol. 119, Academic Press, Orlando, 1985.

[2] W. Feller, *Semigroups of transformations in general weak topologies*, Ann. Math. **57** (1953), 287-308.

[3] D.H. Fremlin, *Consequences of Martin's Axiom*, Cambridge University Press, Cambridge, 1984.

[4] P. Meyer-Nieberg, *Banach lattices*, Springer Verlag, Berlin-Heidelberg-New York, 1991.

[5] R. Nagel (ed.), *One-parameter semigroups of positive operators*, Lect. Notes in Math., vol. 1184, Springer Verlag, Berlin-Heidelberg-New York, 1986.

[6] J.M.A.M. van Neerven *The Adjoint of a Semigroup of Linear Operators*, Lect. Notes in Math., vol 1529, Springer-Verlag, New York-Heidelberg-Berlin, 1992.

[7] J.M.A.M. van Neerven, B. de Pagter, *The adjoint of a positive semigroup*, Comp. Math. (to appear).

[8] B. de Pagter, *A Wiener-Young type theorem for dual semigroups*, Acta Appl. Math. **27** (1992), 101-109.

[9] A. Plessner, *Eine Kennzeichnung der totalstetigen Funktionen*, J. f. Reine u. Angew. Math. **160** (1929), 26-32.

[10] W. Sierpinski, *L'équivalence par décomposition finie et la mésure extérieure des ensembles*, Fund. Math. **37** (1950), 209-212.

[11] M. Talagrand, *Closed convex hull of set of measurable functions, Riemann-measurable functions and measurability of translations*, Ann. Inst. Fourier **32** (1982), 39-69.

[12] M. Talagrand, *Pettis integral and measure theory*, Memoirs of the Am. Math. Soc., vol. 307, 1984.

On Nonlinear Scattering
and Unbounded Surfaces

G. F. Roach University of Strathclyde, Glasgow, Scotland

1. INTRODUCTION

Scattering theory can be thought of, most simply perhaps, as the interaction of an evolutionary process with a non-homogeneous and maybe even non-linear medium. It has played a central role in mathematical physics over the years. Scattering processes have been observed in such varied fields as acoustics, geophysical prospecting, medical diagnosis, non-destructive testing processes, optical switching devices, quantum mechanics, and other quite general remote sensing techniques.

The evolution of a system can be modelled in terms of an initial boundary value problem the solution of which describes the state of the system at any point at some future (or past) time. Specifically, given initial data, f, for a system the state u of the system at any point x and time t can be written, abstractly in the form

$$u(x,t) = U(t)f(x)$$

where $U(t)$ is an operator describing the time evolution of the system.

In principle therefore, using (1.1), we can obtain the state of the system at all points (x,t). However, in practice $U(t)$ is very difficult to compute. Indeed, such a computation, which in many practical situations can soon become virtually unmanageable because of its complexity, is not always something which can be justified bearing in mind the needs of the experimentalists.

This situation can be eased by making recourse to the results and techniques of scattering theory which have been so successfully developed and exploited for problems in quantum mechanics.

Experiments in quantum mechanics are characterised by the fact that the system undergoes a measurement at a time t_o and also at a later time t_1 and that for most of the time interval $(t_1 - t_o)$ there is negligeable interaction between the various components of the system. Indeed components only interact for a small part of the interval $(t_1 - t_o)$. This means that for the greater part of the experiment the component parts are behaving as though they were "free". For this reason, at least, it is unnecessary to solve completely the initial boundary value problem which models the system in order to relate theoretical predictions to the available experimental results.

This philosophy and the associated techniques can be applied to problems in many fields other than quantum mechanics, Lax and Phillips (1967) and Wilcox (1975). Of particular interest is the manner in which these ideas, generated initially to deal with linear problems, can even be extended to provide information about non-linear problems.

Scattering theory, as developed in quantum mechanics is associated, almost entirely, with perturbations of the Laplacian interpreted as an operator in $L_2(\mathbf{R}^n)$. This so-called potential scattering is now well developed and quite well understood. However, for target or obstacle scattering the situation is by no means as satisfactory.

In this article we shall indicate how a scattering theory can be developed in fields other than quantum mechanics and survey the results, particularly for target scattering, which have been obtained recently.

First we clarify what was means by saying that a component in the system was behaving as though it were "free". This is perhaps most readily achieved by considering the motion of two particles. The trajectory of the k-th particle, $k = 1, 2$, can be described by a three dimensional vector-valued function $\mathbf{r}_k(t)$ which represents a curve in $\mathbf{R}^3$.

In classical mechanics a particle is in free motion if its trajectory is defined by

$$\mathbf{r}(t) = \mathbf{r}_0 + \mathbf{v}_t, \quad \ddot{r}(T) = 0.$$

Consequently, when the two particles are a large enough distance apart, so that their interaction can be regarded as virtually negligible, their motion can be considered as being almost along a straight line $\mathbf{r}_k^-(t)$ with uniform velocity $\mathbf{v}_k^-$, $k = 1, 2$.

During the time that particles are close they interact and their motion is, in general, highly complex. Eventually they move apart, they cease to interact and once again their motion can be considered as being almost that of a free particle in the sense that they move almost along a straight line, $\mathbf{r}_k^+(t)$ with uniform velocity $\mathbf{v}_k^+$, $k = 1, 2$.

This approximation process can be made precise in a natural way by requiring, for

$k = 1, 2$

$$\lim_{t \to -\infty} |\ \mathbf{r}_k(t) - (\mathbf{r}_k^- + \mathbf{v}_k^- t)\ | = 0$$

$$\lim_{t \to -\infty} |\ \dot{\mathbf{r}}_k(t) - \mathbf{v}_k^-\ | = 0$$

$$\lim_{t \to +\infty} |\ \mathbf{r}_k(t) - (\mathbf{r}_k^+ + \mathbf{v}_k^+ t)\ | = 0$$

$$\lim_{t \to +\infty} |\ \dot{\mathbf{r}}_k(t) - \mathbf{v}_k^+\ | = 0.$$

The vectors $\mathbf{r}_k^-, \mathbf{v}_k^-$ $(r_k^+, \mathbf{v}_k^+)$ characterise a free state of the k-th particle called the incoming (outgoing) asymptotic state of the k-th particle. One of the main aims of scattering theory is to show that these incoming and outgoing asymptotic states actually exist. In this particular case, for example, if long range interaction is being studied then no such states exist.

When the above limits do exist then we say that the particles are **asymptotically free** as $t \to \pm\infty$.

To illustrate for target scattering the various notions which are required in developing a scattering theory and the results which can be obtained we shall, in the first instance, confine attention to linear acoustic scattering modelled by the following initial boundary value problem in $\mathbf{R}^3$

$$\left(\frac{\partial^2}{\partial t^2} - \Delta \right) u(x,t) = F(x,t), \quad (x,t) \in \Omega \times \mathbf{R} \tag{1.1}$$

$$u(x,0) = \phi_0(x), \quad x \in \Omega \tag{1.2}$$

$$u_t(x,0) = \phi_1(x), \quad x \in \Omega \tag{1.3}$$

$$u(x,t) \in (bc), \quad (x,t) \in \partial\Omega \times \mathbf{R} \tag{1.4}$$

where Ω is an open region in $\mathbf{R}^3$, $\partial\Omega$ denotes the boundary of Ω, the quantities F, ϕ_o, ϕ_1 are given and (1.4) indicates that the solution is required to satisfy certain boundary conditions on $\partial\Omega$. The intention is to investigate the existence and uniqueness of solutions to the problem (1.1)-(1.4) and to determine their asysmptotic behaviour as $t \to \pm\infty$.

In the following sections we shall consider first the case when Ω is an unbounded region exterior to a closed bounded boundary $\partial\Omega$. This will allow us to recall the various techniques which have been successfully developed for investigating scattering by bounded targets. We will then indicate how these various concepts have been modified recently in order to cater for the situation when $\partial\Omega$ is possibly unbounded. Unbounded scatterers appear in a wide range of real life situations including, for example, non-destructive testing of composite materials, geophysical prospecting, optical switching devices and ultrasonic medical diagnosis. Since these various examples could possibly involve non-linear materials we give, in the last section, a brief indication how the structure introduced in previous sections can be used as a vehicle for investigating non-linear problems.

2. Scattering by Bounded Targets

We take as prototype problem a homogeneous form of (1.1)-(1.4).

$$\mathbf{P1}: \qquad \left(\frac{\partial^2}{\partial t^2} - \Delta\right) u_0(x,t) = 0, \quad (x,t) \in \Omega \times \mathbf{R} \tag{2.1}$$

$$u_0(x,0) = \phi_0(x), \quad x \in \Omega \tag{2.2}$$

$$u_{0t}(x,0) = \phi_1(x), \quad x \in (\Omega \tag{2.3}$$

$$u_0(x,t) \in (bc), \quad x \in \partial\Omega \times \mathbf{R} \tag{2.4}$$

We consider first the case when $\Omega = \mathbf{R}^3$ and introduce the operator

$$A_0 : L_2(\mathbf{R}^3) \to L_2(\mathbf{R}^3)$$
$$A_0 u_0 = -\Delta u_0, \quad u_0 \in D(A_0)$$
$$D(A_0) = \{u \in L_2(\mathbf{R}^3) : \Delta u_0 \in L_2(\mathbf{R}^3)\}. \tag{2.5}$$

The following results can be obtained (Wilcox (1975)).

Theorem 2.1

    (i) A_0 is a self-adjoint, non-negative operator on $L_2(\mathbf{R}^3)$

    (ii) A_0 has a unique non-negative square root $A_0^{\frac{1}{2}}$ with domain

$$D(A_0^{\frac{1}{2}}) = L_2(\mathbf{R}^3) \cup \{u : D^\alpha u \in L_2(\mathbf{R}^3), |\alpha| \le 1\}.$$

Here we have used the multi-index notation for derivatives which in $\mathbf{R}^n$ implies

$$\alpha = (\alpha_1, \cdots, \alpha_n), \quad \alpha_j = \text{non} - \text{negative integer} \quad j = 1, 2, \cdots n$$
$$|\alpha| = \sum_{j=1}^{n} \alpha_j$$
$$D^\alpha = D_1^{\alpha_1} D_2^{\alpha_2} \cdots D_n^{\alpha_n}, \quad D_j = \partial/\partial x_j \quad j = 1, 2, \cdots n.$$

We now interpret equation (2.1) as

$$\left(\frac{d^2}{dt^2} + A_0\right) u_0(t) = 0, \quad t \in \mathbf{R} \tag{2.6}$$

where the notation emphasises that u_0, the unknown in **P1**, is understood to be an $L_2(\mathbf{R}^3)$-valued function of $t \in \mathbf{R}$. The initial values (2.2), (2.3) then assume the form

$$u_0(0) = \phi_0 \quad \text{and} \quad u_{0t}(0) = \phi_1 \tag{2.7}$$

where it is assumed that $\phi_0, \phi_1 \in L_2(\mathbf{R}^3)$.

A solution of the initial value problem (2.6), (2.7) can be written in the form

$$u_0(t) = (\cos tA_0^{\frac{1}{2}})\phi_0 + A_0^{-\frac{1}{2}}(\sin tA_0^{\frac{1}{2}})\phi_1. \tag{2.8}$$

If further, ϕ_0, ϕ_1 are real valued functions such that $\phi_0 \in L_2(\mathbf{R}^3)$ and $\phi_1 \in D(A_0^{-\frac{1}{2}})$ and if we define on $L_2(\mathbf{R}^3)$

$$h_0 := \phi_0 + iA_0^{-\frac{1}{2}}\phi_1 \tag{2.9}$$

then the solution (2.8) can be expressed in the form

$$u_0(t) \equiv u_0(\cdot, t) = Re\{v_0(., t)\} \tag{2.10}$$

where

$$v_0(t) \equiv v_0(\cdot, t) = \exp(-itA_0^{\frac{1}{2}})h_0 \tag{2.11}$$

is the complex valued solution in $L_2(\mathbf{R}^3)$ of (2.6), (2.7).

Since A_0 is a self-adjoint operator the coefficients involving $A_0^{\frac{1}{2}}$ appearing in (2.8) and (2.11) can be interpreted by means of the spectral theorem. Specifically, if A_0 has a spectral family denoted by $\{E_0(\lambda)\}$ then we have the spectral representations (Helmberg (1969))

$$A_0 = \int_0^\infty \lambda dE_0(\lambda) \tag{2.12}$$

$$\Phi(A_0) = \int_0^\infty \Phi(\lambda)dE_0(\lambda). \tag{2.13}$$

where Φ is a bounded, Lebesgue measurable function of λ.

Consequently $A_0^{\frac{1}{2}}$ is interpreted by setting $\Phi(\lambda) = \lambda^{\frac{1}{2}}$. The remaining coefficients in (2.8) and (2.11) are interpreted in a similar manner.

A more explicit interpretation can be obtained by introducing the Fourier transform

$$F(f)(p) = \hat{f}(p) = \lim_{M \to \infty} \frac{1}{(2\pi)^{\frac{3}{2}}} \int_{|x| \leq M} e^{-ix \cdot p} f(x)dx \tag{2.14}$$

$$f(x) = F^{-1}(\hat{f})(x) = \lim_{M \to \infty} \frac{1}{(2\pi)^{\frac{3}{2}}} \int_{|p| \leq M} e^{ix \cdot p} \hat{f}(p)dp. \tag{2.15}$$

It can then be shown that

$$\Phi(A_0)f(x) = \lim_{M \to \infty} \frac{1}{(2\pi)^{\frac{3}{2}}} \int_{|p| \leq M} e^{ix \cdot p} \Phi(|\,p\,|^2) \hat{f}(p) dp \qquad (2.16)$$

where Φ is a bounded Lebesgue measurable function. We would emphasise that the above limits are taken in the $L_2(\mathbf{R}^3)$ sense.

We now notice that

$$w_0(x,p) = \frac{1}{(2\pi)^{\frac{3}{2}}} e^{ix \cdot p}, \quad x \in \mathbf{R}^3,\ p \in \mathbf{R}^3 \qquad (2.17)$$

satisfies

$$(\Delta + |\,p\,|^2) w_0(x,p) = 0 \quad \text{for all } x,p \in \mathbf{R}^3. \qquad (2.18)$$

Thus $w_0(x,p)$ is an eigenfunction of A_0 corresponding to the spectral parameter $|\,p\,|^2$. In fact it is a generalised eigenfunction since, as direct calculation shows, $w(\cdot,p) \notin L_2(\mathbf{R}^3)$.

Eigenfunction expansions for A_0 can now be obtained by rewriting (2.14)-(2.16) using (2.17) to obtain (Wilcox (1975))

$$F(f)(p) = \hat{f}(p) = \lim_{M \to \infty} \int_{|x| \leq M} \overline{w_0(x,p)} f(x) dx \qquad (2.19)$$

$$f(x) = F^{-1}(\hat{f})(x) = \lim_{M \to \infty} \int_{|p| \leq M} w_0(x,p) \hat{f}(p) dp \qquad (2.20)$$

$$\Phi(A_0)f(x) = \lim_{M \to \infty} \int_{|p| \leq M} w_0(x,p) \Phi(|\,p\,|^2) \hat{f}(p) dp \qquad (2.21)$$

where as before the limits are taken in an $L_2(\mathbf{R}^3)$ sense.

The function w_0 is a steady state solution of the unforced wave equation which represents a plane wave propagating in the direction p where a time dependence of the form $\exp(-i\,|\,p\,|\,t)$ has been assumed in (1.1).

For problems involving an arbitrary domain $\Omega \subset \mathbf{R}^3$, which will correspond to an obstacle being placed in an incident field of the form studied above, we introduce the operator

$$\begin{aligned}
&A : L_2(\Omega) \to L_2(\Omega) \\
&Au = -\Delta u, \quad u \in D(A) \\
&D(A) = \{u \in L_2(\Omega) : -\Delta u \in L_2(\Omega) \ st \ u \in (bc)\}.
\end{aligned} \qquad (2.22)$$

where $u \in (bc)$ indicates that u is required to satisfy certain boundary conditions, denoted (bc), on $\partial\Omega$ the boundary of Ω. To fix ideas we shall take (bc) to mean Neumann conditions.

The following results are available (Wilcox (1975)).

Theorem 2.2

(i) A is a self-adjoint, non-negative operator on $L_2(\Omega)$

(ii) A has a unique square root $A^{\frac{1}{2}}$ with domain

$$D(A^{\frac{1}{2}}) = L_2(\Omega) \cap \{u : D^\alpha u \in L_2(\Omega), \mid \alpha \mid \leq 1\}.$$

The initial boundary value problem **P1** can now be reformulated as an initial value problem in $L_2(\Omega)$ by requiring to find a function

$$u : \mathbf{R} \to L_2(\Omega) \tag{2.23}$$

satisfying

$$\left(\frac{d^2}{dt^2} + A\right) u(t) = 0 \tag{2.24}$$

$$u(0) = \psi_0 \quad \text{and} \quad u_t(0) = \psi_1 \tag{2.25}$$

with the data satisfying $\psi_0, \psi_1 \in L_2(\Omega)$.

Since A is self adjoint a solution of (2.24), (2.25) can be written in the form

$$u(t) = (\cos tA^{\frac{1}{2}})\psi_0 + (A^{-\frac{1}{2}} \sin tA^{\frac{1}{2}})\psi_1 \tag{2.26}$$

where the coefficients are interpreted by means of spectral theory. Specifically

$$\Phi(A) = \int_0^\infty \Phi(\lambda) dE(\lambda) \tag{2.27}$$

where $\{E(\lambda)\}$ denotes the spectral family of A.

Furthermore, if ψ_0, ψ_1 are real valued functions such that $\psi_0 \in L_2(\Omega)$ and $\psi_1 \in D(A^{-\frac{1}{2}})$ and if we define

$$h := \psi_0 + iA^{-\frac{1}{2}}\psi_1 \tag{2.28}$$

then (2.26) can be written in the form

$$u(t) = u(\cdot, t) = Re\{v(\cdot, t)\} \tag{2.29}$$

where

$$v(t) = v(\cdot, t) = \exp\{-itA^{\frac{1}{2}}\}h \tag{2.30}$$

is the complex valued solution in $L_2(\Omega)$ of (2.24), (2.25).

If now we want to compare the asymptotic behaviour, as $t \to \pm\infty$, of solutions to the free problem (2.6), (2.7) and the perturbed problem (2.24), (2.25) we use the complex forms of the solutions (2.11) and (2.30) first writing them in the more convenient form

$$v_0(t) = \exp(-itA_0^{\frac{1}{2}})h_0 =: U_0(t)h_0 \tag{2.31}$$

$$v(t) = \exp(-itA^{\frac{1}{2}})h =: U(t)h. \tag{2.32}$$

An immediate comparison can be obtained by enquiring whether or not solutions to the perturbed problem are asymptotically equal to solutions to the free problem. This will be the case when

$$\lim_{t \to \pm\infty} \| v(t) - v_0(t) \|_{L_2(\Omega)} = 0. \tag{2.33}$$

If we introduce the mapping

$$J(\Omega) : L_2(\Omega) \to L_2(\mathbf{R}^3) \tag{2.34}$$

defined by

$$J(\Omega)v(x) = \begin{cases} v(x), & x \in \Omega \\ 0, & x \in \mathbf{R}^3 \setminus \Omega \end{cases} \tag{2.35}$$

and use (2.31) and (2.32) then (2.33) is equivalent to the requirement

$$\lim_{t \to \pm\infty} \| J(\Omega)U(t)h - U_0(t)h_0 \|_{L_2(\mathbf{R}^3)} = 0. \tag{2.36}$$

Since $\exp\{-itA_0^{-\frac{1}{2}}\}$ is a unitary operator on $L_2(\mathbf{R}^3)$ we can re-write (2.36) in the form

$$\begin{aligned} 0 &= \lim_{t \to \pm\infty} \| U_0^{-1}(t)J(\Omega)U(t)h - h_0 \|_{L_2(\mathbf{R}^3)} \\ &= \lim_{t \to \pm\infty} \| W(t)h - h_0 \|_{L_2(\mathbf{R}^3)} \\ &= \| W_\pm h - h_0 \|_{L_2(\mathbf{R}^3)} \end{aligned} \tag{2.37}$$

where

$$W_\pm := \lim_{t \to \pm\infty} W(t) \tag{2.38}$$

are the so-called wave operators associated with $A_0^{\frac{1}{2}}, A^{\frac{1}{2}}$ and $J(\Omega)$.

Thus we see that the two systems are asymptotically equal as $t \to \pm\infty$ provided the initial data for the perturbed and free systems are related according to

$$W_\pm h = h_0^\pm. \tag{2.39}$$

where the superscripts on h_0 indicate that different initial values may have to be considered when investigating $t \to +\infty$ and $t \to -\infty$.

Proof of the existence of the limits (2.38) defining the wave operators $W_\pm$ is a considerable undertaking. In the meantime we consider the availability of expansion formulae for A similar to those in (2.19)-(2.21) for A_0.

We have noticed that A_0, the unperturbed operator, has a generalised eigenfunction $w_0(x,p)$ given in (2.17). This eigenfunction is associated with the spectral parameter $\mid p \mid^2$ and corresponds to a plane wave propagating in the p direction. The corresponding eigenfunction for A is defined as the steady-state wave $w(x,p)$ defined as the solution of the boundary value problem

$$(\Delta + \mid p \mid^2)w(x,p) = 0, \quad x \in \Omega \tag{2.40}$$

$$w(x,p) \in (bc), \quad x \in \partial\Omega. \tag{2.41}$$

Following Ikebe (1960) and Wilcox (1975) we set

$$w(x,p) = w_0(x,p) + w^s(x,p) \tag{2.42}$$

where $\mid x \mid \geq r_0$ for some sufficiently large r_0.

Physically this means that the total wave, $w(x,p)$, consists of an incident wave, $w_0(x,p)$, plus a scattered wave $w^s(x,p)$. Such eigenfunctions are referred to as distorted plane waves.

In constructing $w(x,p)$ use will be made of the cut-off function

$$j(x) := \phi_1(\mid x \mid -r_0), \quad x \in \mathbf{R}^3 \tag{2.43}$$

where $\phi_1 \in C^\infty(\mathbf{R})$ and is such that (i) $\phi_1'(\tau) \geq 0, \tau \in \mathbf{R}$, (ii) $\phi_1(\tau) \equiv 0, \tau \leq 0$, (iii) $\phi_1(\tau) \equiv 1, \tau \geq 1$ and where r_0 is chosen so that

$$\Omega_- := \mathbf{R}^3 \setminus \Omega \subset B(r_o)$$

with $B(r_0)$ denoting a ball centred on the origin and of radius r_0.

The eigenfunctions of A are assumed to have the form

$$w(x,p) = j(x)w_0(x,p) + w^s(x,p) \tag{2.44}$$

where the scattered wave is required to satisfy

$$(\Delta + \mid p \mid^2)w^s(x,p) = -(\Delta + \mid p \mid^2)j(x)w_0(x,p). \tag{2.45}$$

It will be seen that there are, in fact, two families of distorted plane waves. These are the so-called outgoing and incoming waves and we emphasise this by writing $w_+(x,p)$ and $w_-(x,p)$ respectively. The scattered wave $w^s(x,p)$ will be similarly subscripted.

Once the distorted plane waves have been constructed we can obtain the required generalised Fourier transform and spectral representation in the following form

Wilcox (1975)

$$(F_\pm f)(p) = \hat{f}_\pm(p) = \lim_{R\to\infty} \int_{B(R)} \overline{w_\pm(x,p)} f(x)dx \qquad (2.46)$$

$$f(x) = (F_\pm^* \hat{f}_\pm)(x) = \lim_{R\to\infty} \int_{|p|\leq R} w_\pm(x,p)\hat{f}_\pm(p)dp \qquad (2.47)$$

$$\Phi(A)f(x) = \lim_{R\to\infty} \int_{|p|\leq R} w_\pm(x,p)\Phi(|\,p\,|^2)\hat{f}_\pm(p)dp \qquad (2.48)$$

where $B(R) = \Omega \cap \{x \in \mathbf{R}^3 :| x | < R\}$ and Φ is a bounded Lebesgue measurable function. The limits in (2.46) to (2.48) are taken in the $L_2(\Omega)$ sense.

It can be shown (Wilcox (1975)) that

$$W_+ \equiv W_+(A_0^{1/2}, A^{1/2}, J(\Omega)) = F^* F_- : L_2(\Omega) \to L_2(\mathbf{R}^3)$$
$$W_- \equiv W_-(A_0^{1/2}, A^{1/2}, J(\Omega)) = F^* F_+ : L_2(\Omega) \to L_2(\mathbf{R}^3)$$

and that

$$E(\lambda) = W_+^* E_0(\lambda)W_+ = W_-^* E_0(\lambda)W_-, \quad \lambda \in \mathbf{R}.$$

The above development relies crucially on details concerning the existence and uniqueness of solutions to equations of the form (2.45). This will be discussed in the next section.

3. On the solutions of exterior problems

Exterior problems can conveniently be investigated in a weighted space setting (Neittaanmaki and Roach (1987) - subsequently denoted PN/GFR 1987). The approach has the virtue that it extends readily to problems involving unbounded surfaces.

To fix ideas we consider the problem

$$(\Delta + k^2)u(x) = f(x), \quad x \in \Omega \qquad (3.1)$$
$$u(x) = 0, \qquad x \in \partial\Omega \qquad (3.2)$$

the aim being to settle questions of existence and uniqueness of solutions to (3.1)–(3.2). It is well known that for unbounded $\Omega \subset \mathbf{R}^n$ this boundary value problem may have no solution in $L_2(\Omega)$ Eidus (1965), Kupradze (1956), PN/GFR 1987 . Furthermore, even when a solution exists it may not be unique [17].

Many of the difficulties associated with having to work in finite subdomains of Ω, with applications of the Poincaré inequality and Sobolev embedding theorems can be eased by working in a weighted Sobolev space structure. To this end we introduce the notation

$$\Omega(R) = \{x \in \Omega : | \; x \; |< R\}$$
$$\Omega(R_1, R_2) = \{x \in \Omega : R_1 <| \; x \; |< R_2\}$$
$$E(R) = \{x \in \mathbf{R} : | \; x \; |> R\}$$
$$B(R) = \{x \in \mathbf{R} : | \; x \; |< R\}$$
$$S(R) = \{x \in \mathbf{R} : | \; x \; |= R\}.$$

For convenience we shall suppose that $\Omega_- := \mathbf{R}^n \setminus \bar{\Omega}$ contains the origin.

We shall use standard complex Sobolev Hilbert spaces $H^k_{loc}(\Omega)$, $H^k(\Omega)$, $k = 0, 1, 2, \cdots$. In particular we shall write $L^2_{loc}(\Omega) = H^0_{loc}(\Omega)$ and $L_2(\Omega) = H^0(\Omega)$. The inner product and norm on $H^k(\Omega)$ are denoted by

$$(u, v)_k = \sum_{|\alpha| \le k} \int_\Omega \overline{D^\alpha u} D^\alpha v \, dx \quad \| \; u \; \|^2_k = (u, u)_k.$$

It will be convenient to introduce

$$H^1_{0*}(\Omega) := \{u \in L^2_{loc}(\Omega) : \phi u \in H^1_0(\Omega), \quad \phi \in \mathcal{D}(\Omega)\}$$

where $\mathcal{D}(\Omega)$ denotes the space of infinitely often continuously differentiable functions having compact support in Ω and the subspace $H^1_0(\Omega)$ of $H^1(\Omega)$ corresponds to the homogeneous Dirichlet condition (3.2).

We shall use weight functions of the form

$$p_\delta(x) := (1+ | \; x \; |)^{-\frac{1}{2}} (\log(e+ | \; x \; |))^{-\frac{1}{2}(\delta+1)}$$
$$q_\delta(x) := (\log(e+ | \; x \; |))^{-\frac{\delta}{2}}$$
$$\rho_\delta(x) := (1+ | \; x \; |)(\log(e+ | \; x \; |))^{\frac{\delta}{2}} \tag{3.3}$$

where $\delta \ge 0$ is a fixed number.

For any weight function $g(x) > 0$ we define the norm

$$\| \; u \; \|^2_{k,g} := \sum_{|\alpha| \le k} \int_\Omega g(x)^2 \; | \; D^\alpha u(x) \; |^2 \; dx. \tag{3.4}$$

Finally we introduce

$$L^2_g(\Omega) := \{u \in L^2_{loc}(\Omega) : gu \in L_2(\Omega)\}$$
$$H^k_g(\Omega) := \{u \in H^k_{loc}(\Omega) : gD^\alpha u \in L_2(\Omega), | \; \alpha \; | \le k\}. \tag{3.5}$$

A precise formulation of the radiation problem can now be given.

Definition 3.1

Let $f \in L^2_{loc}(\Omega)$ be given. An element $u \in H^1_{0*}(\Omega)$ is said to be a solution of the radiation problem

$$(\Delta + k^2)u = f \quad \text{in } \Omega \tag{3.6}$$

$$u = 0 \quad \text{in } \partial\Omega \tag{3.7}$$

if

  (i) u satisfies

$$(\nabla u, \nabla \phi)_0 - k^2(u, \phi)_0 = -(f, \phi)_0 \quad \text{for all } \phi \in \mathcal{D}(\Omega) \tag{3.8}$$

  (ii) the radiation condition

$$\frac{\partial u}{\partial r} - iku \in L_2(\Omega), \quad r = |\, x \,|, \tag{3.9}$$

is satisfied.

It can be shown (PN/GFR 1987):

Theorem 3.2

The radiation problem (3.6), (3.7) has at most one solution.

To prove existence of radiating solutions we consider the equation

$$(\Delta + k^2 + i\epsilon)u_\epsilon(x) = f_\epsilon(x), \quad x \in \Omega \tag{3.10}$$

where $u_\epsilon \in H^1_0(\Omega)$, $f_\epsilon \in L_2(\Omega)$.

For this equation the standard L_2 solution theory is available and the following can be established (PN/GFR 1987).

Theorem 3.3

Let $k \in \mathbf{R}$, $k \neq 0$ and $0 < \epsilon \leq 1$. Then for every $f_\epsilon \in L_2(\Omega)$ the equation (3.10) has a unique solution $u_\epsilon \in H^1_0(\Omega)$ and moreover

$$\epsilon \,\| \, u_\epsilon \, \|_1 \leq C_1 \, \| \, f_\epsilon \, \|_0 \tag{3.11}$$

where the constant C_1 is independent of u_ϵ, f_ϵ and ϵ.

It can then be proved that in the sense of some weighted norm the limit

$$\lim_{\epsilon \to 0+} u_\epsilon = u \tag{3.12}$$

exists, and furthermore, the limit function u is the solution of the radiation problem (3.6), (3.7).

To prove the convergence (3.12) the following a priori estimate has been established (PN/GFR 1987).

Theorem 3.4

Let $u_\epsilon \in H_0^1(\Omega)$, $0 < \epsilon \le 1$, be the solution of (3.10) with $f_\epsilon \in L_{\rho_\delta}^2$. Then

$$\| \nabla(e^{-ikr}u_\epsilon) \|_0 + \| u_\epsilon \|_{0,p_\delta} \le C\{\| f_\epsilon \|_{0,\rho_\delta} + \| u_\epsilon \|_1 (\Omega(R_1))\}$$

for some fixed constant $R_1 > 0$ and for $C > 0$ independent of $\epsilon, \delta, u_\epsilon$ and f_ϵ. Here $\| \cdot \|_1 (\Omega(R_1))$ denotes that $\Omega(R_1)$ rather than Ω is used in computing the norm.

The Limiting Absorption Principle (LAB) is now used to prove the following existence theorem for radiating solutions (PN/GFR 1987).

Theorem 3.5

Let Theorem 3.2 be valid and let $\Omega \subset \mathbf{R}^n$ be an exterior domain with the segment property. The radiation problem has, for every $f \in L_{\rho_\delta}^2(\Omega)$, $0 < \delta < \frac{1}{2}$ fixed, a unique solution $u \in H_{0p_\delta}^1(\Omega)$ and

$$\| \nabla(e^{-ikr}u) \|_0 + \| u \|_{0,p_\delta} \le C\{\| f \|_{0,\rho_\delta} + \| u \|_1 (\Omega(R_1))\}$$

for some fixed constants $R_1 > 0$ and $C > 0$ as in Theorem 3.4.

4. The Case of Infinite Interfaces

When $\partial\Omega$ is unbounded fewer results are available. Although Dirichlet and Neumann type boundary value problems have been considered the more natural and physically realistic problem would seem to be the transmission problem. As before the aims are as in the case of bounded $\partial\Omega$ to settle questions of existence and uniqueness. In this connection we would mention the pioneering works of Rellich (1943) and Jones (1953) together with the survey article by Eidus (1969).

Definition 4.1

Let Ω_1, Ω_2 be two unbounded half spaces in $\mathbf{R}^n$ separated by an unbounded surface $\Gamma \in C^2$. By a transmission problem we shall mean:

Find $u = (u_1, u_2) \in C^2(\Omega_1) \times C^2(\Omega_2)$ such that

$$-\Delta u_j - k_j^2 n_j(x)u_j = f_j, \quad x \in \Omega_j, \quad j = 1,2 \tag{4.3}$$

$$\alpha u_1 - u_2 = g_1, \quad x \in \Gamma \tag{4.4}$$

$$\beta\frac{\partial u_1}{\partial \nu} - \frac{\partial u_2}{\partial \nu} = g_2, \quad x \in \Gamma \tag{4.5}$$

$$u \in (RC) \tag{4.6}$$

where $\alpha, \beta \in \mathbf{C} \setminus \{0\}$, $k_j^2 > 0$ $(j = 1, 2)$, and $n_j(x)$, $(j = 1, 2)$ is real valued and tends to unity as $r = \mid x \mid \to \infty$. The relation (4.6) denotes that the solutions are required to satisfy certain, as yet unspecified, radiation conditions (RC).

Throughout we shall assume

A1: $n_j \in C^1(\Omega_j)$ and satisfies

$$\mid n_j(x) - 1 \mid + \mid r\partial_r n_j(x) \mid = 0(r^{-1}(\log r)^{-1-\delta}), \quad r = \mid x \mid \to \infty$$

where $0 < \delta \leq 1$, $j = 1, 2$, $\partial_r g = \partial g / \partial r$.

A2: There exists $R_0 > 0$ such that $\nu(x) = (\nu_1(x), \cdots, \nu_n(x))$ the unit normal to Γ at $x \in \Gamma$ drawn in the direction Ω_1 to Ω_2 satisfies

$$x.\nu(x) = 0, \quad x \in \Gamma, \quad \mid x \mid \geq R_0.$$

A3: $\bar{\alpha}\beta$ is a positive constant.

A4: $g_1 \in H^{\frac{1}{2}}(\Gamma(R))$, $g_2 \in H^{-\frac{1}{2}}(\Gamma(R)) \forall R > 0$

$$g_1 = g_2 = 0, \quad x \in \Gamma, \quad \mid x \mid \geq R_0$$

where

$$\Gamma(R) = \{x \in \Gamma : \mid x \mid < R\}.$$

We shall also need the abbreviations

$$\Omega_j(R) := \{x \in \Omega_j : \mid x \mid < R\}$$
$$E_j(R) := \{x \in \Omega_j : \mid x \mid > R\}$$
$$S_j(R) := \{x \in \Omega_j : \mid x \mid = R\}$$
$$\Omega_j(R_1, R_2) := \{x \in \Omega_j : R_1 < \mid x \mid < R_2\}$$
$$\Gamma(R_1, R_2) := \{x \in \Gamma : R_1 < \mid x \mid < R_2\}$$
$$\Gamma_e(R) := \{x \in \Gamma : \mid x \mid > R\}$$
$$\Omega(x, R) := \{y \in \mathbf{R}^n : \mid x - y \mid < R\}$$
$$\Omega(R) = \Omega(0, R).$$

We shall need certain special function spaces. To this end let $V \subset \mathbf{R}^n$ be an open set, $R^0 = 0$, $R^i = 2^{i-1}$, $i \geq 1$ and $V_i = V \cap \{x \in \mathbf{R}^n : R^{i-1} < \mid x \mid < R^i\}$.

We shall say that $u \in B(V)$ if

(i) $u \in L_2(V \cap \Omega(R)) \quad \forall R > 0$

(ii) $\parallel u \parallel_{B(V)} = \sum_{i=1}^{\infty} (R^i)^{\frac{1}{2}} \{\int_{V_i} \mid u(x) \mid^2 dx\}^{\frac{1}{2}} < \infty.$

We shall say $u \in B^*(V)$ if

(i) $u \in L_2(V \cap \Omega(R))$ $\forall R > 0$

(ii) $\| u \|^2_{B^*(V)} = \sup_{R>1} \frac{1}{R} \{ \int_{\substack{|x|<R \\ x \in V}} | u(x) |^2 \, dx \} < \infty.$

We shall say that $u \in L^{2,\alpha}(V)$ if

(i) $u \in L_2(V \cap \Omega(R))$ $\forall R > 0$

(ii) $\| u \|^2_{\alpha,V} = \int_V \{(1+ | x |)^{1/|\alpha|}(\log(e+ | x |))^{1+1/|\alpha|}\}^\alpha | u(x) |^2 \, dx < \infty.$

Let

$$\rho_\delta(x) := (1+ | x |)^{(n+1)/2}(\log(e+ | x |))^{1+\delta}.$$

We define

$$L^2_{\rho_\delta}(V) := \{u \in L_2(V) \cap \Omega(R))\forall R > 0 : \rho_\delta u \in L_2(V)\}$$

with norm

$$\| u \|^2_{0,\rho_\delta,V} := \int_V \rho_\delta^2 | u |^2 \, dx.$$

To obtain a weak formulation for the transmission problem, which will parallel that used in Definition 3.1 for the radiation problem, we introduce the following bilinear and linear functionals for

$$u = (u_1, u_2), v = (v_1, v_2) \in H^1(\Omega_1) \times H^1(\Omega_2)$$
$$a_i(u_i, v_i) = \int_{\Omega_i} \{\nabla u_i \cdot \nabla \bar{v}_i - k_i^2 n_i(x)u_i \cdot \bar{v}_i\}dx_i, \quad i = 1, 2$$
$$a(u, v) = \bar{\alpha}\beta a_1(u_1, v_1) + a_2(u_2, v_2)$$
$$F(v) = \bar{\alpha}\beta \int_{\Omega_1} f_1 \bar{v} \, dx + \int_{\Omega_2} f_2 \bar{v}_2 dx + \int_\Gamma g_2 \bar{v}_2 ds.$$

and the sets

$$H_e^* = \{u = (u_1, u_2) : u_i \in H^1(\Omega_i(R)), \ i = 1, 2, \ \alpha u_1 - u_2 = g_1, \ x \in \Gamma(R) \quad \forall R > 0\}$$
$$H_e = \{u = (u_1, u_2) : u_i \in H^1(\Omega_i(R)), \ \alpha u_1 - u_2 = 0, \ x \in \Gamma(R) \quad \forall R > 0\}.$$

With this notation we have the following weak formulation of the transmission problem.

Definition 4.2

Find

$$u = (u_1, u_2) \in H_e^* \cap \{B^*(\Omega_1) \times B^*(\Omega_2)\} \text{ such that}$$

$$a(u, \phi v) = F(\phi v) \qquad \forall v \in H_e, \quad \forall \phi \in C_0^\infty(\mathbf{R}^n). \tag{4.7}$$

$$\lim_{R \to \infty} \sum_{j=1}^{2} (\bar\alpha\beta)^{2-j} \{ \frac{1}{R} \int_{\Omega_j(1,R)} [|\, D_{r,k_j}(u_j)\,|^2 + r^{-2} \,|\, \nabla_1 u_j \,|^2] dx +$$

$$+ 2(-1)^{j-1} \frac{1}{R} \int_{\Gamma(1,R)} Im\{ r k_j \bar{u}_j \frac{\partial u_j}{\partial \nu} \} ds \} = 0 \tag{4.8}$$

$$\int_{\Omega_j} (1+r)^{-1} \{ |\, \nabla u_j \,|^2 - |\, \partial_r u_j \,|^2 \} dx < +\infty, \quad j = 1,2 \tag{4.9}$$

where

$$\nabla_1 := r(\nabla - \hat{x}\partial_r) \tag{4.10}$$

$$D_{r,\mu}(u_j) = \frac{\partial u_j}{\partial r} + \frac{(n-1)}{2r} u_j - i\mu u_j, \quad \mu \in \mathbf{C} \tag{4.11}$$

and we refer to $D_{r,u}$ as the radiation operator.

The relations (4.8), (4.9) are the required radiation conditions (RC) indicated in (4.6). These are different from the usual Rellich-Sommerfeld Radiation condition if $k_1 \neq k_2$. However, when $k_1 = k_2$ the surface integral over the part $\Gamma(1,R)$ of the interface Γ vanishes by virtue of the transmission conditions Roach and Zhang (1992).

The following uniqueness of solution is obtained in Roach and Zhang (1992).

Theorem 4.3

The transmission problem (4.7) to (4.9) has at most one solution.

Existence of solution to the problem (4.7) to (4.9) is proved by means of the LAB. In this case we make use of the damped equations

$$-\Delta u_{j\epsilon} - (1 + i\epsilon)k_j^2 n_j(x) u_{j\epsilon} = f_j, \quad \text{in } \Omega_j, \quad j = 1,2 \tag{4.12}$$

where $u_{j\epsilon} \in H^1(\Omega_j)$ and satisfies the transmission conditions (4.4), (4.5).

For the transmission problem centered on (4.12) the standard L_2-solution theory is available. To see this we introduce the following notations for

$$u = (u_1, u_2), \; v = (v_1, v_2) \in H^1(\Omega_1) \times H^1(\Omega_2)$$
$$a_{j\epsilon}(u_j, v_j) = a_j(u_j, v_j) - i\epsilon k_j^2 (n_j u_j, v_j)_0$$
$$a_\epsilon(u, v) = \bar\alpha\beta a_{1\epsilon}(u_1, v_1) + a_{2\epsilon}(u_2, v_2)$$

where $0 < \epsilon \leq 1$ and $j = 1, 2$,

$$H^* = \{ u = (u_1, u_2) : u_j \in H^1(\Omega_j)(j = 1,2), \alpha u_1 - u_2 = g_1 \text{ on } \Gamma(R) \; \forall \, R > 0 \}$$
$$H = \{ u = (u_1, u_2) : u_j \in H^1(\Omega_j)(j = 1,2), \alpha u_1 - u_2 = 0 \text{ on } \Gamma(R) \; \forall \, R > 0 \}.$$

The transmission problem centred on (4.12) is equivalent to the following problem:

Find $u_\epsilon = (u_{1\epsilon}, u_{2\epsilon}) \in H^*$ satisfying

$$a_\epsilon(u_\epsilon, v) = F(v) \quad \forall v \in H. \tag{4.13}$$

The following existence result is then available in Roach and Zhang (1992):

Theorem 4.4

Let $f_j \in L_2(\Omega_j)$ $(j = 1, 2)$ and $0 < \epsilon \leq 1$, then the problem (4.13) has a unique solution $u_\epsilon \in H^*$.

To settle the question of existence for the transmission problem we first make the following additional assumption.

A5: $\mid n_j(x) - 1 \mid = 0(r^{-(n+2))/2}(\log r)^{-2(1+\delta)})$, $\quad r = \mid x \mid \to \infty$, $\quad j = 1, 2.$

The LAB can now be established in the form [23].

Theorem 4.5

Let

    (i) A1 and A5 hold

    (ii) $f_j \in L^2_{\rho_\delta}(\Omega_j) \subset B(\Omega_j)$ $\quad (j = 1, 2)$

    (iii) $u_\epsilon \in H^*$, $0 < \epsilon \leq 1$ be a solution of (4.13)

then there is a $u \in B^*(\Omega_1) \times B^*(\Omega_2)$ such that

$$\lim_{\epsilon \to 0+} u_{j\epsilon} = u_j$$

exists in $W(B^*(\Omega_j), B(\Omega_j))$ $(j = 1, 2)$, that is for every $g \in B(\Omega_j)$

$$\lim_{\epsilon \to o+} \int_{\Omega_j} u_{j\epsilon} \bar{g} dx = \int_{\Omega j} u_j \bar{g} dx$$

and

$$D_{r,k_j}(u_j) \in B^*(E_j(1)) \quad (j = 1, 2).$$

Also u_j is the strong limit of u_{j_ϵ} in $L^{2,-\delta}(\Omega_j)$ $(j = 1, 2)$, that is

$$\lim_{\epsilon \to o+} u_{j\epsilon} = u_j$$

in the norm of $L^{2,-\delta}(\Omega_j)$ $(j = 1, 2)$. Furthermore, u is a solution of the transmission problem (4.3) to (4.6).

The following existence and uniqueness result can now be obtained for the transmission problem (4.3) to (4.6),

Theorem 4.6

Let

    (i) A1 and A5 hold

    (ii) $f_j \in L^2_{\rho_\delta}(\Omega_j) \ (h = 1, 2)$.

Then the transmission problem (4.3) to (4.6) has a unique solution $u \in H^*_e \cap \{B^*(\Omega_1) \times B^*(\Omega_2)\}$ and $D_{r,k_j}(u_j) \in B^*(E_j(1)) \ (j = 1, 2)$.

5. On time dependent problems

In this section we shall mention results which can be obtained concerning the asymptotic behaviour, as $t \to \pm\infty$, of solutions to the problem of determining $w := (w_1, w_2) = w(x, t)$ satisfying

$$(\frac{\partial^2}{\partial t^2} - \mu(x)^{-1}\Delta)w_j = f_j e^{-i\omega t}, (x, t) \in \Omega_j \times \mathbf{R} \tag{5.1}$$

$$\alpha w_1 = w_2, \beta\frac{\partial w_1}{\partial \nu} = \frac{\partial w_2}{\partial \nu}, (x, t) \in \Gamma \times \mathbf{R} \tag{5.2}$$

$$w_j(x, 0) = \frac{\partial w_j}{\partial t}(x, 0) = 0, x \in \Omega_j \tag{5.3}$$

where $\Omega_j \ (j = 1, 2)$ are two half spaces in $\mathbf{R}^n \ (n \geq 2)$ separated by an unbounded interface $\Gamma \in C^2$ with unit normal $\nu(x) = (\nu_1 x), \cdots, \nu_n(x))$ at $x \in \Gamma$ drawn from Ω_1 to Ω_2 and $\alpha, \beta \in \mathbf{C} \setminus \{0\}$. The data f_j are assumed to belong to suitably weighted L_2 spaces on Ω_j. The coefficient $\mu(x) = a(x)^{-1}$ tends to $m_j > 0$ in Ω_j as $r =\mid x \mid \to \infty$ where $a(x)$ is a wave propagation speed. Finally ω is some given positive constant.

    In Roach and Zhang (1992) conditions are given which allow the possibility of proving that $w := (w_1, w_2)$ has the property

$$w_j = u_j \exp(-i\omega t) + o(1), \quad j = 1, 2, \quad t \to +\infty \tag{5.4}$$

in an appropriate topology, where $u := (u_1, u_2)$ satisfies

$$(-\Delta - \omega^2\mu(x))u_j = \mu(x)f_j \quad \text{in } \Omega_j, \quad (j = 1, 2) \tag{5.5}$$

$$\alpha u_1 = u_2, \quad \beta\frac{\partial u_1}{\partial \nu} = \frac{\partial u_2}{\partial \mu} \quad \text{in } \Gamma. \tag{5.6}$$

If (5.4) is valid, where w solves (5.1) to (5.3) with u some solution of (5.5), then we say that the Limiting Amplitude Principle (LAM) holds. The validity of LAM implies that w tends to a stationary solution as $t \to +\infty$. However, we would remark that (5.4) need not always hold (cf Eidus (1965)).

    The LAM has been proved by many authors for various exterior boundary value problems [1, 17]. Results for problems with unbounded boundaries are given in

Ikebe (1960), Eidus (1969), Eidus and Vinnik (1974), Eidus (1986), Weder (1985) and Martin (1988). A particular novelty in the work [24] is that α and β are different.

Eidus (1969) developed a general theory for studying the asymptotic behaviour, for large time, of solutions to the Cauchy problem for the abstract wave equation. Briefly, this can be described as follows. Let A be a linear, self–adjoint operator in $L_2(\Omega)$ where Ω is an exterior domain. To obtain the asymptotic behaviour as $t \to \infty$ of solutions $w(x,t)$ of the Cauchy problem for the forced wave equation

$$\left(\frac{\partial^2}{\partial t^2} + A\right) w(x,t) = e^{-i\omega t} f(x) \tag{5.7}$$

we first prove the existence of the limit

$$\lim_{z \to \lambda+0i} u_z = u_\lambda, \quad \lambda > 0 \tag{5.8}$$

where u_z is assumed to satisfy

$$Au_z = zu_z + f, \quad Im\,z \neq 0 \tag{5.9}$$

with data $f \in L_2(\Omega)$ and the limit function, u_λ, is assumed to satisfy

$$Au_\lambda = \lambda u_\lambda + f, \quad \lambda \in \mathbf{R}^+. \tag{5.10}$$

This of course is equivalent to proving the validity of LAB for $\lambda > 0$. Next we prove that the limit function u_λ satisfies a Hölder condition as a function of λ. Finally we examine the behaviour of u_λ, that is the resolvent of A, as $\lambda \to 0$.

In Roach and Zhang (1993) this strategy is worked in detail for wave propagation in penetrable imhomogeneous media containing unbounded interfaces, that is for problem (5.1) to (5.3). The validity of LAM is proved and the behaviour of the resolvents of the associated reduced wave equations are studied.

The spectral analysis and scattering theory associated with an equation of the form (5.1) have been investigated by a number of authors. We would mention in particular Lax and Phillips (1967), Wilcox (1975) and Leis (1986) and the references cited. Unbounded domains with unbounded boundaries have been studied in Il'in (1989) and Lyford (1979) and perturbed stratified media in Weder (1985, 1986) . Recently the work in Roach and Zhang (1993) has been extended to allow a spectral analysis and scattering theory to be developed for wave propagation problems in inhomogeneous, penetrable media with unbounded interfaces. This work will be communicated separately but the details are as follows. The propagation problem is given a Hilbert space formulation involving a self-adjoint operator A. The LAB obtained in Roach and Zhang (1993) is used to discuss existence and uniqueness of solution to the problem and the spectral family of A determined. Two families of generalised Fourier transforms are constructed and used to obtain generalised eigenfunction expansions for wave propagation in inhomogeneous penetrable media with an unbounded interface. These eigenfunction expansions are used in conjunction with the method of stationary phase to prove the existence of the wave operators and to obtain the representation of the wave operators in terms of the generalised Fourier transforms.

6. Non-linear Problems

The philosophy and analytical structure introduced in the previous sections can also be used as a vehicle for investigating non-linear problems.

As a prototype we consider the non-linear problem **P2**:

$$(\frac{\partial^2}{\partial t^2} - \mu(x)^{-1}\Delta)w_j(x,t) = F_j(w_j(x,t), \quad (x,t) \in \Omega_j \times \mathbf{R}, \; j = 1,2$$

$$\alpha w_i = w_2, \quad \beta\frac{\partial w_1}{\partial \nu} = \frac{\partial w_2}{\partial \nu}, \quad (x,t) \in \Gamma \times \mathbf{R}$$

$$w_j(x,0) = \phi_{0j}(x), \quad \frac{\partial w_j}{\partial t}(x,0) = \phi_{ij}(x), \quad x \in \Omega_j, \; j = 1,2$$

where F_j is a not necessarily linear function of w_j on Ω_j and all the other notation is as for the linear problem.

Associated with **P2**, which we shall regard as the perturbed (hard) problem, is an unperturbed (easy) or free problem denoted by **P3**. The problem **P3** has the same form as **P2** save that the 'perturbation' F_j is now zero and the initial conditions are (say) $\bar{\phi}_{mj}$, $m = 0,1$, $j = 1,2$.

We proceed as before and obtain a problem **P4** of the form

$$\frac{d^2}{dt^2}w + Aw = F(w), \quad w(0) = \phi_0, \quad w_t(0) = \phi_1$$

where

$$A : H_0(\Omega) \supset D(A) \rightarrow H_0(\Omega)$$

$$H_0(\Omega) = L_2(\Omega, b\mu dx), \quad b = (\overline{\alpha}\beta)^{2-j}, \; j = 1,2$$

$$F(w)(x) = F_j(w_j)(x), \quad x \in \Omega_j, \; j = 1,2.$$

the norms on $H_0(\Omega)$ and $L_2(\Omega)$ can be shown to be equivalent, Roach and Zhang (1992). The unknown w is regarded as an $H_0(\Omega)$ valued function of t.

The second order problem **P4** can be reduced to the first order problem **P5**

$$\psi_t + iG\psi = N(\psi), \quad \psi(0) = \psi_0$$

where

$$w_t = v, \quad v_t + Aw = F(w)$$

$$\psi = \begin{bmatrix} w \\ v \end{bmatrix}, \quad \psi(0) = \begin{bmatrix} \phi_0 \\ \phi_i \end{bmatrix} = \psi_o, \quad iG = \begin{bmatrix} 0 & -I \\ A & 0 \end{bmatrix}, \quad N(\psi) = \begin{bmatrix} 0 \\ F(W) \end{bmatrix}.$$

Introduce $H_D(\Omega)$ the closure of $C_0^\infty(\Omega)$ with respect to

$$\| f \|_D^2 = \int_\Omega b(x) \mid \nabla f(x) \mid^2 dx$$

and the energy space

$$H_E(\Omega) = H_D(\Omega) \oplus H_0(\Omega)$$

with inner product

$$(f,g)_E = (f_1,g_1)_D + (f_2,g_2)_0$$

$$f = (f_1,f_2) \in H_E(\Omega), \quad g = (g_1,g_2) \in H_E(\Omega).$$

We now regard G in **P5** as the operator

$$G : H_E(\Omega) \supset D(G) \to H_E(\Omega)$$
$$D(G) = \{f = (f_1,f_2) \in H_E(\Omega) : \begin{bmatrix} 0 & -I \\ A & 0 \end{bmatrix} \begin{bmatrix} f_1 \\ f_2 \end{bmatrix} \in H_E(\Omega)\}$$
$$= \{f = (f_1,f_2) \in H_E(\Omega) : Af_1 \in H_0(\Omega), \quad f_2 \in H_D(\Omega)\}.$$

Direct computation indicates that G is self-adjoint on $H_E(\Omega)$. Consequently by Stone's Theorem we have that $\{U(t)\}_{t\in\mathbf{R}}$, where

$$U(t) := \exp(-itG),$$

is a well defined group of operators.

The problem **P5** can be reformulated as the integral equation

$$\psi(t) = U(t)\psi_\tau + \int_\tau^t U(t-s)N(\psi(s))ds. \qquad\qquad \textbf{P6}$$

We notice that the first term on the right hand side of this equation solves the unperturbed problem with data $U(\tau)\psi_\tau$ at $t = \tau$.

The problems **P5** and **P6** are similar to those which have been studied by Segal (1963), Strauss (1974), Reed (1976), Goldstein and others. The results obtained by these authors can now be used to settle questions of existence uniqueness and well-posedness of non-linear problems associated with unbounded surfaces; the particular features of the unbounded surface being contained in the definition of the operator G as outlined in earlier sections.

References

Eidus, D. M., *The principal of limiting absorption*, Amer. Math. Soc. Transl. Series 2, 47, 1965, 157-191.

Eidus, D. M., *The principle of limit amplitude*, Russian Maths. Surveys 24, 1969, 97-167.

Eidus, D. M., *The limiting absorption and amplitude principles for the diffraction problem with two unbounded media*, Comm. Math. Phys. 107, 1986, 29-38.

Eidus, D. M. and Vinnik, A. A., *On radiation conditions for domains with infinite boundaries*, Dokl. Akad. Nauk. SSR. 214 (1), 1974, 12-15.

Goldstein, J.A., *Semigroups of linear operators and applications*, Oxford University Press, 1985.

Helmberg, G., *Introduction to spectral theory in Hilbert space*, North Holland Publishing Co. London, 1969.

Ikebe, T., *Eigenfunction expansions associated with the Schrödinger operator and their applications to scattering theory*, Arch. Rational Mech. Anal. 5, 1960, 1-34.

Il'in, E.M., *Scattering by unbounded obstacles for elliptic operators of second order*, Proc. Steklov Inst. Math. (2), 1989, 85-107.

Jones, D.S., *The eigenvalues of $\Delta u + \lambda u = 0$ when the boundary conditions are given on semi-infinite domains*, Proc. Camb. Phil. Soc. 49, 1953, 668-684.

Kato, T., *Growth properties of solutions of the reduced wave equation with a variable coefficient*, Communs. Pure Appl. Math. 12, 1959, 403-425.

Kleinman, R.E. and Martin, P.A., *On single integral equations for the transmission problem of acoustics*, SIAM J. Appl. Math. 48, 1988, 307-325.

Kress, R. and Roach, G.F., *Transmission problems for the Helmholtz equation*, J. Math. Phys. 19, 1978, 1433-1437.

Kristensson, G., *A uniqueness theorem for Helmholtz equation penetrable media with an infinite interface*, SIAM J. Math. 11 (6), 1980, 1104-1117.

Kupradze, W. D., *Randwertaufgaben der Schwingungstheorie und Integralgleichungen*, Berlin: Deutsch, Vert. d. Wiss, 1956.

Lax, P.D. and Phillips, R.S., *Scattering Theory*, Academic Press, 1967.

Leis, R., *Initial boundary value problems in mathematical physics*, John Wiley, 1986.

Lyford, W. C., *Asymptotic energy propagation and scattering of waves in waveguides and cylinders*, Math. Ann. 219, 1979, 193-212.

Neittaanmäki, P. and Roach, G.F., *Weighted Sobolev spaces and exterior problems for the Helmholtz equation*, Proc. Roy. Soc. Lond. A 410, 1987, 373-383.

Odeh, F.M., *Uniqueness theorem for the Helmholtz equation in domains with infinite boundaries*, J. Math. Mech. 12, 1963, 857-868.

Ramm, A., *Scattering by a penetrable body*, J. Maths. Phys. 25, 1984, 469-471.

Ramm, A. and Werner, P., *On the limiting amplitude principle for a layer*, J. Reine Angew. Math. 360, 1985, 19-46.

Reed, M., *Abstract non-linear wave equations*, Lecture Notes in Mathematics No. 507, Springer Verlag, Berlin, 1976.

Rellich, F., *Über das asymptotische Verhalten der Lösungen von $\Delta u + \lambda u = 0$ in unendiche Gebieten*, Jber. dt. Mat. Verein 53, 1943, 57-65.

Rellich, F., *Studies and essays presented to R. Courant on his 60th birthday*, Interscience, New York, 1948, 329-344.

Roach, G.F. and Zhang, B., *A transmission problem for the reduced wave equation in inhomogeneous media with an infinite interface*, Proc. Roy. Soc. Lond. A 436, 1992, 121-140.

Roach, G. F. and Zhang, B., *The limiting amplitude principle for wave propagation problems with two unbounded media*, Proc. Camb. Phil. Soc. (to appear).

Saito, Y.M., *A remark on the limiting absorption principle for the reduced wave equation with two unbounded media*, Pacific J. Math. 136, 1989, 183-208.

Segal, I, *Non-linear semigroups*, Ann. Math. 78, 1963, 339-364.

Sommerfeld, A., *Vorlesungen über theoretische Physik VI*, Partielle Differentialgleichungen der Physik vol. 6, Leipzig, Geest and Portig, Akademisehe Verlag, 1966.

Strauss, W., *Non-linear scattering theory*, Scattering Theory in Mathematical Physics (Ed. J. A. Lavita, J.-P. Marchand, Reidel Publishing, Holland, 1974.

Vladimirov, V. S., *Equations of mathematical physics*, Marcel Dekker, Inc., New York, 1971.

Weder, R. *Spectral and scattering theory in perturbed stratified fluids*, J. Maths. Pures et Appl. 64, 1985, 149-173.

Weder, R. *Spectral and scattering theory in perturbed stratified fluids II. Transmission problems and exterior domains*, J. Differential Equations 64, 1986, 103-131.

Wilcox, C. *Scattering theory for the d'Alembert equation in exterior domains*, Lecture Notes in Mathematics, 442, Springer-Verlag, 1975.

Compactness and Asymptotic Stability for Solutions of Functional Differential Equations with Infinite Delay

W. M. Ruess   Universität Essen, Essen, Germany

1   INTRODUCTION

The object of this paper is to study the asymptotic behavior of solutions $x_\varphi : \mathbb{R} \to X$ to the following functional differential equation with infinite delay:

$$\text{(FDE)} \quad \begin{cases} \dot{x}(t) + (\alpha I + B)x(t) \ni F(x_t) \, , \ \ t \geq 0 \\ x\,|_{\mathbb{R}^-} = \varphi \in \hat{E}. \end{cases}$$

Here, α is a real constant, X a Banach (state) space, $B \subset X \times X$ a (generally) nonlinear and multivalued accretive operator in X, E a "suitably" chosen Banach space of continuous initial history functions $\varphi : \mathbb{R}^- \to X$, and $F : \hat{E} \subset E \to X$ a Lipschitz continuous mapping from a subset $\hat{E}$ of E into X. As usual, for a function $x : \mathbb{R} \to X$ and $t \geq 0$, the function $x_t : \mathbb{R}^- \to X$ is defined by $x_t(s) := x(t + s)$, $s \in \mathbb{R}^-$.

Equations of type (FDE) arise in the modeling of time-dependent processes for which the time rate of change at time t is a function both of the state at time t and the history of the process up to time t – such as temperature control in materials with thermal memory, or population dynamics. Examples where the state-responsive operator B typically is multivalued will be recorded below (Example 1.1) and in section 4.

Our analysis will be based on the solution operator semigroup associated to (FDE) in the initial history space E. This approach has long since been well developed for the case of single valued state-responsive operators B, cf. [4, 5, 6, 16, 17, 18, 23, 32, 38, 39, 40], but extended to multivalued operators $B \subset X \times X$ only recently [34, 36]. (For parallel direct approaches to existence and uniqueness via the 'method of lines' and further methods in the single valued nonautonomous case, we refer to [24, 25, 26, 27, 28].)

Section 2 is devoted to the solution operator semigroup for $B \subset X \times X$ multivalued. Along the way, we slightly extend the functional analytic tools as developed for this case in [34]. In sections 3 and 4, we develop applications of this result to the qualitative analysis of solutions to (FDE) in the general multivalued case. The *leitmotiv* for this part will be the following: assume that, in (FDE), the damping dominates the influence of the history, i.e., $\alpha \geq M =$ Lipschitz constant of the history-responsive function F; then the solutions to (FDE) qualitatively enjoy the same asymptotic behavior as the solutions to the corresponding undelayed Cauchy problem

$$\text{(CP)} \quad \begin{cases} \dot{x}(t) + Bx(t) \ni 0 , \ t \geq 0 \\ x(0) = x_0 \in X. \end{cases}$$

Results in this direction heavily depend on the choice of the initial history space E : while $E = BUC(\mathbb{R}^-, X) =$ bounded uniformly continuous functions from $(-\infty, 0]$ into X with sup-norm behaves "poorly" in this respect (cf. [37, Examples 4.1]), positive results can be achieved for weighted sup-norm spaces E_v for weights $v : \mathbb{R}^- \to (0,1]$ that suitably vanish at (minus) infinity (see section 2). For $E = E_v$ with weights v of this kind we show in section 3:

1. If the resolvent $J_\lambda^B = (I + \lambda B)^{-1}$, $\lambda > 0$, of B is compact, then bounded solutions to (FDE) have relatively compact range and, under slight restrictions on the weight v, they are asymptotically almost periodic (Theorems 3.1 and 3.2).

2. If, in addition, the Lipschitz constant of F for a coarser weighted sup-norm on $\hat{E}$ is strictly less than α, then the solutions to (FDE) are asymptotically stable: $\lim_{t\to\infty} \left\| x_\varphi(t) - x_\psi(t) \right\| = 0$, $\varphi, \psi \in \hat{E}$ (Theorem 3.3).

While, for $B : D(B) \subset X \to X$ single valued and linear, results of type 1. and 2. above can be deduced from suitably generalized variation-of-constants formulas for solutions to (FDE), the general case of $B \subset X \times X$ nonlinear and multivalued requires a more detailed analysis of the solution operators associated to (FDE).

In section 4, we apply the foregoing results to several examples of (perturbations of) partial differential operators arising from diffusion/absorption processes in physics and chemistry.

Notation and Terminology

Given a subset D of a Banach space Y, $co\,D$ will denote its convex hull, and $cl\,D$ its (norm-) closure in Y. Recall [10] that a subset $C \subset Y \times Y$ is said to be *accretive* in Y if for each $\lambda > 0$ and each pair $[x_i, y_i] \in C, i \in \{1, 2\}$, we have

$$\|(x_1 + \lambda y_1) - (x_2 + \lambda y_2)\| \geq \|x_1 - x_2\|,$$

maximal accretive if there exists no accretive subset of $Y \times Y$ strictly larger than C, and *m-accretive* in Y if, in addition, $R(I + \lambda C) = Y$ for all $\lambda > 0$. Moreover, for $\lambda > 0$, $J_\lambda = (I + \lambda C)^{-1}$, and $C_\lambda = \lambda^{-1}(I - J_\lambda)$. Finally, we recall ([8]) that if $R(I + \lambda C) \supset D(C)$ for all $\lambda > 0$, the generalized domain $\hat{D}(C)$ of C is given by $\hat{D}(C) = \{\varphi \in cl\,D(C) \mid$

$\lim_{\lambda \to 0+} \|C_\lambda \varphi\| < \infty\}$. For all these notions and the general theory of accretive sets and evolution equations, the reader is referred to [3, 9, 10, 30].

Example 1.1

One of the prototype examples of processes leading to multivalued operators $B \subset X \times X$ is the so-called 'thermostat problem', see [15, Ch. 2.3.2].

Given Ω bounded and open (and with smooth boundary) in $\mathbb{R}^N$, and prescribed temperatures $h_1(x) \leq h_2(x)$, $x \in \Omega$, the temperature $u(t,x)$ at (time $t \geq 0$ and) $x \in \Omega$ is to deviate as little as possible from the interval $[h_1(x), h_2(x)]$ through regulation by heat sources with heat flux $-\tilde{g}$, restricted by $-\tilde{g} \in [g_1, g_2]$ with $0 \in [g_1, g_2]$. In case heat is injected or extracted proportionally to the difference of the actual heat to the bounds $h_1(x)$ and $h_2(x)$, the function $-\tilde{g} = \beta(u)$ is piecewise linear and monotonically non-decreasing (see [15, Ch. 2.3.1]). However, for the ideal case of keeping $u(t,x)$ within a fixed interval $[h_1, h_2]$ at every time, one has to choose (cf. [15, Ch. 2.3.1])

$$\beta(r) = \begin{cases} (-\infty, 0] & , \quad r = h_1 \\ 0 & , \quad h_1 < r < h_2 \\ [0, \infty) & , \quad r = h_2 \end{cases}$$

The process as described above is then governed by the following multivalued operator B in $L^2(\Omega)$:

$$B = (-\Delta + \tilde{\beta}) \qquad \text{with} \qquad D(B) = W^{2,2}(\Omega) \cap W_0^{1,2}(\Omega) \cap D(\tilde{\beta}),$$

where

$$\begin{cases} D(\tilde{\beta}) & = \{u \in L^2(\Omega) \mid \exists v \in L^2(\Omega) : v(\omega) \in \beta(u(\omega)) \text{ a.e. } \omega \in \Omega\}, \\ \tilde{\beta}(u) & = \{v \in L^2(\Omega) \mid v(\omega) \in \beta(u(\omega)) \text{ a.e. } \omega \in \Omega\}, \qquad u \in D(\tilde{\beta}). \end{cases}$$

For this special operator B, (FDE) thus models temperature control (in the interior) for materials with a suitable 'thermal memory'. Further concrete examples will be considered in section 4.

2 THE SOLUTION SEMIGROUP FOR (FDE)

Throughout this paper, the initial history spaces will be weighted sup-norm spaces of the type $E_v = \{\varphi \in C(\mathbb{R}^-, X) \mid v\varphi \in BUC(\mathbb{R}^-, X)\}$, with norm $\|\varphi\|_v := \sup \{v(s) \|\varphi(s)\| \mid s \in \mathbb{R}^-\}$, where the (weight-) function $v : \mathbb{R}^- \to (0, 1]$ has the following properties:

(v1) v is continuous, nondecreasing, and $v(0) = 1$;

(v2) $\lim_{u \to 0-} \dfrac{v(s + u)}{v(s)} = 1$ uniformly over $s \in \mathbb{R}^-$.

The Banach space E_v is sometimes called a UC_g- space ($v = 1/g$), and has been considered by various authors; e.g., see [2], [21, 22], and the further references listed therein. However, we have chosen to view E_v within the more familiar framework associated with the general theory of weighted sup-norm spaces. Eventually (in sections 3 and 4), the following further restrictions on v will play a role:

(v3) $\lim_{s\to-\infty} v(s) = 0$;

(v3*) $\lim_{t\to\infty} \sup_{s\le -t} \dfrac{v(s)}{v(s+t)} = 0.$

We consider (FDE) both

(a) in the global context, i.e., $F : \hat{E} = E_v \to X$ is globally (defined and) Lipschitz continuous, and $B \subset X \times X$ is m-accretive, and

(b) in the local context, i.e., F is only (defined and) Lipschitz continuous on a closed convex subset $\hat{E}$ of E, and $B \subset X \times X$ is accretive such that conditions (A1) – (A3) of [36, 2. B] are fulfilled. In either case, if the operator A in E_v is defined by

$$(2.1) \quad \begin{cases} D(A) &= \{\varphi \in \hat{E} \mid \varphi' \in E,\ \varphi(0) \in D(B),\ \varphi'(0) \in F(\varphi) - (\alpha I + B)\varphi(0)\} \\ A\varphi &= -\varphi', \end{cases}$$

then $\hat{E} \subset R(I + \lambda A)$ for all $\lambda > 0$, and $-A$ generates a strongly continuous semigroup $(S(t))_{t\ge0}$ of type γ on $clD(A)$, with $\gamma = max\{0, M - \alpha\}$ (where $M =$ Lipschitz constant of F); i.e., $\|S(t)\varphi - S(t)\psi\|_v \le e^{\gamma t} \|\varphi - \psi\|_v$ for all $t \ge 0$ and all $\varphi, \psi \in clD(A)$. Moreover, $(S(t))_{t\ge0}$ acts as a translation, i.e., if $\varphi \in clD(A)$, then $S(t)\varphi = (\ x_\varphi)_t$ for all $t \ge 0$, where

$$(2.2) \qquad\qquad x_\varphi(t) = \begin{cases} \varphi(t) & ,\ t \le 0 \\ (S(t)\varphi)(0) & ,\ t \ge 0. \end{cases}$$

The proofs of these facts in the global $(F : \hat{E} = E_v \to X)$ and single valued $(B : D(B) \subset X \to X)$ context (cf. [4, 16, 17, 18, 32, 38, 40]) can be adapted to the above general local and multivalued context, see [36, Thm. 2.1]. This is no longer true for the final step, namely to show that the function x_φ of (2.2) actually is the (unique) solution to (FDE). Here, a continuous function $x : \mathbb{R} \longrightarrow X$ is said to be a *solution* to (FDE) if

(a) $x(s) = \varphi(s)$ for $s \in \mathbb{R}^-$;

(b) $x|_{\mathbb{R}^+}$ is locally absolutely continuous and differentiable a.e.;

(c) $x_t \in \hat{E}$, $x(t) \in D(B)$, and $\dot{x}(t) + \alpha x(t) - F(x_t) \in -B(x(t))$ for a.e. $t \in \mathbb{R}^+$.

We shall now present the techniques to show that for X^* uniformly convex, $B \subset X \times X$ maximal accretive and $\varphi \in \hat{D}(A)$, the function x_φ of (2.2) solves (FDE). More generally, we shall assume that X is a (real) reflexive Banach space whose norm is Fréchet-differentiable at any $x \in X\backslash\{0\}$. The latter condition is equivalent to the duality mapping J of X being single-valued and norm-norm-continuous, and is implied by X^* being locally uniformly convex (cf. [12, Chapter two, §2]). Recall that $J : X \longrightarrow 2^{X^*}$ is defined by $J(x) = \{x^* \in X^* \mid \langle x, x^* \rangle = \|x\|^2 = \|x^*\|^2\}$, $x \in X$.

THEOREM 2.1 *In the context of (FDE) as set forth above, if X is a (real) reflexive Banach space whose norm is Fréchet-differentiable at any $x \in X\backslash\{0\}$, and $B \subset X \times X$ is maximal accretive, then, for every $\varphi \in \hat{D}(A)$, the function x_φ of (2.2) is the unique solution to (FDE).*

(See [36, Thm. 2.5 (a)] and, for the more general nonautonomous case, [34, Thm.3.1].)

PROOF: It is known that, for $\lambda > 0$, the operator $-A_\lambda$ generates a strongly continuous semigroup of operators $(S_\lambda(t))_{t\geq 0}$ on $clco\,D(A)$ such that for $\varphi \in cl\,D(A)$,

$$(2.3) \qquad \lim_{\lambda\to 0+} S_\lambda(t)\varphi = S(t)\varphi \qquad \text{uniformly on bounded t-intervals,}$$

and $u_\lambda(t) = S_\lambda(t)\varphi$ is the unique continuously differentiable solution of

$$\begin{cases} \dot{u}_\lambda(t) + A_\lambda u_\lambda(t) = 0\,, & t \geq 0 \\ u_\lambda(0) = \varphi, \end{cases}$$

cf. [9], so that

$$(2.4) \qquad S_\lambda(t)\varphi(0) = \varphi(0) - \int_0^t A_\lambda S_\lambda(\tau)\varphi(0)\,d\tau, \qquad t \geq 0.$$

Let $T > 0$, and choose any sequence $0 < \lambda_n \to 0$ with $\lambda_n\gamma \leq c < 1$ for all $n \in \mathbb{N}$. Since $\varphi \in \hat{D}(A)$, (2.3) and further results of [10], particularly Lemmas 1.1, 1.3 and 1.4, imply that

$$(2.5) \qquad \begin{cases} (S_{\lambda_n}(\cdot)\varphi)_n,\ (J_{\lambda_n}S_{\lambda_n}(\cdot)\varphi)_n,\ (F(J_{\lambda_n}S_{\lambda_n}(\cdot)\varphi))_n \qquad \text{and} \\ (A_{\lambda_n}S_{\lambda_n}(\cdot)\varphi)_n \qquad \text{are uniformly bounded on } [0,T], \qquad \text{and} \\ \lim_{\lambda\to 0+} \|J_{\lambda_n}S_{\lambda_n}(\tau)\varphi - S(\tau)\varphi\| = 0 \qquad \text{for all } \tau \in [0,T]. \end{cases}$$

We pause to sketch the argument of proof for the case that B is single valued (compare [26]). Then, (2.4) and the fact that $A_\lambda = AJ_\lambda$ (A is single valued), and the definition of A and $D(A)$ in (2.1) imply that, for $t \in [0,T]$, and $n \in \mathbb{N}$,

$$(2.6) \qquad S_{\lambda_n}(t)\varphi(0) = \varphi(0) + \int_0^t [F(J_{\lambda_n}S_{\lambda_n}(\tau)\varphi) - (\alpha I + B)J_{\lambda_n}S_{\lambda_n}(\tau)\varphi(0)]\,d\tau\,.$$

Taking the limit as n tends to infinity, (2.3) and (2.5) together with Lebesgue's Dominated-Convergence theorem imply that

$$(S(t)\varphi)(0) = \varphi(0) + \int_0^t [F(S(\tau)\varphi) - (\alpha I + B)(S(\tau)\varphi)(0)]\,d\tau\,, \qquad t \in [0,T],$$

and thus, according to (2.2), complete the proof for this case if it is possible to show that the sequence $(BJ_{\lambda_n}S_{\lambda_n}(\tau)\varphi(0))_n$ converges weakly to $BS(\tau)\varphi(0)$, $\tau \in [0,T]$. However, for X^* uniformly convex and B single valued, this follows easily from the following Lemma by Kato (for details, cf. [40]).

LEMMA 2.2 ([29, Lemma 2.5]) *Assume that $C \subset X \times X$ is an m-accretive operator in a Banach space X with uniformly convex dual X^* and that $(x_n)_n$ and $(y_n)_n$ are sequences in X such that $[x_n, y_n] \in C$, $n \in \mathbb{N}$. If $(x_n)_n$ converges in norm to $x \in X$ and $(y_n)_n$ converges weakly to $y \in X$ then $[x,y] \in C$.*

Obviously, for the general case of $B \subset X \times X$ multivalued, the above step from (2.4) to (2.6) is no longer possible, and a new technique is required to complete the proof. The following parametrized version of Kato's Lemma above turns out to be the crucial tool.

LEMMA 2.3 *Assume that X is a (real) Banach space with single valued and norm-continuous duality map $J : X \to X^*$, and assume that (Ω, Σ, μ) is a finite measure space, and that $\{B(\omega) \mid \omega \in \Omega\}$ is a family of (generally, nonlinear and multivalued) operators $B(\omega) \subset X \times X$ such that, for $\omega \in \Omega$, $(B(\omega) + \alpha(\omega)I)$ is maximal accretive for some $\alpha(\omega) \in \mathbb{R}$.*
Let $x, x_n : \Omega \to X, n \in \mathbb{N}$, be functions from Ω into $X, y \in L^1(\mu, X)$ and $(y_n)_n \subset L^1(\mu, X)$ be such that
$$\begin{aligned}
&(i) &&\|x_n(\omega) - x(\omega)\| \longrightarrow 0 \quad a.e. \ \omega \in \Omega,\\
&(ii) &&y_n \rightharpoonup y \quad weakly \ in \ L^1(\mu, X), \ and\\
&(iii) &&[x_n(\omega), y_n(\omega)] \in B(\omega) \quad a.e. \ \omega \in \Omega \ for \ all \ n \in \mathbb{N}.
\end{aligned}$$

Then $[x(\omega), y(\omega)] \in B(\omega)$ a.e. $\omega \in \Omega$.

(This is a slight extension of the closely related result of Lemma 2.1 of [34].)

Deferring the proof of this Lemma for the moment, the proof of Theorem 2.1 can be completed in the following way. Define
$$\begin{aligned}
x &: [0, T] \to X : &&x(\tau) = (S(\tau)\varphi)(0)\\
x_n &: [0, T] \to X : &&x_n(\tau) = J_{\lambda_n} S_{\lambda_n}(\tau)\varphi(0)\\
y_n &: [0, T] \to X : &&y_n(\tau) = A_{\lambda_n} S_{\lambda_n}(\tau)\varphi(0) + F(J_{\lambda_n} S_{\lambda_n}(\tau)\varphi)
\end{aligned}$$

Then we have :
$$\begin{aligned}
&(i) &&\|x_n(\tau) - x(\tau)\| \longrightarrow 0 \quad \text{(from (2.5)), and}\\
&(ii) &&[x_n(\tau), y_n(\tau)] \in (\alpha I + B) \ \text{for all } n \in \mathbb{N}, \tau \in [0, T] \quad \text{(from (2.1)).}
\end{aligned}$$

Moreover, $(y_n)_n$ is bounded in $L^\infty([0, T]; X)$. As X is supposed to be reflexive, the vector version of Dunford's classical L^1- weak compactness criterion [13, Ch. IV.2, Theorem 1] reveals that $(y_n)_n$ is weakly relatively compact in $L^1([0, T]; X)$, and thus has a subsequence $(y_{n_k})_k$ converging to some $y \in L^\infty([0, T]; X)$ weakly in $L^1([0, T]; X)$. Lemma 2.3 above shows that

$$(2.7) \qquad\qquad [x(t), y(t)] \in (\alpha I + B) \quad a.e. \ t \in [0, T].$$

From (2.4) we have:

$$(2.8) \qquad S_{\lambda_{n_k}}(t)\varphi(0) = \varphi(0) - \int_0^t y_{n_k}(\tau)\,d\tau + \int_0^t F(J_{\lambda_{n_k}} S_{\lambda_{n_k}}(\tau)\varphi)\,d\tau.$$

Applying (2.5) and weak convergence of $(y_{n_k})_k$ in $L^1([0, T]; X)$ to y with respect to the functionals $\chi_{[0,t)} \otimes x^*$ on $L^1([0, T]; X), t \in [0, T], x^* \in X^*$, we conclude from (2.8) (together with (2.2)) that

$$x_\varphi(t) = \varphi(0) - \int_0^t y(\tau)\,d\tau + \int_0^t F((\,x_\varphi)_\tau)\,d\tau.$$

This shows that x_φ is locally absolutely continuous and differentiable a.e. on $[0, \infty)$, and, according to (2.7), fulfills

$$\dot{x}_\varphi(t) - F((\,x_\varphi)_t) = -y(t) \in -(\alpha I + B)\, x_\varphi(t)$$

for a.e. $t \in [0, \infty)$. Hence, x_φ is the solution to (FDE), and the proof of Theorem 2.1 is complete.

It remains to prove Lemma 2.3: First let us note that a reduction to Kato's Lemma 2.2 is not possible since weak convergence of $(y_n)_n$ to y in $L^1(\mu, X)$ does not imply that

$(y_n(\omega))_n$ converges to $y(\omega)$ weakly in X, $\omega \in \Omega$, not even for $X =$ reals; this is put to rest by the sequence $(r_n)_n$ of the Rademacher functions on $[0,1]$. Instead, we proceed as in part (b) of the proof of [34, Lemma 2.1]: as $(y_n)_n$ is $L^1(\mu, X)-$ weakly convergent to y, there is a sequence $(g'_n)_n$ with $g'_n \in co\{y_l | l \geq n\}$ for $n \in \mathbb{N}$, converging to y in $L^1(\mu, X)-$ norm, and hence a subsequence $(g_n)_n$ of $(g'_n)_n$ converging to y in norm pointwise a.e. on Ω. Note that $g_n \in co\{y_l | l \geq n\}$ for all $n \in \mathbb{N}$ as well, so that

$$g_n = \sum_{i=1}^{r_n} \alpha_i^n y_{l(i,n)},$$

for appropriate $r_n \in \mathbb{N}, \alpha_i^n \geq 0$ and $l(i,n) \geq n$ for all $i \in \{1, ..., r_n\}$, and $\sum_{i=1}^{r_n} \alpha_i^n = 1$. Let $E \in \Sigma$ be a $\mu-$ nullset such that $\|g_k(\omega) - y(\omega)\| \to 0$, $\|x_k(\omega) - x(\omega)\| \to 0$ as $k \to \infty$, and $[x_n(\omega), y_n(\omega)] \in B(\omega)$ for all $\omega \in \Omega \setminus E$, and all $n \in \mathbb{N}$. Given $\omega \in \Omega \setminus E$, we have, for all $[u, v] \in B(\omega)$,

$$\langle y(\omega) - v, J(x(\omega) - u)\rangle$$

$$= \lim_{n\to\infty} \langle g_n(\omega) - v, J(x(\omega) - u)\rangle = \lim_{n\to\infty} \sum_{i=1}^{r_n} \alpha_i^n \langle y_{l(i,n)}(\omega) - v, J(x(\omega) - u)\rangle$$

$$= \lim_{n\to\infty} \left\{ \sum_{i=1}^{r_n} \alpha_i^n \langle y_{l(i,n)}(\omega) - v, J(x(\omega) - u) - J(x_{l(i,n)}(\omega) - u)\rangle \right.$$

$$\left. + \sum_{i=1}^{r_n} \alpha_i^n \langle y_{l(i,n)}(\omega) - v, J(x_{l(i,n)}(\omega) - u)\rangle \right\}$$

$$\geq \lim_{n\to\infty} \left\{ -\alpha(\omega) \sum_{i=1}^{r_n} \alpha_i^n \left\| x_{l(i,n)}(\omega) - u \right\|^2 \right\} = -\alpha(\omega) \left\| x(\omega) - u \right\|^2.$$

Maximal accretivity of $(B(\omega) + \alpha(\omega)I)$ implies that $[x(\omega), y(\omega)] \in B(\omega)$ for all $\omega \in \Omega \setminus E$. This completes the proof of Lemma 2.3.

REMARKS 2.4   1. For the special case of $X = H =$ Hilbert space and $B(\omega) \equiv B$, Lemma 2.3 is due to Brézis [7, Lemma 3, p. 126] (with a different method of proof).
2. Dorroh [14, Lemma 3] has given a result related to Lemma 2.3 for $B(\omega) \equiv B$ in the setting of $L^2([0,1]; X)$ for X^* uniformly convex: mainly, the sequence $(x_n)_n$ is supposed to be in $C([0,1]; X)$ and to converge uniformly to $x \in C([0,1]; X)$, and the sequence $(y_n)_n$ is supposed to converge to y weakly in $L^2([0,1]; X)$ rather than in $L^1([0,1]; X)$. However, an inspection of the proof reveals that the result holds as well for X as in Lemma 2.3 and with $(x_n)_n$ only bounded in $C([0,1]; X)$ and converging pointwise to x. Thus, though weaker than Lemma 2.3 (for finite measure spaces, L^2-weak convergence implies L^1- weak convergence), Dorroh's result, extended to the more general setting as indicated above, could as well have served to complete the proof of Theorem 2.1; the details are rather obvious and are left to the interested reader. (For the more general nonautonomous case [34, Thm.3.1], however, the full force of Lemma 2.3 with regard to time-dependence of the operators $B(\omega)$ is needed.)

At the expense of requiring the operator B to be m-accretive, the geometric restrictions on the Banach space X in Theorem 2.1 can be relaxed to obtain the following variant for Banach spaces with the Radon-Nikodym property [13].

THEOREM 2.5 *In the context of (FDE) as set forth above, if X is a (real) Banach space with the Radon-Nikodym property, and $B \subset X \times X$ is m-accretive, then, for every $\varphi \in \hat{D}(A)$, the function x_φ of (2.2) is the unique solution to (FDE).*

PROOF: Given $\varphi \in \hat{D}(A)$, and $T > 0$, define $f : [0, T] \to X$ by $f(t) = F(S(t)\varphi)$, $t \in [0, T]$. Since f is Lipschitz continuous ([8, Corollary 1]), it is easy to check that the assumptions of [10, Theorem 5.1] are fulfilled for the operator $(\alpha I + B)$ and the function f. Moreover, $\varphi(0) \in \hat{D}(B)$ by [36, Step 1 of the proof of Proposition 3.3]. Hence, [10, Theorem 5.1 (iv)] applies to show that the function

$$U(t) = \lim_{n \to \infty} \prod_{k=1}^{n} \left(I + \frac{t}{n}[(\alpha I + B) - f(\frac{kt}{n})] \right)^{-1} \varphi(0)$$

is the strong solution of

$$\begin{cases} \dot{u}(t) + (\alpha I + B)u(t) \ni f(t), & 0 \le t \le T \\ u(0) = \varphi(0). \end{cases}$$

(Here we have used the fact that in [10, Theorem 5.1 (iv)] the restriction on the Banach space X to be reflexive can actually be weakened to X having the Radon-Nikodym property.) In order to prove Theorem 2.5, we thus need only show $S(t)\varphi(0) = U(t)$ for $t \in [0, T]$. We fix the following notation: given $\lambda > 0$ (sufficiently small), we denote by J_λ and J_λ^B the resolvents of the operators A of (2.1) and of B, respectively, and, for $t \in [0, T]$, we let $J_\lambda(t) = (I + \lambda[(\alpha I + B) - f(t)])^{-1}$. We have to show

(2.9)
$$\lim_{n \to \infty} \left\| (J_{\frac{t}{n}}^n \varphi)(0) - \prod_{k=1}^{n} J_{\frac{t}{n}}(\frac{kt}{n})(\varphi(0)) \right\| = 0$$

for all $t \in (0, T)$. Given $\lambda > 0$, we use the identity $J_\lambda(t)x = J_{\frac{\lambda}{1+\lambda\alpha}}^B [\frac{1}{1+\lambda\alpha}(x + \lambda f(t))]$ and induction to show that

$$\left\| (J_\lambda^n \varphi)(0) - \prod_{k=1}^{n} J_\lambda(k\lambda)(\varphi(0)) \right\| \le \frac{\lambda}{(1 + \lambda\alpha)^{n+1}} \sum_{k=1}^{n} \left\| F(J_\lambda^k \varphi) - F(S(k\lambda)\varphi) \right\| (1 + \lambda\alpha)^k.$$

If, for given $\lambda, \tau > 0$, $[\tau/\lambda]$ denotes the largest integer less than or equal to τ/λ, we thus arrive at

$$\left\| (J_\lambda^n \varphi)(0) - \prod_{k=1}^{n} J_\lambda(k\lambda)(\varphi(0)) \right\| \le (1 + \lambda\alpha)^{-\rho} \int_\lambda^{(n+1)\lambda} \left\| F(J_\lambda^{[\tau/\lambda]} \varphi) - F(S([\tau/\lambda]\lambda)\varphi) \right\| d\tau,$$

where $\rho = 1$ for $\alpha \ge 0$, and $\rho = (n + 1)$ for $\alpha < 0$. In either case, given $t \in (0, T)$, choosing $\lambda = t/n$ and letting $n \to \infty$, Lebesgue's Dominated-Convergence theorem serves to establish (2.9). This completes the proof.

3   COMPACTNESS, ALMOST PERIODICITY AND STABILITY

It is well known that, in the context of the Cauchy problem (CP), if the resolvent $J_\lambda^B = (I + \lambda B)^{-1}$, $\lambda > 0$, of B is compact, then bounded solutions to (CP) have relatively compact range, and thus are asymptotically almost periodic and, for $t \to \infty$, converge to an almost periodic solution to (CP) (cf. [11, 35]). Moreover, in special cases, such as for the perturbation $B = (-\Delta + \tilde\beta)$ of the Dirichlet - Laplacian in $L^2(\Omega)$ as in the 'thermostat problem' of Example 1.1, solutions to (CP) even have a limit in norm at infinity.

In this section, we first show that, in the context of (FDE), if the resolvent $J_\lambda^B = (I + \lambda B)^{-1}$, $\lambda > 0$, of B is compact, and $M \leq \alpha$, then bounded solutions to (FDE) have relatively compact range (Theorem 3.1). This result, in conjunction with Theorems 2.1 and 2.5 above and results of [37], will serve to transfer the above types of asymptotic behavior for solutions from the Cauchy problem (CP) to its delayed counterpart (FDE) (Theorems 3.2 and 3.3) in case the damping dominates the influence of the history, i.e., if $\alpha \geq M$ (compare the *leitmotiv* of the Introduction).

THEOREM 3.1 *Assume that the resolvent $J_\lambda^B = (I + \lambda B)^{-1}$, $\lambda > 0$, of B is compact [respectively, weakly compact] and that $\alpha \geq M$. Then, for $\varphi \in cl\, D(A)$, if $x_\varphi|_{\mathbb{R}^+}$ is bounded, it has relatively compact [respectively, weakly relatively compact] range.*

PROOF: Let $J_\lambda = (I + \lambda A)^{-1}$, $\lambda > 0$, denote the resolvent of the operator A from (2.1), and let $\varphi \in cl\, D(A)$ such that $x_\varphi|_{\mathbb{R}^+} : \mathbb{R}^+ \to X$ is bounded. Then, according to [37, Proposition 2.3], $S(\cdot)\varphi : \mathbb{R}^+ \to E_v$ is bounded as well. Moreover, we recall from [36, Proof of Theorem 2.1] that

$$(J_\lambda S(t)\varphi)(0) = (I + \frac{\lambda}{1 + \lambda\alpha}B)^{-1}\left[\frac{1}{1 + \lambda\alpha}((S(t)\varphi)(0) + \lambda F(J_\lambda S(t)\varphi))\right].$$

Taken together, these facts show that, under the assumptions of the theorem, the set $\{J_\lambda S(t)\varphi(0) \mid t \geq 0\}$ is relatively [weakly] compact, provided $x_\varphi(\mathbb{R}^+)$ is bounded. From [30, Ch. 2, Theorem 3.1] we read that, for $\lambda > 0$,

$$\|(J_\lambda S(t)\varphi)(0) - (S(t)\varphi)(0)\| \leq \|J_\lambda S(t)\varphi - S(t)\varphi\| \leq \frac{4}{\lambda}\int_0^\lambda \|S(s)\varphi - \varphi\|\, ds.$$

This estimate serves to show that, given any $\epsilon > 0$, the set $\{ x_\varphi(t) \mid t \geq 0\} = \{S(t)\varphi(0) \mid t \geq 0\}$ is $\epsilon-$ close (in norm) to the set $\{J_\lambda S(t)\varphi(0) \mid t \geq 0\}$, and thus is relatively [weakly] compact. This completes the proof of Theorem 3.1.

Theorems 2.1 and 3.1 can now be combined with results of [37] to derive [asymptotic] almost periodicity or stability of solutions to (FDE) for general multivalued operators $B \subset X \times X$ in case the weight v for the initial history space E_v satisfies additional properties (besides the standing hypotheses $(v1)$ and $(v2)$). We recall [19, 20] that a function $f \in C_b(\mathbb{R}^+, X)$ is said to be *asymptotically almost periodic* if its set $T(f) = \{f_t \mid t \geq 0\}$ of translates is sup-norm relatively compact. (Here, $f_t(s) = f(t + s), s, t \geq 0$.)

THEOREM 3.2 *In addition to the assumptions of Theorem 3.1, assume that the weight function v for the initial history space E_v satisfies (v3) (of section 2). Then, if $\varphi \in cl\,D(A)$ and either (i) $\lim_{s\to\infty} v(s)\,\|\varphi(s)\| = 0$, or (ii) v satisfies (v3*), we have: If $x_\varphi|_{\mathbb{R}^+}$ is bounded, then $x_\varphi|_{\mathbb{R}^+} : \mathbb{R}^+ \to X$ is asymptotically almost periodic, and there exists a unique element $\psi \in cl\,D(A)$ such that (i) $x_\psi|_{\mathbb{R}^+}$ is almost periodic, and (ii) $\lim_{t\to\infty} \left\| x_\varphi(t) - x_\psi(t) \right\| = 0$. Moreover, if $\varphi \in \hat{D}(A)$, then so is ψ.*

This follows from combining Theorem 3.1 with [37, Theorem 2.9].

A much stronger conclusion can be achieved - namely, stability of solutions to (FDE), in case the history controlling function F has a Lipschitz constant strictly less than the damping α for a coarser weighted norm on $\hat{E}$.

THEOREM 3.3 *Consider (FDE) in the setting of $\hat{E} = \hat{E}_v \subset E_v$, and assume that the resolvents $J_\lambda^B = (I + \lambda B)^{-1}$, $\lambda > 0$, of B are compact, and that $M < \alpha$. Moreover, assume that there exists a weight w satisfying (v1) and (v2) such that*

$$(i)\ w \le v,\quad (ii)\ \lim_{r\to-\infty} \frac{w(r)}{v(r)} = 0,\ and \quad (iii)\ \|F\varphi - F\psi\| \le M_w\,\|\varphi - \psi\|_w$$

for all $\varphi, \psi \in \hat{E}_v$, with a Lipschitz constant $0 \le M_w < \alpha$.
Let $\varphi, \psi \in cl\,D(A)$ such that the corresponding functions x_φ, x_ψ of (2.1) solve (FDE). Then we have:
If $x_\varphi|_{\mathbb{R}^+}$ and $x_\psi|_{\mathbb{R}^+}$ are bounded, then $\lim_{t\to\infty} \left\| x_\varphi(t) - x_\psi(t) \right\| = 0$.
In particular, if $0 \in D(B), 0 \in B(0)$, and $0 \in \hat{E}_v, F(0) = 0$, then $\lim_{t\to\infty} \left\| x_\varphi(t) \right\| = 0$ for all solutions x_φ to (FDE), $\varphi \in cl\,D(A)$.

This follows from combining Theorem 3.1 above with [37, Theorem 3.1].

We close this section by recalling that, with respect to Theorem 3.3, the stronger conclusion of even exponential stability of solutions to (FDE) follows without compactness assumptions from $\alpha > M$ alone in case the weight v is an exponential function of the form $v_\mu(s) = e^{\mu s}$ for some $\mu > 0$ ([33, Corollary 4.1]) or at least such that $v(\cdot)v_{-\mu}(\cdot)$ is nondecreasing on $\mathbb{R}^-$ ([37, Theorem 3.6 and Corollary 3.7]).

4 APPLICATIONS

In this section, we combine the results of sections 2 and 3 to determine the asymptotic behavior of various infinite-memory problems of type (FDE). We consider the following examples of (perturbations of) partial differential operators.
Throughout, Ω will denote a bounded open subset of $\mathbb{R}^N$ with smooth boundary Γ, and $\beta \subset \mathbb{R} \times \mathbb{R}$ will be a maximal monotone subset of $\mathbb{R}^2$ such that $0 \in \beta(0)$. $\tilde{\beta} \subset L^2(\Omega) \times L^2(\Omega)$ will be its realization in $L^2(\Omega)$, viz.,

$$\begin{cases} D(\tilde{\beta}) &= \{u \in L^2(\Omega) \mid \exists v \in L^2(\Omega) : v(\omega) \in \beta(u(\omega)) \text{ a.e. } \omega \in \Omega\}, \\ \tilde{\beta}(u) &= \{v \in L^2(\Omega) \mid v(\omega) \in \beta(u(\omega)) \text{ a.e. } \omega \in \Omega\}, \qquad u \in D(\tilde{\beta}). \end{cases}$$

4.1 The Dirichlet - Laplacian (modeling the 'thermostat problem', see Example 1.1).

$$\begin{cases} D(B_D) &= W_0^{1,2}(\Omega) \cap W^{2,2}(\Omega) \cap D(\tilde{\beta}), \\ B_D u &= -\Delta u + \tilde{\beta}(u), \qquad u \in D(B_D). \end{cases}$$

4.2 The Neumann - Laplacian.

If A denotes the operator in $L^2(\Omega)$ defined by

$$\begin{cases} D(A) &= \{u \in W_0^{2,2}(\Omega) \mid -\dfrac{\partial u}{\partial n} = 0 \quad \text{a.e. on } \Gamma\} \\ Au &= -\Delta u, \qquad u \in D(A), \end{cases}$$

then the operator B_N is defined by

$$\begin{cases} D(B_N) &= D(A) \cap D(\tilde{\beta}), \\ B_N u &= Au + \tilde{\beta}(u), \qquad u \in D(B_N). \end{cases}$$

4.3 The $r-$Laplacian $\Delta_r, 1 \leq r < \infty$ (modeling diffusion/absorption for non-Newtonian fluids [1]).

The $r-$Laplacian Δ_r in $L^p(\Omega), 1 \leq p < \infty, 1 \leq r < \infty$, is defined to be the closure $\bar{A}_r$ in $L^p(\Omega)$ of the operator A_r defined by

$$\begin{cases} D(A_r) &= \{u \in W_0^{1,r}(\Omega) \cap L^\infty(\Omega) \mid -div(\,|grad\,u|^{\,r-1}\,grad\,u) \in L^\infty(\Omega)\}, \\ A_r u &= -div(\,|grad\,u|^{\,r-1}\,grad\,u), \qquad u \in D(A_r). \end{cases}$$

Define the operator B_r by

$$\begin{cases} D(B_r) &= D(\bar{A}_r) \cap D(\tilde{\beta}), \\ B_r u &= \bar{A}_r u + \tilde{\beta}(u), \qquad u \in D(B_r). \end{cases}$$

In the following, let B denote any of the operators of 4.1 to 4.3. It is known that, in all three cases, B is m-accretive with compact resolvents $J_\lambda^B = (I + \lambda B)^{-1}$, $\lambda > 0$, and that $\lim_{t \to \infty} \|x(t)\| = 0$ for the solutions $x(\cdot)$ to the Cauchy problem (CP) associated to B (where, for the case 4.2, the additional assumption $\beta(0) = 0$ is required). This follows from [3, Ch. II.3,Proposition 3.7] and [7] for 4.1, from [31, §7] for 4.2, and from [41, 42] for 4.3.
With the help of Theorems 2.1 and 3.1 to 3.3 above, we can now draw the corresponding conclusions for the associated delayed problem (FDE).

THEOREM 4.1 *Let B denote any of the operators of 4.1 to 4.3 (with restricting p to $1 < p < \infty$ in 4.3), and consider the associated delayed problem (FDE). Moreover, assume that $0 \in \hat{E}, F(0) = 0$, and that $M = M_v \leq \alpha$.*
Then we have for all $\varphi \in \hat{D}(A)$:
(a) The solution x_φ of (FDE), restricted to $\mathbb{R}^+$, has relatively compact range.

(b) If, in addition, the weight v satisfies (v3) and either
    (i) $\lim_{s\to\infty} v(s)\,\|\varphi(s)\| = 0$, or
    (ii) v satisfies (v3),*
then the solution x_φ of (FDE), restricted to $\mathbb{R}^+$, is asymptotically almost periodic, and there exists a solution x_ψ to (FDE) for some $\psi \in \hat{D}(A)$ such that
(i) $x_\psi|_{\mathbb{R}^+}$ is almost periodic, and   (ii) $\lim_{t\to\infty}\left\| x_\varphi(t) - x_\psi(t)\right\| = 0$.
(c) If, in addition, there exists a weight w satisfying (v1) and (v2) such that
(i) $w \le v$,   (ii) $\lim_{r\to-\infty}\dfrac{w(r)}{v(r)} = 0$, and (iii) $\|F\varphi - F\psi\| \le M_w\,\|\varphi - \psi\|_w$
for all $\varphi, \psi \in \hat{E}_v$ for a Lipschitz constant $0 \le M_w < \alpha$, then $\lim_{t\to\infty}\left\| x_\varphi(t)\right\| = 0$.

PROOF: First, since in each of the cases 4.1 to 4.3, $B \subset X \times X$ is m-accretive and X^* is uniformly convex, Theorem 2.1 above reveals that for any $\varphi \in \hat{D}(A)$, the function x_φ of (2.1) is the unique solution to (FDE). Then, the fact that $0 \in B(0)$ and the extra assumption $0 \in \hat{E}$ and $F(0) = 0$ imply that the zero function is a solution to (FDE) (for $\varphi \equiv 0$) and hence, since $M \le \alpha$, i.e., $\gamma = 0$, that every solution to (FDE) with $\varphi \in cl\,D(A)$ is bounded. As the resolvents $J_\lambda^B = (I + \lambda B)^{-1}$, $\lambda > 0$, of B are compact, the proof of Theorem 4.1 can now be completed by applying Theorems 3.1 to 3.3.

Acknowledgement: The author is grateful to the referee for constructive remarks.

References

[1] G. Astarita and G. Marruci, *Principles of non-Newtonian fluid mechanics*, McGraw-Hill, New York 1974

[2] F.V. Atkinson and J. R. Haddock, *On determining phase spaces for functional differential equations*, Funkcial. Ekvac. 31 (1988), 331-347

[3] V. Barbu, *Nonlinear semigroups and differential equations in Banach spaces*, Noordhoff, Leyden 1976

[4] D.W. Brewer, *A nonlinear semigroup for a functional differential equation*, Trans. Amer. Math. Soc. 236 (1978), 173-191

[5] D.W. Brewer, *The asymptotic stability of a nonlinear functional differential equation of infinite delay*, Houston J. Math. 6 (1980), 321-330

[6] D.W. Brewer, *Locally Lipschitz continuous functional differential equations and nonlinear semigroups*, Illinois J. Math. 26 (1982), 374-381

[7] H. Brézis, *Monotonicity methods in Hilbert spaces and some applications to nonlinear partial differential equations.* In: Contributions to nonlinear functional analysis, E. Zarantonello (Ed.), Acad.Press 1971, 101-156

[8] M.G. Crandall, *A generalized domain for semigroup generators*, Proc. Amer. Math. Soc. 37 (1973), 434-440

[9] M.G. Crandall and T.M. Liggett, *Generation of semi-groups of nonlinear transformations on general Banach spaces*, Amer. J. Math. 93 (1971), 265-298

[10] M.G. Crandall and A. Pazy, *Nonlinear evolution equations in Banach spaces*, Israel J. Math. 11 (1972), 57-94

[11] C.M. Dafermos and M. Slemrod, *Asymptotic behavior of nonlinear contraction semigroups*, J. Funct. Anal. 13 (1973), 97-106

[12] J. Diestel, *Geometry of Banach spaces - selected topics*, Lecture Notes Math. 485, Springer 1975

[13] J. Diestel and J.J. Uhl, *Vector measures*, Math. Surveys 15, Amer. Math. Soc. 1977

[14] J.R. Dorroh, *A nonlinear Hille-Yosida-Phillips Theorem*, J. Functional Analysis 3 (1969), 345-353

[15] G. Duvaut and J.L. Lions, *Inequalities in Mechanics and Physics*, Grundlehren math. Wiss., vol. 219, Springer, Berlin, Heidelberg, New York, 1976

[16] J. Dyson and R. Villella-Bressan, *Functional differential equations and non-linear evolution operators*, Proc. Royal Soc. Edinburgh 75A (1975/76), 223-234

[17] J. Dyson and R. Villella-Bressan, *Semigroups of translation associated with functional and functional differential equations*, Proc. Royal Soc. Edinburgh 82A (1979), 171-188

[18] H. Flaschka and M.J. Leitman, *On semigroups of nonlinear operators and the solution of the functional differential equation* $\dot{x}(t) = F(x_t)$, J. Math. Anal. Appl. 49 (1975), 649-658

[19] M. Fréchet, *Les fonctions asymptotiquement presque-périodiques continues*, C.R. Acad. Sci. Paris 213 (1941), 520-522

[20] M. Fréchet, *Les fonctions asymptotiquement presque-périodiques*, Rev. Sci. 79 (1941), 341-354

[21] J.R. Haddock and W.E. Hornor, *Precompactness and convergence in norm of positive orbits in a certain fading memory space*, Funkcial. Ekvac. 31 (1988), 349-361

[22] J.K. Hale and J. Kato, *Phase space for retarded equations with infinite delay*, Funkcial. Ekvac. 21 (1978), 11-41

[23] F. Kappel and W. Schappacher, *Some considerations to the fundamental theory of infinite delay equations*, J. Differential Equations 37 (1980), 141-183

[24] A.G. Kartsatos, *A direct method for the existence of evolution operators associated with functional evolutions in general Banach spaces*, Funkc. Ekvacioj 31 (1988), 89-102

[25] A.G. Kartsatos and M.E. Parrott, *Global solutions of functional evolution equations involving locally defined Lipschitzian perturbations*, J. London Math. Soc. (2), 27 (1983), 306-316

[26] A.G. Kartsatos and M.E. Parrott, *Convergence of the Kato approximants for evolution equations involving functional perturbations*, J. Differential Equations 47 (1983), 358-377

[27] A.G. Kartsatos and M.E. Parrott, *A method of lines for a nonlinear abstract functional evolution equation*, Trans. Amer. Math. Soc. 286 (1984), 73-89

[28] A.G. Kartsatos and M.E. Parrott, *The weak solution of a functional differential equation in a general Banach space*, J. Differential Equations 75 (1988), 290-302

[29] T. Kato, *Nonlinear semigroups and evolution equations*, J. Math. Soc. Japan 19 (1967), 508-520

[30] N. Pavel, *Nonlinear evolution operators and semigroups*, Lecture Notes Math. 1260, Springer 1987

[31] A. Pazy, *Strong convergence of semigroups of nonlinear contractions in Hilbert space*, J. Analyse Math. 34 (1978), 1- 35

[32] A.T. Plant, *Nonlinear semigroups of translations in Banach space generated by functional differential equations*, J. Math. Anal. Appl. 60 (1977), 67-74

[33] A.T. Plant, *Stability of nonlinear functional differential equations using weighted norms*, Houston J. Math. 3 (1977), 99-108

[34] W.M. Ruess, *The evolution operator approach to functional differential equations with delay*, Proc. Amer. Math. Soc. 119 (1993), 783-791

[35] W.M. Ruess and W.H. Summers, *Minimal sets of almost periodic motions*, Math. Ann. 276 (1986), 145-186

[36] W.M. Ruess and W.H. Summers, *Operator semigroups for functional differential equations with delay*, to appear in Trans. Amer. Math. Soc.

[37] W.M. Ruess and W.H. Summers, *Almost periodicity and stability for solutions to functional differential equations with delay*, to appear.

[38] G.F. Webb, *Autonomous nonlinear functional differential equations and nonlinear semigroups*, J. Math. Anal. Appl. 46 (1974), 1-12

[39] G.F. Webb, *Functional differential equations and nonlinear semigroups in L^p-spaces*, J. Differential Equations 20 (1976), 71-89

[40] G.F. Webb, *Asymptotic stability for abstract nonlinear functional differential equations*, Proc. Amer. Math. Soc. 54 (1976), 225-230

[41] P. Wittbold, *Nonlinear evolution equations and completely accretive operators*, Diplomarbeit, Essen 1992

[42] P. Wittbold, *Asymptotic behaviour of certain nonlinear evolution equations in L^1*, in: Progress in partial differential equations: the Metz surveys 2 (M. Chipot, Ed.), Pitman Res. Notes Math. Series 296, 216 -230; Longman, Essex 1993

Singularity Formation in Smooth Solutions
for a Modified Fourier Law

Katarzyna Saxton   Loyola University, New Orleans, Louisiana

1. INTRODUCTION

The purpose of this paper is to investigate the possibility of formation of singularities
for rigid heat conductors. The model used here admits finite speed of propagation of heat
disturbances, by replacing the classical Fourier law with a modification introduced by
Kosinski [4]. Physical motivation for the model is given in Cimmelli and Kosinski [2].
The Fourier law for heat flux has been modified in a number of papers, for example,
Chester [1], Gurtin and Pipkin [3]. An advantage of the current approach is in retaining
proportionality between heat flux and a form of temperature gradient.

In this model, the concept of a new temperature scale β is introduced, which is
related to the absolute temperature ϑ by the initial-value problem

$$\beta_t = f(\vartheta, \beta) \tag{1.1}$$

$$\beta(t_0) = \beta_0 \quad , \quad \beta \in (0, \infty) \ .$$

The heat flux $\mathbf{q}$ is related to $\nabla \beta$ by:

$$\mathbf{q} = d \, \nabla \beta \ , \tag{1.2}$$

Supported by LEQSF Grant, Program R & D, (1991–1994) –RD–A–22.

where for rigid materials d could be in general a function of ϑ. Constitutive equations for nonlinear heat conductors form a set of equations for the following quantities: the Helmholtz free energy ψ, specific entropy η, the heat flux $\mathbf{q}$ in terms of the temperature $\vartheta > 0$, the new temperature gradient $\nabla\beta$, where

$$\psi = \widehat{\psi}(\vartheta, \nabla\beta)$$
$$\eta = \widehat{\varepsilon}(\vartheta, \nabla\beta) = -\partial_\vartheta \widehat{\psi}(\vartheta, \nabla\beta) \tag{1.3}$$

and

$$\mathbf{q} = d\nabla\beta .$$

For the explicit function f in Eq. (1.1) the Maxwell-Cattaneo relation can be recovered: $\tau\mathbf{q}_t + \mathbf{q} = -k\nabla\vartheta$, τ-relaxation time. We assume that the free energy function $\widehat{\psi}$ is given by

$$\widehat{\psi}(\vartheta, \nabla\beta) = \widehat{\psi}_1(\vartheta) + \frac{1}{2}\widehat{\varepsilon}_2(\nabla\beta)^2 , \tag{1.4}$$

where $\widehat{\varepsilon}_2 = k_0\tau_0/(\rho_0\vartheta_0)$, and τ_0, k_0 are characteristic material constants representing the relaxation time and thermal conductivity at ϑ_0, ρ_0 is a mass density. Additionally, we assume that the function in Eq. (1.2) is a constant, $d = k_0$.

We introduce two variables for a one-dimensional homogeneous rigid body,

$$\beta_t = w$$
$$\beta_x = p . \tag{1.5}$$

From the energy balance law, one obtains the following quasilinear hyperbolic system (cf. Kosinski and Saxton [5]),

$$-C(\beta, w)w_t + bpw_x + ap_x + H(\beta, w) = 0$$
$$\beta_t - w = 0 \tag{1.6}$$
$$p_t - w_x = 0 ,$$

where

$$C(\beta, w) = \rho_0\tau_0\vartheta c_V(\vartheta)$$
$$H(\beta, w) = \rho_0\vartheta c_V(\vartheta)f_0'(\beta)w$$
$$a = k_0\vartheta_0 \tag{1.7}$$
$$b = \tau_0 k_0 ,$$

and

$$\tau_0 f(\vartheta, \beta) = \vartheta_0 \log(\vartheta_0/\vartheta) + f_0(\beta)$$
$$f_0(\beta) = \tau_0 f(\vartheta_0, \beta) . \tag{1.8}$$

Here $c_V = \vartheta\partial_\vartheta\nabla > 0$ is the specific heat at constant volume.

LEMMA: Let $|r_0| = \sup_{x \in \mathbb{R}} |r_0(x)|$, $|s_0| = \sup_{x \in \mathbb{R}} |s_0(x)|$. Then, as long as smooth solution exist, we have

$$\sup_{x \in \mathbb{R}} |r(x,t)| + \sup_{x \in \mathbb{R}} |s(x,t)| \le (|r_0| + |s_0|)e^{\frac{2}{\tau_0}t} .$$

Proof: We rewrite the function $F(p) = r - s$ in Eq.(2.9)$_1$ in more convenient variables ζ, and z defined by

$$\zeta = \frac{b_0}{2a_0}(r - s) \quad , \quad z = \frac{b_0}{\sqrt{4a_0}}p . \tag{2.11}$$

Then

$$\zeta = \mathcal{F}(z) = z\sqrt{z^2 + 1} + \log\left(z + \sqrt{z^2 + 1}\right) .$$

Straightforward calculations lead to the observation

$$|z| \le \sqrt{|\zeta|} , \tag{2.12}$$

and by (2.11),

$$|p| \le \sqrt{\frac{2}{b_0}|r - s|} . \tag{2.13}$$

Integrating both Eq's in (2.10) along their corresponding characteristic yields

$$r(x_2(t,\delta),t) = r_0(\delta) - \frac{1}{2\tau_0}\int_0^t \left\{ r(x_2(\tau,\delta),\tau) + s(x_2(\tau,\delta),\tau) - \frac{b_0}{2}p^2(x_2(\tau,\delta),\tau) \right\} d\tau$$

$$s(x_1(t,\gamma),t) = s_0(\gamma) - \frac{1}{2\tau_0}\int_0^t \left\{ r(x_1(\tau,\gamma),\tau) + s(x_1(\tau,\gamma),\tau) - \frac{b_0}{2}p^2(x_1(\tau,\gamma),\tau) \right\} d\tau , \tag{2.14}$$

where by integrating (2.5), characteristic curves are given by

$$x_2(t,\delta) = \delta + \int_0^t \lambda^2 \, d\tau$$

$$x_1(t,\gamma) = \gamma + \int_0^t \lambda^1 \, d\tau . \tag{2.15}$$

Equations (2.14) and (2.13) imply the inequalities

$$|r(x_2(t,\delta),t)| \le |r_0(\delta)| + \frac{1}{\tau_0}\int_0^t \{|r(x_2(\tau,\delta),\tau)| + |s(x_2(\tau,\delta),\tau)|\} \, d\tau$$

$$|s(x_1(t,\gamma),t)| \le |s_0(\gamma)| + \frac{1}{\tau_0}\int_0^t \{|r(x_1(\tau,\gamma),\tau)| + |s(x_1(\tau,\gamma),\tau)|\} \, d\tau . \tag{2.16}$$

Defining $R(t) = \sup_{x \in \mathbb{R}} |r(x,t)|$, $S(t) = \sup_{x \in \mathbb{R}} |s(x,t)|$, it follows from (2.16) that

$$R(t) \le |r_0| + \frac{1}{\tau_0}\int_0^t \{R(\tau) + S(\tau)\} \, d\tau$$

$$S(t) \le |s_0| + \frac{1}{\tau_0}\int_0^t \{R(\tau) + S(\tau)\} \, d\tau . \tag{2.17}$$

Then

$$R(t) + S(t) \le |r_0| + |s_0| + \frac{2}{\tau_0} \int_0^t \{R(\tau) + S(\tau)\}\, d\tau \;,$$

and finally by Gronwall's inequality

$$R(t) + S(t) \le (|r_0| + |s_0|)\, e^{\frac{2}{\tau_0} t} \;,$$

which establishes the lemma.

THEOREM: If $r_{0,x}$ or $s_{0,x}$ are sufficiently large for some $x \in \mathbb{R}$, then any $C^1(\mathbb{R})^2$ solution (w, p) to system (2.1) only exists for finite time.

Proof: The functions $\lambda^1(p)$ and $\lambda^2(p)$ can be expressed as functions of $(r - s)$ by (2.9). Thus $\lambda^1 = \lambda^1(r - s)$, $\lambda^2 = \lambda^2(r - s)$. Differentiating Eq's (2.8) with respect to x implies

$$
\begin{aligned}
r'_x + \lambda^2_r r_x^2 + \lambda^2_s r_x s_x + \frac{1}{\tau_0} w_x &= 0 \\
s_x + \lambda^1_s s_x^2 + \lambda^1_r r_x s_x + \frac{1}{\tau_0} w_x &= 0 \;.
\end{aligned}
\tag{2.18}
$$

Now we use the relations

$$r_x = \frac{1}{\lambda^1 - \lambda^2}(r - s), \quad s_x = \frac{1}{\lambda^1 - \lambda^2}(r' - s')
\tag{2.19}$$

together with

$$w_x = \frac{\lambda^1}{\lambda^1 - \lambda^2} s_x - \frac{\lambda^2}{\lambda^1 - \lambda^2} r_x \;.
\tag{2.20}$$

We will proceed with the proof based on Eq. $(2.18)_1$, concentrating on the value of r_x. The argument applied here could be applied in a similar way to Eq. $(2.18)_2$ for s_x.

Multiplying both sides of $(2.18)_1$ by the integrating factor $\mu = \mu(r - s)$ defined by

$$\mu(r - s) = exp\left(-\int_0^{r-s} \left(\frac{\lambda^2_r}{\lambda^1 - \lambda^2}\right)(\zeta)\, d\zeta\right)
\tag{2.21}$$

gives

$$(\mu r_x)' + \lambda^2_r \mu r_x^2 - \frac{\lambda^2_r}{\tau_0(\lambda^1 - \lambda^2)} \mu r_x + \frac{\lambda^1}{\tau_0(\lambda^1 - \lambda^2)^2} \mu(r' - s') = 0 \;.
\tag{2.22}$$

We now follow Slemrod [9]. Let us define

$$f(r - s) = -\int_0^{r-s} \left(\frac{\lambda^1}{(\lambda^1 - \lambda^2)^2} \mu\right)(\zeta)\, d\zeta
\tag{2.23}$$

and

$$\theta = \mu r_x \;,
\tag{2.24}$$

so that Eq. (2.22) becomes

$$\theta' = -\lambda_r^2 \mu r^{-1} \theta^2 + \frac{\lambda^2}{\tau_0(\lambda^1 - \lambda^2)}\theta + \frac{1}{\tau_0}f' \ . \tag{2.25}$$

Setting

$$\widehat{\theta} = \theta - \frac{1}{\tau_0}f$$

$$k_1(r - s) = -\lambda_r^2 \mu^{-1} > 0 \tag{2.26}$$

$$k_2(r - s) = -\frac{\lambda^2}{\lambda^1 - \lambda^2} > 0$$

gives us

$$\widehat{\theta}' = k_1\widehat{\theta}^2 + \frac{1}{\tau_0}k_2\left[2\frac{k_1}{k_2}f - 1\right]\widehat{\theta} + \frac{k_1}{\tau_0^2}f^2 - \frac{k_2}{\tau_0^2}f \ . \tag{2.27}$$

Defining

$$\varepsilon = \inf k_1 \geq 0, \quad M = \sup\left|\frac{1}{\tau_0}k_2\left[2\frac{k_1}{k_2}f - 1\right]\right|$$

$$K = \inf\left(-\frac{k_2}{\tau_0^2}f\right) \tag{2.28}$$

implies

$$\frac{1}{\tau_0}k_2\left[2\frac{k_1}{k_2}f - 1\right]\widehat{\theta} \geq -\frac{1}{2}\left(\varepsilon\widehat{\theta} + \frac{M^2}{\varepsilon}\right) \tag{2.29}$$

and

$$k_1\widehat{\theta}^2 + \frac{1}{\tau_0}k_2\left[2\frac{k_1}{k_2}f - 1\right]\widehat{\theta} + \frac{k_1}{\tau_0^2}f^2 - \frac{k_2}{\tau_0^2}f \geq \frac{\varepsilon}{2}\widehat{\theta}^2 - \frac{M}{2\varepsilon} + K \ . \tag{2.30}$$

Without loss of generality, suppose $K \leq 0$.

Next we compare two solutions $\widehat{\theta}$ of Eq. (2.27) and $\widetilde{\theta}$ for which $\widehat{\theta}(x,0) = \widetilde{\theta}(x,0) = \theta_0$, where

$$\frac{d}{dt}\widetilde{\theta} = \frac{\varepsilon}{2}\widetilde{\theta}^2 - \frac{M}{2\varepsilon} + K \tag{2.31}$$

for which $\widehat{\theta}(t) \geq \widetilde{\theta}(t)$ for some $t \in [0, d)$.

The solution $\widetilde{\theta}(t)$ of Eq. (2.31) is given by

$$\widetilde{\theta}(t) = \sqrt{\frac{A}{B}}\frac{1 + D_0 e^{2\sqrt{AB}t}}{1 - D_0 e^{2\sqrt{AB}t}} \tag{2.32}$$

where

$$A = \frac{\varepsilon}{2} \quad , \quad B = \frac{M}{2\varepsilon} - K \quad , \quad D_0 = \frac{\theta_0 - \sqrt{\frac{A}{B}}}{\theta_0 + \sqrt{\frac{A}{B}}} \quad , \tag{2.33}$$

$$\theta_0 = \mu(r_0 - s_0)r_{0,x} - \frac{1}{\tau_0}f(r_0 - s_0) \ .$$

From this, it follows that $\widetilde{\theta}(x_2(t,\delta),t)$, and so $\widehat{\theta}(x_2(t,\delta),t)$ tends to infinity in finite time t^* if $0 < D_0 < 1$ or, equivalently,

$$0 < \frac{2G}{\mu(r_0 - s_0)r_{0,x} - \frac{1}{\tau_0}f(r_0 - s_0) + G} < 1 \tag{2.34}$$

where $G = \sqrt{\frac{A}{B}}$.

Eq. (2.34) is satisifed for $r_{0,x}$ sufficiently large. Since $\mu(r - s)r_x = \widehat{\theta} + \frac{1}{\tau_0}f$, thus $\mu(r - s)r_x$ goes to infinity. From the lemma, $\mu(r - s)$ is bounded for $t \le t^*$, and so r_x goes to infinity.

REFERENCES

[1] M. Chester, Second sound in solids, Phys. Rev. 131, 2103–2015 (1962).

[2] V. A. Cimmelli and W. Kosinski, Heat waves in continuous media at low temperatures, Quaderni del Dipartimento di Matematica, n.6/1992, Universaita della Basilicata, Potenza.

[3] M. E. Gurtin and A. C. Pipkin, A general theory of heat conduction with finite wave speeds, Arch. Rational Mech. Anal. 31, 113–126 (1968).

[4] W. Kosinski, Elastic waves in the presence of a new temperature scale, Elastic wave propagation, M. F. McCarthy and M. A. Hayes, eds., Elsevier Science (North–Holland), Amsterdam, 1989, p. 629.

[5] W. Kosinski and K. Saxton, The effect on finite time breakdown due to modified Fourier laws, Quart. Appl. Math. 51, 55–68 (1993).

[6] P. D. Lax, Development of singularities of solutions of nonlinear hyperbolic partial differential equations, J. Mathematical Physics. 5, 611–613 (1964).

[7] L. Longwei and Z. Yongshu, Existence and non-existence of global smooth solutions for quasilinear hyperbolic systems, Chin, Ann. of Mah. 9B, 372–377 (1988).

[8] T. Nishida, Nonlinear hyperbolic equations and related topoics influid dynamics, Publications Mathematiques D'orsay, Universite de Paris-Sud, Department de Mathematique, 78.02 (1978).

[9] M Slemrod, Instability of steady shearing flow in a nonlinear viscoelastic fluid, Arch. Rational Mech. Anal. 68, 111–225 (1978).

Blow-up at the Boundary of Solutions to Nonlinear Evolution Equations

Ralph Saxton   University of New Orleans, New Orleans, Louisiana

1   INTRODUCTION

In [7], by using Riemann invariants and appropriate ordinary differential equations defined along the characteristic curves of the system, Lax obtained an estimate for the breakdown time of solutions to the one-dimensional equation of compressible elasticity,

$$u_{tt} - \sigma'(u_x)u_{xx} = 0 \tag{1}$$

under the conditions of genuine nonlinearity and strict hyperbolicity. Such a result was later obtained without the use of Riemann invariants by F. John, ([2]), for genuinely nonlinear systems of n first order equations. He showed that, in this case, uncoupling into n simple waves occurs from initial data with compact support, provided breakdown has not previously taken place. Proof of finite time breakdown then follows using simple wave properties, ([3]).

Klainerman and Majda, ([6]), reexamined (1) while allowing both characteristic families to become linearly degenerate on a curve in phase space. Their approach was again based on Riemann invariants and admitted periodic initial data. Keyfitz and Kranzer, ([5]), subsequently broke the requirement of strict hyperbolicity by examining a related system of conservation laws in which both characteristics need not always remain distinct. In this nonstrictly hyperbolic case, one eigenvalue is linearly degenerate while the other is genuinely nonlinear, and breakdown follows as a result of the behaviour of the genuinely nonlinear mode.

383

In §2, we start by considering the behaviour of solutions to scalar conservation laws in n spatial dimensions under conditions related to linear degeneracy. This provides a simple regularity result which allows for the existence of at least one maximal characteristic on which a solution must remain smooth, provided the data is correspondingly smooth, and it also gives some basic information on the domain of analyticity associated with the Cauchy-Kowalevski Theorem when applicable in this situation. Next, in §3, we examine equation (1) in the context of possible linear degeneracy and nonstrict hyperbolicity. This example extends work by Sakamoto, ([9]), for initial-boundary value problems with characteristic boundary, and also ([10]). In §4, we briefly examine a higher dimensional problem motivated by (1) and the full equations of motion for three dimensional nonlinear elasticity. A simple extension of linear degeneracy and nonstrict hyperbolicity is presented for that situation, and we show that the high-dimensional breakdown results presented for the genuinely nonlinear, nonstrictly hyperbolic situation in [10] need no longer apply. Finally, in §5, we examine a more unusual problem involving an algebraic constraint, derived from the context of liquid crystals having high inertial masses, ([1]), ([11]).

2   SCALAR CONSERVATION LAWS

In this section, we briefly examine scalar conservation laws of the form

$$u_t + (\mathbf{a}(u).\nabla)u = 0, \ \mathbf{x} \in I\!\!R^n, \tag{2}$$

$$u(\mathbf{x}, 0) = u_0(\mathbf{x}). \tag{3}$$

Here $u : I\!\!R^n \times I\!\!R_+ \to I\!\!R$, and $\mathbf{a}(.) : I\!\!R \to I\!\!R^n$. It is well known, ([8]), that equation (2) has solutions given in implicit form by

$$u(\mathbf{x}, t) = u_0(\mathbf{x} - \mathbf{a}(u(\mathbf{x}, t))t), \tag{4}$$

that are constant along characteristics

$$\frac{d\mathbf{x}}{dt} = \mathbf{a}(u_0(\boldsymbol{\alpha})), \ \mathbf{x}(0) = \boldsymbol{\alpha}, \tag{5}$$

in which $\boldsymbol{\alpha}$ denotes a (vector-valued) 'Lagrangian' reference coordinate. Taking the spatial gradient of (4) leads to the relation

$$\nabla u(\mathbf{x}, t) = (1 + t\mathbf{a}'(u_0(\boldsymbol{\alpha})).\nabla_\alpha u_0(\boldsymbol{\alpha}))^{-1}\nabla_\alpha u_0(\boldsymbol{\alpha}), \tag{6}$$

which demonstrates that finite time breakdown of solutions is always possible by choosing data such that $\mathbf{a}'(u_0(\boldsymbol{\alpha})).\nabla_\alpha u_0(\boldsymbol{\alpha}) < 0$, unless $\mathbf{a}'(u_0(\boldsymbol{\alpha})) \equiv 0$.

Definition The α-characteristic is *linearly degenerate of order k* with respect to the initial data if $\mathbf{a}^{(i)}(u_0(\boldsymbol{\alpha})) \equiv 0, i = 1, 2..., k$. In particular, if $k = 1$, this means linear degeneracy in the usual sense for scalar conservation laws, and if $k = \infty$, we call the characteristic *completely* linearly degenerate.

Definition If $\mathbf{a}^{(j)}(u_0(\boldsymbol{\alpha})) \neq 0$ for some $j \geq 1$, the α-characteristic is called *nondegenerate of order j*. If $j = 1$, this reduces to the usual definition of genuine nonlinearity, ([8]).

From equation (6), it is clear that ∇u, as well as u, remains constant along any linearly degenerate characteristic. In the following results, it is seen that linear degeneracy is sufficient to guarantee smoothness of solutions along a characteristic, linear degeneracy of order k keeps the first k derivatives constant along the characteristic, while complete linear degeneracy leads to preservation of all derivatives. In particular, 'higher order' breakdown is not possible in any case. The results apply to 'maximal' characteristics, prior to any interaction with, for example, shocks generated by neighbouring, genuinely nonlinear characteristic fields since global regularity of the solution on the chosen characteristic may depend nonlocally on the initial data.

Lemma 2.1 *Let u be any solution to (2) defined on a characteristic, $\alpha = \alpha^*$, which is linearly degenerate. Then regularity of u is conserved over the characteristic.*

Proof Computing the second spatial gradient of equation (2) gives

$$\nabla\nabla u_t + (\mathbf{a}(u).\nabla)\nabla\nabla u$$
$$+ \quad (\mathbf{a}'(u).\nabla)(\nabla u \nabla u) + \nabla\nabla u(\mathbf{a}'(u).\nabla)u + \nabla u \nabla u(\mathbf{a}''(u).\nabla)u = 0. \tag{7}$$

Along the linearly degenerate characteristic, $\alpha = \alpha^*$, this reduces to

$$(\partial_t + (\mathbf{a}(u).\nabla))\nabla\nabla u + \aleph = 0, \tag{8}$$

where $\aleph \equiv \nabla u \nabla u(\mathbf{a}''(u).\nabla)u$ stays constant. Thus, on the characteristic $\mathbf{x} = \mathbf{a}(u_0(\alpha^*))t + \alpha^*$, one has

$$\nabla\nabla u(\mathbf{x}, t) = \nabla\nabla u(\alpha^*) - c\aleph t. \tag{9}$$

where c is a constant.
Similarly, by taking one time derivative, it follows that

$$(\partial_t + (\mathbf{a}(u).\nabla))u_t + u_t(\mathbf{a}'(u).\nabla)u = 0, \tag{10}$$

which reduces to

$$(\partial_t + (\mathbf{a}(u).\nabla))u_t = 0, \tag{11}$$

and so u_t remains constant on $\alpha = \alpha^*$, with its value determined in terms of the initial data through the original conservation law. Applying the gradient operator to (10),

$$\nabla u_{tt} + (\mathbf{a}(u).\nabla)\nabla u_t$$
$$+ \quad (\mathbf{a}'(u).\nabla)(u_t\nabla u) + \nabla u_t(\mathbf{a}'(u).\nabla)u + u_t\nabla u(\mathbf{a}''(u).\nabla)u = 0, \tag{12}$$

reduces to

$$(\partial_t + (\mathbf{a}(u).\nabla))\nabla u_t + \aleph' = 0, \tag{13}$$

where, by (11), $\aleph' \equiv u_t\nabla u(\mathbf{a}''(u).\nabla)u$ stays constant on $\alpha = \alpha^*$. This gives an analogous result to (9) for ∇u_t. The same situation occurs for u_{tt}.

Computation of higher order derivatives proceeds by a simple induction argument, with at most algebraic growth in time possible at each stage.

Corollary 2.2 *Let u be any solution to (2) defined on a characteristic which is completely linearly degenerate. Then all derivatives of u are conserved over the characteristic. For C^ω data, the solution remains C^ω.*

Proof If the α^*-characteristic above is linearly degenerate of order 2, then $\aleph = 0$, and $\nabla\nabla u$ remains constant. The result follows in a similar fashion for every derivative if the characteristic is completely linearly degenerate. It remains to prove that analyticity is conserved along the characteristic. However since each derivative on the characteristic is the same as the corresponding derivative evaluated initially, the same majorant series can be used anywhere on the characteristic as in the Cauchy-Kowalevski Theorem for the initial value problem and the result follows.

Theorem 2.3 *Let u be any solution to (2) defined on a characteristic, $\alpha = \alpha^*$, which is linearly degenerate. Then for C^∞ data, the solution remains C^∞ along the characteristic.*

Proof This is clear from the Lemma.

3   THE ELASTICITY EQUATION

In this section we outline differences, under varying hypotheses, in the possible behaviour of solutions to the wave equation

$$u_{tt} - \sigma'(u_x)u_{xx} = 0, \ x \in \Omega \subseteq \mathbb{R}, \tag{14}$$

and

$$u(x,0) = u_0(x), \ u_t(x,0) = u_1(x), \ x \in \overline{\Omega} \text{ or } x \in \mathbb{R}. \tag{15}$$

It is a classical result ([7]) that in the strictly hyperbolic case, $\sigma' > 0$, together with the condition of genuine nonlinearity, $\sigma'' \neq 0$, one obtains finite time breakdown of solutions. A very brief sketch of the reason for this is the following. If one defines the operator, d_R of differentiation along a right characteristic by $d_R = \partial_t + \sigma'^{1/2}(u_x)\partial_x$, and a Riemann invariant, r, along the right characteristic by $r = u_t - \int^{u_x} \sigma'^{1/2}(z)dz$, then it is clear that $d_R r = 0$. That is, r is constant along curves satisfying $\frac{dx(t)}{dt} = \sigma'^{1/2}(u_x(x(t),t))$. However it turns out ([7]) that the derivative r_x satisfies an ordinary differential equation $d_R(\sigma'^{1/4}(u_x)r_x) + (\sigma'^{-1/4}(u_x))'(\sigma'^{1/4}(u_x)r_x)^2 = 0$ on the same curve. From this it is straightforward to show that $r_x \to \infty$ in finite time in general, however this argument relies both on $\sigma' \neq 0$ and on σ'' having a distinguished sign. An extension to the case where both characteristic families are linearly degenerate, $\sigma'' = 0$, but still strictly hyperbolic, $\sigma' > 0$, was made in [6], in which it was found that for small, periodic data the breakdown time was determined asymptotically as a *nonlocal* function of the initial data. Finally, in [10] it was shown that in the context of an initial boundary value problem one could again obtain breakdown in finite time when a characteristic family existed which satisfied $\sigma' = \sigma'' = 0$, together with $\sigma''' > 0$. We will interpret this condition and its consequences more fully, next.

As a first order system, equation (14) can be written as

$$\begin{pmatrix} v \\ w \end{pmatrix}_t = \begin{pmatrix} 0 & \sigma'(w) \\ 1 & 0 \end{pmatrix} \begin{pmatrix} v \\ w \end{pmatrix}_x$$

in which the matrix has eigenvalues $\lambda_\pm = \pm\sigma'^{1/2}(w)$ and corresponding eigenvectors

$$\mathbf{r}_\pm = \begin{pmatrix} \sigma'^{1/2}(w) \\ \mp 1 \end{pmatrix}$$

Setting $\nabla_U = (\partial_v, \partial_w)$ implies that $\mathbf{r}_\pm . \nabla_U \lambda_\pm = -(\sigma'^{1/2}(w))'$, and the system will be linearly degenerate or genuinely nonlinear at some point of phase space according to whether or not $\mathbf{r}_\pm . \nabla_U \lambda_\pm = 0$ there. Therefore in a nonstrictly hyperbolic region, $\lambda_+ = \lambda_-$, we need to consider the ratio $\sigma'^{-1/2}\sigma''$ where σ' tends to 0. For definiteness, assume that $\sigma'(w) = \sigma''(w) = 0$ at $w = 0$.

In the following, we will consider the behaviour of solutions to equation (14) both for bounded or unbounded domains. In the bounded case, we may examine the evolution of solutions at the boundary itself, ([10]), subject to Neumann boundary conditions or, as in the unbounded case, we can examine a characteristic starting from a point, say at $x = \alpha$, at which the initial data satisfy $u_0'(\alpha) = u_1'(\alpha) = 0$. In either case, consider the time dependence of the spatial derivative of the solution restricted to the 'vertical' line $x = \beta$, where β denotes either α or a boundary point. (In the latter case the following argument is unnecessary since the Neumann boundary conditions make the boundary of Ω characteristic automatically.) Differentiating (14) with respect to x and integrating twice in t implies that on any interval of existence, ([9]), $t < t^*$ of bounded C^3-solutions, $u_x(\beta, t)$ satisfies

$$u_x(\beta, t) = \int_0^t (t - \eta)(\sigma''(u_x(\beta, \eta))u_{xx}^2(\beta, \eta) + \sigma'(u_x(\beta, \eta))u_{xxx}(\beta, \eta))d\eta, \qquad (16)$$

and so if we suppose for example σ'' to be locally Lipschitz continuous, an application of Grönwall's inequality implies that $u_x(\beta, t) = 0$ for $t < t^*$. This means that the line $x = \beta$ is characteristic since $\frac{dx(t)}{dt} = \sigma'^{1/2}(u_x(x(t), t))$ and we have assumed that $\sigma'(0) = 0$.

Note that equation (16) fails to provide a Grönwall estimate beyond $t = t^*$ since one or both of the higher order derivative terms may become unbounded, and the line $x = \beta$ need no longer remain characteristic thereafter. In fact, under the assumptions $\sigma'(0) = \sigma''(0) = 0$, the line $x = \beta$ must remain characteristic until at least $min(t^*, t_h^*)$, where t_h^* is the time to breakdown of solutions lying in the hyperbolic region of the phase space and t^* is the time to breakdown of solutions computed formally in the nonstrictly hyperbolic region. Since our interest is in the latter region we suppose, for simplicity, that data for the hyperbolic region has been chosen in such a way that $t^* \leq t_h^*$. This restriction can be lifted using a more detailed analysis, but we make the main features of breakdown in the nonstrictly hyperbolic case more evident by its use.

The following simple estimate will be useful.

Lemma 3.1 *Let* $f(x) \in C^2$ *satisfy* $f(x), f''(x) \geq 0$, *and* $f(0) = f'(0) = 0$. *Then* $\frac{f'(0)}{f^{1/2}(0)} = 0$ *if and only if* $f''(0) = 0$.

Proof Since $f(0) = f'(0) = 0$, it follows from the Mean Value Theorem that for some $0 < \theta, \psi < 1$

$$f'(x) = \int_0^x f''(y)dy = xf''(\theta x) \ , f(x) = \int_0^x (x-y)f''(y)dy = x^2(1-\psi)f''(\psi x), \qquad (17)$$

so

$$\frac{f'(x)}{f^{1/2}(x)} = sgn(x)(1-\psi)^{-1/2}\frac{f''(\theta x)}{(f''(\psi x))^{1/2}}. \qquad (18)$$

The result follows immediately on letting $x \to \pm 0$.

Theorem 3.2 *Let $u(.,t)$ be any solution to equations (14), (15) defined on a nonstrictly hyperbolic characteristic, $x = \beta$. Then in the linearly degenerate case the characteristic is regularity preserving. If the characteristic is genuinely nonlinear, then finite time breakdown is possible if and only if $\sigma'''(0) > 0$.*

Proof Set $\sigma'''(0) = \mu$. Differentiating equation (14) twice implies

$$u_{ttxx} = \sigma'(u_x)u_{xxxx} + 3\sigma''(u_x)u_{xx}u_{xxx} + \sigma'''(u_x)u_{xx}^3, \qquad (19)$$

and so in particular, for $t < t^*$,

$$u_{ttxx}(\beta,t) = 3\sigma''(0)u_{xx}(\beta,t)u_{xxx}(\beta,t) + \mu u_{xx}^3(\beta,t). \qquad (20)$$

by failure of strict hyperbolicity. Assuming that on $x = \beta$, $|\mathbf{r}_\pm.\nabla_U\lambda_\pm| < \infty$, it follows that $\sigma'^{-1/2}\sigma''$ tends to a finite limit as σ' tends to zero, and so $\sigma''(0) = 0$. Thus (20) reduces to

$$u_{ttxx}(\beta,t) = \mu u_{xx}^3(\beta,t). \qquad (21)$$

It is now simple to apply an induction argument, based essentially on successive differentiation of (19) and on the elementary fact that solutions to (21) can only break down in finite time when $\mu > 0$, to show that $u(\beta,.) \in C^\infty(0,t_h^*)$ when $\mu \le 0$. In particular, using the above Lemma applied to $f = \sigma'$, it follows that a linearly degenerate characteristic preserves the regularity of solutions for a maximal time, determined only by the behaviour of the solution off the characteristic. On the other hand, if the characteristic is genuinely nonlinear, then it follows from the Lemma that $\sigma'''(0) = \mu \neq 0$, and so a nonstrictly hyperbolic, genuinely nonlinear characteristic leads to finite time breakdown only when, in addition, $\sigma'''(0) > 0$, ([10]).

4   A SECOND ORDER EQUATION IN $I\!R^n$

Here we examine a class of initial-boundary value problems of vector-valued solutions defined over a bounded domain in $I\!R^n$,

$$\mathbf{u}_{tt} - \Delta\Psi(\mathbf{u}) = \mathbf{0}, \ \mathbf{x} \in \Omega \subseteq I\!R^n, \tag{22}$$

$$\mathbf{u}(\mathbf{x}, t) = \mathbf{0}, \ \mathbf{x} \in \partial\Omega, \ t \geq 0 \tag{23}$$

and

$$\mathbf{u}(\mathbf{x}, 0) = \mathbf{u}_0(\mathbf{x}), \ \mathbf{u}_t(\mathbf{x}, 0) = \mathbf{u}_1(\mathbf{x}), \ \mathbf{x} \in \overline{\Omega}. \tag{24}$$

Although the equations are set in an arbitrary number of spatial dimensions, it has been shown possible to predict a finite interval of existence for C^1-solutions in this setting, ([10]), under the following conditions for Ψ:-

    1)   Let $\Psi(\mathbf{u}) = \mathbf{u}\mathcal{F}(\frac{1}{2} \mid u \mid^2)$, where
    2) $\mathcal{F} \in C^3(I\!R_+; I\!R)$,
    3) $\mathcal{F}(0) = 0$, and
    4) $\mathcal{F}'(0) = \mu \in (-\infty, \infty)$.

As in the previous section, $\mu > 0$ is a necessary condition for finite time breakdown. Here we give a brief summary of the argument in ([10]) to show this does not take place when $\mu \leq 0$. Restricting attention to the boundary of Ω and using the equation (22) enables one to show that

$$\nabla\mathbf{u} : \nabla\mathbf{u}_{tt} = \mu(\mid \nabla\mathbf{u} \mid^4 + 2 \mid \nabla\mathbf{u}\nabla\mathbf{u}^T \mid^2) \tag{25}$$

and

$$\nabla\mathbf{u}_t : \nabla\mathbf{u}_{tt} = \frac{\mu}{2}(\mid \nabla\mathbf{u} \mid^2 \partial_t \mid \nabla\mathbf{u} \mid^2 + \partial_t \mid \nabla\mathbf{u}\nabla\mathbf{u}^T \mid^2). \tag{26}$$

Equation (25) and an integration with respect to time of (26) together lead to the relation

$$(\mid \nabla\mathbf{u} \mid^2)_{tt} = 3\mu(\mid \nabla\mathbf{u} \mid^4 + 2 \mid \nabla\mathbf{u}\nabla\mathbf{u}^T \mid^2) + 2 \mid \nabla\mathbf{u}_1 \mid^2$$
$$-\mu(\mid \nabla\mathbf{u}_0 \mid^4 + 2 \mid \nabla\mathbf{u}_0\nabla\mathbf{u}_0^T \mid^2). \tag{27}$$

Now, for any tangent vector τ defined on $\partial\Omega$, $\nabla\mathbf{u}.\tau \equiv \mathbf{0}$. Consequently, $\nabla\mathbf{u}$ has rank 1 on the boundary, and takes the form $\alpha(t, \mathbf{x}) \otimes \mathbf{n}(\mathbf{x})$ where $\alpha : I\!R_+ \times I\!R^n \to I\!R^n$ and $\mathbf{n}$ denotes a unit normal on $\partial\Omega$. Therefore $\mid \nabla\mathbf{u}\nabla\mathbf{u}^T \mid = \mid \alpha \mid^2$ and (27) reduces to

$$(\mid (\alpha \mid^2)_{tt} = 9\mu \mid \alpha \mid^4 + c_0, \tag{28}$$

for some function, $c_0(\mathbf{x})$. Multiplying equation (28) by $(\mid \alpha \mid^2)_t$ and integrating in time gives

$$\frac{1}{2}((\mid \alpha \mid^2)_t)^2 = 3\mu \mid \alpha \mid^6 + c_0 \mid \alpha \mid^2 + c_1. \tag{29}$$

for some $c_1(\mathbf{x})$. From this and the assumption that $\mu \leq 0$, the boundedness of α follows directly, and so C^1 solutions $\mathbf{u}$ remain bounded in time.

5   A CONSTRAINED PROBLEM

In this section we examine a particular system motivated by certain ideas in the theory of liquid crystals. It has been shown in [1] that breakdown of solutions to the "geometric optics" approximations of the equations of motion for highly inertial liquid crystals undergoing planar oscillations can take place in finite time. Since these equations possess a Hamiltonian structure but are of nonconservative form, it is interesting to note that the absence of shock structure can be noted during the formation of singularities. Because of the two-dimensional nature of the equations investigated, there is a question of whether such singularities would be stable features of the full three-dimensional equations, or whether regularity could be recovered by an "escape" to the extra dimension. Here we consider a related question by observing that the behaviour of solutions on the edge of a bounded, smooth, three-dimensional domain is of course fundamentally two-dimensional, and one might in certain circumstances expect similar behaviour as in the planar case to occur there.

The system we analyse is derived in basically the same way as in [1] except that we no longer employ geometric optics but work on the problem directly. However as was done in [11], we will set a constitutive term to zero, but unlike [1] and [11] we now work with the full three-dimensional equations. This has the effect of introducing a type of degeneracy not applicable in the case of the final model analysed in [1], but at the same time it makes clear how to interpret the resulting blow-up.

Let $\mathbf{u}(\mathbf{x}, t) : \Omega \subset I\!\!R^3 \times I\!\!R_+ \to S^2$, the unit ball in $I\!\!R^3$, and suppose $v = \nabla.\mathbf{u}$. The equations of motion for the problem, subject to a potential energy density of $\frac{1}{2}v^2$ reduce, after elimination of a Lagrange multiplier, to

$$\mathbf{u}_{tt} - \nabla v + \mathbf{u}(|u_t|^2 - (u.\nabla)v) = 0, \tag{30}$$

$$\mathbf{u}(\mathbf{x}, 0) = \mathbf{u}_0(\mathbf{x}) \in S^2, \ \mathbf{u}_t(\mathbf{x}, 0) = \mathbf{u}_1(\mathbf{x}) \perp \mathbf{u}_0(\mathbf{x}), \ \mathbf{x} \in \overline{\Omega} , \tag{31}$$

$$\mathbf{u}|_{\partial\Omega} = -\mathbf{n}, \tag{32}$$

where $\mathbf{n}$ denotes an outward unit vector on $\partial\Omega$.

Here the energy density $\frac{1}{2}v^2$ differs from that for the wave equation on S^2, $\frac{1}{2}|\nabla\mathbf{u}|^2$, which leads to a harmonic mapping problem in Minkowski space-time. Equation (32) is commonly referred to as a "strong-anchoring" boundary condition, and we will assume for simplicity that $\partial\Omega$ contains a "flat" portion, $\partial\Omega_f$. So on such a portion, one has $\frac{\partial \mathbf{u}}{\partial \tau_i} = \frac{\partial \mathbf{n}}{\partial \tau_i} = 0$, $i = 1, 2$, where τ_i , $i = 1, 2$, denote orthogonal tangent vectors.

Theorem 5.1 *Let $\mathbf{u}(., t)$ be any $C^3(\overline{\Omega})$-solution to equations (30)-(32), $0 \le t \le T$. Then, in general, $T < \infty$. Further, $T \le t^*$, where t^* depends only on $\mathbf{u}_0|_{\partial\Omega}$ and $\mathbf{u}_1|_{\partial\Omega}$.*

Proof We compute successively the divergence and gradient-divergence of (30) to obtain

$$v_{tt} - \Delta v + (u.\nabla + v)(|u_t|^2 - (u.\nabla)v) = 0, \tag{33}$$

and

$$\nabla v_{tt} - \nabla \Delta v + \nabla(u.\nabla + v)(|u_t|^2 - (u.\nabla)v) = 0. \tag{34}$$

In local coordinates on $\partial\Omega_f$, then $\mathbf{u} = (-1,0,0)$ and so since $\mathbf{u}$ is a unit vector, any derivative satisfies $\partial|u|^2 = 0$. In particular, this implies $\mathbf{u}.\mathbf{u}_n = 0$, and so on $\partial\Omega_f$, $\mathbf{u}_n = (0,\alpha,\beta)$ for some α,β, which completes the specification we need for the limiting gradient $\nabla\mathbf{u}|_{\partial\Omega_f}$,

$$\begin{pmatrix} | & | & | \\ \mathbf{u}_n & \mathbf{u}_{\tau_1} & \mathbf{u}_{\tau_2} \\ | & | & | \end{pmatrix} = \begin{pmatrix} 0 & 0 & 0 \\ \alpha & 0 & 0 \\ \beta & 0 & 0 \end{pmatrix}.$$

In particular, then $\nabla.\mathbf{u}|_{\partial\Omega_f} = v|_{\partial\Omega_f} = 0$. Therefore $\nabla v|_{\partial\Omega_f} = \mathbf{n}\frac{\partial v}{\partial n}$, $\Delta v|_{\partial\Omega_f} = \mathbf{n}.\frac{\partial}{\partial n}(\mathbf{n}\frac{\partial v}{\partial n}) + \sum_{i=1}^{2}\tau_i.\frac{\partial}{\partial\tau_i}(\mathbf{n}\frac{\partial v}{\partial n}) = \frac{\partial^2 v}{\partial n^2}$, and $\nabla\Delta v|_{\partial\Omega_f} = (\mathbf{n}\frac{\partial}{\partial n} + \sum_{i=1}^{2}\tau_i\frac{\partial}{\partial\tau_i})(\frac{\partial^2 v}{\partial n^2}) = \mathbf{n}\frac{\partial^3 v}{\partial n^3} + \sum_{i=1}^{2}\tau_i\frac{\partial^3 v}{\partial\tau_i\partial n^2}$. Similarly, considering equation (34) one finds that on $\partial\Omega_f$, $\nabla(u.\nabla)(u.\nabla)v = (\mathbf{n}\frac{\partial}{\partial n} + \sum_{i=1}^{2}\tau_i\frac{\partial}{\partial\tau_i})(-\frac{\partial}{\partial n})(-\frac{\partial}{\partial n})v$ and so the high order expression $-\nabla\Delta v + \nabla(u.\nabla)(u.\nabla)v$ vanishes on $\partial\Omega_f$. Now, dotting equation (34) with $\mathbf{n}$ implies that on $\partial\Omega_f$,

$$\frac{\partial}{\partial n}v_{tt} + \frac{\partial}{\partial n}(-\frac{\partial}{\partial n} + v)|u_t|^2 + \frac{\partial}{\partial n}(v(-\frac{\partial v}{\partial n})) = 0, \tag{35}$$

or,

$$\frac{\partial^2}{\partial t^2}v_n - \frac{\partial^2}{\partial n^2}|u_t|^2 + v_n|u_t|^2 - v_n^2 = 0, \tag{36}$$

since $v|_{\partial\Omega_f} = 0$. But $\mathbf{u}|_{\partial\Omega_f}$ is constant and so on $\partial\Omega_f$, $\mathbf{u}_t = \mathbf{0} \Rightarrow |u_t|^2 = 0$ and $\frac{\partial^2}{\partial n^2}|u_t|^2 = 2(|\frac{\partial}{\partial n}u_t|^2 + \mathbf{u}_t.\frac{\partial^2 \mathbf{u}_t}{\partial n^2}) = 2v_t^2 = 0$ again because $v|_{\partial\Omega_f} = 0$. Finally then, the normal derivative v_n satisfies

$$\frac{\partial^2}{\partial t^2}v_n(\mathbf{x},t) - v_n^2(\mathbf{x},t) = 0, \mathbf{x} \in \partial\Omega_f. \tag{37}$$

Suppose now, for example, that $v_n(\mathbf{x},0) \neq 0$ and $\frac{\partial}{\partial t}v_n(\mathbf{x},0) > 0$ for some $\mathbf{x} \in \partial\Omega_f$. Then under this type of condition it is easy to show that there exists a corresponding time at which the solution to the ordinary differential equation (37) blows up. Letting t^* denote the infimum of all such time estimates as $\mathbf{x}$ varies over $\partial\Omega_f$ gives the result.

Remark. In fact, the previous result can be further interpreted by examining the geometry of the director configuration, in which case the presence or absence of blow-up becomes quite suggestive as to the effect the boundary conditions impose on the solution. However we do not pursue this further here.

References

[1] Hunter, J., and Saxton, R. (1991). *Dynamics of Director Fields*, SIAM J. Appl. Math., 51, 6, 1498-1521.

[2] John, F. (1976). *Delayed singularity formation in solutions of nonlinear wave equations in higher dimensions*, Comm. Pure Appl. Math., Vol. XXIX, 649-682.

[3] John, F. (1982). *"Partial Differential Equations"*, Fourth Edition, Applied Mathematical Sciences, 1, Springer-Verlag.

[4] Klainerman, S. (1980). *Global existence for nonlinear wave equations*, Comm. Pure Appl. Math., Vol. XXXIII, 43-101.

[5] Keyfitz, B. L., and Kranzer, H. C. (1983). *Non-Strictly Hyperbolic Systems of Conservation Laws: Formation of Singularities*, Contemporary Mathematics, 17, 77-90.

[6] Klainerman, S. and Majda, A. (1980). *Formation of Singularities for Wave Equations Including the Nonlinear Vibrating String*, Comm. Pure Appl. Math., Vol. XXXIII, 241-263.

[7] Lax, P. D. (1964). *Development of singularities of solutions of nonlinear hyperbolic partial differential equations*, J. Mathematical Physics, 5, 611-613.

[8] Majda, A. (1984). *"Compressible Fluid Flow and Systems of Conservation Laws in Several Space Variables"*, Applied Mathematical Sciences, 53, Springer-Verlag.

[9] Sakamoto, R. (1989). *Local existence theorems and blow-up of solutions for quasi-linear hyperbolic problems in a domain with characteristic boundary generated by initial data*, Publ. RIMS, Kyoto Univ. 25, 381-405.

[10] Saxton, R. (1992). *Formation of Singularities for a Class of Nonlinear, Hyperbolically Degenerate Initial-Boundary Value Problems*, Appl. Math. Lett., 5, 3, 73-75.

[11] Saxton, R. (1992). *Finite Time Boundary Blowup for a Degenerate, Quasilinear Cauchy Problem, "* Partial Differential Equations " eds. J. K. Hale and J. Wiener, Pitman Research Notes in Mathematics Series, 273, 212-215, Longman.

On N-times Integrated C-Cosine Functions

Sen-Yen Shaw and Yuan-Chuan Li   National Central University, Chung-Li, Taiwan, Republic of China

0. INTRODUCTION

Let X be a sequentially complete locally convex space and C be a linear injection on X. A strongly continuous family $\{C(t); t \geq 0\}$ of continuous linear operators on X is called an *n-times integrated C-cosine function* ($n \geq 1$) ([2], [4], [5]) if $CC(\cdot) = C(\cdot)C$, $C(0) = 0$, and

$$
\begin{aligned}
2C(t)C(s)x = \frac{1}{(n-1)!} \Bigg\{ & (-1)^n \int_0^{|s-t|} (|s-t| - r)^{n-1} C(r)Cx\,dr \\
& + \left[\int_0^{s+t} - \int_0^t - \int_0^s \right] (t + s - r)^{n-1} C(r)Cx\,dr \\
& + \int_0^t (s - t + r)^{n-1} C(r)Cx\,dr + \int_0^s (t - s + r)^{n-1} C(r)Cx\,dr \Bigg\}
\end{aligned}
$$

for all $x \in X$, and $s, t \geq 0$. It is called a (0-times integrated) *C-cosine function* if $C(0) = C$ and $2C(t)C(s) = [C(t + s) + C(|t - s|)]C$ $s, t \geq 0$.

$C(\cdot)$ is said to be *nondegenerate* if $C(t)x = 0$ for all $t > 0$ implies $x = 0$. Note that a C-cosine function automatically is nondegenerate and commutes with C. $C(\cdot)$ is said to be *exponentially equicontinuous* if there is a $w \geq 0$ such that $\{e^{-wt}C(t); t \geq 0\}$ is equicontinuous, and is said to be *exponentially Lipschitz continuous* if $\{e^{-w(t+h)}[C(t + h) - C(t)]/h; t, h \geq 0\}$ is equicontinuous.

Research supported in part by the National Science Council of Taiwan.

When $C = I, n = 0$, $C(\cdot)$ reduces to an ordinary strongly continuous cosine operator function ([1], [9]).

In this paper we discuss some properties of n-times integrated C-cosine functions. The paper consists of the following sections: 1) generators and basic properties; 2) generation theorems; 3) the abstract Cauchy problem; 4) $(r^2 - A)^{-n}C$-cosine function; 5) connection with ordinary cosine operator functions; 6) Perturbation theorems; 7) Representation formulas for C-cosine functions.

1. GENERATORS AND BASIC PROPERTIES

Under the assumption that $C(\cdot)$ is nondegenerate, one can define its generator.

DEFINITION 1.   The integral generator A of a nondegenerate n-times integrated C-cosine function $C(\cdot)$ is defined as:

$$x \in D(A) \text{ and } Ax = y \Leftrightarrow C(t)x - t^n Cx/n! = \int_0^t (t - s)C(s)y\, ds \text{ for } t \geq 0.$$

The assumption that $C(\cdot)$ is nondegenerate implies that the operator A is well-defined. It is also clear that A is linear and closed.

PROPOSITION 1.1.   *The integral generator A of a nondegenerate n-times integrated C-cosine function $C(\cdot)$ is a closed linear operator with the following properties:*
   (i) $C(t)x \in D(A)$ *and* $AC(t)x = C(t)Ax$ *for* $x \in D(A)$ *and* $t \geq 0$;
   (ii) $\int_0^t (t - r)C(r)x\, dr \in D(A)$ *and*
$$A \int_0^t (t - r)C(r)x\, dr = C(t)x - t^n Cx/n! \quad \text{for} \quad x \in X, t \geq 0.$$
   (iii) $C(\cdot)$ *is uniquely determined by* A.

Proof.   (i) follows from the commutativity of the family $C(\cdot)$ and the definition of A.
   (ii) By the definition of A we need to show that for all $s, t \geq 0$

$$[C(s) - \frac{s^n}{n!}C] \int_0^t (t - r)C(r)x\,dr = \int_0^s (s - r)C(r)[C(t) - \frac{t^n}{n!}C]x\, dr,$$

or equivalently, to show that the function

$$f(s) := \int_0^t (t - r)C(s)C(r)x\,dr - \int_0^s (s - r)C(r)C(t)x\,dr + \frac{t^n}{n!}C \int_0^s (s - r)C(r)x\,dr$$

is equal to $\frac{s^n}{n!}C \int_0^t (t - r)C(r)x\, dr$. By substituting the expression for $C(s)C(r)$ into $f(s)$ and then taking derivatives we obtain for $2 \leq j \leq n - 1$

$$2f^{(j)}(s)$$
$$= \frac{1}{(n - j - 1)!}C \int_0^t (t - r)[(-1)^n \int_0^{|s-r|} (\text{sgn}(s - r))^j (|s - r| - \tau)^{n-j-1}C(\tau)x d\tau$$
$$+ (\int_0^{s+r} - \int_0^s - \int_0^\tau)(s + r - \tau)^{n-j-1}C(\tau)x d\tau$$

$$+(-1)^j \int_0^s (r-s+\tau)^{n-j-1}C(\tau)x d\tau$$

$$+\int_0^r (s-r+\tau)^{n-j-1}C(\tau)x d\tau]dr$$

$$-\frac{1}{(n-j+1)!}C[(-1)^n \int_0^{|t-s|}(\operatorname{sgn}(s-t))^j(|t-s|-\tau)^{n-j+1}C(\tau)x d\tau$$

$$+(\int_0^{s+t}-\int_0^s-\int_0^t)(s+t-\tau)^{n-j+1}C(\tau)x d\tau$$

$$+(-1)^j\int_0^s(t-s+\tau)^{n-j+1}C(\tau)x d\tau$$

$$+\int_0^t(s-t+\tau)^{n-j+1}C(\tau)x d\tau].$$

Finally, one obtains

$$2f^{(n)}(s)=C[\int_0^t(t-r)[(-1)^n(\operatorname{sgn}(s-r))^nC(|s-r|)x dr$$

$$-(-1)^n\int_0^{|s-t|}(\operatorname{sgn}(s-t))^n(|s-t|-\tau)C(\tau)x d\tau$$

$$-(-1)^n\int_0^s(t-s+\tau)C(\tau)x d\tau+2\int_0^t(t-\tau)C(\tau)x d\tau]$$

$$=2\int_0^t(t-\tau)C(\tau)x d\tau.$$

Since $f^{(j)}(0)=0$ for $j=0,1,...,n-1$, the function $f(s)$ is equal to $\frac{s^n}{n!}C\int_0^t(t-r)C(r)x\,dr$.

(iii) follows from a routine argument which we omit here.

COROLLARY 1.2 [5, Corollary 3.3]. (i) *For all $x\in X$ and $t\geq 0$ we have $C(t)x\in D(A)$.*

(ii) *$C(\cdot)x$ is right-sided differentiable at $t\geq 0$ if and only if $\int_0^t C(s)x\,ds\in D(A)$. In that case,*

$$\frac{d}{dt}C(t)x=\begin{cases} A\int_0^t C(s)x\,ds+t^{n-1}Cx/(n-1)! & \text{if } n\geq 1,\\ A\int_0^t C(s)x\,ds & \text{if } n=0.\end{cases}$$

(iii) *$C(\cdot)x$ is twice right-sided differentiable at $t\geq 0$ if and only if $C(t)x\in D(A)$. In that case,*

$$\frac{d^2}{dt^2}C(t)x=\begin{cases} AC(t)x+t^{n-2}Cx/(n-2)! & \text{if } n\geq 2,\\ AC(t)x & \text{if } n\leq 1.\end{cases}$$

Next we consider the case that $C(\cdot)$ is exponentially equicontinuous. Then one can take the Laplace transform: $L(\lambda)x:=\int_0^\infty \lambda^n e^{-\lambda t}C(t)x dt$, $x\in X$, $\lambda>w$.

THEOREM 1.3 ([5, Propositions 2.2 and 2.3]). *Let $n \geq 0$. An exponentially equicontinuous function $C(\cdot)$ which commutes with C and has value $C(0) = C$ in case $n = 0$, and $C(0) = 0$ in case $n \geq 1$ is an n-times integrated C-cosine function if and only if $L(\cdot)$ satisfies*

$$(1.1) \qquad (\lambda^2 - \mu^2)L(\mu)L(\lambda) = [\lambda L(\mu) - \mu L(\lambda)]C \quad \text{for all } \lambda, \mu > w.$$

If we put $R(\lambda) := \lambda^{-1/2}L(\lambda^{1/2})$, then (1.1) implies that $R(\cdot)$ is a *C-pseudoresolvent*. That is, $R(\cdot)C = CR(\cdot)$ and

$$(1.2) \qquad (\lambda - \mu)R(\mu)R(\lambda) = R(\mu)C - R(\lambda)C \quad (\lambda > w).$$

THEOREM 1.4 ([3, Theorem 3.4]). *Let $R(\cdot) : \Omega \to L(X)$ be a C-pseudo-resolvent with C an injection.*

(i) *The null space $N(R(\lambda))$ of $R(\lambda)$ is independent of λ.*

(ii) *There exists some closed operator B such that*
$$(1.3) \qquad R(R(\lambda)) \subset D(B) \text{ and } R(\lambda)(\lambda - B) \subset (\lambda - B)R(\lambda) = C \quad \text{for all } \lambda \in \Omega$$
if and only if $N(R(\lambda)) = \{0\}$.

(iii) *When $R(\lambda)$ is injective, the operator A defined by*

$$(1.4) \qquad \begin{cases} D(A) & := C^{-1}[R(R(\lambda))] = \{x \in X; Cx \in R(R(\lambda))\}, \\ Ax & := (\lambda - R(\lambda)^{-1}C)x \text{ for } x \in D(A) \end{cases}$$

is independent of $\lambda > w$ and is a closed operator satisfying (1.3)

(iv) *If B satisfies (1.3), then $B \subset C^{-1}BC = A$. In particular, $C^{-1}AC = A$.*

Notice that when C is the identity operator I, Theorem 1.4 reduces to the well known properties of a *pseudoresolvent* as can be found in [10, Chapter VIII].

Since $\{R(\lambda); \lambda > w^2\}$ is a C-pseudoresolvent (Theorem 1.3), from (i) of Theorem 1.4, the uniqueness of the Laplace transform, and the assumption that $C(\cdot)$ is nondegenerate, one easily deduces that $N(R(\lambda)) = 0$ for all $\lambda > w^2$, and so Theorem 1.4 justifies the following definition.

DEFINITION 2.   When $C(\cdot)$ is nondegenerate and exponentially equicontinuous, there are closed operators B such that $\lambda^2 - B$ is injective and

$$R(L(\lambda)) \subset D(B) \text{ and } L(\lambda)(\lambda^2 - B) \subset (\lambda^2 - B)L(\lambda) = \lambda C \text{ for } \lambda > w.$$

Such closed operators B are called *subgenerators* of $C(\cdot)$. The closed operator A defined by

$$(1.5) \qquad \begin{cases} D(A) & := \{x \in X; \ Cx \in R(L(\lambda))\}; \\ Ax & := \lambda(\lambda - L(\lambda)^{-1}C)x \text{ for } x \in D(A), \end{cases}$$

is independent of $\lambda > w$ and extends all subgenerators. It is called the *Laplace generator* of $C(\cdot)$. As Theorem 1.4 has asserted, $B \subset C^{-1}BC = A$ for all subgenerators B, and in particular, $C^{-1}AC = A$

PROPOSITION 1.5 ([5, Proposition 3.2]). *The Laplace generator A has the following properties:*

(1.6) $\quad R(L(\lambda)) \subset D(A)$ *and* $L(\lambda)(\lambda^2 - A) \subset (\lambda^2 - A)L(\lambda) = \lambda C$ *for* $\lambda > w$;

(1.7) $\quad C(t)x \in D(A)$ *and* $AC(t)x = C(t)Ax$ *for* $x \in D(A)$ *and* $t \geq 0$;

$$(1.8) \qquad \int_0^t (t - r)C(r)x\, dr \in D(A) \text{ and}$$

$$A \int_0^t (t - r)C(r)x\, dr = C(t)x - t^n Cx/n! \text{ for } x \in X,\, t \geq 0.$$

(1.9) $\quad C^{-1}AC = A.$

PROPOSITION 1.6. *The Laplace generator of a nondegenerate, exponentially equi-continuous n-times integrated C-cosine function coincides with the integral generator.*

Proof. Let us temporarily denote by A_1 and A_2 the integral generator and the Laplace generator, respectively. By (1.7), (1.8), and the closedness of A_2 we have $C(t)x - t^n Cx/n! = \int_0^t (t - s)C(s)A_2 x\, ds$ for all $x \in D(A_2)$ and $t \geq 0$. Hence by definition we have $A_2 \subset A_1$. It remains to show that A_1 is a subgenerator, that is,

$$R(L(\lambda)) \subset D(A_1) \text{ and } L(\lambda)(\lambda^2 - A_1) \subset (\lambda^2 - A_1)L(\lambda) = \lambda C \text{ for } \lambda > w.$$

Since A_1 is closed and $A_1 \int_0^t (t-s)C(s)x\, ds = C(t)x - t^n Cx/n!$ is continuous (Proposition 1.1), we have for all $x \in X$

$$A_1 L(\lambda)x = A_1 \int_0^\infty \lambda^n e^{-\lambda t} C(t)x\, dt = A_1 \int_0^\infty \lambda^{n+2} e^{-\lambda t} \int_0^t (t-s)C(s)x\, ds\, dt$$

$$= \int_0^\infty \lambda^{n+2} e^{-\lambda t}(C(t)x - t^n Cx/n!)dt = \lambda^2 L(\lambda)x - \frac{\lambda^{n+2}}{n!} \int_0^\infty e^{-\lambda t} t^n Cx\, dt$$

$$= \lambda^2 L(\lambda)x - \lambda Cx.$$

If $x \in D(A_1)$, then $A_1 C(t)x = C(t)A_1 x$, so that, by the closedness of A_1, we also have $L(\lambda)A_1 x = A_1 L(\lambda)x$, and the proof is complete.

For a *nondegenerate (0-times) C-cosine function* we can give the following definition of a generator.

DEFINITION 3. Let $C(\cdot)$ be a nondegenerate C-cosine function. The operator A defined by

$$(1.10) \qquad \begin{cases} D(A) & := \{x \in X;\ \lim_{t \to 0+} 2t^{-2}[C(t)x - Cx] \in R(C)\} \\ Ax & := C^{-1}\lim_{t \to 0+} 2t^{-2}[C(t)x - Cx] \text{ for } x \in D(A), \end{cases}$$

is called the *infinitesimal generator* of $C(\cdot)$.

When $C = I$, this definition reduces to the well known definition of the infinitesimal generator of a strongly continuous cosine operator function (cf. [1], [9]).

COROLLARY 1.7 *For a nondegenerate, exponentially equicontinuous C-cosine function, the infinitesimal generator, the Laplace generator, and the integral generator all coincide.*

Proof.   In view of Proposition 1.6, we need only to show that the infinitesimal generator A_3 coincides with the Laplace generator A. To prove $A_3 \subset A$ let $x \in D(A_3)$. Then by (1.8) and the closedness of A we have $Cx \in D(A)$ and

$$CA_3x = \lim_{t \to 0+} 2t^{-2}[C(t)x - Cx] = A \lim_{t \to 0+} 2t^{-2} \int_0^t \int_0^s C(r)x \, dr \, ds = ACx.$$

Hence $x \in D(C^{-1}AC) = D(A)$ and $Ax = C^{-1}ACx = C^{-1}CA_3x = A_3x$, by (1.9). Conversely, let $x \in D(A)$. Then by (1.7), (1.8), the closedness of A, and $C(0) = C$, we have $\lim_{t \to 0+} 2t^{-2}[C(t)x - Cx] = \lim_{t \to 0+} 2t^{-2} \int_0^t \int_0^s C(r)Ax \, dr \, ds = CAx$ so that $x \in D(A_3)$ and $A_3x = Ax$. Hence $A = A_3$.

We end this section with two simple examples.

EXAMPLE 1.   For $t \geq 0$ let $C(t)$ be the operator on $X = C_0(\mathbf{R})$ defined by

$$(C(t)f)(s) := \frac{1}{2}\left\{\exp[t(2\chi_{[0,\infty)}(s) - 1)] + \exp[-t(2\chi_{[0,\infty)}(s) - 1)]\right\} \sin(s)f(s), \; f \in C_0(\mathbf{R}),$$

where $\chi_{[0,\infty)}$ denotes the characteristic function of the set $[0,\infty)$. Let C be the injection defined by $Cf := \sin(\cdot)f \; (f \in C_0(\mathbf{R}))$. Then $\{C(t); \; t \geq 0\}$ is a nondegenerate, exponentially bounded C-cosine function whose generator A is the restriction of the identity operator I on $D(A) = \{f \in C_0(\mathbf{R}); \; f(0) = 0\}$.

EXAMPLE 2.   Let $X = C_b(\mathbf{R}) =$ the space of all bounded continuous functions on $\mathbf{R}$. For $t \geq 0$ let $C(t)$ be the operator on $C_b(\mathbf{R})$ defined by

$$(C(t)f)(s) := \frac{1}{2} \int_0^t [f(s+r) + f(s-r)] \, dr, \; s \in \mathbf{R}, \; f \in C_b(\mathbf{R}).$$

Then $\{C(t); \; t \geq 0\}$ is a once integrated cosine function with generator $A = \frac{d^2}{ds^2}$. Note that $C(\cdot)$ is not the integral of a 0-times cosine function because the function $t \to f(\cdot + t) + f(\cdot - t)$ is in general not continuous in $\|\cdot\|_\infty$, i.e. $f(\cdot)$ is not necessarily uniformly continuous on $\mathbf{R}$.

2. GENERATION THEOREMS

In this section we state some generation theorems for nondegenerate $(n + 1)$-times integrated C-cosine functions.

THEOREM 2.1 ([5, Theorem 3.8]).   *Let $C \in L(X)$ be an injection and let A be a closed linear operator. Then we have:*

   (i) *A is the generator of a nondegenerate $(n+1)$-times integrated C-cosine function $C(\cdot)$ such that $\{e^{-wt}[C(t + \tau) - C(t)]/\tau; \; \tau \in (0,1], t \geq 0\}$ is equicontinuous (equivalently, $C(\cdot)$ is exponentially Lipschitz continuous) if and only if A satisfies the following conditions:*

(a) $C^{-1}AC = A$;

(b) $\lambda^2 - A$ is injective and $R(C) \subset D((\lambda^2 - A)^{-k-1})$ for $\lambda > w$ and $k \geq 0$;

(c) $(\lambda^2 - A)^{-1}C$ is infinitely differentiable for $\lambda > w$ and $\{(\lambda - w)^{k+1} D^k[\lambda^{1-n}(\lambda^2 - A)^{-1}C]/k!; \lambda > w, k \geq 0\}$ is equicontinuous.

(ii) $C(\cdot)x$ is twice continuously differentiable on $[0, \infty)$ for $x \in X_1 := \overline{D(A)}$, and the family $\{C_1(t) := \frac{d}{dt}C(t)|_{X_1}; t \geq 0\}$ is an exponentially equicontinuous n-times integrated C_1-cosine function of operators on X_1 with generator A_1, where $C_1 = C|X_1$ and A_1 is the part of A in X_1, i.e. $D(A_1) = \{x \in D(A); Ax \in X_1\}$ and $A_1x = Ax$ for $x \in D(A_1)$.

COROLLARY 2.2 ([5, Corollary 3.9]). *Let $C \in L(X)$ be an injection. A densely defined closed operator is the generator of a nondegenerate, exponentially equicontinuous, n-times integrated C-cosine function if and only if it satisfies conditions (a)-(d).*

For practical application the following theorem is more convenient.

THEOREM 2.3 ([5, Theorem 3.10]). *Let $C \in L(X)$ be an injection, A be a closed operator, and let $C(\cdot) : [0, \infty) \to L(X)$ be a strongly continuous funuction such that $\{e^{-wt}C(t); t \geq 0\}$ is equicontinuous for some $w \geq 0$ and such that $C(\cdot)C = CC(\cdot)$. $C(\cdot)$ is an n-times integrated C-cosine function with A as its generator if and only if $C^{-1}AC = A$, $\lambda^2 - A$ is injective and*

$$(2.1) \qquad (\lambda^2 - A)^{-1}Cx = \lambda^{n-1} \int_0^\infty e^{-\lambda t}C(t)x\, dt$$

for all $x \in X$ and $\lambda > w$.

In Theorem 4.6 we shall formulate another criterion for generators of $2(n+1)$-times integrated C-cosine functions.

3. THE ABSTRACT CAUCHY PROBLEM

Let A be the generator of a nondegenerate n-times integrated C-cosine function $C(\cdot)$ and let $f \in C([0, b], X)$. We consider the following initial value problem:

$$\mathrm{P}(x, y, f) \qquad \begin{cases} u''(t) & = A\,u(t) + f(t), \quad 0 \leq t \leq b \\ u(0) & = x, \quad u'(0) = y. \end{cases}$$

By a solution of $\mathrm{P}(x, y, f)$ we mean a function $u \in C^2([0, b], D(A))$ (i.e. twice continuously differentiable and having values in $D(A)$) which satisfies $\mathrm{P}(x, y, f)$. Let $v_{x,y} : [0, b] \to X$ be the function given by

$$(3.1) \qquad v_{x,y}(t) := C(t)x + \int_0^t C(s)y\, ds + \int_0^t \int_0^s C(r)f(t - s)\, dr\, ds, \quad 0 \leq t \leq b.$$

THEOREM 3.1 ([5, Theorem 4.1]). *For given $x, y \in X$, $\mathrm{P}(x, y, f)$ has a solution if and only if $v_{x,y}^{(n)}([0, b]) \subset R(C)$ and $C^{-1}v_{x,y}^{(n)} \in C^2([0, b], X)$. In this case, the function $u_{x,y} = C^{-1}v_{x,y}^{(n)}$ is the unique solution of $P(x, y, f)$.*

THEOREM 3.2 ([5, Corollary 4.3]). *Suppose the function $g(t) := \int_0^t \int_0^s C(r)\, f(t - s)\, dr\, ds, t \in [0, b]$, satisfies $g^{(n)}([0, b]) \subset R(C)$ and $C^{-1} g^{(n)} \in C^2([0, b], x)$. Then for every pair $x, y \in C(D(A^{m+1}))$, where $m = [(n + 1)/2]$, the problem $P(x, y, f)$ has the unique solution*

$$
\begin{aligned}
u_{x,y}(t) = C^{-1} &\left\{ \frac{1}{(2m - n - 1)!} \int_0^t (t - s)^{2m-n-1} C(s) A^m x\, ds \right. \\
&\left. + \frac{1}{(2m - n)!} \int_0^t (t - s)^{2m-n} C(s) A^m y\, ds + g^{(n)}(t) \right\} \\
&+ \sum_{j=0}^{m-1} \left[\frac{1}{(2j)!} t^{2j} A^j x + \frac{1}{(2j + 1)!} t^{2j+1} A^j y \right].
\end{aligned}
$$

In particular, if A is the generator of a C-cosine function $C(\cdot)$, then for every pair $x, y \in C(D(A)$ the function $u(t) = C^{-1} C(t) x + C^{-1} \int_0^t C(s) y\, ds$ is the unique solution of $P(x, y, 0)$.

4. $(r^2 - A)^{-n} C$-COSINE FUNCTION

In this section we show that a $2n$-times integrated C-cosine function corresponds to a $(r^2 - A)^{-n} C$-cosine function.

THEOREM 4.1. *Let A be a closed operator such that $r^2 - A$ is injective and $R(C) \subset D((r^2 - A)^{-n})$ for some $r > 0$. If there exist $w < r$ and some $0 \le m \le n$ such that A is the generator of a $2m$-times integrated $(r^2 - A)^{-n+m} C$-cosine function $C_{2m}(\cdot)$ and $\{e^{-wt} C_{2m}(t); t \ge 0\}$ is equicontinuous, then for every $0 \le k \le n$ and $\lambda > w$, A generates a $2k$-times integrated $(\lambda^2 - A)^{-n+k} C$-cosine function $C_{2k}(\cdot)$. In particular, A is the generator of a $2n$-times integrated C-cosine function $C_{2n}(\cdot)$ if and only if for $\lambda > w$, A is the generator of a $(\lambda^2 - A)^{-n} C$-cosine function $C_0(\cdot)$. $C_0(\cdot)$ and $C_{2n}(\cdot)$ are related to each other in the following way:*

$$(4.1) \qquad C_0(t) x = D^{2n} C^{-1} C_{2n}(t)(\lambda^2 - A)^{-n} C x$$

and

$$(4.2) \qquad C_{2n}(t) x = (\lambda^2 - A)^n \frac{1}{(2n - 1)!} \int_0^t (t - u)^{2n-1} C_0(u) x\, du,$$

for all $x \in X$ and $t \ge 0$. Moreover, $C_{2n}(\cdot)$ is exponentially Lipschitz continuous if and only if $C_0(\cdot)$ is so.

To prove Theorem 4.1 we shall divide it into the following two propositions.

PROPOSITION 4.2. *Let A be a closed linear operator such that $r^2 - A$ is injective and $R(C) \subset D((r^2 - A)^{-n})$ for some $r > 0$. If A is the generator of a $(r^2 - A)^{-n}C$-cosine function $C_0(\cdot)$ such that $\{e^{-wt}C_0(t); t \geq 0\}$ is equicontinuous, then for each $0 \leq m \leq n$, A is also the generator of the $2m$-times (resp. $(2m+1)$-times) integrated $(r^2 - A)^{-n+m}C$-cosine function $C_{2m}(\cdot)$ (resp. $C_{2m+1}(\cdot)$) defined by*

$$(4.3) \qquad C_{2m}(t)x = (r^2 - A)^m \frac{1}{(2m-1)!} \int_0^t (t-s)^{2m-1} C_0(s)x\,ds$$

$$= (r^2 T^2 - I)^m C_0(t)x + \sum_{j=0}^{m-1} (r^2 T^2 - I)^{m-1-j} \frac{t^{2j}}{(2j)!}(r^2 - A)^{-n+j}Cx,$$

$$(4.4) \qquad C_{2m+1}(t)x = TC_{2m}(t)x = (r^2 T^2 - I)^m T C_0(t)x$$

$$+ \sum_{j=0}^{m-1} (r^2 T^2 - I)^{m-1-j} \frac{t^{2j+1}}{(2j+1)!}(r^2 - A)^{-n+j}Cx,$$

where T is the operator defined by $Tf(t) := \int_0^t f(s)ds$, and $\sum\limits_{j=0}^{-1}$ stands for 0. Moreover, if $C_0(\cdot)$ is exponentially Lipschitz continuous, then so are $C_k(\cdot), 0 \leq k \leq 2n+1$.

Proof. Since $C_0(\cdot)$ is a $(r^2 - A)^{-n}C$-cosine function with generator A, Theorem 2.1 holds with C replaced by $(r^2 - A)^{-n}C$, and so we have

$$(4.5) \qquad [(r^2 - A)^{-n+m}C]^{-1}A[(r^2 - A)^{-n+m}C]$$

$$= [(r^2 - A)^{-n}C]^{-1}(r^2 - A)^{-m}A(r^2 - A)^m[(r^2 - A)^{-n}C]$$

$$= [(r^2 - A)^{-n}C]^{-1}A[(r^2 - A)^{-n}C] = A.$$

Also application of (2.1) gives that $R((r^2 - A)^{-n}C) \subset R((\lambda^2 - A)^{-1})$, and for $x \in X$ and $\lambda > w$

$$(4.6) \qquad \lambda(r^2 - A)^{-n}Cx = (\lambda^2 - A) \int_0^\infty e^{-\lambda t}C_0(t)x\,dt$$

$$= (\lambda^2 - A) \int_0^\infty \lambda^{2m} e^{-\lambda t} \frac{1}{(2m-1)!} \int_0^t (t-s)^{2m-1} C_0(s)x\,ds\,dt.$$

We shall prove the assertion by induction on m. Assume that $C_{2(m-1)}(\cdot)$ is well-defined and has been shown to be a $2(m-1)$-times integrated $(r^2 - A)^{-n+m-1}C$-cosine function and that the second identity in (4.3) holds with m replaced by $m-1$. Then by (1.8) and the closedness of A we see that $C_{2m}(t)$ is well-defined and

$$C_{2m}(t)x = (r^2 - A) \int_0^t (t-s)C_{2(m-1)}(s)x\,ds$$

$$= r^2 T^2 C_{2(m-1)}(t)x - [C_{2(m-1)}(t)x - \frac{t^{2(m-1)}}{(2(m-1))!}(r^2 - A)^{-n+m-1}Cx]$$

$$= (r^2 T^2 - I)C_{2(m-1)}(t)x + \frac{t^{2(m-1)}}{(2(m-1))!}(r^2 - A)^{-n+m-1}Cx$$

$$= (r^2 T^2 - I)[(r^2 T^2 - I)^{m-1} C_0(t)x + \sum_{j=0}^{m-2} (r^2 T^2 - I)^{m-2-j}$$

$$\cdot \frac{t^{2j}}{(2j)!}(r^2 - A)^{-n+j} Cx] + \frac{t^{2(m-1)}}{(2(m-1))!}(r^2 - A)^{-n+m-1} Cx$$

$$= (r^2 T^2 - I)^m C_0(t)x + \sum_{j=0}^{m-1} (r^2 T^2 - I)^{m-1-j} \frac{t^{2j}}{(2j)!}(r^2 - A)^{-n+j} Cx.$$

This expression shows that $C_{2m}(\cdot)$ is strongly continuous on $[0,\infty)$ and is exponentially equicontinuous.

Next, we apply $(r^2 - A)^m$ on both sides of (4.6) and obtain

$$\lambda(r^2 - A)^{-n+m} Cx = (r^2 - A)^m (\lambda^2 - A) \int_0^\infty \lambda^{2m} e^{-\lambda t} \frac{1}{(2(m-1))!}$$

$$\cdot \int_0^t (t-s)^{2m-1} C_0(s)x \, ds \, dt$$

$$= (\lambda^2 - A) \int_0^\infty \lambda^{2m} e^{-\lambda t} C_{2m}(t)x \, dt$$

for $x \in X$ and $\lambda > w$. This and (4.5) show that $C_{2m}(\cdot)$ is a $2m$-times integrated $(r^2 - A)^{-n+m} C$-cosine function with generator A.

Since the integral of an exponentially equicontinuous function is exponentially Lipschitz continuous, it is easily seen from identities in (4.3) and (4.4) that the exponential Lipschitz continuity of $C_0(\cdot)$ implies the same of $C_{2m}(\cdot)$ and $C_{2m+1}(\cdot)$.

PROPOSITION 4.3. *If A is the generator of a $2n$-times integrated C-cosine function $C_{2n}(\cdot)$ such that $\{e^{-wt} C_{2n}(t); t \geq 0\}$ is equicontinuous, then for any $0 \leq k \leq n$ and $r > w$, A is also the generator of the $2(n-k)$-times integrated $(r^2 - A)^{-k} C$-cosine function $C_{2(n-k)}(\cdot)$ defined by*

$$(4.7) \qquad C_{2(n-k)}(t)x = D_t^{2k} C^{-1} C_{2n}(t)(r^2 - A)^{-k} Cx \qquad (x \in X, t \geq 0).$$

Moreover, if $C_{2n}(\cdot)$ is exponentially Lipschitz continuous, then so are $C_{2(n-k)}(\cdot)$, $0 \leq k \leq n$.

For the proof we need the next two lemmas.

LEMMA 4.4 ([5, Lemma 3.7]). *For any $n \geq 0$, if A is the generator of an n-times integrated C-cosine function $C_n(\cdot)$, then $R(C) \subset D((\lambda^2 - A)^{-k})$ and $L^{(k-1)}(\lambda) = p_k((\lambda^2 - A)^{-1})C$ holds for all $\lambda > w$ and $k \geq 1$, where $p_1(t) = \lambda t, p_2(t) = -2\lambda^2 t^2 + t$, and $p_k(t) = -[2k\lambda t p_{k-1}(t) + k(k-1)t p_{k-2}(t)]$ for $k \geq 3$.*

LEMMA 4.5 . *Under the assumption of Lemma 4.4 the following assertions are true for all $k \geq 0$ and $\lambda > w$.*

(i) $C_n(t)(\lambda^2 - A)^{-k} Cx \in R(C)$ for all $x \in X$ and $t \geq 0$, and the function $F_k(\cdot) := C^{-1} C_n(\cdot)(\lambda^2 - A)^{-k} C$ is $2k$-times continuously differentiable on $[0,\infty)$, and $F_k^{(2k)}(\cdot)$ is exponentially equicontinuous.

(ii) If $C_n(\cdot)$ is exponentially Lipschitz continuous, then so is $F_k^{(2k)}(\cdot)$.

Proof. (i) It is easily seen by induction that $p_k(t) = (-1)^{k-1}2^{k-1}((k-1)!)\lambda^k t^k + q_k(t)$ with $\deg q_k \le k-1$. Hence $(\lambda^2 - A)^{-k}C$ can be expressed as a linear combination of $(\lambda^2 - A)^{-k+1}C$, $(\lambda^2 - A)^{-k+2}C,\cdots$, $(\lambda^2 - A)^{-1}C$, C, and $L^{(k-1)}(\lambda)$. Let $G_k(\cdot) := C_n(\cdot)(\lambda^2 - A)^{-k}C$, $k \ge 0$. Then $G_k(\cdot)$ is a linear combination of $G_{k-j}(\cdot)$, $1 \le j \le k$, and $C_n(\cdot)L^{(k-1)}$. Since both $G_0(t)x = C_n(t)Cx$ and

$$C_n(\cdot)L^{(k-1)}x = D_\lambda^{(k-1)} \int_0^\infty \lambda^n e^{-\lambda s} C_n(t)C_n(s)x\, ds$$

lie in $R(C)$, and since $C^{-1}G_0(\cdot)x$ and $C^{-1}C_n(\cdot)L^{(k-1)}x$ are continuous on $[0,\infty)$, by induction we have that $R(G_k(t)) \subset R(C)$ and the function $F_k(\cdot)$ is strongly continuous on $[0,\infty)$.

Next we prove by induction that $F_k(\cdot)$ is $2k$-times continuously differentiable on $[0,\infty)$. The case $k = 0$ is trivial. For $k \ge 1$ we use (1.8) to write

$$(4.8) \qquad F_k(t)x = C^{-1}\Big[\int_0^t (t-s)C_n(s)A(\lambda^2 - A)^{-k}Cx\,ds + \frac{t^n}{n!}C(\lambda - A)^{-k}Cx\Big]$$

$$= \int_0^t (t-s)C^{-1}C_n(s)[\lambda^2(\lambda^2 - A)^{-k} - (\lambda^2 - A)^{-k+1}]Cx\,ds$$

$$+ \frac{t^n}{n!}(\lambda^2 - A)^{-k}Cx$$

$$= \lambda^2 \int_0^t (t-s)F_k(s)x\,ds - \int_0^t (t-s)F_{k-1}(s)x\,ds$$

$$+ \frac{t^n}{n!}(\lambda^2 - A)^{-k}Cx \quad (x \in X).$$

Assume that $F_{k-1}(\cdot)$ is $2(k-1)$-times continuously differentiable and $F_{k-1}^{(2(k-1))}(\cdot)$ is exponentially equicontinuous (and hence so are $F_{k-1}^{(j)}$, $0 \ge j \ge k-1$). Then the function $h(t) := \int_0^t (t-s)F_{k-1}(s)x\,ds - \frac{t^n}{n!}(\lambda^2 - A)^{-k}Cx$ is $2k$-times continuously differentiable. By iterative substitutions, we write

$$F_k(t)x = \lambda^2 \int_0^t (t-s)F_k(s)x\,ds - h(t)$$

$$= \lambda^2 \int_0^t (t-s)\Big[\lambda^2 \int_0^s (s-u)F_k(u)x\,du - h(s)\Big]ds - h(t)$$

$$\cdots$$

$$= \frac{\lambda^{2k}}{(2k-1)!}\int_0^t (t-s)^{2k-1}F_k(s)x - \sum_{j=1}^{k-1}\frac{\lambda^{2j}}{(2j-1)!}\int_0^t (t-s)^{2j-1}h(s)ds - h(t),$$

Which shows that $F_k(\cdot)x$ is $2k$-times continuously differentaible. Also it can be seen that $F_k^{(2k)}(\cdot)$ is exponentially equicontinuous.

Proof of Proposition 4.3. The hypothesis implies that $R(C) \subset D((\lambda^2 - A)^{-1})$ and $\lambda Cx = (\lambda^2 - A)\int_0^\infty \lambda^{2n}e^{-\lambda t}C_{2n}(t)x\,dt$ for all $x \in X$ and $\lambda > w$. Replacing x by

$(r^2 - A)^{-k}Cx$, noting that $C^{-1}AC = A$, and using integration by parts, we obtain

$$\lambda(r^2 - A)^{-k}Cx = C^{-1}(\lambda^2 - A)\int_0^\infty \lambda^{2n}e^{-\lambda t}C_{2n}(t)(r^2 - A)^{-k}Cx\,dt$$

$$= C^{-1}(\lambda^2 - A)C\int_0^\infty \lambda^{2(n-k)}e^{-\lambda t}D_t^{2k}C^{-1}C_{2n}(t)(r^2 - A)^{-k}Cx\,dt$$

$$= (\lambda^2 - A)\int_0^\infty \lambda^{2(n-k)}e^{-\lambda t}C_{2(n-k)}(t)x\,dt,$$

where $C_{2(n-k)}(\cdot)$ is defined as in (4.7). Since in Lemma 4.5 $C_{2(n-k)}(\cdot)$ has been shown to be strongly continuous and exponentially equicontinuous, it follows from Theorem 2.3 that $C_{2(n-k)}(\cdot)$ is a $2(n-k)$-times integrated $(r^2 - A)^{-k}C$-cosine function with generator A.

The following characterization of generators of $2(n+1)$-times integrated C-cosine functions is deduced as a consequence of Theorems 2.1 and 4.1.

COROLLARY 4.6. *The following statements are equivalent:*

    (i) A is the generator of a nondegenerate, exponentially Lipschitz continuous $2(n+1)$-times integrated C-cosine function.

    (ii) A satisfies conditions (a), (b) in Theorem 2.1 and

(c) there is an $r > w$ such that $\{(\lambda - w)^{k+1}D^k(\lambda^2 - A)^{-1}(r^2 - A)^{-n}C/k!; \lambda > w, k \geq 0\}$
is equicontinuous.

Proof. By Theorem 4.1 (with n replaced by $n+1$), we see that (i) is equivalent to that (a), (b) hold, and A generates an exponentially Lipschitz continuous twice integrated C_1-cosine function with $C_1 = (r^2 - A)^{-n}C$. By Theorem 2.1 (with $n = 1$), this latter condition is equivalent to the condition that $C_1^{-1}AC_1 = A$, $\lambda - A$ is injective, $R(C_1) \subset D((\lambda - A)^{-k-1})$ for $k \geq 0$, and $\{(\lambda - w)^{k+1}D^k(\lambda^2 - A)^{-1}C_1/k!; \lambda > w, k \geq 0\}$ is equicontinuous, which is precisely statement (ii).

5. CONNECTION WITH ORDINARY COSINE OPERATOR FUNCTIONS

The next theorem relates C-cosine functions and ordinary cosine operator functions.

THEOREM 5.1 ([7, Theorem 1). *A closed linear operator A on a Banach space X is the generator of an exponentially bounded C-cosine function if and only if the following conditions are satisfied: (a) $C^{-1}AC = A$; (b) there exists a Banach space Y with norm $|\cdot|$ such that $R(C) \subset Y \subset X$, $\|y\| \leq M_1|y|$ for $y \in Y$, $|y| \leq M_2\|C^{-1}y\|$ for $y \in R(C)$, and the part A_Y of A in Y generates a strongly continuous cosine operator function (i.e. I-cosine function) on Y, where $M_i, i = 1, 2$, are nonnegative constants.*

Combining this theorem and Theorem 4.1, we obtain the following generalization.

COROLLARY 5.2. *A closed linear operator A on a Banach space X is the generator of an exponentially bounded $2n$-times integrated C-cosine function if and only if the following conditions are satisfied: (a) $C^{-1}AC = A$; (b) there is an $r > 0$ such that*

$r - A$ is injective and $R(C) \subset D((r - A)^{-n})$; and (c) there exists a Banach space Y_n with norm $|\cdot|$ such that $R((r - A)^{-n}C) \subset Y_n \subset X$, $\|y\| \leq M_1|y|$ for $y \in Y_n$, $|y| \leq M_2\|C^{-1}(r - A)^n y\|$ for $y \in R((r - A)^{-n}C)$, and the part A_{Y_n} of A in Y_n generates a strongly continuous cosine operator function on Y_n.

6. PERTURBATION OF INTEGRATED C-COSINE FUNCTIONS

THEOREM 6.1. *Let A be the generator of an exponentially bounded $2n$-times integrated C-cosine function on a Banach space X. Let B be a linear operator satisfying:*
(a) *$A + B$ is closed, $D(A) \subset D(B)$ and $C^{-1}BC = B$;*
(b) *In case $n \geq 1$, there is a $r > w$ such that $r - A - B$ is injective, $R(C) \subset D((r - A - B)^{-n})$ and $R((r - A - B)^{-n}C) = R((r - A)^{-n}C)$;*
(c) *$BY_n \subset Y_n$ and $B|_{Y_n} \in L(Y_n)$, where Y_n is the Banach space cited in Corollary 5.2.*
Then $A + B$ is also the generator of some exponentially bounded n-times integrated C-cosine function on X.

Proof. Since Corollary 5.2 asserts that A_{Y_n} generates a cosine operator function on Y, the classical perturbation theorem implies that $A_{Y_n} + B|_{Y_n} (= (A + B)_{Y_n})$ also generates a cosine operator function on Y_n. Next, (b) and the definition of Y_n imply that

$$R((r - A - B)^{-n}C) = R((r - A)^{-n}C) \subset Y_n \subset X$$

and that $((r - A)^{-n}C)^{-1}(r - A - B)^{-n}C$ is a bounded operator on X. If $y \in R((r - A - B)^{-n}C)$, then using Corollary 5.2 we have

$$
\begin{aligned}
|y| &\leq M_2\|C^{-1}(r - A)^n y\| \\
&= M_2\|C^{-1}(r - A)^n(r - A - B)^{-n}C[(r - A - B)^{-n}C]^{-1}y\| \\
&\leq M_2\|(r - A)^{-n}C)^{-1}(r - A - B)^{-n}C\|\|C^{-1}(r - A - B)^n y\|.
\end{aligned}
$$

Thus $A + B$ satisfies conditions (a), (b) and (c) in Corollary 5.2 and hence generates an exponentially bounded $2n$-times integrated C-cosine function on X.

In particular, if B is a bounded operator from X into $D(C^{-1}(r - A)^n) = R((r - A)^{-n}C)$, then $C^{-1}(r - A)^n B \in L(X)$, and by (c) of Corollary 5.2 we get the following estimate:

$$
\begin{aligned}
|By| &\leq M_2\|C^{-1}(r - A)^n By\| \leq M_2\|C^{-1}(r - A)^n B\|\|y\| \\
&\leq M_2\|C^{-1}(r - A)^n B\|M_1|y|, \qquad y \in Y_n.
\end{aligned}
$$

Thus condition (c) of Theorem 6.1 is satisfied, and the next corollary follows.

COROLLARY 6.2. *Let A be the generator of an exponentially bounded $2n$-times integrated C-cosine function on a Banach space X, and let B be a bounded linear operator on X satisfying:*
(a) *$BC = CB$;*
(b) *there is a $r > w$ such that $r - A - B$ is injective in case $n \geq 1$, $R(C) \subset D((r - A - B)^{-n})$ and $R((r - A - B)^{-n}C) = R((r - A)^{-n}C)$;*
(c) *$R(B) \subset D(C^{-1}(r - A)^n)$.*
Then $A + B$ is also the generator of an exponentially bounded $2n$-times integrated C-cosine function on X.

When $n = 0$, condition (b) in Corollary 6.2 is redundant, and (c) becomes $R(B) \subset R(C)$. Thus Corollary 6.2 reduces to

COROLLARY 6.3. *If A is the generator of an exponentially bounded C-cosine function on X, then $A + B$ generates an exponentially bounded C-cosine function for every $B \in L(X, R(C))$ which commutes with C.*

7. REPRESENTATION FORMULAS FOR C-COSINE FUNCTIONS

Using Theorem 5.1 one can deduce some representation formulas for exponentially bounded C-cosine functions from those formulas [6] for ordinary cosine operator functions. In the following we only display three of them.

THEOREM 7.1 ([7, Theorem 5]). *Let $C(\cdot)$ be an exponentially bounded C-cosine function with generator A. The following formulas hold for all $x \in X$ and $t \geq 0$.*

$$C(t)x = \lim_{n \to \infty} \sum_{k=0}^{n} \sum_{j=0}^{k} \binom{2n}{2k} \binom{k}{j} (-1)^{k-j} \left[1 - (t/2n)^2 A \right]^{-(2n-k+j)} Cx;$$

$$C(t)x = \lim_{n \to \infty} \sum_{k=0}^{n} \sum_{j=0}^{k} \binom{2n+1}{2k} \binom{k}{j} (-1)^{k-j} \left[1 - (\frac{t}{2n+1})^2 A \right]^{-(2n+1-k+j)} Cx;$$

$$C(t)x = \lim_{n \to \infty} e^{-nt} \sum_{m=0}^{\infty} \sum_{k=0}^{m} \sum_{j=0}^{k} \frac{(nt)^{2m}}{(2m)!} \binom{2m}{2k} \binom{k}{j} (-1)^{k-j}$$
$$\cdot \left[1 + \frac{nt}{2m - 2k + 1}(1 - n^{-2}A)^{-1} \right] (1 - n^{-2}A)^{-(2m-k+j)} Cx,$$

the convergence being uniform on bounded intervals.

REFERENCES

1. H. O. Fattorini, Ordinary differential equations in linear topological spaces. I, *J. Differential Equations* **5** (1968), 72-105.

2. Y.-C. Li, Integrated C-semigroups and C-cosine functions of operators on locally convex spaces, Ph.D. dissertation, National Central University, 1991.

3. Y.-C. Li and S.-Y. Shaw, Integrated C-semigroups and the abstract Cauchy problem, preprint (1991).

4. Y.-C. Li and S.-Y. Shaw, On generators of integrated C-semigroups and C-cosine functions, *Semigroup Forum* **47** (1993), 29-35.

5. Y.-C. Li and S.-Y. Shaw, Integrated C-cosine functions and the abstract Cauchy problem, preprint (1991).

6. Shaw, S.-Y., Lee, C.-S. and Chiou, W.-L., Representation formulas for cosine and sine functions of operators II, Aequationes Math. **31** (1986), 64-75.

7. S.-Y. Shaw, On exponentially bounded C-cosine functions, preprint (1992).

8. S.-Y. Shaw and Y.-C. Li, Representation formulas for C-semigroups, *Semigroup Forum* **46** (1993), 123-125.

9. M. Sova, Cosine operator functions, *Rozprawy Math.* **49** (1966), 1-47.

10. K. Yosida, "Functional Analysis," ed. 6, Springer-Verlag, New York/ Heidelberg/ Berlin, 1980.

Hopf Bifurcation and Stability for a Quasilinear Reaction–Diffusion System

Gieri Simonett   University of California at Los Angeles, Los Angeles, California

1. INTRODUCTION

In this note we consider a parameter-dependent quasilinear reaction-diffusion system on a bounded domain of $\mathbb{R}^n$. We will study the dynamic behavior of solutions to this equation. In particular, we are interested in the behavior of the solution set if the parameter is varied. It is shown that a unique one-parameter family of periodic solutions bifurcates from an equilibrium. Finally, we will be concerned with the question, whether the bifurcating orbits are stable or unstable (at least in a simplified situation). We will rely on results carried out in [**14**], where the existence of finite dimensional (parameter-dependent) center manifolds, containing all the local recurrence of small solutions, has been established. We thus make use of a 'reduction by invariant manifolds,' allowing to apply results on Hopf bifurcation for ordinary differential equations.

Let $\Omega \subset \mathbb{R}^n$ be a bounded domain having a smooth boundary. Then we consider the parameter-dependent system of *strongly coupled reaction-diffusion equations*

supported by Schweizerischer Nationalfonds

$$\begin{cases} \partial_t u - \partial_j(a(u,\lambda)\partial_j u) = f(u,\lambda) & \text{in} \quad \Omega \times (0,\infty) , \\ a(u,\lambda)\partial_\nu u = g(u,\lambda) & \text{on} \quad \partial\Omega \times (0,\infty) , \end{cases} \tag{1.1}_\lambda$$

where $u = (u_1, u_2)$, $\lambda \in \Lambda := (-\lambda_0, \lambda_0)$ for some $\lambda_0 > 0$, and where $\partial_\nu := \nu^j \partial_j$ denotes the derivative along the outer normal field $\nu = (\nu^1, \cdots \nu^n)$ on $\partial\Omega$. We assume that the matrix $a(\eta, \lambda)$ is given by

$$a(\eta, \lambda) = \begin{bmatrix} a_{11}(\eta,\lambda) & a_{12}(\eta,\lambda) \\ -a_{21}(\eta,\lambda) & 0 \end{bmatrix} , \tag{1.2}$$

where

$$a_{11},\ a_{12},\ a_{21} \in C^\infty(G \times \Lambda, \mathbb{R}) , \tag{1.3}$$

with G being an open convex neighborhood of 0 in $\mathbb{R}^2$. Moreover, we ask that

$$a_{11}(\eta,\lambda) > 0, \quad a_{12}(\eta,\lambda)a_{21}(\eta,\lambda) > 0, \quad (\eta,\lambda) \in G \times \Lambda , \tag{1.4}$$

which will ensure the parabolicity of the quasilinear system of partial differential equations in $(1.1)_\lambda$. Let us assume that the function f is given by

$$f(u,\lambda) = (\psi(u_1,\lambda)u_1, 0) , \tag{1.5}$$

with ψ being a nonlocal function, satisfying

$$[(v,\lambda) \mapsto \psi(v,\lambda)] \in C^\infty(C(\bar{\Omega}) \times \Lambda, \mathbb{R}) . \tag{1.6}$$

Finally, we suppose that

$$g \in C^\infty(G \times \Lambda, \mathbb{R}^2) , \quad g(0,\cdot) = 0 . \tag{1.7}$$

We will first show that the quasilinear system $(1.1)_\lambda$ has a unique (classical) solution for each initial value belonging to an appropriate function space. In addition, we will even obtain that the solutions generate a (smooth) local semiflow. Note that this property forms the basis of a geometric theory for the quasilinear reaction diffusion system $(1.1)_\lambda$, allowing to give a thorough meaning to the concepts of stability, invariant manifolds, bifurcation, etc.

Let $p > n$ and set

$$V := \{u \in H_p^1(\Omega, \mathbb{R}^2); u(\bar{\Omega}) \subset G\} . \tag{1.8}$$

It follows from Sobolev's embedding theorem and the fact that $\bar{\Omega}$ is compact that V is a well-defined open subset of the Sobolev space $H^1(\Omega, \mathbb{R}^2)$. Then we have the following result on the existence, uniqueness, and smooth dependence of solutions of $(1.1)_\lambda$:

THEOREM 1.1. *Given any initial value $u_0 \in V$, there exists a unique maximal classical solution*

$$u(\cdot, u_0, \lambda) \in C([0, t^+(u_0, \lambda), V) \cap C^\infty(\bar{\Omega} \times (0, t^+(u_0, \lambda)), \mathbb{R}^2) \tag{1.9}$$

of $(1.1)_\lambda$ satisfying $u(0, u_0, \lambda) = u_0$. The mapping

$$(t, u_0) \mapsto u(t, u_0, \lambda) \tag{1.10}$$

defines a smooth semiflow on V depending smoothly on $\lambda \in \Lambda$ and having 0 as a rest-point.

PROOF Let $(v, \lambda) \in V \times \Lambda$ be given. It follows from (1.4) and [**1**, Theorem 4.4] that the boundary value system $(-\partial_j(a(v, \lambda)\partial_j), a(v, \lambda)\partial_\nu)$ is normally elliptic in the sense of [**1,3,4**]. The assertion (1.9) is then a consequence of (1.3)-(1.7) and the results proven in [**1**, Corollary 9.4], (at least for $g = 0$). Indeed, it is clear that the 'Standard Situation' of [**1**, Proposition 7.1 and Corollary 9.4] is satisfied for the matrix a. It is not difficult to see that the function f fulfills [**1**, Assumption (**A3**) on p. 44.], showing that Theorem 7.3, Theorem 9.2 and Corollary 9.3 of the very same paper hold true. As an immediate consequence of (1.5)-(1.6) we obtain that

$$f_1 \in C^\infty(C^\sigma(\bar{\Omega}) \times \Lambda, C^\sigma(\bar{\Omega})) , \quad \sigma \in \mathbb{R}^+ ,$$

where $C^k(\bar{\Omega})$ has the usual meaning for $k \in \mathbb{N}$. $C^{k+\alpha}(\bar{\Omega})$, $k \in \mathbb{N}, \alpha \in (0, 1)$, then denotes the Banach space of all functions in $C^k(\bar{\Omega})$ whose derivatives of order k are α-Hölder continuous. An inspection of the proof of [**1**, Corollary 9.4] then gives the assertion (1.9). The case $g \neq 0$ can be reduced to the situation above, see [**1**, formulas (3) and (4)]. Next, we can infer from [**1**, Corollary 7.4 and Theorem 10.5] that (1.10) holds true. Note that Theorem 10.5 is satisfied, thanks to [**1**, Remark 10.6] and our assumptions (1.5)-(1.6). Finally, it is easily seen that $u \equiv 0$ is an equilibrium for the quasilinear system $(1.1)_\lambda$. $\square$

To simplify some computations, we assume that there exists a positive constant b so that

$$g(u, \lambda) = -ba(u, \lambda)u . \tag{1.11}$$

It should be mentioned that a similar example as $(1.1)_\lambda$ has been considered in [**2**]. If we also introduce the flux vectors

$$\begin{aligned}
j_1(u, \lambda) :&= -a_{11}(u, \lambda)\nabla u_1 - a_{12}(u, \lambda)\nabla u_2 , \\
j_2(u, \lambda) :&= a_{21}(u, \lambda)\nabla u_1 ,
\end{aligned} \tag{1.12}$$

we can rewrite $(1.1)_\lambda$ as

$$\partial_t u_r + \mathrm{div}\, j_r(u,\lambda) = f_r(u,\lambda) \quad \text{in} \quad \Omega \times (0,\infty)\,, \qquad r = 1,2\,, \tag{1.13}$$

and

$$-(j_r(u,\lambda)\,|\,\nu) = g_r(u,\lambda) \quad \text{on} \quad \partial\Omega \times (0,\infty)\,, \qquad r = 1,2\,. \tag{1.14}$$

Note that $(1.1)_\lambda$ can be interpreted as a two populations model, see [2]. Indeed, assuming that $a_{12}(u,\lambda) > 0$, "the first quantity in (1.13) means that the flux of the first population goes in the direction where the density of its own species decreases and also to places where the density of the second population (the enemies) is low. Thus $a_{11}(u,\lambda)\nabla u_1$ can be interpreted as a social friction term, which prevents overcrowdedness. The second quantity means that the second population runs in the direction where there is a higher density of the first species. The boundary condition (1.14) means that the first population tends to stay away from those parts of the boundary, where there is already a lot of its own species or a lot of enemies, whereas the second population wants to go to those places on the boundary, where there is a lot of the first species". The function $f_1(u,\lambda) = \psi(u_1,\lambda)u_1$ describes some device to 'produce' the first species. (If we expect periodic behavior, the loss of the first population caused by diffusion has to be overcome).

We will now consider the spectrum of the linearization of $(1.1)_\lambda$ at 0, i.e. we study the elliptic eigenvalue problem

$$\begin{aligned} a(0,\lambda)\Delta v + \partial_1 f(0,\lambda)v &= \mu(\lambda)v && \text{in} \quad \Omega\,, \\ a(0,\lambda)(\partial_\nu v + bv) &= 0 && \text{on} \quad \partial\Omega\,. \end{aligned} \tag{$1.15)_\lambda$}$$

Observe that $(1.15)_\lambda$ is well-posed since the L_p-realization has a compact resolvent. It is a consequence of (1.4) that $\sigma(a(0,\lambda)) \subset [\mathrm{Re}\, z > 0]$, i.e. all of the eigenvalues of $a(0,\lambda)$ are contained in the right half plane. Thus, $(1.15)_\lambda$ is equivalent to the system

$$\begin{aligned} a(0,\lambda)\Delta v + \partial_1 f(0,\lambda)v &= \mu(\lambda)v && \text{in} \quad \Omega\,, \\ \partial_\nu v + bv &= 0 && \text{on} \quad \partial\Omega\,. \end{aligned} \tag{$1.16)_\lambda$}$$

We now consider the scalar eigenvalue problem

$$\begin{aligned} \Delta\varphi &= \kappa\varphi && \text{in} \quad \Omega\,, \\ \partial_\nu\varphi + b\varphi &= 0 && \text{on} \quad \partial\Omega\,. \end{aligned} \tag{1.17}$$

Let

$$0 > \kappa_0 > \kappa_1 \geq \kappa_2 \geq \dots \tag{1.18}$$

denote the sequence of eigenvalues of (1.17), counted according to its multiplicity. It follows that the eigenvalues of $(1.16)_\lambda$ are given by

$$\sigma(\lambda) := \{\mu_{k,j}(\lambda);\ k \in \mathbb{N},\ j = 1,2\}\,, \tag{1.19}$$

where $\mu_{k,1}(\lambda)$ and $\mu_{k,2}(\lambda)$ are the eigenvalues of the matrices

$$a(0,\lambda)\kappa_k + \partial_1 f(0,\lambda) \in \mathcal{L}(\mathbb{R}^2) , \quad k \in \mathbb{N} ,$$

see [2]. Due to (1.5), we obtain that

$$a(0,\lambda)\kappa_k + \partial_1 f(0,\lambda) = \begin{bmatrix} a_{11}(0,\lambda)\kappa_k + \psi(0,\lambda) & a_{12}(0,\lambda)\kappa_k \\ -a_{21}(0,\lambda)\kappa_k & 0 \end{bmatrix} \tag{1.20}$$

for $k \in \mathbb{N}$. It is now easy to see that

$$\sigma(\lambda) \subset [Re\, z < 0] \quad \text{if} \quad \psi(0,\lambda) < -a_{11}(0,\lambda)\kappa_0 \tag{1.21}$$

and that

$$\sigma(\lambda) \subset [Re\, z \le 0] , \quad \sigma(\lambda) \cap i\mathbb{R} = \{\pm i\omega_0(\lambda)\} \tag{1.22}$$

where $\omega_0(\lambda) = |\kappa_0|\sqrt{a_{12}(0,\lambda)a_{21}(0,\lambda)}$, provided $\psi(0,\lambda) = -a_{11}(0,\lambda)\kappa_0$.

Whenever (1.21) holds true, we infer from the principle of linearized stability that 0 is an asymptotically stable critical point for the semiflow induced by $(1.1)_\lambda$, see [9], or [13] for a more general situation. Hence, the dynamical behavior of small solutions of $(1.1)_\lambda$ remains essentially the same as long as (1.21) is satisfied. But there is a drastic change if some of the eigenvalues hit or cross the imaginary axis.

We assume that

$$\psi(0,\lambda) < -a_{11}(0,\lambda)\kappa_0 \quad \text{if} \quad \lambda < 0 , \quad \psi(0,0) = -a_{11}(0,0)\kappa_0 . \tag{1.23}$$

We then infer from (1.22) that $\sigma(0) = \sigma_s \cup \{\pm i\omega_0\}$, where $\sigma_s \subset [Re\, z < 0]$ and

$$\omega_0 = |\kappa_0|\sqrt{a_{12}(0,0)a_{21}(0,0)} . \tag{1.24}$$

Finally, let

$$\frac{d}{d\lambda}(a_{11}(0,\lambda)\kappa_0 + \psi(0,\lambda))|_{\lambda=0} > 0 . \tag{1.25}$$

We are now ready to state the following theorem on the existence and stability of bifurcating periodic solutions for the quasilinear reaction-diffusion system $(1.1)_\lambda$.

THEOREM 1.2. *Suppose that (1.23) and (1.25) are satisfied. Then there exists a unique one-parameter family $\{\Gamma(s)\, ; \, 0 < s < \varepsilon\}$ of nontrivial periodic orbits of the system $(1.1)_\lambda$ in a neighborhood of $(0,0) \in V \times \Lambda$.*

More precisely, there exists $\varepsilon > 0$ and a map

$$(\lambda(\cdot), T(\cdot), u(\cdot)) \in C^k((-\varepsilon,\varepsilon),\, \mathbb{R} \times \mathbb{R} \times V)$$

with

$$(\lambda(0), T(0), u(0)) = (0, 2\pi/\omega_0, 0)$$

and such that

$$\Gamma(s) := \Gamma(u(s))$$

is a nontrivial $T(s)$-periodic orbit of the quasilinear reaction-diffusion system $(1.1)_{\lambda(s)}$ passing through $u(s) \in V$. If $0 < s_1 < s_2 < \varepsilon$, then $\Gamma(s_1) \neq \Gamma(s_2)$.

Moreover, the family $\{\Gamma(s); 0 < s < \varepsilon\}$ contains every nontrivial periodic orbit of $(1.1)_\lambda$ lying in a neighborhood of $(0, T(0), 0) \in \Lambda \times \mathrm{IR} \times V$.

If

$$s\dot{\lambda}(s) > 0 \quad for \quad s > 0 \qquad (\text{`supercritical bifurcation'}),$$

then each orbit $\Gamma(s)$ is asymptotically stable in V.

In the case of

$$s\dot{\lambda}(s) < 0 \quad for \quad s > 0 \qquad (\text{`subcritical bifurcation'}),$$

each of the orbits $\Gamma(s)$, $0 < s < \varepsilon$, is unstable in V.

PROOF   If follows from (1.17) and (1.23) that $\{\pm i\omega_0\}$ are simple eigenvalues of the eigenvalue problem $(1.16)_0$, and that all of the remaining eigenvalues are contained in $[Re\, z < 0]$. Let $\mu_0(\lambda)$ be the unique continuation of the eigenvalue $i\omega_0$ of $(1.16)_0$ (for λ in a neighborhood of 0). It follows that

$$Re\,\mu_0(\lambda) = \frac{1}{2}\left(a_{11}(0, \lambda)\kappa_0 + \psi(0, \lambda)\right)$$

for λ sufficiently small. Assumption (1.25) then yields the transversality condition $\frac{d}{d\lambda}(Re\,\mu_0(\lambda))|_{\lambda=0} > 0$. The Theorem now follows from [**14**, Theorem 1]. (An obvious modification of [**14**, formula (6.18)] takes care of the present situation). $\square$

It should be noticed that the existence of a unique one-parameter family of bifurcating periodic orbits for a similar system of quasilinear reaction-diffusion equations has also been obtained in [**2**], using a different approach. However, our result contains, in addition, information on the qualitative behavior of the bifurcating orbits.

2. STABILITY CONDITIONS

We will now give conditions which allow to decide, whether supercritical or subcritical bifurcation occurs. For doing this, we will follow an algorithm which has been described in [14, Section 5]. In general, it can be rather difficult to implement each of the steps of the algorithm to a given quasilinear system of partial differential equations [14, Remark 5.5]. If we impose an additional assumption on the coefficients of the 'diffusion matrix' $a(u,0)$, the computations turn out to be fairly easy.

Suppose that

$$\partial_r a_{11}(0,0,0) = \partial_r a_{12}(0,0,0) = \partial_r a_{21}(0,0,0) = 0 \ , \quad r = 1,2 \ . \tag{2.1}$$

Let

$$\beta(z) := \psi(z\varphi_0, 0) \ , \quad z \in \mathbb{R} \ , \tag{2.2}$$

where φ_0 denotes the eigenfunction of (1.17) corresponding to the eigenvalue κ_0. It follows from (1.6) that β is a smooth real function, i.e. $\beta \in C^\infty(\mathbb{R}, \mathbb{R})$.

PROPOSITION 2.1. *Let (2.1) and (2.2) be satisfied. If*

$$\beta''(0) < 0 \ ,$$

the occurring Hopf bifurcation for the quasilinear system $(1.1)_\lambda$ *of reaction-diffusion equations is supercritical and thus, each orbit* $\Gamma(s)$ *is asymptotically stable in V.*

In the case of

$$\beta''(0) > 0 \ ,$$

the appearing Hopf bifurcation is subcritical.

PROOF We will rely on results given in [14, Sections 5-6].

Let $\{\xi_1, \xi_2\}$ be a basis of $\mathbb{R}^2$, given by

$$\xi_1 := (1,0) \ , \quad \xi_2 := (0, -\sqrt{a_{21}(0,0)/a_{12}(0,0)}) \ . \tag{2.3}$$

(1.20) and the second part of (1.23) yield

$$[a(0,0)\kappa_0 + \partial_1 f(0,0)]\xi_1 = -\omega_0\xi_2 \ , \quad [a(0,0)\kappa_0 + \partial_1 f(0,0)]\xi_2 = \omega_0\xi_1 \ , \tag{2.4}$$

showing that the matrix $[a(0,0)\kappa_0 + \partial_1 f(0,0)]$ admits the representation

$$\begin{bmatrix} 0 & \omega_0 \\ -\omega_0 & 0 \end{bmatrix} \tag{2.5}$$

with respect to the basis $\{\xi_1, \xi_2\}$.

Let $H^2_{p,\mathcal{B}}(\Omega) := \{u \in H^2_p(\Omega, \mathbb{R}^2); \partial_\nu u + bu = 0\}$ and, moreover, let φ_0 be the eigenfunction of (1.17) corresponding to the eigenvalue κ_0. It follows from (1.17) and (2.4) that the linear differential operator

$$-L : H^2_{p,\mathcal{B}}(\Omega) \to L_p(\Omega) , \quad v \mapsto a(0,0)\Delta v + \partial_1 f(0,0)v \qquad (2.6)$$

leaves the two-dimensional space $X^c := span\{\varphi_0\xi_1, \varphi_0\xi_2\}$ invariant. Let then $-L_c := L|X^c$ be the part of $-L$ in X^c. Thanks to (2.5), we get that

$$-L_c = \begin{bmatrix} 0 & \omega_0 \\ -\omega_0 & 0 \end{bmatrix} \qquad (2.7)$$

with respect to the basis $\{\varphi_0\xi_1, \varphi_0\xi_2\}$. In the following, we will identify each element of X^c with its coordinates in the basis $\{\varphi_0\xi_1, \varphi_0\xi_2\}$, i.e.

$$(z_1, z_2) \leftrightarrow z_1\varphi_0\xi_1 + z_2\varphi_0\xi_2 . \qquad (2.8)$$

Given $u \in H^2_{p,\mathcal{B}}(\Omega)$, we define the nonlinear function ϕ

$$\phi(u) := \partial_j(a(u,0)\partial_j u) + f(u,0) . \qquad (2.9)$$

By performing the derivative $\partial_j := \partial_{x_j}$ we get

$$\phi(u) = [\partial_1 a(u_1, u_2, 0)\partial_j u_1 + \partial_2 a(u_1, u_2, 0)\partial_j u_2]\partial_j u + a(u,0)\Delta u + f(u,0) , \qquad (2.10)$$

where $[\partial_r a(u_1, u_2, 0)\partial_j u_r]$ stands for the matrix

$$\begin{bmatrix} \partial_r a_{11}(u_1, u_2, 0)\partial_j u_r & \partial_r a_{12}(u_1, u_2, 0)\partial_j u_r \\ -\partial_r a_{21}(u_1, u_2, 0)\partial_j u_r & 0 \end{bmatrix} , \quad r = 1, 2 .$$

For the remainder, we will sometimes use the notation

$$\xi_r = \mu_r e_r , \quad \mu_1 := 1 , \quad \mu_2 := -\sqrt{a_{21}(0,0)/a_{12}(0,0)} , \qquad (2.11)$$

where $\{e_1, e_2\}$ denotes the standard basis of $\mathbb{R}^2$. We now consider the case where u belongs to X^c, and thus is given as $u = z_1\varphi_0\xi_1 + z_2\varphi_0\xi_2$. Using (2.2) and the identification (2.8), the definition of φ_0 and (2.11), (2.10) then becomes

$$\phi(z_1, z_2) = [\partial_1 a((z_1, \mu_2 z_2)\varphi_0, 0)z_1 + \partial_2 a((z_1, \mu_2 z_2)\varphi_0, 0)\mu_2 z_2](z_1, \mu_2 z_2)|\nabla\varphi_0|^2$$
$$+ \kappa_0[a((z_1, \mu_2 z_2)\varphi_0, 0)](z_1, \mu_2 z_2)\varphi_0 + (\beta(z_1)z_1\varphi_0, 0) .$$

We will now decompose the function $\phi(z_1, z_2)$ in a part belonging to X^c and a remaining part (belonging to a complemented space, say a complemented subspace in $L_p(\Omega)$). In fact, we can identify

$$\phi(z_1, z_2) \leftrightarrow (\phi^1(z_1, z_2), \phi^2(z_1, z_2), \phi^3(z_1, z_2)) \in \mathbb{R} \times \mathbb{R} \times L_p(\Omega, \mathbb{R}^2)$$

by means of

$$\phi(z_1, z_2) = \phi^1(z_1, z_2)\varphi_0\xi_1 + \phi^2(z_1, z_2)\varphi_0\xi_2 + \phi^3(z_1, z_2) \ , \tag{2.12}$$

where $v := \phi^3(z_1, z_2) \notin X^c$. This can be accomplished by taking

$$\begin{aligned}
\phi^1(z_1, z_2) &= \ \ \kappa_0 a_{11}(0,0)z_1 + \beta(z_1)z_1 + \kappa_0 a_{12}(0,0)\mu_2 z_2 \ , \\
\phi^2(z_1, z_2) &= -\kappa_0 a_{21}(0,0)\mu_2^{-1} z_1 \ .
\end{aligned} \tag{2.13}$$

The function $\phi^3(z_1, z_2)$ consists of linear combinations of terms of the form

$$\partial_r a_{kl}((z_1, \mu_2 z_2)\varphi_0, 0)\mu_r z_r \mu_s z_s e_i \, |\nabla\varphi_0|^2 \ , \quad i, r, s \in \{1, 2\} \ , \tag{2.14}$$

(where $a_{kl} = a_{11}, a_{12}$, or a_{21}) and, in addition, of the term

$$\kappa_0 \left[\begin{matrix} a_{11}((z_1, \mu_2 z_2)\varphi_0, 0) - a_{11}(0,0) & a_{12}((z_1, \mu_2 z_2)\varphi_0, 0) - a_{12}(0,0) \\ -a_{21}((z_1, \mu_2 z_2)\varphi_0, 0) - a_{21}(0,0) & 0 \end{matrix} \right] (z_1, \mu_2 z_2)\varphi_0 \ . \tag{2.15}$$

In order to see that the function $v := \kappa_0 \left[a((z_1, \mu_2 z_2)\varphi_0, 0) - a(0,0,0)\right](z_1, \mu_2 z_2)\varphi_0$ does not belong to $X^c = span\{\varphi_0\xi_1, \varphi_0\xi_2\}$, we can use Taylor expansion (or the mean value theorem) for each of the coefficients in $[a((z_1, \mu_2 z_2)\varphi_0, 0) - a(0,0,0)]$. This will produce an additional term φ_0, so that

$$\kappa_0 \left[a((z_1, \mu_2 z_2)\varphi_0, 0) - a(0,0,0)\right](z_1, \mu_2 z_2)\varphi_0 = \left[\alpha((z_1, \mu_2 z_2)\varphi_0, 0)\right](z_1, \mu_2 z_2)\varphi_0^2 \ ,$$

with an appropriate matrix $[\alpha((z_1, \mu_2 z_2)\varphi_0, 0)]$. As a consequence of (2.1), the second order partial derivatives of (2.14) and (2.15), evaluated at $(z_1, z_2) = (0,0)$, vanish. Indeed, it is easy to see that

$$\partial_j \partial_k \left(\partial_r a_{11}(z_1, \mu_2 z_2)\mu_r\mu_s z_r z_s\right)\big|_{z_1 = z_2 = 0} = 0$$

for each combination of $j, k \in \{1, 2\}$ and $r, s \in \{1, 2\}$. The very same holds true for the functions a_{12} and a_{21}. Moreover, we obtain from (2.1) that

$$\partial_j \partial_k \, \kappa_0 \left[a((z_1, \mu_2 z_2)\varphi_0, 0) - a(0,0,0)\right](z_1, \mu_2 z_2)\varphi_0\big|_{z_1 = z_2 = 0} = (0,0)$$

for $j, k \in \{1, 2\}$. Hence, we have shown that

$$\partial_j \partial_k \phi^3(0,0) = 0 \ , \quad j, k \in \{1, 2\} \ . \tag{2.16}$$

According to [14, Sections 5-6], we now have to compute the expression

$$\begin{aligned}
\delta := \omega_0 \left[\partial_1^3\phi^1 + \partial_1\partial_2^2\phi^1 + \partial_1^2\partial_2\phi^2 + \partial_2^3\phi^2 \right] &- \partial_1^2\phi^1 \, \partial_1\partial_2\phi^1 + \partial_1^2\phi^1\partial_1^2\phi^2 \\
&- \partial_1\partial_2\phi^1 \, \partial_2^2\phi^1 - \partial_2^2\phi^1 \, \partial_2^2\phi^2 + \partial_1^2\phi^2 \, \partial_1\partial_2\phi^2 + \partial_1\partial_2\phi^2 \, \partial_2^2\phi^2 \ .
\end{aligned} \tag{2.17}$$

Due to (2.13), we end up with

$$\delta = 3\omega_0 \beta''(0) . \tag{2.18}$$

We will now show that the computations of this section reflect the statements of [**14**, Sections 5-6], which are formulated in a somewhat more abstract (and general) setting. In fact, it should be noted that the nonlinear boundary conditions in $(1.1)_\lambda$ can be reduced to a homogeneous linear boundary condition, due to (1.11) and (1.4). For a given $(v, \lambda) \in V \times \Lambda$, let

$$a(v, \lambda)(w, u) := \int_\Omega (\partial_j w \,|\, a(v, \lambda)\partial_j u) \, dx + \int_{\partial\Omega} (w \,|\, bu) \, d\sigma \tag{2.19}$$

be the Dirichlet form associated to the boundary value system $(-\partial_j(a(v, \lambda)\partial_j\,), \partial_\nu + b)$. We can perform the same steps as in [**14**, Section 6], obtaining a linear operator $A(v, \lambda)$, induced by the Dirichlet form $a(v, \lambda)$, and a function $F(v, \lambda)$, induced by $f(v, \lambda)$. Let then ϕ be defined as in [**14**, formula (5.2)] and let $L := L(0)$ be the linear operator in [**14**, formula (4.8)]. It can be shown that the restriction of L and ϕ to the space $H^2_{p,\mathcal{B}}$ coincide with (2.6) and with (2.9), respectively. [**14**, Proposition 5.2, Remark 5.5b) and Theorem 5.6] then imply the statements of Proposition 2.1. $\square$

REMARKS 2.2.

$a)$ Observe that the 'reaction' term in $(1.1)_\lambda$, i.e. the function $f(u_1, u_2, \lambda)$, is decoupled. Thus, Hopf bifurcation is caused by the strong coupling of the principal part (by 'cross diffusion'). The function f describes a device to "produce the first species in order to compensate for the loss caused by diffusion". In fact, what we have in mind is some device governed by

$$f(u, \lambda) := \Big(\int_\Omega \mu(u_1(x), \lambda) \, dx \, u_1, 0 \Big) , \quad \mu \in C^\infty(\mathbb{R} \times \Lambda, \mathbb{R}) . \tag{2.20}$$

Note that f satisfies the assumption (1.5). The function β, introduced in (2.2), then becomes $\beta(z) := \int_\Omega \mu(z\varphi_0(x), 0) \, dx$. It follows that

$$\beta''(0) = \partial_1^2 \mu(0, 0) \int_\Omega \varphi_0^2(x) \, dx . \tag{2.21}$$

$b)$ It is not difficult to see that there really exist functions $\mu \in C^\infty(\mathbb{R} \times \Lambda, \mathbb{R})$ such that ψ - defined in (2.20) - satisfies the assumptions (1.23) and (1.25). In addition, we can always find a function μ so that the first condition of Proposition 2.1 holds true. It should be clear that μ is related to the diffusion coefficient a_{11} .

$c)$ (2.1) means that the diffusion coefficients should not vary to fast for small solutions. Note that this condition was necessary in order to use a simple version of the stability algorithm, see [**14**, Remark 5.5 b)].

The result of Proposition 2.1 can now be read as follows

COROLLARY 2.3. *There exists a function* $\mu \in C^\infty(\mathbb{R} \times \Lambda, \mathbb{R})$ *such that*

$$f(u, \lambda) = (\int_\Omega \mu(u_1(x), \lambda) \, dx \, u_1, 0)$$

'stabilizes' each of the bifurcating orbits $\Gamma(s) = \Gamma(f, s)$, $0 < s < \varepsilon$, *provided (2.1) is satisfied.*

Using invariant (center) manifolds is a well-known approach to study bifurcation of small solutions, such as periodic solutions or steady states. The method of invariant manifolds is common in dynamical systems and in ordinary differential equations and has also been used in the context of semilinear evolution equations [12,10,5]. However, not much is done in the case of quasilinear equations. In fact, we rely on [13,14]. We would also like to draw attention to [8]. Hopf bifurcation has been widely studied for ordinary differential equations and semilinear evolution equations. Extensions to more general nonlinear equations have been given in [2,7]. In fact, there are quite different approaches to the study of Hopf bifurcation. Instead of using reduction via invariant manifolds, one can also use the so called Ljapunov-Schmidt reduction. This has been done in [2], where the existence of bifurcating solutions for quasilinear reaction-diffusion equations has been proven. However, the author does not study stability conditions; see the remarks at the end of his article. On the other hand, there is a functional analytic approach which was introduced in [6] and later also used by [7] for fully nonlinear equations. The authors in the latter paper obtain the existence of bifurcating periodic solutions even for fully nonlinear equations, but they also do not consider stability. In [8], the authors prove the existence of center manifolds for fully nonlinear equations and then state a result on Hopf bifurcation, and in [11], the stability of periodic solutions for a large class of nonlinear equations is studied.

REFERENCES

[1] H. Amann, *Dynamic theory of quasilinear parabolic equations II. Reaction diffusion systems*, Differential and Integral Equations **3** (1990), 13–75.

[2] H. Amann, *Hopf bifurcation in quasilinear reaction diffusion systems*, in S.Busenberg, M. Martelli, editors, Delay Differential Equations and Dynamical Systems, Proceedings, Claremont 1990, Springer Verlag, LNM 1475, New York, 1991, 53-63.

[3]   H. Amann, *Nonhomogeneous Linear and Quasilinear Elliptic and Parabolic Boundary Value Problems*, preprint.

[4]   H. Amann, Linear and Quasilinear Parabolic Problems, book in preparation.

[5]   J. Carr, Applications of Centre Manifold Theory, Appl. Math. Sc. 35, Springer, New York, 1981.

[6]   M.G. Crandall and P. Rabinowitz, *The Hopf bifurcation theorem in infinite dimensions*, Arch. Rat. Mech. Anal. **67** (1977/78), 53–72.

[7]   G. Da Prato and A. Lunardi, *Hopf bifurcation for fully nonlinear equations in Banach spaces*, Ann. Inst. Henri Poincaré **3** (1986), 315–329.

[8]   G. Da Prato and A. Lunardi, *Stability, instability and center manifold theorem for fully nonlinear autonomous parabolic equations in Banach space*, Arch. Rat. Mech. Anal. **101** (1988), 115–144.

[9]   A.-K. Drangeid, *The principle of linearized stability for quasilinear parabolic evolution equations*, Nonlinear Analysis, TM&A. **13** (1989), 1091–1113.

[10]   D. Henry, Geometric Theory of Semilinear Parabolic Equations, Springer, LNM 840, Berlin, 1981.

[11]   A. Lunardi, *Stability of the periodic solutions to fully nonlinear parabolic equations in Banach spaces*, Differential and Integral equations **1** (1988), 253–279.

[12]   J.E Marsden and M. McCracken, The Hopf Bifurcation and Its Applications, Appl. Math. Sc. 19, Springer New York, 1976.

[13]   G. Simonett, *Center manifolds for quasilinear reaction-diffusion systems*, Differential and Integral Equations. To appear.

[14]   G. Simonett, *Invariant manifolds and bifurcation for quasilinear reaction-diffusion systems*, Nonlinear Anal., TM&A. To appear.

On the Stability of Steady States, Finite Time Blow-up, and Decay Rates of Solutions to a Semilinear Heat Equation in $\mathbb{R}^n$

Xuefeng Wang   Tulane University, New Orleans, Louisiana

Consider the following Cauchy problem

$$\begin{cases} u_t = \Delta u + u^p \, , & u = u(x,t), \quad x \in \mathbb{R}^n, \quad t > 0, \\ u(x,0) = \varphi(x) \, , \end{cases} \tag{1}$$

where $p > 1$ and φ is a nonnegative bounded continuous function in $\mathbb{R}^n$, $\varphi \not\equiv 0$. It is well-known that there exists $0 < T_\varphi \leq \infty$ so that (1) has a unique classical solution u which is bounded on any strip $\mathbb{R}^n \times [0, T']$ for $0 < T' < T_\varphi$, and if T_φ is finite, $\|u(\cdot, t)\|_{L^\infty(\mathbb{R}^n)} \to \infty$ as $t \to T_\varphi$ from below. When $T_\varphi = \infty$, we say that u is a global solution; while when $T_\varphi < \infty$, we say that u blows up in finite time.

The study of the global existence for (1) goes back to the fundamental work of H. Fujita [Fu] in 1966. Since then, (1) and its various generalizations have been attacked by many authors, as model problems for the mathematical study of pathological phonomena. But our understanding of (1), in spite of its simple appearance, is still incomplete. In

---

Supported in part by NSF Grant DMS 9305658.

419

this short expository paper, we shall concentrate on some recent progress on the question of stability of nonnegative steady states, sufficient conditions on the initial value φ for finite time blow-up, and on the decay rates (in time t) of solutions of (1). For results concerning the role of exponents played in blow-up theorems, the readers are referred to [L] for a survey; for results concerning blow-up profiles, they are referred to the expository paper [HV] and the references therein.

In [Fu], Fujita proved that when $1 < p < (n+2)/n$, the local solution u of (1) blows up in finite time, while when $p > (n+2)/n$ and φ is "small", say, dominated by a small multiple of the Gaussian $\exp(-|x|^2)$, the solution u of (1) is global and $\|u(\cdot,t)\|_{L^\infty(\mathbb{R}^n)}$ decays like $t^{-n/2}$ as $t \to \infty$. Subsequently, Hayakawa [H], Kobayashi, Siaro and Tanaka [KST] proved that in the borderline case $p = (n+2)/n$, u also blows up in finite time. Simpler proofs of this last fact were given by Aronson and Weinberger [AW] and Weissler [We].

From the point of view of stability, the above results imply that for $1 < p \le (n+2)/n$ the rest state $u_0 \equiv 0$ is unstable in any reasonable topology, but for $p > (n+2)/n$, u_0 is stable in some sense. While the results for the case $1 < p \le (n+2)/n$ is complete, in case $p > (n+2)/n$ (which we shall always assume throughout this paper), one may ask in what topology (or norm) the rest state u_0 is stable. Also, by the eigenfunction method of Kaplan [K] (also see [Fu]), the solution u of (1) blows up in finite time if the initial value φ is large on a bounded domain. So when $p > (n+2)/n$ the attraction domain of $u_0 \equiv 0$ cannot be too large and there must exist a "threshold" of attraction. Then one may demand a description of this threshold. In particular, one asks how fast φ has to decay at $x = \infty$ so that the solution u of (1) exists globally and converges to u_0 as $t \to \infty$. In this direction, Weissler's integral condition (Theorem 3 in [We]) allows, as one can check, polynomial decay of $\varphi(x) \sim |x|^{-a}$ with a large, improving the earlier result of Fujita. In [LN], Lee and Ni showed that the critical decay of φ for global existence is $\varphi(x) \sim |x|^{-2/(p-1)}$. More precisely, they proved that (i) for any $k > 0$, there exists a small positive $b = b(p,n,k)$ such that if $\varphi(x) \le b(k + |x|^2)^{-1/(p-1)}$, then the solution u of (1) is global with its $L^\infty(\mathbb{R}^n)$ norm decaying like $t^{-1/(p-1)}$ as $t \to \infty$; (ii) if the solution u is global, then necessarily we have

$$\liminf_{x \to \infty} |x|^{\frac{2}{p-1}} \varphi(x) \le (\lambda(B_1))^{\frac{1}{p-1}} , \tag{2}$$

where $\lambda(B_1)$ is the first eigenvalue of $-\Delta$ with Dirichlet boundary condition on the boundary of the unit ball B_1 in $\mathbb{R}^n$.

Now one natural question (as an effort in describing the attraction domain of u_0) is

(Q) How large can the b in Lee and Ni's sufficient condition be if we want the solution u of (1) to be global?

Note by (2), b cannot be larger than $(\lambda(B_1))^{1/(p-1)}$. Recently, an attempt to answer this question was made by Wang [Wa] and a partial answer is the consequence of the following result proved in [Wa].

THEOREM 1 *Suppose $n \geq 3$.*

(i) *When $p > n/(n-2)$ ($> (n+2)/n$), if $\varphi \leq \lambda u_s$ for some $\lambda \in (0,1)$, where*

$$u_s(x) = L|x|^{-2/(p-1)} \text{ with } L = \left(\frac{2(n-2)}{(p-1)^2}\left(p - \frac{n}{n-2}\right)\right)^{\frac{1}{(p-1)}}$$

(which is a singular steady state of (1)), then the solution u of (1) is global and satisfies $0 \leq u \leq \lambda u_s$ in $\mathbb{R}^n \times [0,\infty)$, and

$$\|u(\cdot,t)\|_{L^\infty(\mathbb{R}^n)} \leq \left((\lambda^{1-p} - 1)(p-1)t\right)^{-1/(p-1)}, \quad t > 0 . \tag{3}$$

(ii) *When $n/(n-2) < p < p_c$, where*

$$p_c = \begin{cases} \frac{(n-2)^2 - 4n + 8\sqrt{n-1}}{(n-2)(n-10)} , & n > 10 , \\ \infty , & 3 \leq n \leq 10 , \end{cases}$$

if $\varphi \leq u_s$ in $\mathbb{R}^n$, then u is global with $0 \leq u \leq u_s$ and $\|u(\cdot,t)\|_{L^\infty(\mathbb{R}^n)} \to 0$ as $t \to \infty$.

(iii) *If $p \geq p_c$ and $\liminf_{x\to\infty}(\varphi/u_s)(x) > 1$, then u blows up in finite time.*

(More criteria for global existence and finite time blow-up can be found in [Wa].)

From Theorem 1, we see that *when $n \geq 3$ and $n/(n-2) < p < p_c$, b in Lee and Ni's condition can be as large as L, and that when $n \geq 3$ and $p \geq p_c$, b can be arbitrarily close to L and the least upper bound of b is L.* (The least upper bound in the case $(n+2)/n < p < p_c$ has not yet been found even now!) Moreover, if we introduce the weighted L^∞ norm $|u|_a = \|u(\cdot) \cdot |\cdot|^a\|_{L^\infty(\mathbb{R}^n)}$, then (i) of Theorem 1 implies that for $p > n/(n-2)$, $B_\lambda^+ = \{\varphi \in C(\mathbb{R}^n) \mid \varphi \geq 0 \text{ and } |\varphi|_a < \lambda L, a = \frac{2}{p-1}\}$ is an invariant set of the dynamical system for every $0 < \lambda < 1$. (i) of Theorem 1 also yields that as long as $\varphi \in B_1^+$, the "heat flow" u (initiated at φ) converges to the rest state u_0 in the norms $|\cdot|_a$ with $a < \frac{2}{p-1}$ as $t \to \infty$. Thus for $p > \frac{n}{n-2}$, u_0 is stable in the weighted norm $|\cdot|_a$ with $a = \frac{2}{p-1}$ and is actually "weakly asymptotically stable" in this norm (the precise meaning of this term will be given momentarily). Since the "radius" of attraction domain of u_0 can be as large as L (and equals L in case $p \geq p_c$ by (iii) of Theorem 1), and since for fixed p, L increases to ∞ as $n \to \infty$, it seems that u_0 is "increasingly stable" as the dimension n increases (for more points supporting this assertion, see [Wa]). The intuitive explanation of this is as follows: For fixed p, when n is large there is more room in which "heat" can dissipate away to ∞ so that for larger initial values the "temperature" u ultimately drops down to zero, while for smaller n, "heat" cannot dissipate fast enough and blow-up occurs – by the results of [Fu], [H] and [KST], in low dimensions blow-up always takes place.

Now let us return to the question of the topologies (norms) for stability of the rest state $u_0 \equiv 0$. Recall that u_0 is stable in $|\cdot|_a$ with $a = \frac{2}{p-1}$ for $p > n/(n-2)$ and $n \geq 3$. In an on-going joint project of Gui, Ni and Wang [GNW2], we prove the stability of u_0 in a scale of slightly different weighted L^∞ norms for the full range $p > (n+2)/n$. We show that *if the topology is too fine or too rough*, i.e., *if the weight at ∞ is too heavy or too light, then u_0 is unstable*. Actually Lee and Ni's result (see (2)) already implies the instability of u_0 if the weight is too light. To describe our result more precisely, we introduce a scale of weighted L^∞ norms

$$\|u\|_a = \sup_{x \in \mathbb{R}^n} |(1 + |x|)^a u(x)|, \quad a > 0 \ .$$

A steady state u_α of (1) is said to be *stable* in the norm $\|\cdot\|_a$ if $\forall \ \varepsilon > 0$, $\exists \ \delta > 0$ such that for $\|\varphi - u_\alpha\|_a < \delta$, we have $\|u(\cdot, t) - u_\alpha\|_a < \varepsilon$ for all $t \geq 0$. u_a is said to be *weakly asymptotically stable* in $\|\cdot\|_a$ if it is stable in this norm and if $\exists \ \delta > 0$ such that for $\|\varphi - u_\alpha\|_a \leq \delta$, we have $\|u(\cdot, t) - u_\alpha\|_{a'} \to 0$ as $t \to \infty$ for all $a' < a$. In the standard definition of stability of steady states, the heat flow $u(\cdot, t)$ is required to be in the function space (in our case, the weighted L^∞ space). We think that this requirement is not essential and the point of introducing function norms is to measure the deviation of the heat flow from the steady state. The following is proved in [GNW2].

THEOREM 2 *The rest state u_0 is stable in $\|\cdot\|_a$ if and only if $2/(p-1) \leq a < n$ and $p > (n+2)/n$. In this case, u_0 is actually weakly asymptotically stable.*

We remark that the asymptotic stability of u_0 in the usual sense is impossible. The proof of Theorem 2 may be sketched as follows. In view of the blow-up results of [Fu], [H] and [KST], it is obvious that $p > (n+2)/n$ is necessary for the stability of u_0. Also, as mentioned above, by (2) we see that $a \geq 2/(p-1)$ is necessary. To show the instability of u_0 in $\|\cdot\|_a$ for $a > n$, we prove that for $a > n$, $\|e^{t\Delta}\varphi\|_a(t)$ is unbounded if $\liminf_{x \to \infty} |x|^a \varphi(x) > 0$. When proving the stability of u_0 in $\|\cdot\|_a$ for $2/(p-1) \leq a < n$ and $p > (n+2)/n$, we make use of positive self-similar solutions $v(x, t) = t^{-1/(q-1)} f_q(x/\sqrt{t})$ of

$$v_t = \Delta v + v^q \ , \quad x \in \mathbb{R}^n, \quad t > 0 \ ,$$

where $(n+2)/n < q \leq p$. It is known that for $q > (n+2)/n$, $f_q > 0$ in $\mathbb{R}^n$, $\lim_{x \to \infty} |x|^{2/(q-1)} f_q(x)$ exists and is positive (see [HW]). Define $\overline{u}(x, t) = \lambda v(x, t+1)$, $\lambda > 0$. It is straightforward to check that $\overline{u}$ is a super solution of the differential equation in (1) if λ is small. For any $a \in [2/(p-1), n)$, there exists $q \in ((n+2)/n, p]$ such that $a = 2/(q-1)$. Now the stability of u_0 in $\|\cdot\|_a$ follows from the comparison principle.

We now turn to the stability question of positive steady states of (1), i.e., solutions of the Lane-Emden equation

$$\Delta u + u^p = 0 \ , \quad u = u(x) > 0 \ , \quad \text{in } \mathbb{R}^n \ . \tag{4}$$

where $p > 1$. The study of (4) dates back to the fundamental work of Fowler early in this century. The following is well-known (see, e.g., [CGS], [CL], [Li] and [N]).

PROPOSITION 3   (i) *For $1 < p < (n+2)/(n-2)^+$ ($= \infty$ if $n = 1, 2$), (4) has no solution;*

(ii) *For $p = (n+2)/(n-2)$, all solutions of (4) are given by (up to translation)*

$$u_\alpha(x) = \left(\frac{\lambda\sqrt{n(n-2)}}{\lambda^2 + |x|^2}\right)^{\frac{n-2}{2}},$$

where $u_\alpha(0) = \alpha$ with $\alpha = \left(\lambda^{-1}\sqrt{n(n-2)}\right)^{\frac{n-2}{2}}$.

(iii) *For $p > (n+2)/(n-2)$, the set of all radial solutions of (4) is a one-parameter family $\{u_\alpha\}_{\alpha>0}$ with $u_\alpha(0) = \alpha$, $u_\alpha(r)$ decreasing in $r = |x|$, $u_\alpha(r) = \alpha u_1(\alpha^{(p-1)/2}r)$ and $r^{2/(p-2)}u_\alpha(r) \to L$ as $r \to \infty$ (L is as given in the statement of Theorem 1).*

(So far no nontrivial nonradial solution of (4) has been found.)

From now on, we shall assume $n \geq 3$ and $p \geq (n+2)/(n-2)$. The following result is proved in [Wa].

THEOREM 4   *Let u_α's be the positive radial steady states of (1). Then the following statements hold.*

(i) *If $\varphi \leq \lambda u_\alpha$ for some $\lambda \in (0,1)$ and some $\alpha > 0$, then the solution u of (1) is global, satisfies that $0 \leq u \leq \lambda u_\alpha$, and the decay rate (3) holds.*

(ii) *If $\varphi \geq \lambda u_\alpha$ for some $\lambda > 1$ and some $\alpha > 0$, then the solution u of (1) blows up in finite time.*

This result may be viewed as yet another description of the domain of attraction for the rest state $u_0 = 0$. It may also be viewed as an instability result for every steady state u_α: it implies that u_α is not only unstable in L^∞ norm, but also unstable in the more refined weighted norm $\|\cdot\|_a$ where $a = 2/(p-1)$ if $p > (n+2)/(n-2)$, and $a = n - 2$ if $p = (n+2)/(n-2)$. It turns out that the stability question for u_α is much more delicate: *when $p > p_c$, u_α is unstable in any reasonable topology, while when $p \geq p_c$, the topologies mentioned above are too rough to detect the stability of u_α.* More precisely, Gui, Ni and Wang [GNW1] proved the following two theorems.

THEOREM 5   *Supose $(n+2)/(n-2) \leq p < p_c$.*

(i) *If $\varphi \leq u_\alpha$ and $\varphi \not\equiv u_\alpha$ for some $\alpha > 0$, then the solution u of (1) is global with $\|u(\cdot, t)\|_{L^\infty(\mathbb{R}^n)} \to 0$ as $t \to \infty$;*

(ii) *If $\varphi \geq u_\alpha$ and $\varphi \not\equiv u_\alpha$ for some $\alpha > 0$, then u blows up in finite time.*

THEOREM 6   *Let $\lambda_1 \leq \lambda_2$ be the positive roots of $P(z) = z^2 - (n-2-2m)z + 2(n-2-m)$, $m = 2/(p-1)$, $p \geq p_c$.*

(i) *If $p = p_c$ (in this case $\lambda_1 = \lambda_2$), any positive radial steady state u_α of (1) is weakly asymptotically stable in the norm $\|\cdot\|_{m+\lambda_2}$.*

(ii) *If $p > p_c$ (in this case $\lambda_1 < \lambda_2$), u_α is stable in the norm $\|\cdot\|_{m+\lambda_1}$, and weakly asymptotically stable in the norm $\|\cdot\|_{m+\lambda_2}$.*

From Theorem 5 and Theorem 6, we see that the transition from instability to stability occurs at $p = p_c$. This is closely related to the following drastic change in the mutual behavior of all radial steady states of (1): when $(n+2)/(n-2) \leq p < p_c$, any two radial steady states must intersect each other, while when $p \geq p_c$, no two of them can intersect each other.

As we have already seen, the rest state u_0 is unstable if the topology used to measure the deviation of the heat flow from u_0 is too rough or too fine. *This is still the case for any positive radial steady state u_α when $p \geq p_c$* (recall by Theorem 5, u_α is unstable in any reasonable topology when $p < p_c$). The following theorem was proved in the current joint project of Gui, Ni and Wang [GNW2].

THEOREM 7 *For $p > p_c$, any radial positive steady state u_α of (1) is weakly asymptotically stable in $\|\cdot\|_a$ with $m + \lambda_1 < a \leq m + \lambda_2$. Moreover, u_α is unstable in $\|\cdot\|_a$ if $a < m + \lambda_1$ or $a \geq n$.*

A similar result is also obtained in [GNW2] for the case $p = p_c$. Recall that by Theorem 6 u_α is stable in $\|\cdot\|_a$ with $a = m + \lambda_1$. The question of the stability of u_α in $\|\cdot\|_a$ when $m + \lambda_2 < a < n$ remains unsolved so far and we are still working on this.

In [GNW2], we also proved the following theorem concerning global existence, finite time blow-up and decay rates of solutions of (1).

THEOREM 8 (i) *Suppose $p \geq p_c$. If the initial value $\varphi(x) \leq L/|x|^m \equiv u_s(x)$ for $x \in \mathbb{R}^n$ and*

$$\varphi(x) \leq \frac{L}{|x|^m} - \frac{a}{|x|^{m+\sigma}} \quad at \quad x = \infty \tag{5}$$

for some $a > 0$ and some $\sigma \in (0, \lambda_1)$, then the solution u of (1) exists globally with $\|u(\cdot, t)\|_{L^\infty(\mathbb{R}^n)} \to 0$ as $t \to \infty$. Furthermore, this decaying solution must decay (in time) slower than $t^{-1/(p-1)}$, provided that φ is "large" at infinity in the sense that φ satisfies (5) with reversed inequality sign and with a possibly larger a. On the other hand, if

$$\varphi(x) \geq \frac{L}{|x|^m} + \frac{a}{|x|^{m+\sigma}} \quad at \quad x = \infty$$

for some $a > 0$ and some $\sigma \in (0, \lambda_1)$, then the solution u of (1) blows up in finite time.

(ii) *Suppose $n/(n-2) < p < p_c$. If $\varphi \leq \psi$ in $\mathbb{R}^n$ for some radial continuous weak super-solution ψ of the Lane-Emden equation (4) with*

$$\limsup_{r \to \infty} r^{2/(p-1)} \psi(r) \leq L , \qquad (6)$$

then the solution u of (1) is global and $\|u(\cdot,t)\|_{L^\infty(\mathbb{R}^n)}$ decays like $Ct^{-1/(p-1)}$.

The last part of (i) of this theorem improves (iii) of Theorem 1, while the first part of (i) of Theorem 8 is a complement of (ii) of Theorem 1. Furthermore, (ii) of Theorem 8 implies that the $L^\infty(\mathbb{R}^n)$ norm of the global solutions u in (ii) of Theorem 1 and (i) of Theorem 5 decays at least as fast as $t^{-1/(p-1)}$ when $t \to \infty$, because by the proofs of Theorem 1 and Theorem 5, it is always possible to find a super-solution ψ satisfying the property in the statement of Theorem 8. If the condition "$\varphi \leq \psi$" is replaced by "$\varphi \leq \lambda\psi$" for some $\lambda \in (0,1)$, then (ii) of Theorem 8 is proved in [Wa] (Theorem 4.1) without requiring $p < p_c$ and (6). At this moment, we do not know the decay rate of decaying solution u in (i) of Theorem 8. However, we believe that this is the first time that the solutions of (1) decaying slower than $t^{-1/(p-1)}$ are proved to exist. The proof of Theorem 7, (i) of Theorem 8 (as well as the proof of Theorem 6) heavily rely on the precise asymptotic expansions of positive radial solutions of

$$\Delta u + K(|x|)u^p = 0 \quad \text{in } \mathbb{R}^n , \qquad (7)$$

where $K \equiv 1$ near ∞. The study of such asymptotic expansions was initiated by Yi Li [Li]. The following, which is a slight extension of Li's original result, is proved in [GNW1].

PROPOSITION 9 (i) *Suppose $p \geq p_c$. Then there exist $(p_c =)p_1(n) < p_2(n) < \ldots < p_N(n) < \infty$, where N is an integer depending on n, so that if u is a positive radial solution of (7) with $\lim_{r \to \infty} r^m u(r) > 0, m = 2/(p-1)$, the following statements hold.*
(i) *For $p = p_k(n)$, $k = 1, \ldots, N$, we have $\lambda_2 = k\lambda_1$ and at $r = \infty$,*

$$u(r) = \frac{L}{r^m} + \frac{a_1}{r^{m+\lambda_1}} + \ldots + \frac{a_{k-1}}{r^{m+(k-1)\lambda_1}} + \frac{a_k \log r}{r^{m+k\lambda_1}} + \frac{b_1}{r^{m+\lambda_2}} + \ldots + O\left(\frac{1}{r^{n-2+\varepsilon}}\right).$$

(ii) *For $p_k(n) < p < p_{k+1}$, $k = 1, \ldots, N$, $p_{N+1} = \infty$, we have $k\lambda_1 < \lambda_2 < (k+1)\lambda_1$ and at $r = \infty$,*

$$u(r) = \frac{L}{r^m} + \frac{a_1}{r^{m+\lambda_1}} + \ldots + \frac{a_k}{r^{m+k\lambda_1}} + \frac{b_1}{r^{m+\lambda_2}} + \frac{c}{r^{m+(k+1)\lambda_2}} + \ldots + O\left(\frac{1}{r^{n-2+\varepsilon}}\right).$$

The coefficients $a_2, \ldots, a_N$ are uniquely determined once a_1 is determined. Moreover, once a_1 and b_1 are determined, then all coefficients in the expansion are uniquely determined.

By the above result, we see that for $p \geq p_c$ and for every positive radial solution u_α of the Lane-Emden equation (4),

$$\frac{L}{|x|^m} - \frac{a}{|x|^{m+\sigma}} < u_\alpha(x) < \frac{L}{|x|^m} + \frac{a}{|x|^{m+\sigma}} \quad \text{at} \quad x = \infty,$$

where $a > 0$, $\sigma \in (0, \lambda_1)$. This observation is crucial in the proof of (i) of Theorem 8.

In the rest of this paper, we shall sketch the proof of Theorem 7. The instability of u_α in $\| \cdot \|_a$ for $a \geq n$ follows easily from the proof of Theorem 2. To show that u_α is unstable in $\| \cdot \|_a$ for $a < m + \lambda_1$, we just need to use (i) of Theorem 8. To show the stability of u_α in $\| \cdot \|_a$ for $a \in (m + \lambda_1, m + \lambda_2)$ (the case $a = m + \lambda_2$ is covered by Theorem 6), it is decisive to construct one-parameter families of radial subsolutions $\{\underline{u}_\delta\}_{\delta>0}$ and radial super-solutions $\{\overline{u}_\delta\}_{\delta>0}$ of (4), so that

   (i) $\underline{u}_\delta < u_\alpha < \overline{u}_\delta$ in $\mathbb{R}^n$;

   (ii) $\lim_{\delta \to 0} \underline{u}_\delta = u_\alpha = \lim_{\delta \to 0} \overline{u}_\delta$;

   (iii) $\lim_{x \to \infty} |x|^a (\overline{u}_\delta - u_\alpha) = \lim_{x \to \infty} |x|^a (u_\alpha - \underline{u}_\delta) = \delta$.

Since the coefficient a_1 in the expansion of every positive radial solution of (4) is negative (see [GNW1], Lemma 4.13), by Proposition 9 we see that u_α is the *only* radial steady state of (1) pinched between $\underline{u}_\delta$ and $\overline{u}_\delta$ for any $\delta > 0$ (recall $a > m + \lambda_1$). This will lead to the desired weak asymptotic stability of u_α via a standard parabolic argument. We shall only describe the construction of $\overline{u}_\delta$ – the construction of $\underline{u}_\delta$ is similar. Define $\overline{v}_\delta(r) = u_\alpha(r) + \delta r^{-a}$. By a direct computation, we have that $\overline{v}_\delta$ is a super-solution of (4) at $r = \infty$. By using Proposition 9, it is possible to show $\overline{v}_\delta < u_\beta$ at $r = \infty$ for every $\beta > \alpha$. Now the desired $\overline{u}_\delta$ can be obtained by "gluing" $\overline{v}_\delta$ and u_β together with $\beta > \alpha$ and close to α so that the last intersection of $\overline{v}_\delta$ and u_β occurs sufficiently far away.

REFERENCES

[AW] D. G. Aronson and H. F. Weinberger, Multidimensional nonlinear diffusion arising in population genetics, <u>Adv. Math.</u> 30, 1978, pp. 33–76.

[CGS] L. Caffarelli, B. Gidas and J. Spruck, Asymptotic symmetry and local behavior of semilinear elliptic equations with critical Sobolev growth, <u>Comm. Pure Appl. Math.</u> 42, 1989, pp. 271–297.

[CL] W. Chen and C. Li, Classification of solutions of some nonlinear elliptic equations, <u>Duke Math. J.</u> 52, 1985, pp. 485–506.

[Fu] H. Fujita, On the blowing up of solutions of the Cauchy problem for $u_t = \Delta u + u^{1+\alpha}$, <u>J. Fac. Sci. Univ. Tokyo</u> (I) 13, 1966, pp. 109–124.

[GNW1] C. Gui, W.-M. Ni and X. Wang, On the stability and instability of positive steady states of a semilinear heat equation in $\mathbb{R}^n$, <u>Comm. Pure Appl. Math.</u> 45, 1992, pp. 1153–1181.

[GNW2] C. Gui, W.-M. Ni and X. Wang, in preparation.

[H] K. Hayakawa, On nonexistence of global solutions of some semilinear parabolic equations, <u>Proc. Japan Acad</u>. 49, 1973, pp. 503–505.

[HV] M. Herrero and J. Velázquez, Some results on blow up for semilinear parabolic problems, IMA Preprint Series 1000, 1992.

[HW] A. Haraux and F. B. Weissler, Non-uniqueness for a semilinear initial value problem, <u>Indiana Univ. Math. J</u>. 31, 1982, pp. 167–189.

[K] S. Kaplan, On the growth of solutions of quasilinear parabolic equations, <u>Comm. Pure Appl. Math</u>. 16, 1963, pp. 305–333.

[KST] K. Kobayashi, T. Siaro and H. Tanaka, On the blowing up problem for semilinear heat equations, <u>J. Math. Soc. Japan</u> 29, 1977, pp. 407–424.

[L] H. Levine, The role of critical exponents in blow up theorems, <u>SIAM Review</u>. 32, 1990, pp. 262–288.

[Li] Yi Li, Asymptotic behavior of positive solutions of equation $\Delta u + K(x)u^p = 0$ in $\mathbb{R}^n$, <u>J. Diff. Eqns</u>. 95, 1992, pp. 304–330.

[LN] T.-Y. Lee and W.-M. Ni, Global existence, large time behavior and life span of solutions of a semilinear parabolic Cauchy problem, <u>Trans. Amer. Math. Soc</u>. 333, 1992, pp. 365–378.

[N] W.-M. Ni, On the elliptic equation $\Delta u + K(x)u^{\frac{n+2}{n-2}} = 0$, <u>Indiana Univ. Math. J</u>. 31, 1982, pp. 493–529.

[Wa] X. Wang, On the Cauchy problem for reaction-diffusion equations. <u>Trans. Amer. Math. Soc</u>., 337, 1993, pp. 549–589.

[We] F. Weissler, Existence and nonexistence of global solutions for a semilinear heat equation, <u>Israel J. Math</u>. 38, 1981, pp. 29–40.

Finite Element Error Analysis of a Quasi-linear Evolution Equation

Dongming Wei   University of New Orleans, New Orleans, Louisiana

1. INTRODUCTION

As supplement to the earlier work [7] where the case $2 \leq p < \infty$ was considered, we show that the Cauchy-Dirichlet problem of a quasi-linear evolution equation governed by the p-Laplacian for $1 < p < 2$, has a unique weak solution which satisfies the equation pointwise with respect to time for certain initial conditions. With this formulation, L^2 error estimates for the error between the true solution and its extended fully-discrete finite element approximations are obtained. However, unlike in the case $2 \leq p < \infty$, we are unable to obtain similar L^2 error estimates for the full range $1 < p < 2$ due to failure of the Sobolev imbedding $W_0^{1,p}(\Omega) \to L^2(\Omega)$ for certain values of p given the dimension n of the domain Ω. Here we mainly state the results and outline their proofs which are analogous to those given in [7].

2. EXISTENCE AND UNIQUENESS OF SOLUTION

Throughout this paper, we shall assume that Ω is a bounded convex domain in R^n with smooth boundary $\partial\Omega$, C a generic constant, and $1 < p < 2$. We also use $u(t)$ or simply u to denote the function $u(x,t)$. We use the notation: $|u|_{1,p} = [\int_\Omega |\nabla u|^p dx]^{\frac{1}{p}}$,

$\| u \|_2 = [\int_\Omega |u|^2 dx]^{\frac{1}{2}}$, where $|u|_{1,p}$ is the semi-norm for $W^{1,p}(\Omega)$ which is a norm for $W_0^{1,p}(\Omega)$ and $\| u \|_2$ the usual $L^2(\Omega)$ norm. For $u \in W^{1,p}(\Omega)$, we denote by $u|_{\partial\Omega}$ the trace of u on $\partial\Omega$ [see e.g. 8].

Let A: $W^{1,p}(\Omega) \to W^{-1,p'}(\Omega)$ be the operator defined by

$$(Au, v) = \int_\Omega |\nabla u|^{p-2} \nabla u \cdot \nabla v \, dx, \forall v \in W^{1,p}(\Omega).$$

For the definition of the Sobolev spaces $W_0^{1,p}(\Omega), W^{1,p}(\Omega)$ and $W^{-1,p'}(\Omega)$, see [1] or [5]. We shall consider the following Cauchy-Dirichlet problem of evolution equation:

$$\frac{du}{dt} + Au = f, x \in \Omega, t \in (0,T], \tag{2.1}$$

$$u(x,t)|_{\partial\Omega} = \phi(x), t \in (0,T], \tag{2.2}$$

$$u(x,0) = u_0(x), x \in \Omega, \tag{2.3}$$

where $u_0 \in W^{1,p}(\Omega)$, $u_0|_{\partial\Omega} = \phi(x)$, and $f : [0,T] \to L^2(\Omega)$ is Lipschitz continuous, i.e., there exists a positive constant C such that $\| f(t) - f(t') \|_2 \leq C|t-t'|$ for any $t, t' \in [0,T]$.

DEFINITION 2.1 Let $u(x,t) : [0,T] \to L^2(\Omega)$. If there exists a function $g(x,t) : [0,T] \to L^2(\Omega)$ such that

$$\lim_{\Delta t \to 0} \| \frac{u(t+\Delta t) - u(t)}{\Delta t} - g(t) \|_2 = 0.$$

Then we say that u is differentiable at t, and $g(x,t)$ the derivative of $u(x,t)$ at t, which is denoted by $\frac{du(x,t)}{dt}$ or simply $\frac{du}{dt}$.

In following $<\cdot, \cdot>$ is understood as the usual inner product in $L^2(\Omega)$ and $(\cdot, \cdot)$ as the duality for a pair in $W^{1,p}(\Omega) \times W^{-1,p'}(\Omega)$.

DEFINITION 2.2 We say that u is a solution of (2.1), (2.2), and (2.3) if $u(x,t) \in W^{1,p}(\Omega) \cap L^2(\Omega)$ for all $t \in (0,T]$,

$$< \frac{du}{dt}, v > + (Au, v) = < f, v >, \tag{2.4}$$

$$< u(0), v > = < u_0, v >, \forall v \in W_0^{1,p}(\Omega) \cap L^2(\Omega), \tag{2.5}$$

and

$$u(x,t)|_{\partial\Omega} = \phi(x), t \in (0,T], \tag{2.6}$$

where $\frac{du(x,t)}{dt}$ is the derivative in the sense of Definition 2.1, $u_0 \in W^{1,p}(\Omega)$, and $u_0|_{\partial\Omega} = \phi(x)$

THEOREM 2.1 Suppose that $u_0 \in W_0^{1,p}(\Omega) \cap L^2(\Omega)$ and $\nabla \cdot (|\nabla u_0)|^{p-2}\nabla u_0) \in L^2(\Omega)$. Then problem (1.1), (1.2), (1.3) has a unique solution u in the sense of Definition 2.2. Furthermore, $u \in C[0,T; W^{1,p}(\Omega) \cap L^2(\Omega)]$.

The proof of this theorem is based on the following property of the operator A for $1 < p < 2$ which is analogous to that for $2 \le p < \infty$ in Lemma 1 of [7]:

$$\alpha \parallel u - v \parallel^2 \le (Au - Av, u - v)(\parallel u \parallel + \parallel v \parallel)^{2-p},$$

$$\parallel Au - Av \parallel \le \beta \parallel u - v \parallel^{p-1},$$

$\forall\, u,v \in W^{1,p}(\Omega)$, where α and β are positive constants independent of u and v (see also [4]).

For simplicity, let m be a positive integer and $\{t_i\}_{i=0}^m$ be an uniform partition of [0,T], $\Delta t = \frac{T}{m}$ and $t_i = i\Delta t$. Consider following recursive quasi-linear elliptic problems:

Given u_{i-1}, find u_i such that

$$< \frac{u_i - u_{i-1}}{\Delta t} > +(Au_i, v) = < f_i, v >, \tag{2.7}$$

$$u_i|_{\partial\Omega} = u_{i-1}|_{\partial\Omega}, \forall v \in W_0^{1,p}(\Omega) \cap L^2(\Omega), \tag{2.8}$$

where $u_i = u(x, t_i), f_i = f(x, t_i), i = 0, ..., m$.

Since, by its property, A is a strictly monotone operator and it satisfies all the conditions in Theorem 29.5 of [3, pp. 242 - 243], for each such partition $\{t_i\}_{i=0}^m$, (2.7), (2.8) generate a unique sequence $\{u_i\}_{i=0}^m$ in $W_0^{1,p}(\Omega) \cap L^2(\Omega)$.

Now, let

$$u_m(t) = \frac{t - t_i}{\Delta t_i} u_{i+1} + \frac{t_{i+1} - t}{\Delta t_i} u_i,$$

$$u_m(0) = u(0), t_i < t \leq t_{i+1}, i = 0, ..., m-1.$$

This is called a semi-discrete solution of the problem. Then it can be shown that

$$\| u(t) - u_m(t) \|_2 \leq C\Delta t, \tag{2.9}$$

where $\Delta t = \displaystyle\max_{1 \leq i \leq m} \Delta t_i$, and C is a positive constant independent of u, u_m.

3. L^2 ERROR ESTIMATES

By a classical theorem on global interpolation error estimates in the finite element theory, ([2], or [6]) we have

LEMMA 3.1 Suppose that $\{T_h\}_h$, where $h > 0$, is a regular family of triangulation of Ω. We then have the following interpolation error estimates:

$$|u - \Pi_h u|_{1,p} \leq Ch|u|_{2,p}, \forall u \in W^{2,p}(\Omega), \tag{3.1}$$

where $1 < p < \infty$, $|u|_{2,p} = [\int_\Omega \sum_{|\alpha|=2} |D^\alpha u|dx]^{\frac{1}{p}}$, and C is a constant independent of u, and h is the parameter of element size for the triangulation $\{T_h\}_h$, and Π_h the finite element interpalotion operator defined in [6, Vol. iv, pp. 63 - 64].

Let $\{u_i\}_{i=0,n}$, be the sequence generated by (2.7), (2.8). Let $S_h(\Omega)$ be the finite element space corresponding to $\{T_h\}_h$ with property (3.1). For each i, consider the following problem:

Find $W_i \in S_h(\Omega)$, such that

$$(AW_i, V) = (Au_i, V), \forall v \in W_0^{1,p}(\Omega) \cap L^2(\Omega), \tag{3.2}$$

$$W_i|_{\partial\Omega} = \Pi_h u_i, |_{\partial\Omega}, i = 0, ..., m. \tag{3.3}$$

By Theorem 29.5 of [3], for each i, (3.2) and (3.3) has a unique solution $W_i, i = 0, ..., m$. It can be shown, similarly to the proof of Lemma 10 of [7], by using the property of A that

$$|u_i - W_i|_{1,p} \leq C(u_0, f)(|u_i - \Pi_h u_i|_{1,p})^{\frac{1}{(3-p)}}, i = 0, ..., m. \tag{3.4}$$

Now, we consider the fully-discrete scheme:

Let $U_0 = W_0$, where W_0 is defined by (3.2) and (3.3). Find $U_i \in S_h(\Omega)$, such that

$$< \frac{U_i - U_{i-1}}{\Delta t}, V > + (AU_i, V) = < f_i, V >, \forall V \in S_h(\Omega) \cap W_0^{1,p}(\Omega), \qquad (3.5)$$

$$U_i|_{\partial\Omega} = W_i|_{\partial\Omega}, i = 1, ..., m. \qquad (3.6)$$

We extend the fully-discrete solution $\{U_i\}_{i=0,...,m}$ to [0,T] by using

$$U_m(t) = \frac{t - t_i}{\Delta t_i} U_{i+1} + \frac{t_{i+1} - t}{\Delta t_i} U_i,$$

$$U_m(0) = U_0, t_i < t \le t_{i+1}, i = 0, ..., m - 1, \qquad (3.7)$$

which is similar to the definition of $u_m(t)$.

It can be verified that

$$\| u_m(t) - U_m(t) \|_2^2 \le C_1 \Delta t + C_2 \max_{1 \le i \le m} \| u_i - W_i \|_2 + \| u_0 - W_0 \|_2^2 . \qquad (3.8)$$

By (2.9) and (3.8), we have

$$\| u(t) - U_m(t) \|_2^2 \le 2(\| u(t) - u_m(t) \|_2^2) + \| u_m(t) - U_m(t) \|_2^2,$$

$$\le C_1 \Delta t + C_2 \max_{1 \le i \le m} \| u_i - W_i \|_2 + 2 \| u_0 - W_0 \|_2^2, \qquad (3.9)$$

where C_1 and C_2 are positive constants depending only on u_0, f and Ω. By [2, Theorem 5.3.2, p.317] and (3.9), without assuming "higher regularity" on solution u, we have

$$\lim_{h \to 0} \| u_i - W_i \|_2 = 0, i = 0, ..., m.$$

Therefore, we have

THEOREM 3.1 Let $u(t)$ be the true solution of problem (2.1), (2.2) and (2.3) obtained in Theorem 2.1, and let $U_m(t)$ be the extended fully-discrete solution defined by (3.7). Then we have

$$\lim_{\Delta t \to 0} (\lim_{h \to 0} \| u(t) - U_n(t) \|_2) = 0$$

without assuming 'higher regularity' on solution u.

By the Sobolev imbedding theorem [1], we have $W_0^{1,p}(\Omega) \to L^2(\Omega)$ if $\frac{2n}{n+2} \leq p$. Therefore, there exists a positive constant C such that

$$\| u_i - W_i \|_2 \leq C|u_i - W_i|_{1,p}, i = 0, ..., m. \tag{3.10}$$

By (3.2), (3.9), (3.10), and Lemma 3.1, we have

COROLLARY 3.1 If in addition, it is assumed that for each i the solution $u_i \in W^{2,p}(\Omega), i = 1, ..., m$ and $\frac{2n}{n+2} \leq p < 2$. Then, we have the following L^2 error estimates

$$\| u(t) - U_m(t) \|_2^2 \leq C_1 \Delta t + C_2 (\max_{1 \leq i \leq m} |u_i|_{2,p}) h^{\frac{1}{3-p}} + C_3 |u_0|_{2,p} h^{\frac{2}{3-p}}.$$

REMARK: For any $n \in N^+$, $\frac{2n}{n+2}$ is always less than 2 and greater than 1 when n is greater or equals to 3, and $\frac{2n}{n+2} \to 2$ as $n \to \infty$. Therefore, there exists a gap between 1 and $\frac{2n}{n+2}$ for $3 \leq n$ where the above estimates are not available and this gap increases as $n \to \infty$. The above results also hold for $p = 2$.

REFERENCES

[1] R. A. Adams, Sobolev spaces, Academic Press, New York, 1978.

[2] P. G. Ciarlet, The finite element method for elliptic problems, North Holland, Amsterdam, 1978.

[3] S. Fučik and A. Kufner, Nonlinear differential equations, Elsevier Scientific Publishing Company, New York, 1980.

[4] R. Glowinski and A. Marroco, Sur l'approximation par éléments finis d'ordre un, et la résolution par penalisation-dualité, d'une classe des problémes de Dirichlet non linéares, RAIRO Aanal Numer. 9, R-2, (1975), 41-76.

[5] V. G. Maz'ja, Sobolev Spaces, Springer-Verlag, New York, 1985.

[6] J. T. Oden and G. F. Carey, Finite elements, Prentice-Hall, Inc., Englewood Cliffs, New Jersey, 1984.

[7] D. Wei, Existence, uniqeness and numerical analysis of solutions of a quasi-linear parabolic problem, SIAM J. Numer. Anal., Vol. 29, No.2, (1992), 484 - 496.

[8] A. Ženíšek, Nonlinear elliptic and evolution problems and their finite element approximations, Academic Press, New York, 1990.

Index